Recent Advances in Parasitomics

Juan David Ramírez González
Editor

Recent Advances in Parasitomics

Implications for Parasite and Vector Research

Springer

Editor
Juan David Ramírez González
Department of Pathology, Molecular and Cell-based Medicine
Icahn School of Medicine at Mount Sinai
New York, United States of America

Faculty of Natural Sciences
Universidad del Rosario
Bogotá, Colombia

ISBN 978-3-031-70593-9 ISBN 978-3-031-70591-5 (eBook)
https://doi.org/10.1007/978-3-031-70591-5

This Springer imprint is published by the registered company Springer Nature Switzerland AG
The registered company address is: Gewerbestrasse 11, 6330 Cham, Switzerland

Contents

Part I

Trypanosomatids

Deployment of Amplicon-Based Sequencing Platforms Using Oxford Nanopore Technologies for the Detection and Surveillance of Trypanosomatids

Lissa Cruz-Saavedra and Juan David Ramírez

Abstract

This chapter delves into the transformative potential of Amplicon-based sequencing (ABS) by Oxford Nanopore Technologies (ONT) for detecting and surveilling Trypanosomatids, a family of challenging protozoan parasites in medical and veterinary domains. Our focus centers on *T. cruzi* and *Leishmania*. Leveraging ONT's real-time sequencing and full-length reads, ABS-ONT underscores the importance of precise genomic target selection, evidence-based primer design, and robust PCR standardization for reliable results. We emphasize the significance of high-fidelity, long-read polymerases in ABS, crucial for capturing subtle genetic variations essential for precise genotyping. In remote areas, the field implementation of ABS-ONT could proves invaluable, facilitating on-site diagnostics and surveillance with high resolution and sensitivity. The method's accuracy offers real-time insights into the prevalence and diversity of Trypanosomatids. We discuss challenges in large-scale ABS implementation, including logistical and ethical considerations, emphasizing the need for collaborative efforts, standardized protocols, and stringent ethical practices. Envisioning ABS as the gold standard for diagnostics and genotyping, we contribute to a comprehensive understanding of Trypanosomatids, guiding data-driven public health decisions. Here, we provide a holistic perspective on advancements in detection and surveillance, showcasing the potential of ONT-ABS platform to revolutionize understanding and pave the way for personalized medicine in a one-health approach in parasitology.

Keywords

Amplicon-based sequencing · Oxford Nanopore · Trypanosomatids · Diagnostic · Genotyping

L. Cruz-Saavedra (✉)
Institute of Parasitology, McGill University, Ste-Anne-de-Bellevue, Québec, Canada

Centro de Investigaciones en Microbiología y Biotecnología-UR (CIMBIUR), Facultad de Ciencias Naturales, Universidad del Rosario, Bogotá, Colombia

J. D. Ramírez
Centro de Investigaciones en Microbiología y Biotecnología-UR (CIMBIUR), Facultad de Ciencias Naturales, Universidad del Rosario, Bogotá, Colombia

Molecular Microbiology Laboratory, Department of Pathology, Molecular and Cell-Based Medicine, Icahn School of Medicine at Mount Sinai, New York, NY, USA

J. D. Ramírez González (ed.), *Recent Advances in Parasitomics*,
https://doi.org/10.1007/978-3-031-70591-5_1

1 Introduction

The deployment of amplicon-based sequencing platforms, particularly leveraging the cutting-edge Oxford Nanopore Technologies, represents a significant advancement in the detection and surveillance of Trypanosomatids. The Trypanosomatidae family, comprising protozoan parasites with substantial medical and veterinary importance, poses unique challenges for effective monitoring and control [1]. By employing amplicon-based sequencing on the Oxford Nanopore platform, this chapter aims to explore innovative methodologies for unraveling the genomic complexities of Trypanosomatids. With a focus on key species such as *Trypanosoma cruzi*, *Trypanosoma brucei*, and *Leishmania*, this approach holds the promise of enhancing our understanding of the genetic diversity, transmission dynamics, and potential zoonotic risks associated with these parasites [2]. The unparalleled advantages of Oxford Nanopore Technologies (ONT), including real-time sequencing capabilities and generation of full-length reads, offer a transformative potential for precision surveillance in both human and animal populations [3, 4]. Through this exploration, we aim to pave the way for improved diagnostics, surveillance, and ultimately, more effective strategies for the management of Trypanosomatid infections.

2 Importance of Trypanosomatids Worldwide

Trypanosomatids, exhibit a fascinating array of life cycles and ecological adaptations. Notable genera within this family include *Trypanosoma*, *Leishmania*, *Crithidia*, and *Phytomonas* [5]. With their distinctive presence of kinetoplast, long flagellum, trypanosomatids are transmitted primarily through the bites of insect vectors, such as tsetse flies, sandflies, and reduviid bugs [6, 7]. This family's significance is underscored by its members' role as causative agents of various diseases affecting humans, animals, and plants [8].

2.1 Phylogeny of Trypanosomatids

The phylogeny of trypanosomatids, a diverse group of parasitic protozoa belonging to the phylum Euglenozoa, offers valuable insights into their evolutionary relationships and genetic diversity. Among the prominent genera within the Trypanosomatidae family are *Trypanosoma*, *Leishmania*, *Phytomonas*, *Crithidia*, and others genus [5].

In the current classification system, the family Trypanosomatidae is organized into six officially recognized subfamilies and 22 genera. The subfamily Leishmaniinae brings together both monoxenous parasites primarily infecting insects (e.g., *Borovskiya*, *Crithidia*, *Leptomonas*) and dixenous parasites capable of infecting both insects and vertebrates (e.g., *Leishmania*, *Paraleishmania*, *Endotrypanum*). Two monogeneric subfamilies, Blechomonadinae and Paratrypanosomatinae, consist of monoxenous parasites of Siphonaptera (genus *Blechomonas*) and the early branching lineage of the genus *Paratrypanosoma*, respectively. The subfamily Strigomonadinae includes genera *Angomonas*, *Kentomonas*, and *Strigomonas*, which harbor bacterial endosymbionts. Another subfamily, Phytomonadinae, encompasses genera *Phytomonas*, *Herpetomonas*, and *Lafontella*. Remarkably, the genus *Trypanosoma*, previously unassigned to any subfamily, is now proposed to be included in the nominotypical subfamily, Trypanosomatinae [9, 10] (Fig. 1).

Two medically significant genera, *Trypanosoma* and *Leishmania*, occupy distinct branches in the phylogenetic tree. The tree reveals the specific monophyletic branch of different *Trypanosoma* species responsible for diseases such as African sleeping sickness (*T. brucei*) and Chagas disease (*T. cruzi*), as well as other *Trypanosoma* gender species important in veterinary such as *Trypanosoma evansi*, *Trypanosoma vivax*, and *Trypanosoma theileri*. Importantly, in a high resolution, phylogenetic is possible to observe the two subspecies of *T. brucei* (*T. brucei gambiense* and *T. brucei rhodesiense*) and also

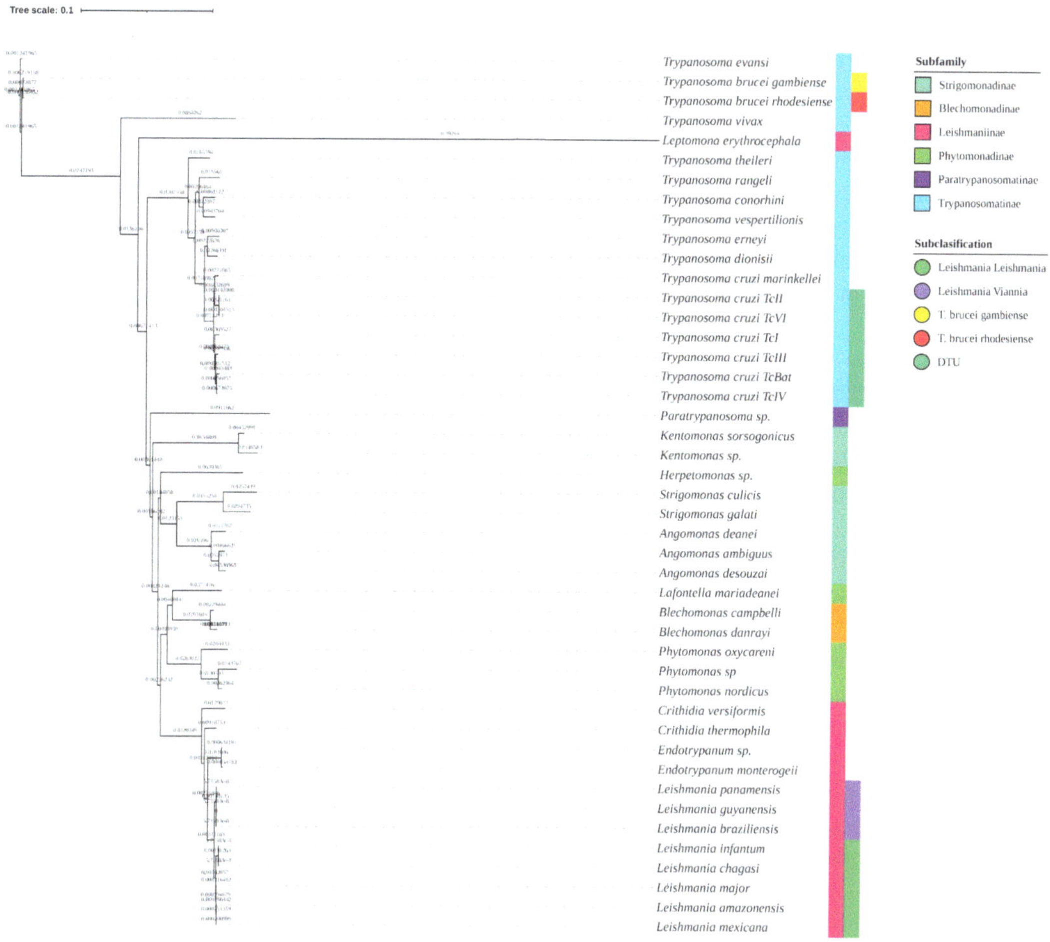

Fig. 1 Phylogeny of trypanosomatids. Phylogenetic relationships among trypanosomatids are inferred from the analysis of the 18S rRNA gene across 46 genera. The color-coded bars represent the six officially recognized subfamilies within the Trypanosomatidae family. Notably, there is a distinct line indicating subgender, subfamily and discrete typing units divisions within the *Trypanosoma* and *Leishmania* genera

the six DTUs of *T. cruzi* (TcI–TcVI) [11, 12]. Similarly, the 53 *Leishmania* species, causing leishmaniasis, display a diverse phylogenetic distribution where it could be observed the two subgenera *Leishmania Leishmania* and *Leishmania Viannia*, reflecting their adaptation to various hosts and vectors (Fig. 1) [13]. The phylogenetics of trypanosomatids provide a comprehensive understanding of the evolutionary relationships among different genera and species within the family. This knowledge is instrumental in unraveling the diversity, adaptation, and pathogenicity of these parasitic protozoa.

2.2 Genomic Markets

The classification of *T. cruzi* relies on various gene markers, including the mini-exon gene, cytochrome *b* (Cyt *b*) gene, Histone H1, GPI loci, LSU rDNA, 24Sα rRNA, 18S rRNA, cruzipain, P7–P8, and gp72. These markers are combined in different ways to determine discrete typing units (DTUs) or genetic lineages of *T. cruzi*. While some markers enable clear identification of individual DTUs, e.g., the intergenic region of the miniexon gene and the DTU TcI, others require a combination of markers for reliable assignment.

Common methods such as PCR-restriction fragment length polymorphism analysis (PCR-RFLP) and nuclear multilocus sequence typing (MLST) are used for strain DTU assignment [14–19]. Next-generation sequencing (NGS) of the mini-exon gene marker reveals both inter-lineage and intra-lineage diversity, crucial for understanding Chagas disease epidemiology and the link between parasite genotype and clinical outcomes [20, 21].

The classification of *Leishmania* species relies on a variety of genetic markers essential for understanding their diversity and epidemiology. These markers include both nuclear and mitochondrial genes such as the internal transcribed spacer (ITS) region of ribosomal DNA, Cyt *b* gene, glycoprotein 63 (gp63), heat shock protein 70 (HSP70), and the mitochondrial kinetoplast DNA (kDNA) minicircles [22–27]. The ITS region, commonly used in phylogenetic studies, offers insights into the evolutionary relationships among *Leishmania* species. The Cyt *b* gene, known for its high variability, aids in distinguishing different strains and species. Meanwhile, gp63 and HSP70 genes provide valuable information on the virulence and adaptation mechanisms of *Leishmania* parasites. The kDNA minicircles, with their high copy numbers and polymorphisms, serve as useful targets for molecular diagnostics and epidemiological studies.

For other trypanosomatids, such as *T. brucei*, genetic markers such as variable surface glycoproteins (VSGs), ITS regions of ribosomal DNA, and mini-exon genes are employed to better understand their phylogenetic relationships and genetic diversity. These markers, along with splice leader (SL) RNA genes and kinetoplast DNA (kDNA) maxicircle genes, contribute to our understanding of the population structure and epidemiology of *T. brucei* parasites, aiding in the development of strategies for disease control and treatment. The genetic analysis of *T. brucei* using techniques such as amplified fragment length polymorphism (AFLP) and mobile genetic elements (MGE), PCR has revealed distinct genetic profiles for different subspecies, such as *T. b. gambiense*, *T. b. brucei*, and *T. b. rhodesiense* [28, 29].

The molecular identification of *Trypanosoma theileri* has been accomplished using PCR and sequencing of specific genes such as cathepsin L-like (CatL) and the ITS of the 18S gene [6, 30–32]. Phylogenetic analysis based on these gene sequences has revealed the presence of different genotypes and lineages of *T. theileri*, including IC, IB, IIB, TthI, TthII, and a putative lineage TthIII. These techniques have allowed for the identification of genetic markers that can differentiate between subspecies and determine their relatedness.

To classify species of *Phytomonas*, *Leptomonas*, and *Crithidia*, scientists primarily rely on genetic markers such as the 18S ribosomal RNA, glyceraldehyde-3-phosphate dehydrogenase (gGAPDH), and mitochondrial genes such as Cyt *b* [33–35]. These genes are crucial for understanding the evolutionary relationships and comparing genetic sequences among different species within these genera. Additionally, specific genetic markers, including ITS and particular proteins, help distinguish strains or lineages within each genus. Unlike the protozoa mentioned earlier, the classification of these trypanosomatids predominantly depends on these genes.

2.3 Clinic and Infection

The Trypanosomatidae family encompasses a diverse group of protozoa with significant zoonotic potential, leading to severe medical and veterinary diseases. Among the approximately 120 identified *Trypanosoma* species affecting animals, roughly 10% are recognized as zoonotic, primarily transmitted by insect vectors. Key species such as *T. cruzi*, *T. vivax*, *T. evansi*, and *T. theileri* play crucial roles in causing diseases in both humans and domestic animals across the Americas. Simultaneously, *T. brucei*, the causative agent of African trypanosomiasis, remains a major public health concern primarily impacting sub-Saharan Africa [9].

Chagas disease, stemming from *T. cruzi* infection, affects approximately 10 million people worldwide and has extended beyond its initial endemic regions due to globalization. It manifests in acute and chronic phases, potentially resulting in Chagas cardiomyopathy or megasymdromes characterized by anomalous growth of the esophagus and colon. Chagas disease has two phases: an acute phase, usually asymptomatic or presenting with nonspecific febrile symptoms, and a chronic phase, which can manifest as indeterminate or determinate forms, with various organ-specific manifestations potentially arising. The incomplete clearance of the parasite by the host immune response contributes to the persistence of the disease [36].

T. brucei is a protozoan parasite responsible for human African trypanosomiasis (HAT), commonly known as sleeping sickness, affecting approximately 70 million people in 36 countries. Transmitted by the tsetse fly, the disease unfolds in two stages. The initial hemolymphatic stage presents with symptoms such as fever, headache, joint pain, and itching, progressing to severe neurological complications if left untreated [37]. The subspecies, *T. b. gambiense* and *T. b. rhodesiense*, cause chronic and acute forms, respectively [28, 38].

Additionally, *T. vivax*, *T. evansi*, and *T. theileri* contribute to animal trypanosomiasis, impacting ruminants, horses, dogs, and other domestic animals. *T. vivax* causes African animal trypanosomiasis (AAT) in various domestic and wild animals, primarily affecting cattle but also impacting other mammals, leading to economic losses in livestock production [39]. Clinical manifestations in infected animals include fever, anemia, weight loss, and weakness. *T. evansi* is responsible for Surra, affecting domestic and wild animals such as horses, camels, and buffaloes [40]. Clinical signs include fever, anemia, weight loss, and neurological symptoms, with disease severity varying among host species. *T. theileri*, a nonpathogenic trypanosome in cattle, lacks significant clinical effects but can complicate the diagnosis of more harmful trypanosome species in cattle populations [31]. Nonpathogenic species such as *T. rangeli* are also part of this diverse family [41].

Leishmaniasis, caused by various *Leishmania* species, affects millions of people globally, with distinct clinical forms including cutaneous and visceral Leishmaniasis [13]. Different *Leishmania* species serve as causative agents for various clinical forms of the disease. *Leishmania* is estimated to cause 700,000 to 1 million new cases annually, described in approximately 90 countries and affecting more than 12 million people. Visceral Leishmaniasis (VL) causes severe systemic symptoms, including fever, weight loss, and hepatosplenomegaly. Cutaneous Leishmaniasis (CL) is the most prevalent form. Understanding the diverse manifestations of Leishmaniasis is crucial for effective diagnosis and treatment strategies.

Crithidia represents a genus of monoxenous trypanosomatids primarily found in insects. While not major contributors to human diseases, *Crithidia* species serve as valuable models for understanding basic biological processes in trypanosomatids. However, this genus has recently been associated with fatal visceral Leishmaniasis, especially in coinfection with *Leishmania infantum*, contributing to our broader understanding of the evolutionary adaptations within this diverse group of protozoa [42–44]. *Phytomonas* is a genus of trypanosomatids uniquely adapted to infect plants, transmitted by phytophagous insects and leading to diseases in various plant species. While not directly associated with human or animal infections, *Phytomonas* highlights the versatility of trypanosomatids in colonizing diverse hosts and ecosystems [45].

These infections result in substantial economic losses and various health complications, including death or misdiagnoses due to false positives. The intricate web of Trypanosomatidae underscores the critical importance of developing effective detection and surveillance methods to curb the spread of these diseases in both human and animal populations.

3 Sequencing Strategies in Amplicon-Based Sequencing

The history of sequencing platforms is marked by significant milestones, beginning with Frederick Sanger's pioneering work in 1977, laying the foundation for DNA sequence decoding. Next-generation sequencing (NGS) platforms, such as Illumina and Ion Torrent, revolutionized genomics in the mid-2000s, enabling large-scale and cost-effective DNA sequencing [46, 47]. Third-generation sequencing technologies, exemplified by Pacific Biosciences and Oxford Nanopore Technologies (ONT), introduced long-read and nanopore sequencing, with ONT providing portability and real-time capabilities [3, 48]. Ongoing advancements in sequencing platforms enhance speed, accuracy, and cost-effectiveness. Amplicon-based sequencing, rooted in Kary Mullis' development of the polymerase chain reaction (PCR) in the 1980s, gained prominence with NGS. This method involves PCR amplification of specific DNA regions followed by high-throughput sequencing, finding applications in targeted resequencing and microbial diversity studies. Illumina platforms initially contributed to its evolution. The advent of third-generation sequencing brought innovations, overcoming limitations associated with short-read technologies [49]. Amplicon-based sequencing remains versatile in genomics, clinical diagnostics, and environmental studies, offering high sensitivity, cost-effectiveness, and focused genomic region analysis. It has made significant contributions to biodiversity assessments, ecosystem health monitoring, clinical diagnostics, cancer genomics, and infectious disease identification. Early studies exploring microbiome analysis and environmental DNA studies played a foundational role in shaping amplicon-based sequencing's trajectory.

Amplicon-based sequencing, particularly leveraging the 16S rRNA gene marker for bacteria and phagos relationship, has been transformative in both bacterial and virus-focused studies. In bacterial research, it unravels the complexities of the human microbiome, providing insights into microbial community structures and their implications for human health. The method tracks antibiotic resistance genes' dynamics, aiding in understanding their evolution and dissemination. It facilitates rapid detection of emerging bacterial pathogens, crucial for prompt response strategies. Beyond clinical use, it is extensively employed in environmental studies, monitoring contamination and revealing bacterial community dynamics in diverse ecosystems. In virus-focused studies, key genes such as capsid, envelope, glycoproteins, and polymerase are targeted to understand viral diversity and evolution. Amplicon sequencing of genes related to viral tropism, host interaction, and immune evasion deciphers the genetic makeup of viral populations, aiding in species identification and community characterization. This method's continuous evolution, coupled with innovative bioinformatics, positions it as crucial in unraveling the intricacies of both bacterial and viral ecosystems, impacting health, agriculture, and ecosystems.

Amplicon-based sequencing, with a focus on specific genes like 18S, has proven indispensable in examining blood samples for detecting *Plasmodium* species causing malaria and filarial parasites. In stool samples, this sequencing approach has demonstrated effectiveness in identifying intestinal protozoa such as *Giardia* and *Cryptosporidium*. The genetic insights gained from these studies significantly contribute to our understanding of parasitic infections, shaping diagnostic methodologies and enriching molecular parasitology. Ongoing research continually expands knowledge in this domain, with recent findings shedding light on the diversity, evolution, and clinical relevance of various parasitic organisms. For the latest and most comprehensive information, exploring recent publications and scientific databases is recommended. In the field of Trypanosomatids, limited studies have explored genes related to 18S, HSP70 (heat shock protein 70), miniexon, dnaK, and cytochrome in human, reservoirs, and insect vectors. Specific genes like tcscd5 have been employed, especially in the study of *T. cruzi* in congenital infections, providing crucial insights into the genetic makeup of this pathogenic organism. Despite the relative scarcity of studies on Trypanosomatids compared to other microbial groups, these investigations carry substantial implications for parasitology, playing a pivotal role in the development of diagnostic tools and targeted interventions against parasitic infections.

4 Oxford Nanopore Technology

In recent years, Oxford Nanopore Technologies (ONT) has been a novel nucleic acid sequencing, pioneering nanopore-based technologies since its establishment in 2005. The introduction of portable sequencers, such as the MinION in 2012, and subsequent releases such as the GridION and PromethION, underscores ONT's commitment to refining and expanding sequencing capabilities, ushering in an era of real-time, long-read genomics [50].

The Oxford Nanopore sequencing workflow begins with end repair, where DNA fragments undergo enzymatic treatment to mend their ends for subsequent adapter ligation, ensuring uniformity in DNA structure and optimizing library preparation efficiency. After end repair, DNA or RNA samples are ligated with specific adapters, allowing interaction with nanopores during sequencing. The nanopore sequencing device, such as the MinION or GridION, draws the sample through nanopores embedded in a membrane while an electric current passes through. Disruptions in the electric current caused by DNA or RNA molecules passing through the nanopores generate distinct signal patterns interpreted by the device's electronics, which are then converted into base calls representing sequence information. Bioinformatics tools analyze the base calls for alignment to a reference genome or assembly into contigs, extracting biological insights (Fig. 2b).

Oxford Nanopore sequencing offers real-time data generation, portability, and scalability, facilitating diverse genomic and transcriptomic applications in research and diagnostics. By detecting voltage disruptions, ONT enables the real-time sequencing of both DNA and RNA in portable devices, significantly enhancing the feasibility and efficiency of nucleic acid sequencing. Nanopore sequencing operates on a groundbreaking principle, threading single-stranded nucleic acid molecules through minuscule pores in a membrane, and observing alterations in ionic currents during the molecule's passage allows for real-time detection and identification of DNA or RNA sequences [51]. This technology, renowned for producing extended sequencing reads and facilitating real-time analyses without prior amplification, has become transformative in genomics, finding versatile applications in genetics, diagnostics, and environmental research [52, 53]. The ability to obtain data simultaneously as sequencing occurs adds to the real-time nature of this approach, further streamlining and expediting genomic analyses [54].

Nanopore sequencing goes beyond individual genome sequencing, offering a versatile approach for the simultaneous analysis of multiple samples, including amplicons derived from various PCR genomic products [4]. Through the integration of unique DNA barcodes during library preparation, this technology facilitates the pooling of amplicons from different sources into a single sequencing library [55]. Whether targeting entire genomes or multiple amplicons, the nanopore sequencing platform allows for a comprehensive and high-throughput analysis in a single sequencing run. This ability to sequence entire genomes or multiple amplicons simultaneously enhances the efficiency and scalability of nanopore sequencing, positioning it as an invaluable tool for diverse applications, from whole-genome sequencing to targeted studies of specific genomic regions (Fig. 2b).

Nanopore amplicon-based sequencing stands out as a powerful tool for virus detection and identification, offering rapid and accurate sequencing of viral genomes across various strains. It has been successfully applied to hepatitis A virus (HAV), Crimean-Congo Hemorrhagic Fever Virus (CCHFV), Jingmen tick virus (JMTV), Seoul orthohantavirus (SEOV), Zika virus, and SARS-CoV-2. With high genome coverage rates and accurate consensus sequences, nanopore sequencing is reliable for molecular

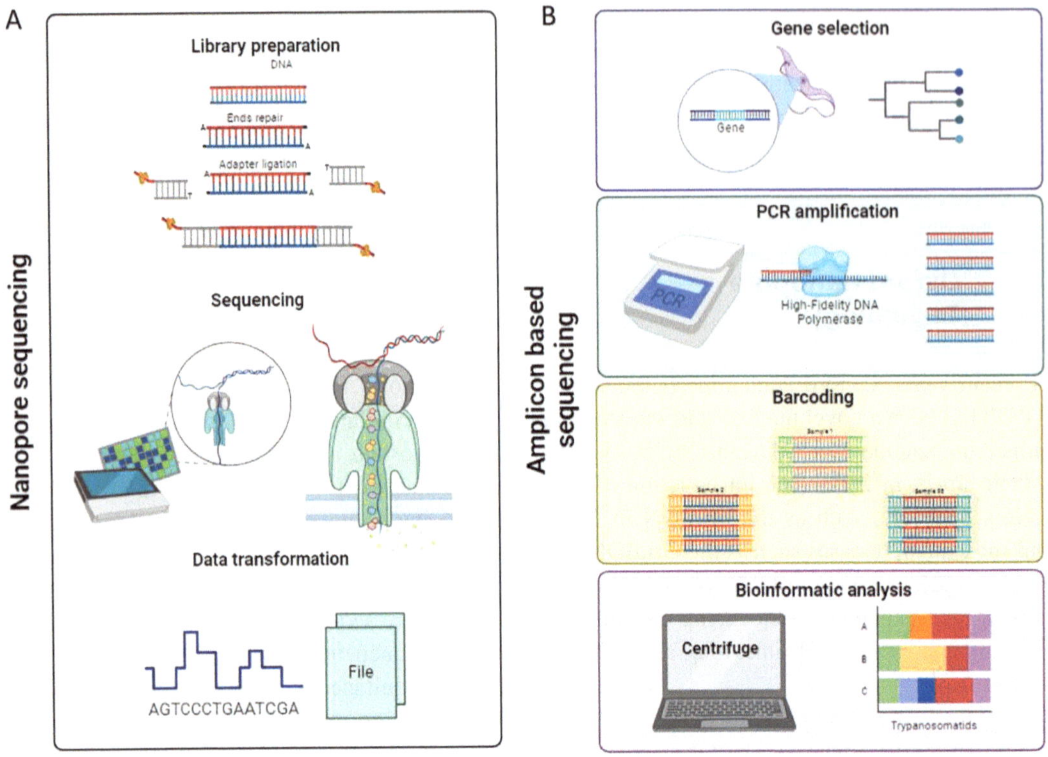

Fig. 2 Oxford Nanopore Technology. (**a**) Steps during Oxford Nanopore Technology—Library preparation: DNA is prepared for sequencing. Adapter ligation: Adapters are attached to DNA fragments. Sequencing: DNA is sequenced using a nanopore device. Data transformation: Sequencing data is processed for analysis. (**b**) Step for amplicon sequencing based on Oxford Nanopore Technology—Gene selection: Specific genes are chosen for analysis. PCR amplification: DNA is amplified using PCR with high-fidelity DNA polymerase. Barcoding: Unique barcodes are added to DNA fragments. Bioinformatic analysis: Data is analyzed using bioinformatics tools like centrifuge for taxonomic classification

diagnosis and epidemiological studies. Its real-time capability, portability, and cost-effectiveness make it suitable for point-of-care testing and outbreak surveillance. Leveraging adaptive sampling and bioinformatics analysis enhances the efficiency and accuracy of nanopore sequencing. However, in contrast to viruses, the application of nanopore amplicon sequencing in protozoa remains underexplored, with limited studies investigating its potential. This gap underscores the need for further research to unlock the full capabilities of nanopore technology in protozoan detection and identification, highlighting an area ripe for exploration and innovation.

Amplicon-based sequencing with Oxford Nanopore Technologies (ONT) holds significant promise for advancing bacterial epidemiological studies. ONT's Flongle sequencing platform enables rapid identification of bacteria and detection of antimicrobial resistance genes (ARGs) directly from blood cultures. Metagenomic sequencing analysis (mNGS) with ONT's nanopore adaptive sampling (NAS) technology enhances microbial detection resolution. Hybrid assembly, combining long-read nanopore sequencing and short-read Illumina sequencing, is the most accurate and cost-effective approach for large-scale genomic epidemiological studies. ONT-based amplicon sequencing is instrumental in studying microbial communities and bacterial serotyping, providing high accuracy in generating full-length 16S rRNA gene amplicon sequences. The real-time capability of ONT sequencers allows for adaptive sampling and realistic software testing. Although cluster detection remains challenging, ONT sequencing reliably identifies multilocus sequence types (MLST), serotypes, and antimicrobial resistance determinants, showing promise for genomic epidemiology and public health investigations. Furthermore, ONT sequencing is effective in rapid genotyping and disease diagnosis in livestock, positioning itself as a valuable tool for epidemiology in bacteria, offering reliable genotyping and disease detection capabilities.

Nanopore amplicon sequencing has emerged as a valuable tool for detecting and identifying protozoa across various scenarios. Despite its proven effectiveness, research in this area remains limited. Patiño et al. employed amplicon-based NGS to detect trypanosomatid species in diverse hosts and vectors. Matsuo introduced a versatile two-step PCR method for nanopore amplicon sequencing, offering customizable primer sets. Early et al. developed the PASEC pipeline to enhance analysis of low-density *Plasmodium* DNA samples. Additionally, Díaz-Viraqué et al. demonstrated improved assembly of the *T. cruzi* Berenice strain using Illumina short reads and MinION long reads. Moreover, our group pioneered an amplicon-based sequencing approach utilizing 18S for trypanosomatid identification and genotyping, broadening nanopore technology's scope in protozoan research. The limited exploration underscores the imperative for further investigation to fully exploit nanopore sequencing's potential in protozoan detection and identification.

5 Development of Amplicon-Based Sequencing Based on Oxford Nanopore Technology in Trypanosomatids

The use of ONT in amplicon-based sequencing is greatly enhanced by its ability to analyze multiple samples simultaneously, streamlining the process and increasing efficiency. Strategies like multilocus sequence typing (MLST) and barcoding, relying on simple sequence markers, have been pivotal in deepening our understanding of pathogens [16, 56, 57]. For instance, MLST has been employed to study the diversity of *T. cruzi* strains, analyzing genes such as gGAPDH and Cyt *b*. Key markers in Trypanosomatids, including Cyt *b*, kDNA, the intergenic region of the miniexon gene, and fragments of 18S rRNA and heat shock protein 70 (HSP70), have played essential roles [12, 33, 58, 59]. Concatenation studies involving these genes have provided valuable insights into genetic diversity and population dynamics [17]. By leveraging these markers and sequencing approaches, researchers have gained a deeper understanding of Trypanosomatids' genetic diversity, evolutionary relationships, and population dynamics. For instance, studies analyzing the diversity of Cyt *b* in Trypanosomatids have revealed distinct genetic lineages and provided insights into their evolutionary history and adaptation strategies.

Despite the significant contributions of strategies such as MLST and barcoding, their limitations have become apparent. These approaches often rely on short gene fragments, making precise classification and detection of infections challenging due to limited capacity and depth. For instance, DTU characterization in *T. cruzi* relied on sequencing these genes, and discrimination between *Leishmania* species currently relies on sequencing the HSP70 gene [12, 24]. Similar sequencing approaches can differentiate species within the genus *Trypanosoma*, such as *T. evansi* and *T. theileri*, affecting animals [24, 60, 61].

Despite the significant contributions of strategies such as MLST and barcoding, which leverage simple sequence markers, their limitations have become apparent. These approaches often rely on short gene fragments, making precise classification and detection of infections challenging due to limited capacity and depth. This complexity is amplified when dealing with uneven abundance of genotypes or species, especially in samples from humans, vectors, and reservoirs [57, 62–64]. Accurate classification is crucial for effective diagnosis, treatment monitoring, and epidemiological studies of Trypanosomatids. Consequently, there is a recognized need for more sophisticated and comprehensive sequencing approaches to overcome these challenges and comprehend the intricate dynamics within trypanosomatid populations.

In response to these challenges, amplicon-based sequencing strategies have gained prominence. Recently, authors have started incorporating ONT as a platform for developing these strategies, showcasing high specificity, sensitivity, utility, and simplicity in implementation. Drawing from the experience in amplicon-based sequencing development, we outline some key considerations for further refinement of this technique.

5.1 Genomic Target Selection

In the intricate realm of Trypanosomatids, where multigene families proliferate and recombination rates soar, selecting genomic targets becomes a pivotal task. Researchers opt for genes less entwined within multigene families to mitigate genetic variability triggered by recombination events. This strategy allows for the identification of more stable and conserved genomic regions, enhancing the precision of genetic diversity assessments across Trypanosomatid species or genotypes. Trypanosomatids undergo genetic mutations, leading to genomic plasticity and adaptation to adverse conditions [65]. Pseudogenes, altered genes losing their original function, offer insights into the evolutionary history of functional genes in trypanosomatids. These parasites showcase unique features, such as intronless polycistronic transcription, trans-splicing, and RNA editing, significantly influencing gene expression and metabolism [66]. Moreover, they employ mechanisms for DNA damage repair and adaptation, including chromosomal rearrangements, gene expansion, and DNA polymerase gene duplication [67]. The emphasis on stability and conservation in genomic targets underscores the quest for reliable and reproducible outcomes in amplicon sequencing studies, especially vital in unraveling the complex genetic dynamics of trypanosomatids. By honing in on regions less susceptible to genetic variability, researchers can unravel evolutionary relationships and population dynamics within Trypanosomatid species, fostering a deeper understanding of their biology and pathogenicity.

Key considerations include resolution, with a focus on genomic regions offering sufficient discriminatory power for distinguishing closely related species or strains. The selection process also addresses copy number variation in target genes, preventing biases in genomic material quantification. In silico analysis of single nucleotide polymorphisms (SNPs) within chosen genomic targets provides insights into genetic variations among different isolates. Additionally, discriminating between species and strains is achieved through phylogenetic analysis based on the chosen targets.

For precise molecular characterization of *Trypanosoma*, *T. cruzi*, and *Leishmania* species, common gene selections include the ITS and 18S ribosomal RNA (18S rRNA) for *Trypanosoma* species, Cyt *b* for *Trypanosoma*, and HSP70 and Miniexon for *T. cruzi*. *Leishmania* relies on gp63, ITS, and cysteine proteinase B (cpb) for strain characterization [20, 24, 30, 60, 61, 68–71]. These gene choices balance variability and precision, tailored to specific taxonomic levels, ensuring accurate identification of species and genotypes (Fig. 3a).

We construct a Venn diagram that provides a glimpse into the genes utilized for classifying various species of Trypanosomatids, including *T. cruzi*, *Leishmania*, *T. brucei*, and others. Notably, the 18S rRNA gene appears universally across all species, underlining its pivotal role in phylogenetic analysis and taxonomy within Trypanosomatids as we showed at the beginning of this chapter. Furthermore, the ITS (internal transcribed spacer) region is shared by *T. cruzi*, *Leishmania* and *T. brucei*, hinting at potential evolutionary connections among these species. However, each species boasts distinct genetic markers: *T. cruzi* features the Mini-exon, *Leishmania* contains gp63, *T. brucei* possesses VSGs, and other Trypanosomatids exhibit unique genes such as gGAPDH and CatL. This genetic diversity underscores the evolutionary trajectories and biological complexities inherent to these parasitic organisms (Fig. 3b).

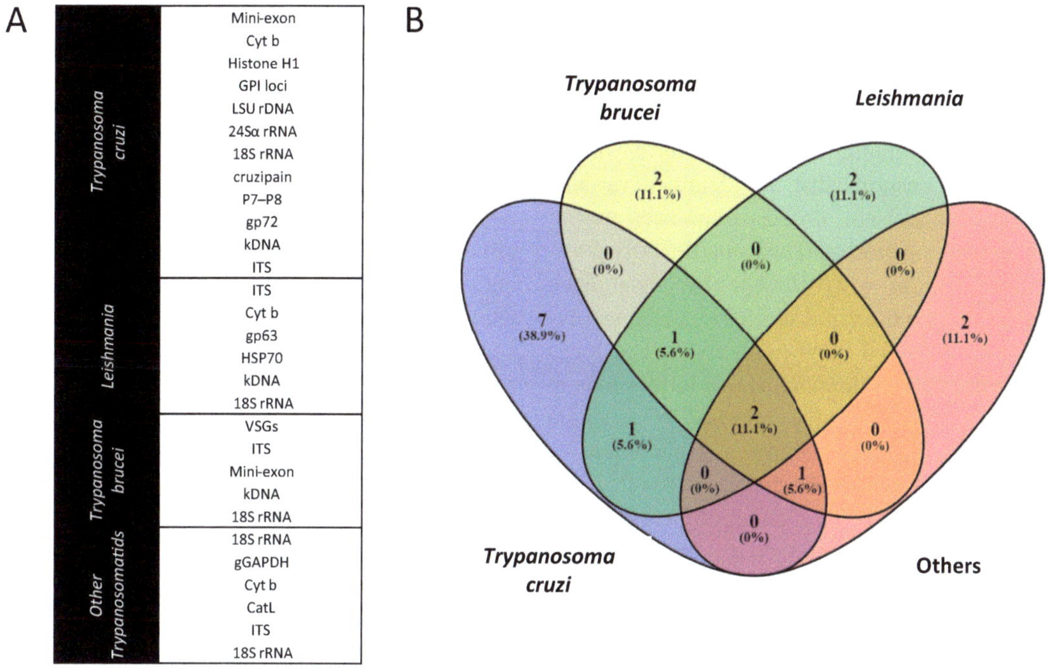

Fig. 3 Gene target for molecular characterization. (**a**) Gene used for molecular characterization in trrypanosomatids. (**b**) Common genes markets used for molecular characterization—Venn diagram. 18S rRNA: universal across all species. ITS: shared by *T. cruzi*, *Leishmania*, and *T. brucei*

5.2 Amplicon and Primers Design

Designing primers for specific and sensitive discrimination of species or genotypes in Trypanosomatids requires an evidence-based approach. Start by obtaining curated sequences from databases for each species or genotype of interest, aligning them using tools like MAFFT, and extracting a consensus sequence [72]. Utilize tools such as BLASTprimer for primer design, considering the expected amplicon sizes. Leverage ONT's capability to sequence entire genes by limiting primer design to expected amplicon sizes. After selecting the primer pair, scrutinize the alignment for polymorphisms, ensuring primer specificity for the target species and review the difference Tm between primer, it should not be more than five. Additionally, evaluate the probability of secondary structures such as hairpin and dimers.

It is essential to address potential risks. Having a minimum of two sequences per species is advisable, and in cases of high variability, such as the TcI DTU in *T. cruzi*, include all known variants. Overlooking polymorphisms or using nonspecific primers can lead to inaccurate results, impacting the precision of species or genotype discrimination. Therefore, a thorough alignment-based approach, complemented by in-depth analysis, is crucial for reliable primer design in Trypanosomatids.

5.3 PCR Standardization

Achieving reliable results in amplicon-based sequencing begins with robust PCR standardization, a process critical for consistent outcomes. The significance of employing high-fidelity, long-read polymerases in amplicon-based sequencing extends to the accurate detection of genetic variants that happened in human and bacteria, especially in scenarios where only a few single nucleotide polymorphisms (SNPs) differentiate closely related trypanosomatid species [73–75]. Given the relatively small variations in their genomic sequences, the ability to precisely capture and distinguish these subtle differences becomes crucial for accurate genotyping and species identification. High-fidelity polymerases contribute to minimizing errors during DNA replication, ensuring that the identified variants are genuine and not artifacts introduced during the sequencing process. This is particularly pertinent in regions where the genetic divergence between trypanosomatid species is subtle, involving only a limited number of SNPs. The long-read capability of the polymerase further aids in spanning these variable regions in a single sequencing read, providing a comprehensive view of the genomic landscape.

Consideration of the number of PCR cycles is crucial in amplicon-based sequencing (ABS) as it can impact the abundance of the test results and include bias [76]. Standardizing a uniform number of PCR cycles across all ABS reactions becomes essential to ensure consistency and reliability in the generated data. This approach aims to mitigate potential biases introduced by variations in PCR cycle numbers, providing a more uniform and comparable basis for evaluating the genomic information derived from different samples. By adopting a standardized number of PCR cycles for the entire amplicon-based sequencing process, researchers enhance the robustness of their results, facilitating accurate comparisons and interpretations across various genomic targets and experimental conditions.

Incorporating reference DNA from each variant under investigation is essential, providing a reliable benchmark for subsequent analyses. The standardization process involves initiating a conventional PCR with selected reference samples to evaluate the expected amplicon length. The goal is to obtain a unique band, ensuring the specificity of the amplification. In instances where the target gene remains unamplified or yields non-specific bands, careful adjustments in the number of cycles and melting temperature (Tm) are made. This iterative refinement continues until the expected and specific results are achieved.

Upon successfully standardizing the PCR conditions, the protocol is applied to all reference samples. It is imperative to exercise caution during this phase, as variations in the PCR conditions may introduce biases or inconsistencies in subsequent sequencing results. Rigorous validation, including assessing the uniformity of amplicon lengths across all variants, is crucial. To corroborate the obtained sequences, Sanger sequencing is employed. This orthogonal method provides an additional layer of validation, ensuring the accuracy of the amplicon sequences. It is vital to note that despite the meticulous standardization, risks such as contamination or technical errors during PCR setup can impact the reliability of the sequencing results. Implementing strict laboratory practices, including separate workspaces for PCR setup, helps mitigate these risks and ensures the integrity of the sequencing data.

5.4 ON Library Preparation and Real-Time Sequencing

Preparing sequencing libraries for ONT involves critical steps, including end-repair, barcode incorporation for multiplexing, and adapter ligation. The end-repair readies the DNA for subsequent processes, while barcodes enable simultaneous analysis of multiple samples. Adapter ligation ensures the PCR products are ready for sequencing on the MinIonMk1C instrument with MinKNOW software.

Considerations for library efficiency and quality are paramount. Including both negative and positive controls is advised. A positive control with a known parasite concentration serves as a reference, and a negative control helps identify cross-contamination between samples [77]. Caution is needed when dealing with samples of varying parasite concentrations, especially in patients with Chagas disease, where low concentrations may pose challenges. Additionally, samples from reservoirs may harbor diverse trypanosomatids, increasing the risk of contamination. In Illumina metagenomics, it is a known fact that samples with high concentrations can contaminate other samples in the same run with lower concentrations [78]. Hence, meticulous planning and selective barcode use are crucial to ensure accurate and reliable sequencing results, particularly in the presence of these complexities. ONT's capability for a 96-sample run emphasizes high-throughput analysis, but controls and strategic barcode use are vital in the face of diverse sample characteristics.

Following library preparation, a crucial step involves base refinement to enhance the accuracy of sequencing data. Guppy, a specialized software, is employed for this purpose. Guppy refines the raw signal data generated during sequencing, improving base-calling accuracy [79]. This refinement is integral in reducing errors inherent in nanopore sequencing, thereby enhancing the overall quality of the data [80]. Achieving a high-quality score is pivotal, as it directly correlates with the reliability of the sequence information obtained. This meticulous base refinement process significantly contributes to the precision and fidelity of downstream analyses, ensuring robust and accurate outcomes.

5.5 Bioinformatic Analysis

The intricate genetic variability among Trypanosomatids presents formidable challenges in diagnostics and genotyping, particularly when utilizing long reads. Amplicon-based sequencing, a valuable tool in this context, relies on bioinformatic analyses to decipher complex genomic variations. Traditional tools like BLAST, while widely used, face limitations in accurately discriminating between closely related variants. This is due to their reliance on local alignment, which may not adequately capture the nuances of genetic diversity, resulting in misassignments between species and discrete typing units (DTUs).

To address these challenges, newer algorithms such as DADA2 have emerged, providing a robust alternative, especially for short-read data. DADA2, by modeling amplicon errors, distinguishes true biological variation from sequencing errors, enhancing the reliability of assignments. However, for long-read data generated by platforms like Oxford Nanopore sequencing, DADA2 may encounter challenges due to increased error rates and complex genomic structures [81].

In this landscape, tools like Centrifuge shine as they offer efficient taxonomic classification, particularly adept at handling long-read data complexities. Centrifuge's rapid and accurate hashing-based algorithm makes it a valuable asset that requires less computational resources, complementing bioinformatic analyses and improving taxonomic assignment accuracy [82, 83].

A crucial aspect for a successful bioinformatic analysis is a well-curated and validated reference database. Utilizing a database that has been thoroughly curated through alignments and phylogenies ensures consistency and reliability in species and DTU assignments. Navigating the intricate genetic landscape of Trypanosomatids requires a nuanced bioinformatic strategy, especially with the advent of long-read technologies. Hybrid approaches, incorporating tools such as BLAST, DADA2, and Centrifuge, offer a comprehensive analysis, leveraging the strengths of each tool. In cases of discrepancies, a meticulous examination of assigned reads unveils errors and enhances assignment accuracy. Given the challenges posed by genetic diversity, a cautious and multi-methodological approach is paramount to building confidence in bioinformatic analyses. Validating assignments through various means, including direct scrutiny of assigned sequence reads, adds robustness to the exploration of the intricate genetic dynamics within Trypanosomatids.

6 Field Implementation

The groundbreaking studies on Trypanosomatids and *Leishmania*, harnessing the capabilities of ONT, have ushered in a new era of understanding and managing these parasitic infections.

The application of ONT in Trypanosomatids research, particularly through the innovative amplicon-based sequencing (ABS) method, signifies a notable breakthrough in diagnostic precision and genotyping accuracy [2]. A recent study targeting the 18S rDNA gene has demonstrated remarkable capabilities, successfully distinguishing between 11 *Trypanosoma* species and six *T.*

cruzi discrete typing units (DTUs). Beyond this, the method has unveiled coinfections and displayed a limit of detection, reaching as low as one parasite per mL. This cutting-edge approach, marked by its high resolution, sensitivity, and accuracy, represents a substantial leap forward in understanding the prevalence and complexity of trypanosomatid infections. Its versatility is evident in its successful application across diverse samples, encompassing humans, animals, and vectors. Importantly, the method holds potential for application in Chagas samples related to oral outbreaks and chronic cases, as well as in reservoirs and domestic animals infected with various non-cruzi trypanosome species. Moreover, its utility extends to vector samples, contributing significantly to the enhanced comprehension and monitoring of trypanosomatid infections and transmission cycles (Data in revision for publication).

Moreover, similar to how a generic 18S gene allowed discrimination among trypanosomatids species, more specific genes such as HSP70 have proven instrumental in distinguishing between *Leishmania* species. HSP70 demonstrates exceptional specificity and discriminatory capabilities. Its application in clinical samples from humans and reservoirs with cutaneous Leishmaniasis (CL) and visceral Leishmaniasis (VL) not only distinguishes between different *Leishmania* species but also identifies instances of coinfections. This study underscores the outstanding specificity and effectiveness of HSP70-based ONT sequencing, offering valuable insights into the dynamics of diverse *Leishmania* species and coinfection events [2].

Moving beyond the laboratory, the field implementation of ONT technology emerges as a game-changer, especially in remote and endemic areas [84]. The portability and logistical efficiency of ONT sequencing make it an invaluable asset for on-site diagnostics and surveillance even under condition of solar panel energy [85]. The amplicon-based sequencing (ABS) method's high resolution, sensitivity, and accuracy empower researchers and healthcare professionals to gain real-time insights into the prevalence and diversity of Trypanosomatids as has been demonstrate in genes such as 18S and HSP70 for those parasites and even in other parasites such as *Blastocystis* [2, 86]. This not only enhances our understanding of these parasitic infections but also catalyzes advancements in field-based research and interventions, ushering in a more effective era of disease management strategies. The integration of ONT technology and tailored sequencing methods represents a significant leap forward in precision medicine and infectious disease control.

7 Challenges and Ethical Considerations

Boarding on the widespread implementation of amplicon-based sequencing (ABS) for trypanosomatids poses challenges and ethical considerations that warrant careful consideration. Scaling up ABS introduces logistical and technical hurdles, with the sheer volume of samples potentially straining resources. It requires optimized workflows to ensure efficiency, addressing issues related to standardization across laboratories and potential variations in sample quality. Additionally, robust infrastructure and skilled personnel are crucial for the accurate and timely analysis of extensive ABS datasets.

To enhance the large-scale implementation of ABS for trypanosomatids, collaborative efforts between research institutions, organizations, and governments are essential. Sharing resources can streamline collective capacity to handle a high volume of samples. Developing standardized protocols adaptable across various laboratories would mitigate challenges related to standardization and variations in sample quality. Investing in training programs for personnel involved in ABS would enhance their skills, ensuring proficient management of the technology and data [87].

Oxford Nanopore Technologies (ONT), renowned for its portable and real-time sequencing capabilities, faces challenges related to potentially higher error rates, especially in homopolymeric regions [88, 89]. The rapid nanopore sequencing process, while efficient, may introduce inaccuracies during base calling, particularly in repetitive nucleotide sequences. Despite these challenges, continuous advancements in base-calling algorithms aim to improve accuracy, solidifying ONT's status as a transformative sequencing technology. Researchers utilizing ONT should be mindful of these nuances, applying appropriate error correction strategies to ensure the reliability of their sequencing results.

To further improve ethical considerations, ongoing education and training programs for researchers and healthcare professionals on ethical best practices in genomics research could be implemented. Regular updates and adherence to evolving ethical guidelines would ensure that research practices align with the latest standards. Collaboration with ethical review boards remains essential, and periodic evaluations of data management practices can identify areas for improvement. Fostering open communication and trust-building efforts with study participants contributes to a more ethically sound implementation of ABS in trypanosomatid research.

8 Results and Future Perspectives

Up to the present moment, the implementation of Oxford Nanopore Technology (ONT) has been limited in comparison to other platforms such as Illumina [57, 90–92]. However, the broader application of amplicon-based sequencing, even on alternative platforms such as Illumina, has demonstrated its utility. The standardization of this method as a gold standard for diagnostics and genotyping could significantly enhance data availability and generalize the currently employed methods for Trypanosomatids. This expansion of available data would contribute to a more comprehensive understanding of these parasites, allowing for data-driven public health decisions that could control transmission and advance treatment of these infections. Currently, the relationship between many species or genotypes and drug resistance is not entirely clear, with information limited primarily to *Leishmania* species. The use of these advanced sequencing techniques could offer broader insights, paving the way for more personalized medicine based on a one-health approach.

Moreover, the analysis of the quantity of generated reads could shed light on the magnitude and frequency of interactions in mixed infections and coinfections. These techniques not only enable a more sensitive diagnosis but also provide access to genotyping, a challenging task with these parasites. Another notable advantage, particularly with ONT, is the portability of the technology. The ability to conduct sequencing in remote and resource-limited areas enhances accessibility and facilitates on-site diagnostics, contributing to a more widespread and efficient application of genomic tools in the field.

In conclusion, the potential benefits of standardizing amplicon-based sequencing, particularly with ONT, extend beyond mere diagnostics. They encompass a deeper understanding of parasite interactions, resistance patterns, and personalized treatment strategies. As these methods become more prevalent, the field holds promise for significant advancements in both research and clinical applications.

References

1. Desquesnes M, Gonzatti M, Sazmand A, Thévenon S, Bossard G, Boulangé A, Gimonneau G, Truc P, Herder S, Ravel S, Sereno D, Jamonneau V, Jittapalapong S, Jacquiet P, Solano P, Berthier D (2022) A review on the diagnosis of animal trypanosomoses. Parasit Vectors 15:64. https://doi.org/10.1186/s13071-022-05190-1
2. Patiño LH, Ballesteros N, Muñoz M, Jaimes J, Castillo-Castañeda AC, Madigan R, Paniz-Mondolfi A, Ramírez JD (2023) Validation of Oxford nanopore sequencing for improved New World Leishmania species identification via analysis of 70-kDA heat shock protein. Parasit Vectors 16:458. https://doi.org/10.1186/s13071-023-06073-9
3. Petersen LM, Martin IW, Moschetti WE, Kershaw CM, Tsongalis GJ (2019) Third-generation sequencing in the clinical laboratory: exploring the advantages and challenges of nanopore sequencing. J Clin Microbiol 58:e01315–e01319. https://doi.org/10.1128/JCM.01315-19
4. Karst SM, Ziels RM, Kirkegaard RH, Sørensen EA, McDonald D, Zhu Q, Knight R, Albertsen M (2021) High-accuracy long-read amplicon sequences using unique molecular identifiers with nanopore or PacBio

sequencing. Nat Methods 18:165–169. https://doi.org/10.1038/s41592-020-01041-y
5. Kaufer A, Ellis J, Stark D, Barratt J (2017) The evolution of trypanosomatid taxonomy. Parasit Vectors 10:287. https://doi.org/10.1186/s13071-017-2204-7
6. Rodrigues JCF, Godinho JLP, de Souza W (2014) Biology of human pathogenic trypanosomatids: epidemiology, lifecycle and ultrastructure. Subcell Biochem 74:1–42. https://doi.org/10.1007/978-94-007-7305-9_1
7. Wheeler RJ, Gluenz E, Gull K (2013) The limits on trypanosomatid morphological diversity. PLoS One 8:e79581. https://doi.org/10.1371/journal.pone.0079581
8. Horn D (2022) A profile of research on the parasitic trypanosomatids and the diseases they cause. PLoS Negl Trop Dis 16:e0010040. https://doi.org/10.1371/journal.pntd.0010040
9. Maslov DA, Opperdoes FR, Kostygov AY, Hashimi H, Lukeš J, Yurchenko V (2019) Recent advances in trypanosomatid research: genome organization, expression, metabolism, taxonomy and evolution. Parasitology 146:1–27. https://doi.org/10.1017/S0031182018000951
10. Votýpka J, Kment P, Yurchenko V, Lukeš J (2020) Endangered monoxenous trypanosomatid parasites: a lesson from island biogeography. Biodivers Conserv 29:3635–3667. https://doi.org/10.1007/s10531-020-02041-2
11. Balmer O, Beadell JS, Gibson W, Caccone A (2011) Phylogeography and taxonomy of Trypanosoma brucei. PLoS Negl Trop Dis 5:e961. https://doi.org/10.1371/journal.pntd.0000961
12. Zingales B, Miles MA, Campbell DA, Tibayrenc M, Macedo AM, Teixeira MMG, Schijman AG, Llewellyn MS, Lages-Silva E, Machado CR, Andrade SG, Sturm NR (2012) The revised Trypanosoma cruzi subspecific nomenclature: rationale, epidemiological relevance and research applications. Infect Genet Evol 12:240–253. https://doi.org/10.1016/j.meegid.2011.12.009
13. Akhoundi M, Kuhls K, Cannet A, Votýpka J, Marty P, Delaunay P, Sereno D (2016) A historical overview of the classification, evolution, and dispersion of Leishmania parasites and sandflies. PLoS Negl Trop Dis 10:e0004349. https://doi.org/10.1371/journal.pntd.0004349
14. Ramírez JD, Duque MC, Montilla M, Cucunubá ZM, Guhl F (2012) Multilocus PCR-RFLP profiling in Trypanosoma cruzi I highlights an intraspecific genetic variation pattern. Infect Genet Evol 12:1743–1750. https://doi.org/10.1016/j.meegid.2012.06.018
15. Sá ARN, Dias GBM, Kimoto KY, Steindel M, Grisard EC, Toledo MJO, Gomes ML (2016) Genotyping of Trypanosoma cruzi DTUs and Trypanosoma rangeli genetic groups in experimentally infected Rhodnius prolixus by PCR-RFLP. Acta Trop 156:115–121. https://doi.org/10.1016/j.actatropica.2016.01.006
16. Ramírez JD, Tapia-Calle G, Guhl F (2013) Genetic structure of Trypanosoma cruzi in Colombia revealed by a high-throughput nuclear multilocus sequence typing (nMLST) approach. BMC Genet 14:96. https://doi.org/10.1186/1471-2156-14-96
17. Ramírez JD, Duque MC, Montilla M, Cucunubá Z, Guhl F (2012) Natural and emergent Trypanosoma cruzi I genotypes revealed by mitochondrial (Cytb) and nuclear (SSU rDNA) genetic markers. Exp Parasitol 132:487–494. https://doi.org/10.1016/j.exppara.2012.09.017
18. Messenger LA, Miles MA, Bern C (2015) Between a bug and a hard place: Trypanosoma cruzi genetic diversity and the clinical outcomes of Chagas disease. Expert Rev Anti-Infect Ther 13:995–1029. https://doi.org/10.1586/14787210.2015.1056158
19. Llewellyn MS Genome-scale multilocus microsatellite typing of Trypanosoma cruzi discrete typing unit I reveals phylogeographic structure and specific genotypes linked to human infection. https://pubmed.ncbi.nlm.nih.gov/19412340/. Accessed 7 Sep 2023
20. Souto RP, Fernandes O, Macedo AM, Campbell DA, Zingales B (1996) DNA markers define two major phylogenetic lineages of Trypanosoma cruzi. Mol Biochem Parasitol 83:141–152. https://doi.org/10.1016/s0166-6851(96)02755-7
21. Herrera C, Guhl F, Falla A, Fajardo A, Montilla M, Adolfo Vallejo G, Bargues MD (2009) Genetic variability and phylogenetic relationships within Trypanosoma cruzi I isolated in Colombia based on miniexon gene sequences. J Parasitol Res 2009:897364. https://doi.org/10.1155/2009/897364
22. Mauricio IL, Stothard JR, Miles MA (2004) Leishmania donovani complex: genotyping with the ribosomal internal transcribed spacer and the mini-exon. Parasitology 128:263–267. https://doi.org/10.1017/s0031182003004578
23. Davila M, Pineda V, Calzada JE, Saldaña A, Samudio F (2021) Evaluation of cytochrome b sequence to identify Leishmania species and variants: the case of Panama. Mem Inst Oswaldo Cruz 116:e200572. https://doi.org/10.1590/0074-02760200572
24. Hernández C, Alvarez C, González C, Ayala MS, León CM, Ramírez JD (2014) Identification of six New World Leishmania species through the implementation of a high-resolution melting (HRM) genotyping assay. Parasit Vectors 7:501. https://doi.org/10.1186/s13071-014-0501-y
25. Kocher A, Valière S, Bañuls A-L, Murienne J (2018) High-throughput sequencing of kDNA amplicons for the analysis of Leishmania minicircles and identification of neotropical species. Parasitology 145:585–594. https://doi.org/10.1017/S0031182017002013
26. Mousavi T, Shokohi S, Abdi J, Naserifar R, Ahmadi M, Mirzaei A (2018) Determination of genetic diversity of Leishmania species using mini-circle kDNA, in Iran-Iraq countries border. Trop Parasitol 8:77–82. https://doi.org/10.4103/tp.TP_3_18
27. Spanakos G, Piperaki E-T, Menounos PG, Tegos N, Flemetakis A, Vakalis NC (2008) Detection and species identification of Old World Leishmania in clinical samples using a PCR-based method. Trans R Soc

Trop Med Hyg 102:46–53. https://doi.org/10.1016/j.trstmh.2007.05.019
28. Simo G, Cuny G, Demonchy R, Herder S (2008) Trypanosoma brucei gambiense: study of population genetic structure of central African stocks using amplified fragment length polymorphism (AFLP). Exp Parasitol 118:172–180. https://doi.org/10.1016/j.exppara.2007.07.010
29. Agbo EEC, Majiwa PAO, Claassen HJHM, Te Pas MFW (2002) Molecular variation of Trypanosoma brucei subspecies as revealed by AFLP fingerprinting. Parasitology 124:349–358. https://doi.org/10.1017/s003118200100138x
30. Cortez AP, Rodrigues AC, Garcia HA, Neves L, Batista JS, Bengaly Z, Paiva F, Teixeira MMG (2009) Cathepsin L-like genes of Trypanosoma vivax from Africa and South America--characterization, relationships and diagnostic implications. Mol Cell Probes 23:44–51. https://doi.org/10.1016/j.mcp.2008.11.003
31. De la Cadena E, Camacho M, Vaca F, Enríquez S, Eleizalde MC, Arrivillaga-Henríquez J, Mendoza M, Navarro JC, Ramírez-Iglesias JR (2023) Molecular identification of Trypanosoma theileri in cattle from the Ecuadorian Amazon. Vet Parasitol Reg Stud Rep 37:100824. https://doi.org/10.1016/j.vprsr.2022.100824
32. Jaimes-Dueñez J, Cantillo-Barraza O, Triana-Chávez O, Mejia-Jaramillo AM (2020) Molecular surveillance reveals bats from eastern Colombia infected with Trypanosoma theileri and Trypanosoma wauwau-like parasites. Prev Vet Med 184:105159. https://doi.org/10.1016/j.prevetmed.2020.105159
33. Borges AR, Engstler M, Wolf M (2021) 18S rRNA gene sequence-structure phylogeny of the Trypanosomatida (Kinetoplastea, Euglenozoa) with special reference to Trypanosoma. Eur J Protistol 81:125824. https://doi.org/10.1016/j.ejop.2021.125824
34. Frolov AO, Kostygov AY, Yurchenko V (2021) Development of Monoxenous Trypanosomatids and Phytomonads in insects. Trends Parasitol 37:538–551. https://doi.org/10.1016/j.pt.2021.02.004
35. Uttaro AD, Altabe SG, Rider MH, Michels PA, Opperdoes FR (2000) A family of highly conserved glycosomal 2-hydroxyacid dehydrogenases from Phytomonas sp. J Biol Chem 275:31833–31837. https://doi.org/10.1074/jbc.M006080200
36. Rassi A, Rassi A, Marin-Neto JA (2010) Chagas disease. Lancet 375:1388–1402. https://doi.org/10.1016/S0140-6736(10)60061-X
37. Büscher P, Cecchi G, Jamonneau V, Priotto G (2017) Human African trypanosomiasis. Lancet 390:2397–2409. https://doi.org/10.1016/S0140-6736(17)31510-6
38. Mudji J, Künzli E, Molyneux D, Blum J (2021) Long-term sequelae of congenital gambiense human African trypanosomiasis. Trans R Soc Trop Med Hyg 115:932–936. https://doi.org/10.1093/trstmh/trab082
39. Odeniran PO, Ademola IO, Macleod ET, Welburn SC (2018) Bovine and small ruminant African animal trypanosomiasis in Nigeria - a review. Vet Parasitol Reg Stud Rep 13:5–13. https://doi.org/10.1016/j.vprsr.2018.03.001
40. Aregawi WG, Agga GE, Abdi RD, Büscher P (2019) Systematic review and meta-analysis on the global distribution, host range, and prevalence of Trypanosoma evansi. Parasit Vectors 12:67. https://doi.org/10.1186/s13071-019-3311-4
41. de Sousa MA, da Silva FT, Dos Santos BN, Dos Santos Pereira SM, Carvalhal C, Hasslocher Moreno AM (2008) Trypanosoma rangeli Tejera, 1920, in chronic Chagas' disease patients under ambulatory care at the Evandro Chagas Clinical Research Institute (IPEC-Fiocruz, Brazil). Parasitol Res 103:697–703. https://doi.org/10.1007/s00436-008-1033-1
42. Fakhar M, Derakhshani-Nia M, Gohardehi S, Karamian M, Hezarjaribi HZ, Mohebali M, Akhoundi B, Sharbatkhori M (2022) Domestic dogs carriers of Leishmania infantum, Leishmania tropica and Crithidia fasciculata as potential reservoirs for human visceral leishmaniasis in northeastern Iran. Vet Med Sci 8:2329–2336. https://doi.org/10.1002/vms3.929
43. Rogerio LA, Takahashi TY, Cardoso L, Takamiya NT, de Melo EV, de Jesus AR, de Oliveira FA, Forrester S, Jeffares DC, da Silva JS, Ribeiro JM, Almeida RP, Maruyama SR (2023) Co-infection of Leishmania infantum and a Crithidia-related species in a case of refractory relapsed visceral leishmaniasis with non-ulcerated cutaneous manifestation in Brazil. Int J Infect Dis 133:85–88. https://doi.org/10.1016/j.ijid.2023.05.012
44. Takamiya NT, Rogerio LA, Torres C, Leonel JAF, Vioti G, de Sousa Oliveira TMF, Valeriano KC, Porcino GN, de Miranda Santos IKF, Costa CHN, Costa DL, Ferreira TS, Gurgel-Gonçalves R, da Silva JS, Teixeira FR, De Almeida RP, Ribeiro JMC, Maruyama SR (2023) Parasite detection in visceral Leishmaniasis samples by dye-based qPCR using new gene targets of Leishmania infantum and Crithidia. Trop Med Infect Dis 8:405. https://doi.org/10.3390/tropicalmed8080405
45. Jaskowska E, Butler C, Preston G, Kelly S (2015) Phytomonas: trypanosomatids adapted to plant environments. PLoS Pathog 11:e1004484. https://doi.org/10.1371/journal.ppat.1004484
46. Hu T, Chitnis N, Monos D, Dinh A (2021) Next-generation sequencing technologies: an overview. Hum Immunol 82:801–811. https://doi.org/10.1016/j.humimm.2021.02.012
47. Akintunde O, Tucker T, Carabetta VJ (2023) The evolution of next-generation sequencing technologies. arXiv:2305.08724v1
48. Kumar KR, Cowley MJ, Davis RL (2019) Next-generation sequencing and emerging technologies. Semin Thromb Hemost 45:661–673. https://doi.org/10.1055/s-0039-1688446
49. Pervez MT, Hasnain MJU, Abbas SH, Moustafa MF, Aslam N, Shah SSM (2022) A comprehensive review of performance of next-generation sequencing platforms. Biomed Res Int 2022:3457806. https://doi.org/10.1155/2022/3457806

50. Deamer D, Akeson M, Branton D (2016) Three decades of nanopore sequencing. Nat Biotechnol 34:518–524. https://doi.org/10.1038/nbt.3423
51. MacKenzie M, Argyropoulos C (2023) An introduction to nanopore sequencing: past, present, and future considerations. Micromachines (Basel) 14:459. https://doi.org/10.3390/mi14020459
52. Lin B, Hui J, Mao H (2021) Nanopore technology and its applications in gene sequencing. Biosensors (Basel) 11:214. https://doi.org/10.3390/bios11070214
53. Wang Y, Zhao Y, Bollas A, Wang Y, Au KF (2021) Nanopore sequencing technology, bioinformatics and applications. Nat Biotechnol 39:1348–1365. https://doi.org/10.1038/s41587-021-01108-x
54. Ying Y-L, Hu Z-L, Zhang S, Qing Y, Fragasso A, Maglia G, Meller A, Bayley H, Dekker C, Long Y-T (2022) Nanopore-based technologies beyond DNA sequencing. Nat Nanotechnol 17:1136–1146. https://doi.org/10.1038/s41565-022-01193-2
55. Whitford W, Hawkins V, Moodley KS, Grant MJ, Lehnert K, Snell RG, Jacobsen JC (2022) Proof of concept for multiplex amplicon sequencing for mutation identification using the MinION nanopore sequencer. Sci Rep 12:8572. https://doi.org/10.1038/s41598-022-12613-7
56. Messenger LA, Llewellyn MS, Bhattacharyya T, Franzén O, Lewis MD, Ramírez JD, Carrasco HJ, Andersson B, Miles MA (2012) Multiple mitochondrial introgression events and heteroplasmy in trypanosoma cruzi revealed by maxicircle MLST and next generation sequencing. PLoS Negl Trop Dis 6:e1584. https://doi.org/10.1371/journal.pntd.0001584
57. Dario MA, Moratelli R, Schwabl P, Jansen AM, Llewellyn MS (2017) Small subunit ribosomal metabarcoding reveals extraordinary trypanosomatid diversity in Brazilian bats. PLoS Negl Trop Dis 11:e0005790. https://doi.org/10.1371/journal.pntd.0005790
58. Maia de Souza R, Ruedas Martins RC, Moyses Franco LA, Tuon FF, de Oliveira Junior IG, Maia da Silva CA, Imamura R, Amato VS (2022) Identification of Leishmania species by next generation sequencing of hsp70 gene. Mol Cell Probes 61:101791. https://doi.org/10.1016/j.mcp.2022.101791
59. Lima L, Espinosa-Álvarez O, Ortiz PA, Trejo-Varón JA, Carranza JC, Pinto CM, Serrano MG, Buck GA, Camargo EP, Teixeira MMG (2015) Genetic diversity of Trypanosoma cruzi in bats, and multilocus phylogenetic and phylogeographical analyses supporting Tcbat as an independent DTU (discrete typing unit). Acta Trop 151:166–177. https://doi.org/10.1016/j.actatropica.2015.07.015
60. Geysen D, Delespaux V, Geerts S (2003) PCR-RFLP using Ssu-rDNA amplification as an easy method for species-specific diagnosis of Trypanosoma species in cattle. Vet Parasitol 110:171–180. https://doi.org/10.1016/s0304-4017(02)00313-8
61. Holland WG, Claes F, My LN, Thanh NG, Tam PT, Verloo D, Büscher P, Goddeeris B, Vercruysse J (2001) A comparative evaluation of parasitological tests and a PCR for Trypanosoma evansi diagnosis in experimentally infected water buffaloes. Vet Parasitol 97:23–33. https://doi.org/10.1016/s0304-4017(01)00381-8
62. Ardila MM, Herrera L, Zabala-Monterroza W, Bedoya-Polo A, Lozano-Arias D, García-Alzate R, Pérez-Doria A (2022) Molecular diagnosis of trypanosomatids in Didelphis marsupialis from Los Montes de María: a first report of Trypanosoma rangeli from Colombian Caribbean region. J Parasit Dis 46:323–327. https://doi.org/10.1007/s12639-021-01459-x
63. Burgos JM, Diez M, Vigliano C, Bisio M, Risso M, Duffy T, Cura C, Brusses B, Favaloro L, Leguizamon MS, Lucero RH, Laguens R, Levin MJ, Favaloro R, Schijman AG (2010) Molecular identification of Trypanosoma cruzi discrete typing units in end-stage chronic Chagas heart disease and reactivation after heart transplantation. Clin Infect Dis 51:485–495. https://doi.org/10.1086/655680
64. Schwabl P, Maiguashca Sánchez J, Costales JA, Ocaña-Mayorga S, Segovia M, Carrasco HJ, Hernández C, Ramírez JD, Lewis MD, Grijalva MJ, Llewellyn MS (2020) Culture-free genome-wide locus sequence typing (GLST) provides new perspectives on Trypanosoma cruzi dispersal and infection complexity. PLoS Genet 16:e1009170. https://doi.org/10.1371/journal.pgen.1009170
65. Abrahim M, Machado E, Alvarez-Valín F, de Miranda AB, Catanho M (2022) Uncovering pseudogenes and intergenic protein-coding sequences in TriTryps' genomes. Genome Biol Evol 14:evac142. https://doi.org/10.1093/gbe/evac142
66. Cabrera G (2016) Genome maintenance in Trypanosomatids. http://www.eurekaselect.com. http://www.eurekaselect.com/chapter/9921. Accessed 29 Jan 2024
67. Genois MM, Paquet ER, Laffitte MC, Maity R, Rodrigue A, Ouellette M, Masson JY (2014) DNA repair pathways in Trypanosomatids: from DNA repair to drug resistance. https://www.ncbi.nlm.nih.gov/pmc/articles/PMC3957735/. Accessed 29 Jan 2024
68. Liarte DB, Murta SMF, Steindel M, Romanha AJ (2009) Trypanosoma cruzi: multiplex PCR to detect and classify strains according to groups I and II. Exp Parasitol 123:283–291. https://doi.org/10.1016/j.exppara.2008.12.005
69. Ramírez JD, Guhl F, Rendón LM, Rosas F, Marin-Neto JA, Morillo CA (2010) Chagas cardiomyopathy manifestations and Trypanosoma cruzi genotypes circulating in chronic Chagasic patients. PLoS Negl Trop Dis 4:e899. https://doi.org/10.1371/journal.pntd.0000899
70. Brisse S, Verhoef J, Tibayrenc M (2001) Characterisation of large and small subunit rRNA and mini-exon genes further supports the distinction of six Trypanosoma cruzi lineages. Int J Parasitol 31:1218–1226. https://doi.org/10.1016/s0020-7519(01)00238-7
71. Duffy T, Bisio M, Altcheh J, Burgos JM, Diez M, Levin MJ, Favaloro RR, Freilij H, Schijman AG

(2009) Accurate real-time PCR strategy for monitoring bloodstream parasitic loads in chagas disease patients. PLoS Negl Trop Dis 3:e419. https://doi.org/10.1371/journal.pntd.0000419
72. Katoh K, Standley DM (2013) MAFFT multiple sequence alignment software version 7: improvements in performance and usability. Mol Biol Evol 30:772–780. https://doi.org/10.1093/molbev/mst010
73. Sze MA, Schloss PD (2019) The impact of DNA polymerase and number of rounds of amplification in PCR on 16S rRNA gene sequence data. mSphere 4:e00163-19. https://doi.org/10.1128/mSphere.00163-19
74. Wu J-Y, Jiang X-T, Jiang Y-X, Lu S-Y, Zou F, Zhou H-W (2010) Effects of polymerase, template dilution and cycle number on PCR based 16 S rRNA diversity analysis using the deep sequencing method. BMC Microbiol 10:255. https://doi.org/10.1186/1471-2180-10-255
75. Filges S, Yamada E, Ståhlberg A, Godfrey TE (2019) Impact of polymerase fidelity on background error rates in next-generation sequencing with unique molecular identifiers/barcodes. Sci Rep 9:3503. https://doi.org/10.1038/s41598-019-39762-6
76. Krehenwinkel H, Wolf M, Lim JY, Rominger AJ, Simison WB, Gillespie RG (2017) Estimating and mitigating amplification bias in qualitative and quantitative arthropod metabarcoding. Sci Rep 7:17668. https://doi.org/10.1038/s41598-017-17333-x
77. Hornung BVH, Zwittink RD, Kuijper EJ (2019) Issues and current standards of controls in microbiome research. FEMS Microbiol Ecol 95:fiz045. https://doi.org/10.1093/femsec/fiz045
78. Jurasz H, Pawłowski T, Perlejewski K (2021) Contamination issue in viral metagenomics: problems, solutions, and clinical perspectives. Front Microbiol 12:745076. https://doi.org/10.3389/fmicb.2021.745076
79. Wick RR, Judd LM, Holt KE (2019) Performance of neural network basecalling tools for Oxford nanopore sequencing. Genome Biol 20:129. https://doi.org/10.1186/s13059-019-1727-y
80. Zhang Y, Akdemir A, Tremmel G, Imoto S, Miyano S, Shibuya T, Yamaguchi R (2020) Nanopore basecalling from a perspective of instance segmentation. BMC Bioinformatics 21:136. https://doi.org/10.1186/s12859-020-3459-0
81. Prodan A, Tremaroli V, Brolin H, Zwinderman AH, Nieuwdorp M, Levin E (2020) Comparing bioinformatic pipelines for microbial 16S rRNA amplicon sequencing. https://www.ncbi.nlm.nih.gov/pmc/articles/PMC6964864/. Accessed 10 Jan 2024
82. Kim D, Song L, Breitwieser FP, Salzberg SL (2016) Centrifuge: rapid and sensitive classification of metagenomic sequences. Genome Res 26:1721–1729. https://doi.org/10.1101/gr.210641.116
83. Watts GS, Thornton JE, Youens-Clark K, Ponsero AJ, Slepian MJ, Menashi E, Hu C, Deng W, Armstrong DG, Reed S, Cranmer LD, Hurwitz BL (2019) Identification and quantitation of clinically relevant microbes in patient samples: comparison of three k-mer based classifiers for speed, accuracy, and sensitivity. PLoS Comput Biol 15:e1006863. https://doi.org/10.1371/journal.pcbi.1006863
84. Pomerantz A, Peñafiel N, Arteaga A, Bustamante L, Pichardo F, Coloma LA, Barrio-Amorós CL, Salazar-Valenzuela D, Prost S (2018) Real-time DNA barcoding in a rainforest using nanopore sequencing: opportunities for rapid biodiversity assessments and local capacity building. Gigascience 7:giy033. https://doi.org/10.1093/gigascience/giy033
85. Gowers G-O, Vince O, Charles J-H, Klarenberg I, Ellis T, Edwards A (2019) Entirely off-grid and solar-powered DNA sequencing of microbial communities during an ice cap traverse expedition. Genes (Basel) 10:902. https://doi.org/10.3390/genes10110902
86. Maloney JG, Molokin A, Santin M (2020) Use of Oxford Nanopore MinION to generate full-length sequences of the Blastocystis small subunit (SSU) rRNA gene. Parasit Vectors 13:595. https://doi.org/10.1186/s13071-020-04484-6
87. Mugambi ML, Peter T, Martins SF, Giachetti C (2018) How to implement new diagnostic products in low-resource settings: an end-to-end framework. BMJ Glob Health 3:e000914. https://doi.org/10.1136/bmjgh-2018-000914
88. Delahaye C, Nicolas J (2021) Sequencing DNA with nanopores: troubles and biases. PLoS One 16:e0257521. https://doi.org/10.1371/journal.pone.0257521
89. Chen J, Xu F (2023) Application of nanopore sequencing in the diagnosis and treatment of pulmonary infections. Mol Diagn Ther 27:685–701. https://doi.org/10.1007/s40291-023-00669-8
90. Maiguashca Sánchez J, Sueto SOB, Schwabl P, Grijalva MJ, Llewellyn MS, Costales JA (2020) Remarkable genetic diversity of Trypanosoma cruzi and Trypanosoma rangeli in two localities of southern Ecuador identified via deep sequencing of mini-exon gene amplicons. Parasit Vectors 13:252. https://doi.org/10.1186/s13071-020-04079-1
91. Patiño LH, Castillo-Castañeda AC, Muñoz M, Jaimes JE, Luna-Niño N, Hernández C, Ayala MS, Fuya P, Mendez C, Hernández-Pereira CE, Delgado L, Sandoval-Ramírez CM, Urbano P, Paniz-Mondolfi A, Ramírez JD (2021) Development of an amplicon-based next-generation sequencing protocol to identify Leishmania species and other Trypanosomatids in Leishmaniasis endemic areas. Microbiol Spectr 9:e0065221. https://doi.org/10.1128/Spectrum.00652-21
92. Hakim JMC, Waltmann A, Tinajeros F, Kharabora O, Málaga Machaca E, Calderon M, Menduiña MDC, Wang J, Rueda D, Zimic M, Verástegui M, Juliano JJ, Gilman RH, Mugnier MR, Bowman NM, Chagas Working Group (2023) Amplicon sequencing reveals complex infection in infants congenitally infected with Trypanosoma cruzi and informs the dynamics of parasite transmission. J Infect Dis 228(6):769–776. https://doi.org/10.1093/infdis/jiad125

Proteomics of *Leishmania* Parasites

Erika M. Costa, Rebeca F. Arruda, and Patricia Cuervo

Abstract

Leishmania parasites are etiological agents of a wide spectrum of diseases in humans, denominated leishmaniasis. Clinical forms of these diseases range from self-healing cutaneous lesions to disseminated, mucocutaneous, and fatal visceral infections affecting over 12 million people worldwide. Both parasite species and host immune responses contribute to the clinical outcome. During the last decades, impressive advances in sequencing the genome of most of the pathogenic *Leishmania* spp. has been accomplished resulting in remarkable improvements in the understanding of parasite biology. Despite these successes, the few differences identified at the genome level among *Leishmania* species do not explain the broad spectrum of clinical manifestations caused by these parasites. Because *Leishmania* regulates gene expression at the posttranscriptional and posttranslational levels, a poor correlation is observed between transcripts and protein levels. In such scenario, proteomics analysis represents an impair tool for understanding the contribution of parasites to clinical outcomes. However, the achievements in large-scale proteomics analyses did not follow the genomic efforts and the *Leishmania* proteome remains largely uncharacterized. Proteomics intends to provide detailed descriptions and integration of protein data to better ascertain protein function in biological systems. Proteomics studies in *Leishmania* spp. have focused on global proteome profiling, identification and quantification of proteins related to the different evolutionary forms, and proteins enriched in different cellular fractions or organelles. Also, relevant contributions have been made to the description of posttranslational modifications, quantification of the abundance of proteins in response to drugs, and those involved in the regulation of the host immune response. These qualitative and quantitative profiles have contributed to the functional annotation of proteins, the identification of molecules involved in virulence and drug resistance, in addition to expand the design of vaccines and therapeutic alternatives. Here, we present a historical review on the contribution of proteomics to the understanding of *Leishmania* biology.

Erika M. Costa and Rebeca F. Arruda contributed equally to this work and share first authorship.

E. M. Costa · R. F. Arruda · P. Cuervo (✉)
Laboratório de Pesquisa em Leishmanioses, Instituto Oswaldo Cruz, FIOCRUZ, Rio de Janeiro, RJ, Brazil
e-mail: erika.costa@ioc.fiocruz.br; rebeca.arruda@ioc.fiocruz.br; patricia.cuervo@fiocruz.br

J. D. Ramírez González (ed.), *Recent Advances in Parasitomics*,
https://doi.org/10.1007/978-3-031-70591-5_2

Keywords

Leishmania spp. proteomics · Developmental proteomics · Posttranslational modifications · Subcellular proteomes · Secretome · Proteomics of drug resistance · *Leishmania* interactome

1 Introduction

The term leishmaniasis describes a large group of human diseases caused by protozoan parasites of the *Leishmania* genus. These parasites are associated with a large spectrum of clinical manifestations ranging from self-healing cutaneous lesions to disseminated, mucocutaneous, and visceral infections [1]). Both asymptomatic infections and those that develop clinical symptoms result from subtle interactions among diverse parasite factors and host immune competences (Murray et al. 2005; [1, 2]) including the virulence traits of parasite species and strains. In this regard, genomic data from more than 16 *Leishmania* species and hundreds of strains and clinical isolates have revealed extensive genomic diversity among parasites, different patterns of aneuploidy, mitotic and meiotic recombination events, and variation in gene copy numbers, all of which contribute to parasite virulence and drug resistance (https://tritrypdb.org/tritrypdb/app, [3–8]). In addition to such impressive contributions, *Leishmania* spp. genome sequencing has paved the way for the analyses of the proteomes of these parasites. Interestingly, less than 2.5% of protein coding genes (~200 genes, including some pseudogenes) are species-specific and therefore, differences among the clinical phenotypes associated with each *Leishmania* species are not fully explained by their genome sequences [7, 9, 10].

Leishmania genome exhibits a polycistronic organization and absence of RNA polymerase II promoters that results in constitutive gene transcription [11, 12]. In these parasites, as in other trypanosomatids, gene regulation occurs mainly at the initiation of transcription involving 3′UTR signatures and at posttranscriptional level by mechanisms of RNA or protein stabilization [13–15]. In addition, gene copy number and mosaic aneuploidy also may determine transcript abundance differences in *Leishmania* [16, 17]. However, it is still poorly understood whether protein abundances reflect such variation in gene and chromosome copy number. Indeed, a poor correlation is observed between transcripts and proteins levels in these parasites [18, 19].

In such scenario, high-resolution proteomic approaches offer even opportunities to gain insights into protein profiles that contribute to specific clinical phenotypes in leishmaniasis. Indeed, proteomics can provide, for example, information concerning cellular localization, posttranslational modifications or protein interactions that cannot be obtained from genome or mRNA data. The proteome is defined as the totality of proteins in a cell, tissue, or organism under specific conditions and at a given point in time. The study of the proteomes intends to provide detailed descriptions and integration of protein data to understand protein function in biological systems.

Proteomic studies of *Leishmania* spp. (revised early by [20]) have focused on qualitative and quantitative descriptions of parasites' proteome, profiling of stage-specific protein abundance, characterization of posttranslational modifications (PTMs) and virulence factors, description of subcellular proteomes, secretomes and extracellular vesicle proteomes, identification and quantitation of proteins involved in drug response and drug resistance, and description of protein interactions, among others (Fig. 1).

Few years after the description by O'Farrell of the two-dimensional electrophoresis (2DE) technique for simultaneous determination of the molecular weight and isoelectric point of proteins [21], the first gel-based protein maps of *Leishmania* spp. using this technique were published [22, 23]. These seminal works on *Leishmania* proteomes intended to identify antigens in *L. tropica* associated with virulence [22], and to differentiate distinct *Leishmania* (*Viannia*) species causing American tegumentary leishmaniasis [23]. However, the lack of large-scale protein identification methods at the time rendered

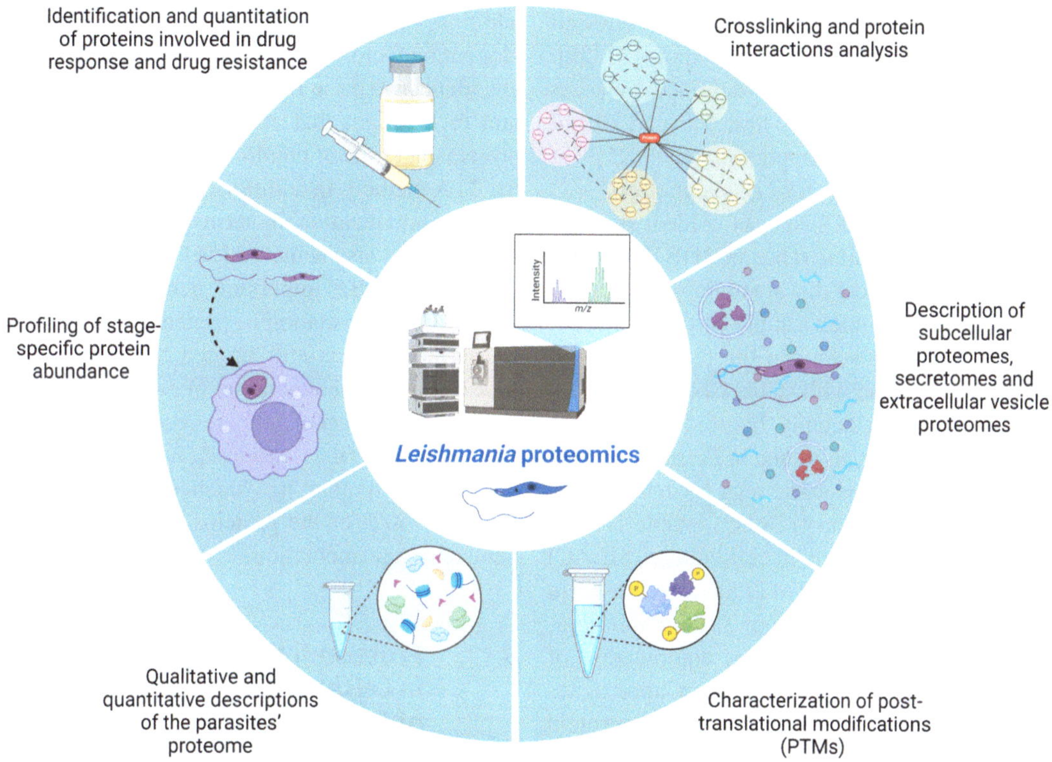

Fig. 1 Biological dimensions of mass spectrometry-based proteomics studies of *Leishmania* parasites. Figure shows some of the biological dimensions of *Leishmania* parasites that have been studied using different quantitative proteomics approaches in the last two decades

those maps essentially descriptive. In the late 1980s, the description of soft ionization techniques for proteins allowed the identification of these molecules by mass spectrometry. Matrix-assisted laser ionization/desorption (MALDI) and electrospray ionization (ESI) in combination with different mass analyzers allowed identification of protein and peptides after analysis by mass spectrometry (MS) [24–26]. The availability of *Leishmania* genome sequences plus bioinformatic tools, in 2005 [27], permitted the beginning of description of *Leishmania* proteomes. Because of their limited reproducibility and resolution, 2DE gel-based approaches are not more used for large-scale analyses and currently liquid chromatography coupled to tandem mass spectrometry (LC-MS/MS) is the leading technique for proteomics studies in this parasite. Today, better sample preparation methods and advances in MS-based proteomics technologies, including depth of identification and label-free quantification, allows thoughtful quantitative exploration of the *Leishmania* proteome under different conditions. In this chapter, we will summarize the different dimensions of proteomics studies in *Leishmania* spp. in the last two decades, focusing on the advancements made until today and perspectives.

2 Describing the Proteome of *Leishmania* spp.

In the early 2000s, the first proteomics studies of *Leishmania* focused on the description of protein maps of the parasite. At first, most of these maps used 2DE to describe a visual array of proteins in different species of *Leishmania* or proteins developmentally regulated in the parasite [28–36]. Combination of 2DE with different staining methods, including fluorescent dyes, allowed detection of 600 to ~3000 protein spots [37–39], but the number of proteins identified by mass spectrometry was still very low, between 20 and 200, which represents approximately ~2.5% of the predicted proteome for *Leishmania* spp.

Still at early 2000s, it was evident that the resolution and dynamic range of 2DE limits protein detection to those that are mainly soluble and abundant in a complex sample [40]. Then, 2DE was gradually replaced and the use of LC-MS/MS in combination with stable isotopes for peptide labeling considerably boosted quantitative comparisons of the proteome in *Leishmania* spp. Labeling based on isobaric tags for relative and absolute quantification (iTRAQ), stable isotope labeling by amino acids in cell culture (SILAC) and tandem mass tags (TMT) have been used in various proteomic analysis of *Leishmania* parasites ([41–50]). More recently, combination of LC-MS/MS with robust label-free approaches including the exponentially modified protein abundance index (emPAI), the total protein approach (TPA), and the "histone ruler" [51–53] have facilitated and democratized proteomics analysis of these parasites [54–58].

Using gel-free analyses, more comprehensive proteomics studies of *Leishmania* were accomplished, identifying up to ~1800 proteins in *L. donovani* and *L. mexicana* that represents ~20% of the predicted proteome [44, 55]. Proteogenomic reanalysis of *L. donovani* proteome identified ~4000 unique proteins [59]. Recently, large-scale and comprehensive quantitative descriptions were achieved for the proteomes of *L. braziliensis*, *L. panamensis*, and *L. guyanensis* with more than 6900 proteins identified, comprising ~85% of the *Leishmania* predicted proteome (~8000 protein-coding genes predicted, considering one protein per gene) [56, 57]. Such in-depth analysis also showed that protein copy number values in these parasites span 6 orders of magnitude with low abundant proteins presenting 50 or less copies per cell, and highly abundant proteins with ~800,000 copies per cell [56]. Protein abundances estimated in *L.* (*Viannia*) datasets by TPA and histone ruler revealed that deep proteomics are able to clearly distinguish the three species; in addition, data showed that histones, tubulin, elongation factor 1-alpha, and HSPs are the most abundant proteins in *Leishmania* [56]. It is important to highlight that absolute concentration of proteins in terms of pmol/mg was obtained using the TPA method; in addition, description of subcellular organization in terms of protein copy numbers was also achieved [56].

Despite those important achievements, the biological interpretation of proteomics data (and transcriptomics) is still limited because of the lack of functional annotation of *Leishmania* spp. After near 25 years of genome sequencing, 30%–55% proteome of *L. major*, *L. infantum*, and *L. braziliensis* remain as "uncharacterized," and about 40–50% of the proteins do not have any gene ontology annotation.

3 Proteomics of Development in *Leishmania* and the Role of Posttranslational Modifications (PTMs) During Differentiation

The life cycle of *Leishmania* is digenetic, involving insect vectors and different vertebrate hosts. The interaction of these parasites with such different hosts requires rapid physiological adaptation that implies rapid modulation of protein abundances for surviving in each environment. Most of the proteomic studies of *Leishmania*' differentiation have been on parasites of the *Leishmania* (*Leishmania*) subgenus, maybe because their culturing in vitro is easier and because of the availability of genomic information for these parasites.

Differentiation from the promastigote to the amastigote stage in *Leishmania* is triggered by elevation of temperature and acidic milieu, by increased exposure to oxygen and nitrogen reactive species, and by nutritional starvation [60–63]. For comparative proteomic studies, axenic parasites are cultivated in vitro under conditions that mimic the vector gut environment (26 °C, pH 7) or the phagolysosome milieu (37 °C, pH 5.5). Proteomics studies of the *Leishmania's* developmental process represent most of the proteomic literature in these parasites. The reason for this may be that besides drastic morphological changes, the transformation from promastigote to amastigotes implies also metabolic adaptation and expression of virulence factors for survival and persistence. Then, identification of molecules involved in such process is a rich source of candidates for therapeutic intervention. The studies quoted below have compared between axenic forms of parasites, different stages of promastigotes during metacyclogenesis, between promastigotes and intracellular amastigotes, or even between different types of amastigotes.

As mentioned before, the first proteomic comparisons between promastigotes and amastigotes were carried out by 2DE. Between 800 and 2000 protein spots were detected in gels of *L. panamensis*, *L. infantum*, *L. donovani*, and *L. mexicana* and about 3–5% of the spots were differentially abundant [29, 32, 34, 36]. Despite the limited protein identification (between 2 and 70 proteins), these studies agreed, in general lines, that the glycolytic pathway is more active in promastigotes, while amastigotes use the fatty acid oxidation and gluconeogenesis as their main energy sources. Such findings were further supported by using an axenic differentiation model combined with an iTRAQ-based proteomic quantitative approach to analyze the changes in protein abundance during seven stages of differentiation of *L. donovani* [44]. Using LC-MS/MS, 21% of the proteome of the parasite was identified and quantified, corroborating that fatty acid oxidation became the main source of metabolic energy of amastigotes whereas enzymes of the glycolytic pathway became diminished. Also, differentiating parasites increased the abundance of proteins involved in gluconeogenesis as well as of those involved in Krebs cycle, respiratory chain, and oxidative phosphorylation [44]. Enrichment of basic proteins from axenic amastigotes and promastigotes of *L. infantum* using free-flow electrophoresis (FFE) revealed ~2400 proteins with pH >7.0 in both stages and increased abundance of basic proteins involved in fatty acid β-oxidation [64]. Increased abundance of enzymes involved in gluconeogenesis, mitochondrial respiration, and stress response was also observed in both intracellular and axenic amastigotes of *L. donovani*. In addition, a significant increase in the vesicular transport, potentially important for survival inside phagolysosomes, was also reported in this species [41]. Similar results were observed in the comparison between promastigotes and amastigotes of *L. infantum* [65].

Leishmania metacyclogenesis is a prerequisite for resistance to complement-mediated lysis and intracellular survival in the mammalian host. Such process consists in the differentiation from noninfective procyclic promastigotes into metacyclic infective parasites in the sandfly gut. This differentiation process is mimicked in vitro during parasites' transformation from logarithmic (log) to stationary (stat) phase of growth in culture. First proteomics studies of metacyclogenesis in *L. infantum* showed increased abundance of cysteine protease and proteins involved in response to oxidative stress in metacyclic promastigotes [34]. Such proteins are crucial for infectivity and parasite survival in host cells. Further studies comparing promastigotes and amastigotes from *L. infantum* supported such findings [66, 67]. In agreement with the general downregulation of protein synthesis reported for stat parasites [68], early proteomics studies showed decreased abundance of proteins involved with synthetic activity in metacyclic promastigotes of *L. major*, whereas proteins involved with parasite motility and cytoskeleton, including flagellar proteins and tubulins, were increased [69]. More recent proteomics studies showed increased abundance of proteins involved in response to oxidative stress during metacyclogenesis in *L. major* and *L. tropica* [70, 71].

A recurrent criticism of using axenic amastigotes is that they may not provide an accurate picture of what happens in host cells, despite displaying biochemical markers of the intracellular form [72]. Studies using intracellular amastigotes are technically more challenging because of the high contamination with proteins of host cells. However, the use of fluorescent transgenic promastigotes of *L. mexicana* for in vivo and in vitro infections made possible the analysis of intracellular amastigotes [55, 73]. After infection, amastigotes were obtained by host cell lysis, and sorting by flow cytometry. Comparison of those amastigotes against promastigotes indicated that adaptation to the intracellular environment requires high metabolic activity of these parasites, and increased abundance of basic protein, which can function as an attempt to buffer the protons released into the phagolysosome and neutralize the total charge of this milieu [55, 73]. Other studies, comparing axenic and intracellular amastigotes of *L. donovani* demonstrated that axenic forms have significant increase in cell size and nuclear diameter and reduced infectivity and survival in the intracellular environment when compared with intracellular amastigotes [39]. Interestingly, 152 of 2152 proteins detected through 2D-DIGE analysis were differentially abundant between axenic and intracellular amastigotes. Proteins involved in antioxidant resistance and growth under nutritionally deprived conditions were increased in the intracellular forms [39].

A recent study used LC-MS/MS and TMT labeling for comparing among *L. mexicana* lesion-derived amastigotes, axenic amastigotes, and amastigotes from macrophages infected in vitro. After filtering off contaminants, 1479 proteins of *L. mexicana* were identified. Significant increase in the abundance of virulence factors such as amastin, cysteine peptidases, glucose transporter, and GP63 were observed in lesion-amastigotes [48]. Also, increased abundance of proteins involved in vesicular trafficking, fatty acid b-oxidation, amino acids metabolism, and oxidative phosphorylation, whereas decreased of those involved in glycolysis was observed in lesion-amastigotes. Because more remarkable differences were observed between axenic amastigotes and lesion-derived ones whereas macrophage-derived amastigotes were mostly intermediate between them, authors suggest that macrophage-derived amastigotes may be more relevant model for experimental studies than axenic amastigotes [48].

Leishmania panamensis, an etiological agent of cutaneous and mucocutaneous leishmaniasis in the New World, is the only species of the *L.* (*Viannia*) subgenus whose differentiation process has been studied using a proteomic approach [36]. Despite the limited protein identification in this seminal 2DE-based study, the results early pointed out to increased amino acid metabolism in amastigotes. Despite the broad range of clinical forms associated to *L.* (*Viannia*) parasites, their genetic diversity, as well as their impact for public health in the Americas, proteomics studies on *L.* (*Viannia*) differentiation is an overlooked topic and there are still many gaps in the knowledge of this process, especially about parasites belonging to the *L.* (*Viannia*) subgenus.

In general, the proteomics data on differentiation of distinct species of *Leishmania* are consistent. These studies support the idea that promastigotes and amastigotes are highly adapted to their respective environments by a strong regulation of their energetic metabolism. Meanwhile, these proteomic studies revealed a small number of proteins (3–9%) that are differentially abundant between stages, suggesting that other mechanisms are involved in the developmental process. Analysis of posttranslational events has shed light on those mechanisms.

As mentioned before, mRNA stability, translation initiation and protein folding are important mechanisms of posttranscriptional and posttranslational regulation of gene expression in *Leishmania*. In addition to these mechanisms, posttranslational modifications (PTMs) play a critical role in the stability of proteins contributing to the control of gene expression. PTMs regulate protein function by altering their localization, molecular interactions, or activities, increasing further the proteome complexity. Phosphorylation, methylation, acetylation, and glycosylation are among the most common PTMs. De novo sequencing analysis of proteins identified in the first proteomic study of *L. braziliensis* demonstrated that ~20% of the identified proteins exhibited PTMs [31].

Analysis of PTMs during *L. donovani* differentiation using a quantitative approach based on iTRAQ and LC-MS/MS detected peptides modified by phosphorylation, methylation, acetylation, and glycosylation [44]. The most significant changes in the PTMs mentioned above occurred during later differentiation times, after promastigotes had initiated their morphological transition into amastigotes [44]. A further analysis using prefractionation with metal affinity chromatography (iMAC) to enrich phosphorylated proteins before 2DE revealed the occurrence of multiple phosphorylation in several proteins. Few proteins were differentially regulated in amastigotes compared to promastigotes, and the *L. donovani* phosphoproteome was considered constitutively expressed [37]. Further, using 2D-DIGE was observed that phosphoproteins such as chaperones and other proteins involved in response to stress had increased abundance in amastigotes of *L. donovani* and could play an important role on infectivity [38].

To extend the coverage of *L. donovani* amastigotes phosphoproteome, protein extracts enriched with iMAC were analyzed by LC-MS/MS. Such strategy resulted in the identification of 445 phosphorylated proteins [74]. Assignment of phosphorylation sites was possible by the enrichment of phosphopeptides on titanium dioxide (TiO_2) columns, showing that phosphorylation occurred mostly on serine residues (86%), and less frequently in threonine (12%) and tyrosine (2%) [74]. Phosphoproteins identified in this study were mainly involved in signal transduction (including protein phosphatases and kinases), protein folding, transcription/translation processes, and RNA/protein turnover [74].

Comparison of the phosphoproteomes of *L. donovani* axenic amastigotes and promastigotes revealed proteins with multiple phosphorylated sites (p-sites) at the distinct stages. Identification of 1614 p-sites corresponding to 627 proteins showed that most of the p-sites were stage-specific and serine phosphorylation in "SF" motif was significantly enriched in amastigotes. Phosphotyrosines (pY) were identified in proteins involved in signal transduction pathways. Interestingly, changes in phosphoprotein abundance did not positively correlate with changes in protein abundance for over half of the phosphorylation events [75]. Furthermore, quantitation of phosphorylation profiles during *L. donovani* development using iTRAQ and LC-MS/MS showed that differentiation signal activates a phosphorylation pathway that initiates parasites transformation. During the phases of signal perception and amastigote morphogenesis, parasites exhibited a significant increase in phosphorylation, whereas during phases of movement cessation/aggregation and maturation, parasites showed increased dephosphorylation. Interestingly, a series of protein kinases and protein phosphatases exhibited phosphorylated sites and phosphorylation of protein kinases occurred early in differentiation, suggesting that they are part of the differentiation initiation pathway [76].

MAPKinase1, the downstream kinase of MAPK signal transduction pathway, is indispensable for survival and infectivity of *L. donovani*. To understand its downstream effector proteins/signaling networks, a label free LC-MS/MS analysis of the phosphoproteome was conducted. Comparison among parasites overexpressing MAPK1 and MAPK1 single replacement mutants with wild-type parasites identified 140 MAPK1 modulated phosphoproteins, mainly involved in pathways such as signal transduction, metabolism, transcriptional, translational, post-translational modification, and regulation of heat shock proteins. This study also revealed that LdMAPK1 interacts directly with only six phosphoproteins, i.e., casein kinase, casein kinase II, HSP83/HSP90, LACK, protein kinase, and serine/threonine protein kinase [77].

To dissect the role of HSP90 in the differentiation from promastigote to amastigote stage, a combination of treatment with a HSP90 inhibitor, phosphoproteome enrichment, and TMT-based labeling approach was used for characterizing the effect of HSP90 inhibition on the phosphoproteome of *L. mexicana* across its differentiation [78]. Decreased phosphorylation of HSP90 and its co-chaperone HSP70 was observed in amastigote. Phosphoproteins involved in kinase activity and microtubule motor activity were negatively affected in promastigotes, whereas ribosomal proteins, protein folding, and proton channel activity were preferentially enriched in the perturbed amastigote phosphoproteome [78].

Together, proteomics studies of differentiation show that parasites rapidly modulate their proteomes to adapt to the different host environments and that the mechanisms underlying these adaptive specificities involve mainly posttranslational modification of proteins by specific phosphorylation pathways.

4 Subcellular Proteomes, Secretome, and Extracellular Vesicle Proteomes

One of the most interesting aspects of subcellular proteomics is the possibility of inferring protein function and cellular organization based on protein localization. This aspect is relevant since around 40–50% of the genes in *Leishmania* spp. remains uncharacterized. In addition, cell homeostasis depends on the discrete biological processes occurring in organellar protein niches [79]. Subcellular proteomes can be obtained by selective purification or specific enrichment of organelles or cellular fractions; furthermore, advances in sample preparation for proteomics and resolution of mass spectrometry allow direct identification of subcellular proteomes in deep-proteomics datasets [56–58, 79, 80].

Seminal proteomics analysis of organelles in *Leishmania* sp. used fractionation with digitonin to purify and enrich the mitochondrial and cytosolic fractions of *L. infantum* [66, 81]. Digitonin is a nonionic detergent that permeabilizes membranes in a selective way, according to their different concentrations of sterol. After serial digitonin extractions, proteins associated with cytosol, mitochondria, endoplasmic reticulum, nucleus, and glycosome were identified. Fluorescence microscopy of GFP-coupled proteins corroborated the cellular localization of several hypothetical proteins identified in cytosolic and mitochondrial fractions and near 70% of the identified proteins had not been detected in previous studies [66], evidencing the relevance of such approach.

Since many proteins involved in the virulence of parasites must be in their surface, i.e., associated with their plasma membrane, this fraction is very relevant for identification of virulence factors and diagnostic molecules. However, due to its high hydrophobicity and low solubility, proteins from the plasma membrane were generally underrepresented in proteomic screens. First proteomics comparisons of this fraction between procyclic and metacyclic promastigotes of *L. infantum* (syn. *L. chagasi*) was made using streptavidin–biotin affinity separation or *n*-octyl glucoside detergent extraction for membrane preparations. Identification by LC-MS/MS revealed that metacyclic promastigotes express stage-specific forms of cysteine peptidases and GP63 not detected in procyclics, while several transporters of nutrient and protons were uniquely present in the membrane fractions of the logarithmic promastigotes [82]. Further studies combining serial centrifugation and free-flow zone electrophoresis (a liquid-based method that separate charged particles according to their isoelectric points) allowed identification of new plasma membrane proteins in *L. infantum* previously unknown [83]. For promastigotes and amastigotes, this protocol allowed the detection of 260 proteins containing at least one transmembrane domain [83]. These proteins were assigned into several functional categories such as transport, metabolism, transcription, translation, surface, protein folding, and proteolysis. Still using 2DE and MALDI-TOF/MS, the proteome of membrane-enriched fractions of *L. donovani* was analyzed. Among the 72 proteins identified, multidrug resistance protein and several surface membrane transporters were detected [84].

Subcellular multistep fractionation of *L. donovani*, using a combination of differential and density-gradient ultracentrifugation techniques plus gel-free and label-free LC-MS/MS, allowed identification of ~3900 proteins, included 645 integral membrane proteins with 1–28 predicted transmembrane domains. Identified membrane proteins corresponded to ~43% of the annotated *L. infantum* membrane proteome. Remarkably, integral membrane proteins including components of the electron transport chain and numerous transporters were localized in several subcellular organellar membranes [85].

Recently, the combination of the filter-aided sample preparation (FASP) method with the consecutive multienzyme digestion [86, 87] followed by high accuracy mass spectrometry, allowed deep coverage of the proteome of *L.* (*Viannia*) species [56]. Based on the gene ontology annotations of cellular component available, the in-depth analysis allowed the identification of ~1100 integral membrane proteins without previous subcellular fractionation [56].

The flagellum in trypanosomatids is a multifunctional organelle involved in locomotion, cell attachment, cell-to-cell communication, cell morphogenesis, and undergoes remarkable structural changes during the parasite's life cycle [88, 89]. Studies on the *L. mexicana* flagellar proteome identified ~2800 proteins, among which, 701 composed the flagellar proteome [90]. Based on protein identifications, a CRISPR-Cas9 knockout library of mutants was produced to screen for flagellar defects. Mutant parasites with altered swimming speed were further used for in vivo mixed infections of the permissive sand fly vector *Lutzomyia longipalpis*. Mutant promastigotes were severely diminished in the sand fly after defecation of the bloodmeal or even did not reach anterior regions of the insect gut, indicating that directional motility is required for successful colonization of sand flies [90].

Leishmania spp., as well as other trypanosomatids, present glycosomes, a specialized peroxisome-related organelle that harbors most of the glycolytic enzymes and exhibits high protein content including enzymes involved in pyrimidine synthesis, nucleotide-sugar biosynthesis, purine salvage and squalene biosynthesis, and enzymes involved in the pentose-phosphate pathway (PPP), among others [91]. Because of the contents, glycosomes are considered potential drug targets. Using sucrose density gradient, glycosomal fraction of *L. donovani* promastigotes was enriched for LC-MS/MS analysis after fractionation by SDS-PAGE. About 1355 proteins were identified in the enriched fraction and near one-third of them were "putative glycosomal" proteins including enzymes of the carbohydrate metabolism, lipid metabolism, thiol metabolism, and purine and pyrimidine metabolism [92].

Physiological processes related to cell–cell communication involve active release of proteins into the extracellular space and proteolytic cleavage of cell surface proteins. Together, actively released proteins and those from proteolytic cleavage and shedding found in the extracellular milieu are referred to as the secretome. Metabolites, enzymes, and virulence factors released/secreted by *Leishmania* can interfere in host cell signaling, modulating the host immune response. Secreted proteins indeed are key factors for the establishment of infection, pathogenesis, and persistence inside host cells. Early studies purified and analyzed the microsomal fraction of *L. major* and *L. amazonensis* with the aim of identifying proteins potentially secreted by the parasites [93]. Proteins such as GP63, GP46, tryparedoxin peroxidase, and cytosolic nucleoside diphosphate kinase were identified, suggesting that the microsomal fraction is rich in vesicles addressed to the cell membrane for secretion. The GTP binding Rab-protein involved in the intracellular traffic and localized mainly in Golgi vesicles was also identified [93].

Mass spectrometry-based analysis of parasites secretome involves technical challenges because abundant proteins present in concentration of g/mL in the serum used for *Leishmania* cultivation hamper the detection of parasites' secreted proteins present at concentrations in a range of low ng/mL [45, 94]. The use of serum-free conditioned medium for parasite culture helps to avoid the dynamic range limitation [95]. Another challenge is the distinction of bona fide secreted proteins from the contaminant proteins from dead parasites, which interfere with the interpretation of the biological significance of the experiment.

A first study of the *L. donovani* secretome used a quantitative approach-based SILAC to differentiate the proteins actively secreted from those associated with the cell body of the parasite [45]. Assays were performed in serum-free conditioned medium containing a nontoxic peptidase inhibitor. Analysis by SDS-PAGE and LC-MS/MS identified 358 proteins in the conditioned medium and, based on SILAC labeling, 151 were considered to be actively secreted by the parasites. Interestingly, 10% of the proteins identified had been previously detected in exosome-like vesicles released from other eukaryotic cells, and only few of the identified had the classical amino-terminal secretion signal peptide, indicating that

Leishmania has nonclassical secretion pathways to regulate the release of proteins [45]. To investigate this possibility, exosomes from *L. donovani* promastigotes were isolated through high-speed centrifugations in different conditions of pH and temperature. Mass spectrometry analysis detected 329 proteins, 52% of them shared with the total parasite' secretome, confirming that exosomes transport much of the material secreted by the parasite [96].

Further studies observed that *L. donovani* exosomes interact with human monocytes in vitro, increasing the production of the anti-inflammatory cytokine IL-10 and then contributing to the establishment of infection [97]. Thus, the presence of immunogenic proteins in *Leishmania* secretome was investigated through the comparison between wild-type and mutant HSP100$^{-/-}$ *L. donovani.* HSP100$^{-/-}$ parasites showed reduced ability to infect macrophages in vitro and their exosomes did not show known virulence factors. In contrast, in exosomes from wild-type parasites, immunogenic proteins such as GP63, elongation factor 1β, kinetoplast membrane protein 11, and a putative homolog of iron superoxide dismutase were detected [97].

An early study on the secretome of *L. braziliensis* using 2DE and MALDI-MS/MS detected 270 proteins from the conditioned medium of secretion. In total, 42 were identified including proteins involved with signal transduction, intracellular survival and modulation of the host immune response such as HSP70, acid phosphatase, activated protein kinase C receptor (LACK), elongation factor 1β and tryparedoxin peroxidase. Data mining showed that ~5% of the identified proteins present a classical secretion signal, while ~57% were secreted following nonclassical secretion mechanisms, indicating that protein export in *L. braziliensis* occurs mainly through unconventional routes including via exosomal vesicles [94].

Based on the observation that temperature shift induces a significant increase in protein release from *L. mexicana*, including increased budding of exovesicles from its surface, a comparative proteomics approach was conducted to identify released proteins. Most of the 72 proteins identified lack a signal peptide. Secreted proteins induced cleavage and activation of the host protein tyrosine phosphatases, specifically SHP-1 and PTP1-B, in a murine bone-marrow-derived macrophage cell line, and inhibition of translocation of the inflammatory transcription factors NF-κB and AP-1. The exoproteome also inhibited nitric oxide production, resulting in attenuation of the inflammatory response and deactivation of the macrophage [98].

A proteomics comparison of the protein content and immunomodulatory properties of vesicles derived from *L. major* wild-type and *L. major* GP63$^{-/-}$ showed that absence of GP63 alters the proteins delivered by parasite exosomes including other important virulence factors such as LACK and tubulin. In addition, this study suggested a role for GP63 in *Leishmania* exosomal protein sorting and also suggest that the GP63 knockout strategy may alter the global protein content of EVs [98]. Indeed, a further study showed that *L. amazonensis*-derived EVs with altered GP63 levels (up- or down-regulated) by episomal expression have similar protein content. This study also showed that GP63 is a primordial component of the ability of these EVs in augmenting the inflammatory cutaneous response in *Leishmania* spp. infection [99]

Further secretome characterization compared between the proteomic profiles of the vesicles and vesicle depleted exoproteomes from logarithmic and stationary *L. infantum* promastigotes. Protein identification showed that GP63 was the most abundant protein in both fractions. However, protein composition of each fraction was distinct: vesicle-derived exoproteome recovered from logarithmic parasites was enriched in ribosomal proteins, whereas vesicles from stationary phase parasites exhibited specific features of apoptotic vesicles. In addition, vesicle depleted exoproteome from stationary parasites was enriched in proteins associated with proteasome, suggesting that exoproteome in this stage may play a role in cell death whereas in log phase parasite has a role in protein turnover [100]. This study also highlighted the biological relevance of the vesicle depleted exoproteomes as this fraction accounts for a significant portion of the total secreted proteins and is rich in potential targets for diagnosis, drug, and vaccine development [46, 47, 100]. In agreement with that, proteomics profiling of secretome from *L. infantum* promastigotes showed that conditioned medium is rich in immunogenic proteins useful to discriminate different clinical forms of canine visceral leishmaniasis [54]. Recently, a study comparing the secretome of logarithmic and stationary *L. infantum* promastigotes corroborated the specific protein content of the exosomes of each stage [101].

Since the composition of extracellular vesicles (EVs) in *Leishmania* can vary according to the state of their parental cell, it was suggested that EVs released by drug-resistant *L. infantum* could contain unique enriched proteins according to the drug-resistance mechanisms involved in the survival of their parental cell line [102]. Proteomes of EVs derived from drug-resistant *L. infantum* cell lines and a wild-type strain were obtained. A core EV proteome highly conserved was observed among strains. However, a few specific proteins enriched differently depending on the drug were identified in the EVs from antimony-, miltefosine-, and amphotericin-resistant parasites. In addition, the study showed changes in the morphology, size, and distribution of *L. infantum* EVs according to the stressful condition [102].

Leishmania braziliensis is a New World parasite associated with localized self-healing cutaneous lesions (LCL), disseminated leishmaniasis (DL), and mucocutaneous lesions (MCL). To further understand the role of *L. braziliensis* secretome in the modulation of host immune response, infection rates and cytokine profiles of macrophages pretreated with secretomes of *L. braziliensis* strains associated with DL and LCL were evaluated. Besides, an iTRAQ-based quantitative proteomics approach was used to compare the abundance of proteins secreted by those strains [43]. Pretreatment with secretome resulted in higher number of infected macrophages, as well as higher number of amastigotes per cell. Additionally, macrophages pretreated with LCL secretome exhibited a proinflammatory cytokine profile, whereas those pretreated with the DL one did not. In line with this, proteomics analysis showed that the DL secretome is rich in molecules involved in macrophage deactivation, while is poor in proteins that activate proinflammatory pathways. Those finding suggested that the secretome of *L. braziliensis* may contribute to the infection and persistence, which are prerequisites for parasite dissemination [43].

Despite the important achievements in the characterization of *Leishmania* secretomes, this field still lack reliable and specific markers for *Leishmania*-derived exosomes. To approach this issue, three fractions of extracellular vesicles from *L. major* were obtained and characterized by nanoparticle tracking analysis (NTA), transmission electron microscopy (TEM), and LC-MS/MS. Proteomic data revealed potential novel protein markers for *L. major*-derived exosomes, including proteins involved in endosomal machinery and the endosomal sorting complex required for transport (ESCRT) pathway, as well as the parasitic phosphatase PRL-1 [103].

The studies on secreted proteins summarized above are consistent, demonstrating that the secreted fraction is an important source of effector molecules involved in parasite–host interaction and modulating host immune responses. Proteomics studies are also consistent showing that only few of the identified proteins present the canonical signal peptide for secretion typical of eukaryotic cells and confirming the existence of unconventional secretion mechanisms in these parasites, mainly mediated by transport through vesicles. Furthermore, these studies showed that a large portion of secreted proteins have previously described intracellular functions, revealing the multifunctionality and multilocalization of proteins in *Leishmania* spp.

5 Proteomics of Drug Resistance in *Leishmania*

In the absence of vaccines, control of leishmaniasis depends largely on chemotherapy treatment. However, the compounds of pentavalent antimony [Sb(V)], meglumine antimoniate, and sodium stibogluconate—first line of choice in the treatment of patients with leishmaniasis until few years ago (PAHO [104])—present high toxicity and severe side effects that include cardiac arrhythmias and kidney damage. Furthermore, after more than 70 years of use, antimony pentavalent-refractory patients and clinically resistant isolates have been reported in different parts of the world [105]. Other medications such as amphotericin B, pentamidine, and the oral medication miltefosine also have high toxicity, questionable efficacy, and cross-resistance, in addition to not being economically viable for many countries where the disease is endemic. Despite such drawbacks, miltefosine, pentamidine, and amphotericin B, alone or in combined schemes and new formulations are now recommended as first-line drugs in various epidemiological scenarios of the world, as well as pentavalent antimonial is still recommended for mucosal leishmaniasis treatment (PAHO [104]). Since reports of failure in the treatment with miltefosine and amphotericin B have grown in the last years, resistance monitoring and surveillance systems are urgently needed [106]. In this context, it is imperative to elucidate the mechanisms involved in resistance, as well as the discovery and validation of new chemotherapy targets.

Proteomic approaches allow to identify proteins involved with drug resistance and new drug targets in in *Leishmania*. Proteomic studies for identifying molecules potentially involved in the drug resistance in these parasites have been carried out using either axenic promastigotes/amastigotes that have been selected to drug resistance in vitro or axenic promastigotes/amastigotes derived from clinical isolates that are considered naturally drug resistant. The compounds evaluated in these studies include drugs currently used for the treatment of leishmaniasis, such as antimonial, amphotericin, and miltefosine, in addition to model drugs for the study of resistance, such as methotrexate (MTX), arsenite, and bicyclic nitro-drugs. The reader can find a comprehensive revision on this topic covering proteomics of drug resistance until 2017 at [107]. More recent studies will be summarized here.

The emergence of resistance and the "resistant genotype" of *Leishmania* parasites have been studied by the state-of-the-art genome sequencing approaches [108–112], but the emergence of resistance phenotypes is poorly understood at the protein level. A recent study on the development of resistance to miltefosine upon experimental selection in vitro in a *L. infantum* strain showed that miltefosine resistance emergence took ~6 months. Label-free quantitative mass spectrometry-based proteomics analyses showed that this process involves remodeling of components of the membrane and mitochondrion, with significant increase in oxidative phosphorylation complexes and energy metabolism mainly dependent on β-oxidation of fatty acids. Proteins canonically involved in ROS detoxification did not contribute to the resistant process obtained in vitro whereas sterol biosynthesis enzymes could have a role in this development. In addition, changes in the abundance of proteins known to be involved in miltefosine resistance such as ABC transporters and phospholipid transport ATPase were detected in *L. infantum* [58].

Fatty acids with medium chain length, including caprylic acid (CA, C8) and lauric acid (LA, C12), have been explored as antimicrobial agents. The antileishmanial efficacy of different saturated medium-chain fatty acids was recently evaluated, showing that caprylic acid was the most potent against *L. donovani* [113]. Proteomics studies of CA-treated parasites showed that the abundance of enzymes of the ergosterol biosynthesis pathway, including acetyl-coenzyme A synthetase, was decreased in those parasites, and confirmed that there was a significant decrease in the abundance of mevalonate kinase in CA-treated cells compared to untreated control. The abundance of enzymes involved in antioxidant response such as cysteine synthase, glutathione synthase, and trypanothione reductase was also reduced in CA-treated *L. donovani*. Thus, ergosterol depletion was the major cause of CA-mediated cell death and the enzyme mevalonate kinase of the ergosterol biosynthesis pathway was the key target of CA [113].

A novel approach based on thermal proteome profiling (TPP) has been proposed to further characterize the mode of action antileishmania drugs and understand the mechanisms of drug resistance deployed by *Leishmania* [114]. TPP is based on the principle that binding of a drug to its protein target can significantly alter the thermal stability of that protein [115]. Phenotypic screening identified a benzothiophene compound with activity against *L. donovani*, and using TPP, oxidosqualene cyclase (OSC), a key enzyme of sterol biosynthesis, was identified as the target of this compound [114]. Also using TPP, multiplexed quantitative mass spectrometry-based proteomics was used to monitor the melting profile of thousands of expressed soluble proteins in WT, antimony-resistant, miltefosine-resistant, and amphotericin B-resistant *L. infantum* parasites in the presence (or absence) of the drugs. Proteins that underwent changes (fold change >4) caused by complexation with a drug were identified unveiling novel drug targets in the *L. infantum* proteome and proteins potentially involved in drug resistance [116].

6 Proteomics Analysis of the Interactome in *Leishmania*

Regulation of gene expression in *Leishmania* is achieved by mechanisms based mainly in RNA-binding proteins (RBPs), but few is known about these proteins. RNA-binding proteins in *Leishmania* control the processing and trafficking of RNA molecules from synthesis to degradation and its identification is relevant to understand the mechanisms of gene regulation in these parasites. A methodological approach involving covalent UV crosslinking of RBPs to mRNA in vivo, followed by interactome capture using oligo(dT) magnetic beads, allowed obtaining the mRNA interactome of *L. donovani* amastigotes. Mass spectrometry-based proteomics analysis of captured proteins identified 79 mRNA interacting proteins, 49 of which had no orthologs or homologs in the human genome and therefore may represent candidates for selective drug targeting [117].

Further optimization of the crosslinking method in association with quantitative mass spectrometry allowed a comprehensive analysis of 1400 mRNA binding proteins (mRBPs) and whole cell proteomes in procyclic promastigotes, metacyclic promastigotes, and amastigotes of *L. mexicana*. Remarkably, the isolated mRNA-bound proteomes are differentially enriched and over 250 RBPs exhibit stage-specific expression. Multiple candidate RBPs were endogenously tagged and found to associate with discrete mRNA target pools in a stage-specific manner [118].

PUF proteins are common RBPs in eukaryotic organisms and might play too an essential role modulating the fate of mRNAs and regulating gene expression in *Leishmania* parasites. To identify putative protein partners of the *L. major* PUF1, a proteomic approach was performed. The PUF1 interactome was characterized by co-immunoprecipitation using *L. major* cellular extracts and an anti-LiPUF1 antibody. Analysis of the co-immunoprecipitated proteins by mass spectrometry identified 90 LmPUF1 candidate partners. Remarkably, many of the identified proteins are other RBPs and/or putative P bodies and mRNA-exporting machinery components [119].

Recently, two RBPs, SCD6, and RBP42 were characterized in *L. braziliensis* [120]. To determine the interactome of SCD6 and RBP42, in vivo UV crosslinking and immunoprecipitation assays were used to obtain protein complexes in which SCD6 or RBP42 were present. Protein mass spectrometry analysis of the immunoprecipitated proteins identified 96 proteins potentially associated with SCD6 and 173 proteins associated with RBP42. Many of the identified proteins were shared in both interactomes, indicating a possible functional relationship between SCD6 and RBP42. Remarkably, interacting proteins were related to RNA metabolism and translation processes, and are components of ribonucleoprotein granules in *Leishmania* and related trypanosomatids, indicating that SCD6 and RBP42 participate in the fate (decay/accumulation/translation) of *L. braziliensis* transcripts [120].

The proteins that interact with nonadenylated RNAs, including noncoding RNAs and pre-mRNAs in *Leishmania*, are mostly unidentified. Using a combination of orthogonal organic phase separation methodology and TMT-labeling-based high resolution quantitative proteomic mass spectrometry, 2417 RBPs were identified, including 1289 putative novel non-poly(A)-RNA-binding proteins in promastigotes and amastigotes of *L. mexicana*. Eight out of 20 *Leishmania* deubiquitinases were exclusively identified in the non-poly(A)-RNA-interactome. Additionally, an increased representation of WD40 repeat domains were observed in the *Leishmania* non-poly(A)-RNA-interactome, uncovering potential involvement of this protein domain in RNA--protein interactions in *Leishmania* [121].

7 Concluding Remarks

Proteomic approaches are particularly useful tools for the identification of proteins and mechanisms involved in several aspects of the biology of *Leishmania* parasites. New roles for already well-characterized proteins as well as the assignment of roles for proteins of unknown function are being defined for several *Leishmania* species using those proteomics tools. Important advances in the deep description and quantitation of *Leishmania* proteomes have been achieved but still ~40% proteins remain uncharacterized and for most of *Leishmania* species more than 50% of the predicted proteome has not experimental evidence. The recurrent identification of the most abundant proteins that precludes the identification of the less abundant proteins and obscures the studied phenomena is being overcome through better sample preparation methods and higher resolution mass spectrometry. The growing availability of *Leishmania*-specific bioinformatic tools and databases will be valuable for a better exploitation and interpretation of the *Leishmania* proteomics data. The characterization of the interactome (protein–protein, RNA–protein) of *Leishmania* is a promising field for discovering new aspects of the biology of these parasites. Finally, the resumption of large-scale proteomic studies on drug resistance is urgent and should be encouraged in the "leishmaniac" research community.

References

1. Burza S, Croft SL, Boelaert M (2018) Leishmaniasis. Lancet 392(10151):951–970. https://doi.org/10.1016/S0140-6736(18)31204-2
2. Murray HW, Berman JD, Davies CR (2005) Saravia NG. Advances in leishmaniasis. Lancet 366(9496):1561–1577. https://doi.org/10.1016/S0140-6736(05)67629-5
3. Franssen SU et al (2020) Global genome diversity of the *Leishmania donovani* complex. eLife 9:e51243. https://doi.org/10.7554/eLife.51243
4. Heeren S et al (2023) Diversity and dissemination of viruses in pathogenic protozoa. Nat Commun 14(1):8343. https://doi.org/10.1038/s41467-023-44085-2
5. Imamura H et al (2016) Evolutionary genomics of epidemic visceral leishmaniasis in the Indian subcontinent. eLife 5. https://doi.org/10.7554/eLife.12613
6. Lypaczewski P, Matlashewski G (2021) *Leishmania donovani* hybridisation and introgression in nature: a comparative genomic investigation. Lancet Microbe 2(6). https://doi.org/10.1016/S2666-5247(21)00028-8
7. Rogers MB et al (2011) Chromosome and gene copy number variation allow major structural change between species and strains of *Leishmania*. Genome Res 21(12):2129–2142. https://doi.org/10.1101/gr.122945.111
8. Van den Broeck F et al (2020) Ecological divergence and hybridization of neotropical Leishmania parasites. Proc Natl Acad Sci USA 117(40):25159–25168. https://doi.org/10.1073/pnas.1920136117
9. Fiebig M, Kelly S, Gluenz E (2015) Comparative life cycle transcriptomics revises *Leishmania mexicana* genome annotation and links a chromosome duplication with parasitism of vertebrates. PLoS Pathog 11(10). https://doi.org/10.1371/journal.ppat.1005186
10. Peacock CS et al (2007) Comparative genomic analysis of three Leishmania species that cause diverse human disease. Nat Genet 39(7):839–847. https://doi.org/10.1038/ng2053
11. Cohen-Freue G et al (2007) Global gene expression in *Leishmania*. Int J Parasitol 37(10):1077–1086. https://doi.org/10.1016/j.ijpara.2007.04.011
12. Leifso K et al (2007) Genomic and proteomic expression analysis of *Leishmania* promastigote and amastigote life stages: the *Leishmania* genome is constitutively expressed. Mol Biochem Parasitol 152(1):35–46. https://doi.org/10.1016/j.molbiopara.2006.11.009
13. Alcoforado-Diniz J et al (2021) Protein methyltransferase 7 deficiency in *Leishmania major* increases neutrophil associated pathology in murine model. PLoS Negl Trop Dis 15(3):e0009230. https://doi.org/10.1371/journal.pntd.0009230
14. Clayton C, Shapira M (2007) Post-transcriptional regulation of gene expression in trypanosomes and

leishmanias. Mol Biochem Parasitol 156(2):93–101. https://doi.org/10.1016/j.molbiopara.2007.07.007

15. Grünebast J, Clos J (2020) *Leishmania*: responding to environmental signals and challenges without regulated transcription. Comput Struct Biotechnol J 18:4016–4023. https://doi.org/10.1016/j.csbj.2020.11.058
16. Iantorno SA et al (2017) Gene expression in *Leishmania* is regulated predominantly by gene dosage. mBio 8(5). https://doi.org/10.1128/mBio.01393-17
17. Ubeda JM et al (2008) Modulation of gene expression in drug resistant Leishmania is associated with gene amplification, gene deletion and chromosome aneuploidy. Genome Biol 9(7):115. https://doi.org/10.1186/gb-2008-9-7-r115
18. Cortazzo da Silva L, Aoki JI, Floeter-Winter LM (2022) Finding correlations between mRNA and protein levels in *Leishmania* development: is there a discrepancy? Front Cell Infect Microbiol 12. https://doi.org/10.3389/fcimb.2022.852902
19. Cuypers B et al (2022) Four layer multi-omics reveals molecular responses to aneuploidy in *Leishmania*. PLoS Pathog 18(9). https://doi.org/10.1371/journal.ppat.1010848
20. Paape D, Aebischer T (2011) Contribution of proteomics of *Leishmania spp.* to the understanding of differentiation, drug resistance mechanisms, vaccine and drug development. J Proteomics 74(9):1614–1624. https://doi.org/10.1016/j.jprot.2011.05.005
21. O'Farrell PH (1975) High resolution two-dimensional electrophoresis of proteins. J Biol Chem 250(10):4007–4021
22. Handman E, Mitchell GF, Goding JW (1981) Identification and characterization of protein antigens of *Leishmania tropica* isolates. J Immunol 126(2):508–512
23. Saravia NG et al (1984) Two-dimensional electrophoresis used to differentiate the causal agents of American tegumentary leishmaniasis. Clin Chem 30(12 Pt 1):2048–2052
24. Fenn JB et al (1989) Electrospray ionization for mass spectrometry of large biomolecules. Science 246(4926):64–71. https://doi.org/10.1126/science.2675315
25. Karas M, Hillenkamp F (1988) Laser desorption ionization of proteins with molecular masses exceeding 10,000 daltons. Anal Chem 60(20):2299–2301. https://doi.org/10.1021/ac00171a028
26. Tanaka K et al (1988) Protein and polymer analyses up to m/z 100000 by laser ionization time of flight mass spectrometry. Rapid Commun Mass Spectrom 2:151–153
27. El-Sayed NM et al (2005) Comparative genomics of trypanosomatid parasitic protozoa. Science 309(5733):404–409. https://doi.org/10.1126/science.1112181
28. Acestor N et al (2002) Establishing two-dimensional gels for the analysis of *Leishmania* proteomes. Proteomics 2(7):877. https://doi.org/10.1002/1615-9861(200207)2:7<877::AID-PROT877>3.0.CO;2-D
29. Bente M et al (2003) Developmentally induced changes of the proteome in the protozoan parasite *Leishmania donovani*. Proteomics 3(9):1811–1829. https://doi.org/10.1002/pmic.200300462
30. Brobey RKB et al (2006) Comparative two-dimensional gel electrophoresis maps for promastigotes of *Leishmania amazonensis* and *Leishmania major*. Braz J Infect Dis 10(1):1–6. https://doi.org/10.1590/S1413-86702006000100001
31. Cuervo P et al (2007) Proteome analysis of *Leishmania (Viannia) braziliensis* by two-dimensional gel electrophoresis and mass spectrometry. Mol Biochem Parasitol 154(1):6–21. https://doi.org/10.1016/j.molbiopara.2007.03.013
32. el Fakhry Y, Ouellette M, Papadopoulou B (2002) A proteomic approach to identify developmentally regulated proteins in *Leishmania infantum*. Proteomics 2(8):1007. https://doi.org/10.1002/1615-9861(200208)2:8<1007::AID-PROT1007>3.0.CO;2-G
33. Góngora R et al (2003) Mapping the proteome of *Leishmania Viannia* parasites using two-dimensional polyacrylamide gel electrophoresis and associated technologies. Biomedica 23(2):153–160
34. Nugent PG et al (2004) Proteomic analysis of *Leishmania mexicana* differentiation. Mol Biochem Parasitol 136(1):51–62. https://doi.org/10.1016/j.molbiopara.2004.02.009
35. Thiel M, Bruchhaus I (2001) Comparative proteome analysis of *Leishmania donovani* at different stages of transformation from promastigotes to amastigotes. Med Microbiol Immunol 190(1–2):33–36. https://doi.org/10.1007/s004300100075
36. Walker J et al (2006) Identification of developmentally-regulated proteins in *Leishmania panamensis* by proteome profiling of promastigotes and axenic amastigotes. Mol Biochem Parasitol 147(1):64–73. https://doi.org/10.1016/j.molbiopara.2006.01.008
37. Morales MA et al (2008) Phosphoproteomic analysis of *Leishmania donovani* pro- and amastigote stages. Proteomics 8(2):350–363. https://doi.org/10.1002/pmic.200700697
38. Morales MA et al (2010) Phosphoproteome dynamics reveal heat-shock protein complexes specific to the *Leishmania donovani* infectious stage. Proc Natl Acad Sci USA 107(18):8381–8386. https://doi.org/10.1073/pnas.0914768107
39. Pescher P et al (2011) Quantitative proteome profiling informs on phenotypic traits that adapt *Leishmania donovani* for axenic and intracellular proliferation. Cell Microbiol 13(7):978–991. https://doi.org/10.1111/j.1462-5822.2011.01593.x
40. Bantscheff M et al (2007) Quantitative mass spectrometry in proteomics: a critical review. Anal Bioanal Chem 389(4):1017–1031. https://doi.org/10.1007/s00216-007-1486-6
41. Biyani N, Madhubala R (2012) Quantitative proteomic profiling of the promastigotes and the

intracellular amastigotes of *Leishmania donovani* isolates identifies novel proteins having a role in Leishmania differentiation and intracellular survival. Biochim Biophys Acta 1824(12):1342–1350. https://doi.org/10.1016/j.bbapap.2012.07.010
42. Kalesh K et al (2021) Quantitative proteomics reveals that Hsp90 inhibition dynamically regulates global protein synthesis in *Leishmania mexicana*. mSystems 6(3). https://doi.org/10.1128/mSystems.00089-21
43. Rodríguez-Vega A et al (2021) Quantitative analysis of proteins secreted by *Leishmania (Viannia) braziliensis* strains associated to distinct clinical manifestations of American Tegumentary Leishmaniasis. J Proteomics 232:104077. https://doi.org/10.1016/j.jprot.2020.104077
44. Rosenzweig D et al (2008) Retooling *Leishmania* metabolism: from sand fly gut to human macrophage. FASEB J 22(2):590–602. https://doi.org/10.1096/fj.07-9254com
45. Silverman JM et al (2008) Proteomic analysis of the secretome of *Leishmania donovani*. Genome Biol 9(2):R35. https://doi.org/10.1186/gb-2008-9-2-r35
46. Esteves BB et al (2022) Characterization of differentially abundant proteins among *Leishmania (Viannia) braziliensis* strains isolated from atypical or typical lesions. Front Cell Infect Microbiol 12:824968. https://doi.org/10.3389/fcimb.2022.824968
47. Esteves S et al (2022) *Leishmania* vesicle-depleted exoproteome: what, why, and how? Microorganisms 10(12):2435. https://doi.org/10.3390/microorganisms10122435
48. Pacakova L et al (2022) Three types of *Leishmania mexicana* amastigotes: proteome comparison by quantitative proteomic analysis. Front Cell Infect Microbiol 12. https://doi.org/10.3389/fcimb.2022.1022448
49. Tandon R et al (2023) Identification of protein biomarkers of attenuation and immunogenicity of centrin or p27 gene deleted live vaccine candidates of *Leishmania* against visceral leishmaniasis. Parasitol Int 92:102661. https://doi.org/10.1016/j.parint.2022.102661
50. Porta EOJ, Isern JA, Kalesh K, Steel PG (2022) Discovery of leishmania druggable serine proteases by activity-based protein profiling. Front Pharmacol 13:929493. https://doi.org/10.3389/fphar.2022.929493
51. Ishihama Y et al (2005) Exponentially modified protein abundance index (emPAI) for estimation of absolute protein amount in proteomics by the number of sequenced peptides per protein. Mol Cell Proteomics 4(9):1265–1272. https://doi.org/10.1074/mcp.M500061-MCP200
52. Wiśniewski JR (2017) Label-free and standard-free absolute quantitative proteomics using the 'total protein' and 'proteomic ruler' approaches. Methods Enzymol 585:49–60. https://doi.org/10.1016/bs.mie.2016.10.002
53. Wiśniewski JR et al (2014) A 'proteomic ruler' for protein copy number and concentration estimation without spike-in standards. Mol Cell Proteomics 13(12):3497–3506. https://doi.org/10.1074/mcp.M113.037309
54. Braga MS et al (2014) Shotgun proteomics to unravel the complexity of the *Leishmania infantum* exoproteome and the relative abundance of its constituents. Mol Biochem Parasitol 195(1):43–53. https://doi.org/10.1016/j.molbiopara.2014.07.001
55. Paape D et al (2010) Gel free analysis of the proteome of intracellular *Leishmania mexicana*. Mol Biochem Parasitol 169(2):108–114. https://doi.org/10.1016/j.molbiopara.2009.10.009
56. Pinho N et al (2020) In-depth quantitative proteomics uncovers specie-specific metabolic programs in *Leishmania (Viannia)* species. PLoS Negl Trop Dis 14(8):e0008509. https://doi.org/10.1371/journal.pntd.0008509
57. Pinho N et al (2022) Nitric oxide resistance in *Leishmania (Viannia) braziliensis* involves regulation of glucose consumption, glutathione metabolism and abundance of pentose phosphate pathway enzymes. Antioxidants 11(2):277. https://doi.org/10.3390/antiox11020277
58. Saboia-Vahia L et al (2022) In-depth quantitative proteomics characterization of in vitro selected Miltefosine resistance in *Leishmania infantum*. Proteomes 10(2):10. https://doi.org/10.3390/proteomes10020010
59. Nirujogi RS et al (2014) Moving from unsequenced to sequenced genome: reanalysis of the proteome of *Leishmania donovani*. J Proteomics 97:48–61. https://doi.org/10.1016/j.jprot.2013.04.021
60. Besteiro S et al (2007) Protein turnover and differentiation in *Leishmania*. Int J Parasitol 37(10):1063–1075. https://doi.org/10.1016/j.ijpara.2007.03.008
61. Clos J, Grünebast J, Holm M (2022) Promastigote-to-amastigote conversion in *Leishmania* spp.—a molecular view. Pathogens 11(9):1052. https://doi.org/10.3390/pathogens11091052
62. Khan YA, Andrews NW, Mittra B (2018) ROS regulate differentiation of visceralizing *Leishmania* species into the virulent amastigote form. Parasitol Open 4. https://doi.org/10.1017/pao.2018.15
63. Mittra B et al (2013) Iron uptake controls the generation of *Leishmania* infective forms through regulation of ROS levels. J Exp Med 210(2):401–416. https://doi.org/10.1084/jem.20121368
64. Brotherton M-C et al (2010) Analysis of stage-specific expression of basic proteins in *Leishmania infantum*. J Proteome Res 9(8):3842–3853. https://doi.org/10.1021/pr100048m
65. Fialho Junior L et al (2021) Proteomic analysis reveals differentially abundant proteins probably involved in the virulence of amastigote and promastigote forms of *Leishmania infantum*. Parasitol Res 120(2):679–692. https://doi.org/10.1007/s00436-020-07020-8
66. Foucher AL, Papadopoulou B, Ouellette M (2006) Prefractionation by Digitonin extraction increases representation of the cytosolic and intracellular

proteome of *Leishmania infantum*. J Proteome Res 5(7):1741–1750. https://doi.org/10.1021/pr060081j
67. McNicoll F et al (2006) A combined proteomic and transcriptomic approach to the study of stage differentiation in Leishmania infantum. Proteomics 6(12):3567–3581. https://doi.org/10.1002/pmic.200500853
68. Sacks DL, Perkins P v (1984) Identification of an infective stage of *Leishmania* promastigotes. Science 223(4643):1417–1419. https://doi.org/10.1126/science.6701528
69. Mojtahedi Z, Clos J, Kamali-Sarvestani E (2008) *Leishmania major*: identification of developmentally regulated proteins in procyclic and metacyclic promastigotes. Exp Parasitol 119(3):422–429. https://doi.org/10.1016/j.exppara.2008.04.008
70. Amiri-Dashatan N et al (2020) Quantitative proteomic analysis reveals differentially expressed proteins in *Leishmania major* metacyclogenesis. Microb Pathog 149:104557. https://doi.org/10.1016/j.micpath.2020.104557
71. Ashrafmansouri M, Amiri-Dashatan N, Ahmadi N (2022) Identification of protein profile in metacyclic and amastigote-like stages of *Leishmania tropica*: a proteomic approach. AMB Express 12(1):142. https://doi.org/10.1186/s13568-022-01481-z
72. Holzer TR, McMaster WR, Forney JD (2006) Expression profiling by whole-genome interspecies microarray hybridization reveals differential gene expression in procyclic promastigotes, lesion-derived amastigotes, and axenic amastigotes in Leishmania mexicana. Mol Biochem Parasitol 146(2):198–218. https://doi.org/10.1016/j.molbiopara.2005.12.009
73. Paape D et al (2008) Transgenic, fluorescent Leishmania mexicana allow direct analysis of the proteome of intracellular amastigotes. Mol Cell Proteomics 7(9):1688–1701. https://doi.org/10.1074/mcp.M700343-MCP200
74. Hem S et al (2010) Identification of *Leishmania*-specific protein phosphorylation sites by LC-ESI-MS/MS and comparative genomics analyses. Proteomics 10(21):3868–3883. https://doi.org/10.1002/pmic.201000305
75. Tsigankov P et al (2013) Phosphoproteomic analysis of differentiating *Leishmania* parasites reveals a unique stage-specific phosphorylation motif. J Proteome Res 12(7):3405–3412. https://doi.org/10.1021/pr4002492
76. Tsigankov P et al (2014) Regulation dynamics of *leishmania* differentiation: deconvoluting signals and identifying phosphorylation trends. Mol Cell Proteomics 13(7):1787–1799. https://doi.org/10.1074/mcp.M114.037705
77. Kaur P et al (2021) Comparative phosphoproteomic analysis unravels MAPK1 regulated phosphoproteins in *Leishmania donovani*. J Proteomics 240:104189. https://doi.org/10.1016/j.jprot.2021.104189
78. Porta EO et al (2023) Inhibition of HSP90 distinctively modulates the global phosphoproteome of *Leishmania mexicana* developmental stages. Microbiol Spectr 11(6):e0296023. https://doi.org/10.1128/spectrum.02960-23
79. Bhushan V, Nita-Lazar A (2024) Recent advancements in subcellular proteomics: growing impact of organellar protein niches on the understanding of cell biology. J Proteome Res. https://doi.org/10.1021/acs.jproteome.3c00839
80. Wiśniewski JR, Rakus D (2014) Multi-enzyme digestion FASP and the 'total protein approach'-based absolute quantification of the Escherichia coli proteome. J Proteomics 109:322–331. https://doi.org/10.1016/j.jprot.2014.07.012
81. Hide M et al (2008) *Leishmania infantum*: tuning digitonin fractionation for comparative proteomic of the mitochondrial protein content. Parasitol Res 103(4):989–992. https://doi.org/10.1007/s00436-008-1062-9
82. Yao C et al (2010) Proteomic examination of *Leishmania chagasi* plasma membrane proteins: contrast between avirulent and virulent (metacyclic) parasite forms. Proteomics Clin Appl 4(1):4–16. https://doi.org/10.1002/prca.200900050
83. Brotherton M-C et al (2012) Analysis of membrane-enriched and high molecular weight proteins in *Leishmania infantum* promastigotes and axenic amastigotes. J Proteome Res 11(8):3974–3985. https://doi.org/10.1021/pr201248h
84. Kumar A et al (2015) Proteomic analyses of membrane enriched proteins of *Leishmania donovani* Indian clinical isolate by mass spectrometry. Parasitol Int 64(4):36–42. https://doi.org/10.1016/j.parint.2015.01.004
85. Jardim A et al (2018) Proteomic profiling of *Leishmania donovani* promastigote subcellular organelles. J Proteome Res 17(3):1194–1215. https://doi.org/10.1021/acs.jproteome.7b00817
86. Wiśniewski JR, Mann M (2012) Consecutive proteolytic digestion in an enzyme reactor increases depth of proteomic and phosphoproteomic analysis. Anal Chem 84(6):2631–2637. https://doi.org/10.1021/ac300006b
87. Wiśniewski JR et al (2009) Universal sample preparation method for proteome analysis. Nat Methods 6(5):359–362. https://doi.org/10.1038/nmeth.1322
88. Gadelha APR, Cunha-e-Silva NL, de Souza W (2013) Assembly of the *Leishmania amazonensis* flagellum during cell differentiation. J Struct Biol 184(2):280–292. https://doi.org/10.1016/j.jsb.2013.09.006
89. Wheeler RJ, Gluenz E, Gull K (2015) Basal body multipotency and axonemal remodelling are two pathways to a 9+0 flagellum. Nat Commun 6(1):8964. https://doi.org/10.1038/ncomms9964
90. Beneke T et al (2019) Genetic dissection of a Leishmania flagellar proteome demonstrates requirement for directional motility in sand fly infections. PLoS Pathog 15(6). https://doi.org/10.1371/journal.ppat.1007828
91. Andrade-Alviárez D et al (2022) Delineating transitions during the evolution of specialised peroxi-

somes: glycosome formation in kinetoplastid and diplonemid protists. Front Cell Dev Biol 10:979269. https://doi.org/10.3389/fcell.2022.979269
92. Jamdhade MD et al (2015) Comprehensive proteomics analysis of glycosomes from *Leishmania donovani*. OMICS 9(3):157–170. https://doi.org/10.1089/omi.2014.0163
93. de Oliveira AHC et al (2006) Subproteomic analysis of soluble proteins of the microsomal fraction from two *Leishmania* species. Comp Biochem Physiol Part D Genomics Proteomics 1(3):300–308. https://doi.org/10.1016/j.cbd.2006.05.003
94. Cuervo P et al (2009) Proteomic characterization of the released/secreted proteins of *Leishmania (Viannia) braziliensis* promastigotes. J Proteomics 73(1):79–92. https://doi.org/10.1016/j.jprot.2009.08.006
95. Chevallet M et al (2007) Toward a better analysis of secreted proteins: the example of the myeloid cells secretome. Proteomics 7(11):1757–1770. https://doi.org/10.1002/pmic.200601024
96. Silverman JM et al (2010) An exosome-based secretion pathway is responsible for protein export from *Leishmania* and communication with macrophages. J Cell Sci 123(6):842–852. https://doi.org/10.1242/jcs.056465
97. Silverman JM et al (2010) *Leishmania* exosomes modulate innate and adaptive immune responses through effects on monocytes and dendritic cells. J Immunol 185(9):5011–5022. https://doi.org/10.4049/jimmunol.1000541
98. Hassani K et al (2011) Temperature-induced protein secretion by *Leishmania mexicana* modulates macrophage signalling and function. PLoS One 6(5):18724. https://doi.org/10.1371/journal.pone.0018724
99. da Silva Lira Filho A et al (2022) *Leishmania* exosomes/extracellular vesicles containing GP63 are essential for enhance cutaneous Leishmaniasis development upon co-inoculation *of Leishmania amazonensis* and its exosomes. Front Cell Infect Microbiol 11. https://doi.org/10.3389/fcimb.2021.709258
100. Santarém N et al (2013) Exoproteome dynamics in *Leishmania infantum*. J Proteomics 84:106–118. https://doi.org/10.1016/j.jprot.2013.03.012
101. Forrest DM et al (2020) Proteomic analysis of exosomes derived from procyclic and metacyclic-like cultured *Leishmania infantum chagasi*. J Proteomics 227:103902. https://doi.org/10.1016/j.jprot.2020.103902
102. Douanne N et al (2020) Unravelling the proteomic signature of extracellular vesicles released by drug-resistant *Leishmania infantum* parasites. PLoS Negl Trop Dis 14(7). https://doi.org/10.1371/journal.pntd.0008439
103. da Silva Lira Filho A et al (2024) Identification of potential novel proteomic markers of *Leishmania spp.*-derived exosomes. Front Cell Infect Microbiol 14. https://doi.org/10.3389/fcimb.2024.1354636
104. Pan American Health Organization (2022) Guideline for the treatment of Leishmaniasis in the Americas, 2nd edn. Pan American Health Organization, Washington, DC. https://doi.org/10.37774/9789275125038
105. Croft SL, Sundar S, Fairlamb AH (2006) Drug resistance in Leishmaniasis. Clin Microbiol Rev 19(1):111–126. https://doi.org/10.1128/CMR.19.1.111-126.2006
106. Hendrickx S et al (2018) Evaluating drug resistance in visceral leishmaniasis: the challenges. Parasitology 145(4):453–463. https://doi.org/10.1017/S0031182016002031
107. Saboia-Vahia L, de Jesus JB, Cuervo P (2018) The role of proteomics in the study of drug resistance. In: Padrón-Nieves M, Ponte-Sucre A (eds) Drug resistance in Leishmania parasites, 2nd edn. Springer, Cham, pp 209–245
108. Hefnawy A et al (2022) Genomic and phenotypic characterization of experimentally selected resistant *Leishmania donovani* reveals a role for dynamin-1-like protein in the mechanism of resistance to a novel antileishmanial compound. mBio 13(1). https://doi.org/10.1128/mbio.03264-21
109. Jara M et al (2023) Unveiling drug-tolerant and persister-like cells in *Leishmania braziliensis* lines derived from patients with cutaneous leishmaniasis. Front Cell Infect Microbiol 13. https://doi.org/10.3389/fcimb.2023.1253033
110. Medina J et al (2021) Comparative analysis of the transcriptional responses of five *Leishmania* species to trivalent antimony. Parasit Vectors 14(1):419. https://doi.org/10.1186/s13071-021-04915-y
111. Patino LH et al (2019) Major changes in chromosomal somy, gene expression and gene dosage driven by SbIII in *Leishmania braziliensis* and *Leishmania panamensis*. Sci Rep 9(1):9485. https://doi.org/10.1038/s41598-019-45538-9
112. Shaw CD et al (2019) Genomic and metabolomic polymorphism among experimentally selected paromomycin-resistant *Leishmania donovani* strains. Antimicrob Agents Chemother 64(1):e00904-19. https://doi.org/10.1128/AAC.00904-19
113. Prasad SR et al (2022) Mechanistic insight into the role of mevalonate kinase by a natural fatty acid-mediated killing of *Leishmania donovani*. Sci Rep 12(1):16453. https://doi.org/10.1038/s41598-022-20509-9
114. Paradela LS et al (2021) Multiple unbiased approaches identify oxidosqualene cyclase as the molecular target of a promising anti-leishmanial. Cell Chem Biol 28(5):711–721.e8. https://doi.org/10.1016/j.chembiol.2021.02.008
115. Jafari M, Shirbazou S, Norozi M (2014) Induction of oxidative stress in skin and lung of infected BALB/C mice with Iranian strain of Leishmania major (MRHO/IR/75/ER). Iran J Parasitol 9(1):60–69
116. Ibarra-Meneses AV et al (2022) Exploring direct and indirect targets of current antileishmanial

drugs using a novel thermal proteomics profiling approach. Front Cell Infect Microbiol 12. https://doi.org/10.3389/fcimb.2022.954144
117. Nandan D et al (2017) Comprehensive identification of mRNA-binding proteins of Leishmania donovani by interactome capture. PLoS One 12(1):e0170068. https://doi.org/10.1371/journal.pone.0170068
118. de Pablos LM et al (2019) The mRNA-bound proteome of *Leishmania mexicana*: novel genetic insight into an ancient parasite. Mol Cell Proteomics 18(7):1271–1284. https://doi.org/10.1074/mcp.RA118.001307
119. Sanchiz Á et al (2020) Proteins interacting with *Leishmania* major PUF1: a proteomic dataset. Data Brief 33:106594. https://doi.org/10.1016/j.dib.2020.106594
120. Nocua PA, Requena JM, Puerta CJ (2021) Identification of the interactomes associated with SCD6 and RBP42 proteins in *Leishmania braziliensis*. J Proteomics 233:104066. https://doi.org/10.1016/j.jprot.2020.104066
121. Kalesh K et al (2022) Transcriptome-wide identification of coding and noncoding RNA-binding proteins defines the comprehensive RNA interactome of *Leishmania mexicana*. Microbiol Spectr 10(1):e0242221. https://doi.org/10.1128/spectrum.02422-21

Decoding Chagas Disease: What Next-Generation Sequencing Has Taught Us

Rachel Clear, Eric Dumonteil, and Claudia Herrera

Abstract

Chagas disease is the most important parasitic disease in the Western Hemisphere, caused by the protozoan parasite *Trypanosoma cruzi*. Understanding and controlling the disease has been challenging due to the extensive complexity of transmission cycles, which involve a wide variety of vectors and host species, highly diverse parasite strains, and clinical presentations. In recent years, the use of next-generation sequencing (NGS) has allowed several breakthroughs to be made in addressing some of these challenges that we review here. Sequencing of parasite genomes and transcriptomic studies have shed light on parasite diversity, its differentiation and responses to drug treatment. *T. cruzi* has the ability to infect any mammalian species, and the elucidation of parasite–host interactions has been enhanced in great part by NGS technologies. Thus, host and parasite genetic diversity has been shown to play a role in Chagas disease pathogenesis, and the host microbiome is also likely involved in these interactions. Finally, NGS has provided insight into the molecular ecology of triatomine vectors and parasite transmission. Bug species, blood feeding sources, *T. cruzi* parasite diversity, and triatomine gut microbiota all play a role in transmission cycles and risk of infection. Characterizing these interactions can lead to improved control. While these studies provide a robust basis to describe the complexity of Chagas disease, their future integration is poised to provide further insight into disease biology and ecology, and lead to new tools and approaches to disease control.

Keywords

Trypanosoma cruzi · Triatomine · Chagas disease · Genome · Transcriptome · Gene expression · Pathogenesis · Transmission cycle · Microbiota

1 Introduction

Chagas disease is caused by the protozoan parasite *Trypanosoma cruzi*, and is transmitted by triatomine bugs. There are an estimated 8 million cases worldwide, and Chagas disease may be responsible for approximately 12,000 deaths per year [1]. It accounts for an estimated loss of US$

R. Clear · E. Dumonteil · C. Herrera (✉)
Department of Tropical Medicine and Infectious Disease, School of Public Health and Tropical Medicine, and Vector-Borne and Infectious Disease Research Center, Tulane University, New Orleans, LA, USA
e-mail: cherrera@tulane.edu

J. D. Ramírez González (ed.), *Recent Advances in Parasitomics*,
https://doi.org/10.1007/978-3-031-70591-5_3

7.2 billion annually [2]. An additional 75 million people are at risk for infection globally [3].

Chagas disease has two distinct phases. The acute phase is characterized by high parasitemia, as well as nonspecific symptoms such as fever, anorexia, nausea or vomiting, diarrhea, and rash; more specific symptoms may also present at the site of infection, such as chagoma (raised tumor-like swellings on the skin) and Romaña's sign (unilateral swelling of the eye) [1]. Without treatment, infected patients will enter the chronic stage, where blood parasites recede to invade tissues—usually in the cardiac or gastrointestinal muscles. Chronic disease symptoms usually manifest more than a decade after the initial infection. Approximately 10–20% of people will develop gastrointestinal disease such as mega-esophagus or megacolon, and 20–40% of people will develop Chagas-related cardiomyopathy [4]. Currently, there is no reliable way to predict which patients are at risk for developing chronic disease and which are likely to remain asymptomatic.

Carlos Chagas, the physician who first described the disease in 1909, himself remarked on the observable differences in clinical presentations and parasite diversity, and stated the need to identify what factors may affect such variability [5]. What Carlos Chagas did not have was the ability to use next-generation sequencing (NGS) to better understand *T. cruzi* infections. Indeed, the first NGS platform was introduced in 2005 and since then, sequencing technology has continued to improve in speed, quality, and ease of use [6]. In recent years, NGS approaches have provided insight into multiple facets of the *T. cruzi* parasite–host–vector triad. This chapter will highlight some of the key advances in our understanding of Chagas disease.

2 *T. cruzi* Parasite Genomics

2.1 Genome Structure and Content

A primary use of NGS has been to better understand the intricacies of the *T. cruzi* genome. Like other members of the kinetoplastida order, it has a unique mitochondrial DNA structure known as kinetoplastid DNA (kDNA) consisting of dozens of maxicircles and thousands of minicircles and making up about 25% of the total genome [7] (Fig. 1). The nuclear genome, on the other hand, comprises 50% of tandemly repeated sequences. This includes satellite DNA, multigene families, and retro-transposable elements [8].

The overall makeup of the *T. cruzi* genome can vary significantly between strains. Accounting for both nuclear and kinetoplast DNA, *T. cruzi* DNA content can differ up to 40%, and genome size can vary from 25 to 53 Mbp in haploid genome size, encoding for 10,000–20,000 genes [9].

The extensive presence of repeat sequences has historically hampered genome sequencing and the first sequence of the CL Brener reference strain suffered high levels of fragmentation and gene collapse [10]. Since then, advances in NGS technologies have allowed the use of combination of short- and long-read sequencing, and a chromosomal level assembly of *T. cruzi* genome is getting closer [11, 12]. To date, at least 42 *T. cruzi* genomes have been assembled and are representative of every discrete typing unit (DTUs) besides TcBat [13].

The *T. cruzi* nuclear genome has been found to be compartmentalized into two regions: The "core" compartment contains conserved genes—genes encoding proteins with known functions—and hypothetically conserved genes—genes of unknown function present in multiple Trypanosomatid species [14]. The "disruptive" compartment contains the three largest multigene families, whereas dozens of other multigene families are dispersed across both compartments [14].

Most genes have at least two copies per haploid genome, but assembly of the Dm28c strain found 5000 single-copy genes, with only about one-fourth having an assigned function [14]. In addition, some genes previously considered to be single-copy have now been shown to exhibit multiple copies, and underestimation of copy number could explain the failure of previous gene knockout experiments [14].

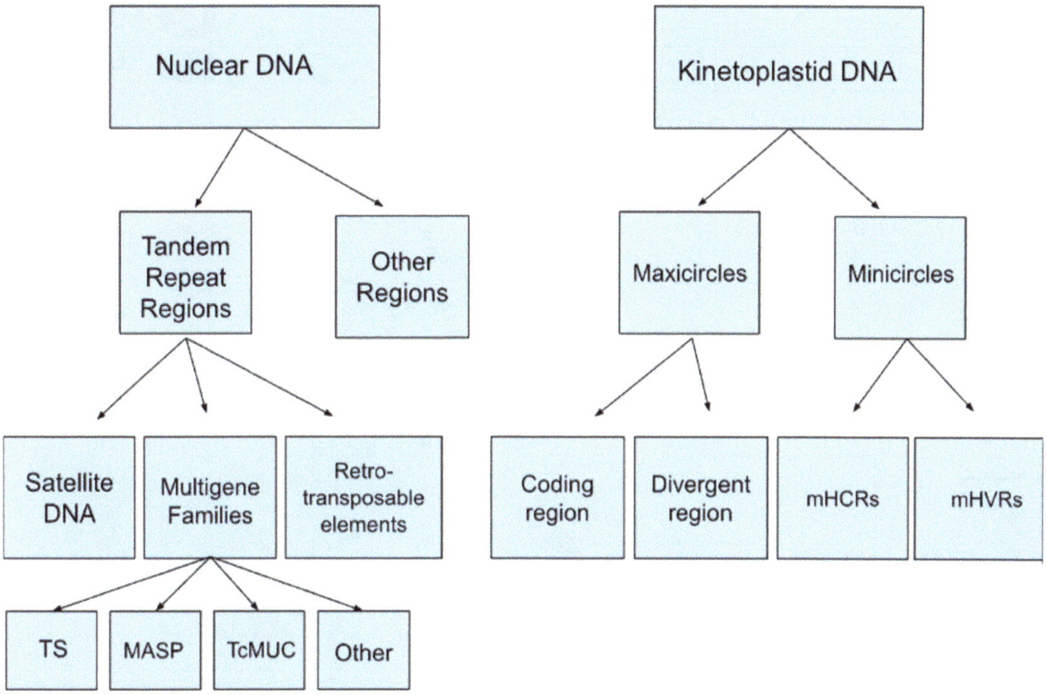

Fig. 1 Organization of *T. cruzi* genome. The nuclear genome is highly repetitive due to satellite DNA, multigene families, and retro-transposable elements. The main multigene families include trans-sialidases (TS), mucin-associated surface proteins (MASP), and *T. cruzi* mucins (TcMUC), among others. The mitochondrial/kinetoplast DNA includes mini- and maxi-circles, which can further be subdivided into conserved and variable regions. See text for additional explanations

T. cruzi strains are customarily divided into seven DTUs (TcI–TcVI and TcBat). Comparative genomics supports this diversity of *T. cruzi* strains, both within and across DTUs. TcI, the most common DTU infecting humans, exhibits particularly high intra-lineage diversity and can be further identified by subgroups TcIa–TcIe [15, 16]. There is notable geographic structuring of TcI across the Americas [17] (Fig. 2). Genetic diversity has also been described among TcII in Brazil, as well as for TcIV between North and South America, suggesting a comparable geographic structuring among other parasite DTUs, but the number of genomes sequenced from these is still too limited to allow for reliable estimates of their genomic diversity [18, 19].

Kinetoplast DNA is also highly informative for assessing evolutionary ancestries and phylogenetic relationships among *T. cruzi* strains. Maxicircles comprise a coding region including 21–25 genes, which are highly conserved across all Trypanosomatid species, as well as a divergent region [20] (Fig. 1). They can be divided into three main clusters corresponding to TcI (cluster A); TcII (cluster C); and TcIII, TcV and TcVI (cluster B), which supports the hypothesis that TcV and TcVI are hybrid strains with a mitochondrial ancestor derived from TcIII. Nonetheless, *T. cruzi* kDNA has also been shown to exhibit introgression and heteroplasmy, representing evidence of genetic exchange among strains [17]. Minicircles are similarly made up of highly conserved regions (mHCRs) and hypervariable regions (mHVRs) (Fig. 1). At least in some strains, minicircles come in a variety of sizes from about 300 bp to about 1500 bp, and the number of mHCRs and mHVRs correlate to size; a new set of mHCR and mHVR occurs about every 300 bp [21].

Remarkably, *T. cruzi* nuclear genome exhibits variable aneuploidy, which is likely associated with adaptative responses to environmental stress

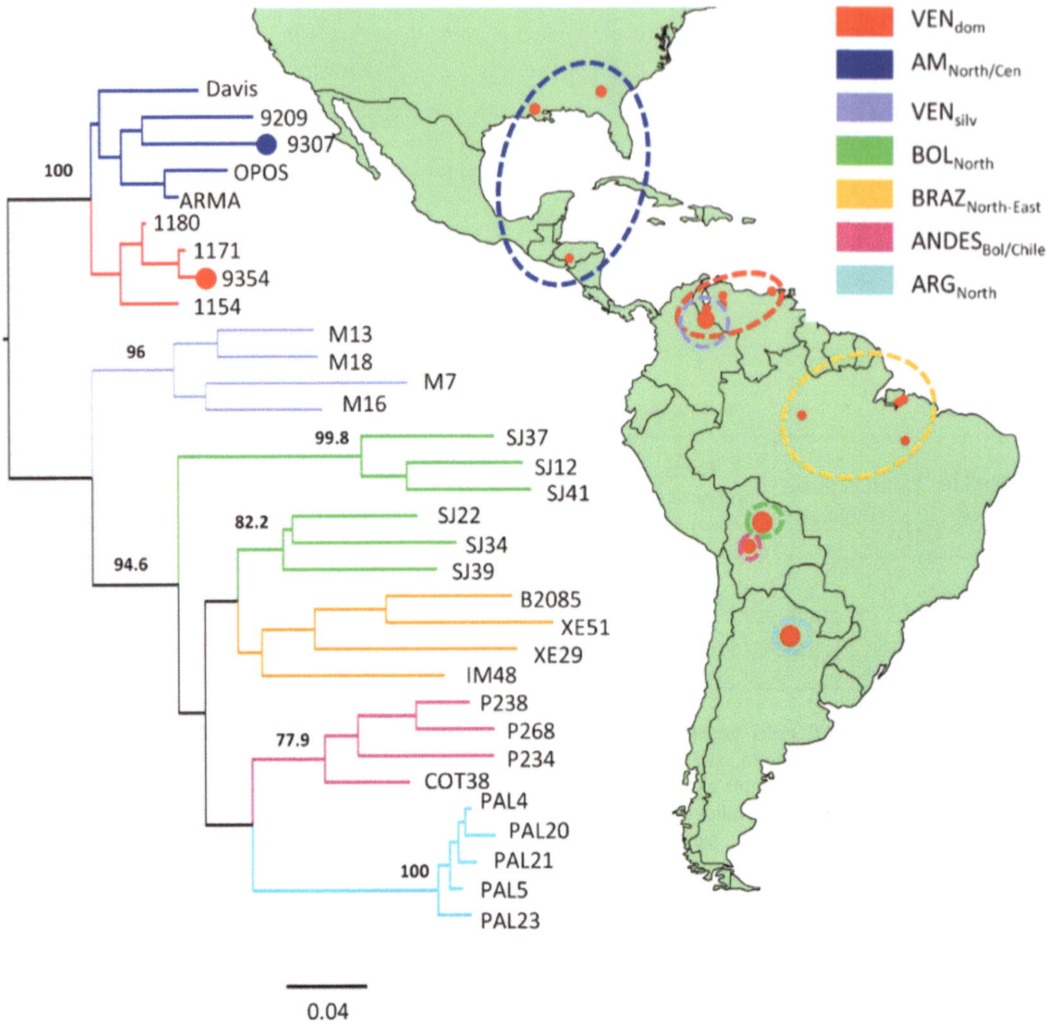

Fig. 2 *T. cruzi* TcI population structure across the Americas. A panel of 32 TcI genomes was assembled for analysis. Origin of individual strains is shown on the map by small red circles. Large red circles correspond to multiple samples, isolated from the same geographical area. Branch colors indicate strain population. The blue and red circles on branches represent inter-lineage introgression events. This relationship is further represented by the structure of the constructed phylogenetic tree. Genetic similarity between strains shows a distinct relationship with the origin of infection. Modified from Messenger et al. [17]

involving gene duplication and segmental aneuploidy [22, 23]. Hybridization is thought to have occurred to give rise to TcV and TcVI DTUs, and has been experimentally observed [24]. Such experimental hybrids are initially tetraploid but show progressive genome erosion with gradual losses of chromosome copies over time [7], highlighting the high tolerance of *T. cruzi* for aneuploidy and strengthening the adaptative role of this process. Meiotic genetic exchange and recombination have also been detected in natural parasite populations [25, 26], but its frequency seems too low to alter the stability of the main parasite DTUs [7, 27].

While comparative genomics is providing a new understanding of *T. cruzi* parasite evolution, it is also key to identify highly conserved antigens that may be used for improved serological diagnostics, as well better understand the impact of parasite diversity in pathogenesis and disease progression [28, 29].

2.2 Multigene Families

One of the hallmarks of the *T. cruzi* genome is its multigene families, some of which are present in several hundreds of copies, which vary among strains and DTUs. The three largest multigene families—trans-sialidases (TS), mucin-associated surface proteins (MASPs), and *T. cruzi* mucins (TcMUCs)—may contribute heavily to parasite–host interactions.

Sequence variability and copy number of these multigene families were determined to be associated with DTU, and TS copy number was strongly correlated with overall genome size [10]. Hybrid strains also maintained higher cluster variability compared to non-hybrid isolates. There is also some evidence that TS and MASP expression plays a role in the progress of chronic infections and evasion of host immune response. The high variability of multigene families may be due to their location on telomeric and subtelomeric regions—thus subjecting them to continuous evolutionary pressure [30]. The two most abundant subgroups of TS are group V (associated with antigenic variation) and group II (implicated in host cell attachment and invasion) [10]. Knockout mutants lacking active TS are still able to infect cells and differentiate into cytosolic amastigotes but are unable to differentiate from amastigotes to trypomastigotes and egress from host cells, resulting in avirulent infections in mouse models [31], highlighting the role of TS as virulence factors. Cruzipain, another virulence factor that is also part of a multigene family, was found to include two large families whose sequence differences in the active site suggest different susceptibility to inhibitory drugs, although the development of broad-spectrum inhibitors may still be feasible [32].

Analysis of nucleotide substitution patterns among the multigene families suggested that these expanded gradually during the diversification of *T. cruzi* as a species, and that the expansion of the dispersed gene families likely happened long before other gene families [11]. Also, proteins from these multicopy gene families appear to evolve faster than core genes, due to a more relaxed purifying selection. Further studies of these multigene families should expand our understanding of their key biological role in *T. cruzi* infection and may lead to new drug targets [33].

2.3 Transcriptomics of *T. cruzi* Parasites

Even though the parasite relies on unique mechanisms for the regulation of gene expression, based on the constitutive polycistronic transcription of protein-coding genes associated with extensive RNA editing and trans-splicing, RNA sequencing has also been used to investigate functional aspects of *T. cruzi* parasites [34]. It is estimated that about half of *T. cruzi* genes are differentially regulated during its life cycle [34–36]. A recent study further described the underlying mechanisms involved in differential gene expression in *T. cruzi*, highlighting major differences between the core and disruptive compartments of the genome [37]. These compartments strongly differ in their level of DNA methylation and chromatin structure, which results in genes from the core compartment having a high and medium expression level, while genes in the disruptive compartment present a low or undetectable expression level, but some of these genes can be upregulated during parasite differentiation [37].

Transcriptomic analysis of *T. cruzi* life cycle stages provided important insight into the mechanisms of parasite differentiation [35]. For example, there are broad transcriptome changes during amastigote differentiation, initiated within hours of trypomastigote entry into host cells, likely in response to the shift from an extracellular to an intracellular environment [35]. Protein surface remodeling and metabolic switches are the foundation for parasite differentiation and can be initiated by transcriptomic changes [38]. A comparison of virulent and non-virulent strains revealed delayed expression of specific genes in the non-virulent strain—genes that encode for surface proteins from multigene families such as TS, mucin, MASP, and GP63 associated with the transition from amastigote to trypomastigote [39].

The functionality of transcriptomes in benznidazole-resistant strains has also been studied. Benznidazole is one of only two drugs that are currently licensed for Chagas disease treatment, so understanding the mechanisms for resistance is imperative to preventing further spread of drug resistance. Drug-resistant parasites had 1819 transcripts that were differently expressed compared to susceptible parasites, most of which (83.7%) presented functional annotations, which allowed to identify 11 biological processes that may be associated with a benznidazole resistance [40]. An additional concern surrounding drug-resistant *T. cruzi* is the recent discovery of amastigote dormancy. Dormant cells exhibit a 50-fold increase in their IC_{50} to BZ, possibly due to limited metabolic activity [41]. A transcriptomic approach was able to identify heterogenous gene expression profile among amastigotes and six distinct subpopulations were identified, including one (2–5% of total amastigotes) that presented characteristics of non-replicative dormant cells [42]. The non-replicative subpopulation displayed several overexpressed genes, which may serve as candidate markers for its identification.

T. cruzi can also regulate gene expression through posttranscriptional mechanisms. *T cruzi* U-rich RNA-binding protein 1 (TcUBP1) is known to modulate gene expression of multiple surface glycoproteins in the infective stage of the parasite, which increases its ability to survive within a host. An RNA sequencing analysis, comparing noninfective parasites with an overexpression of TcUNP1, found that 13% of genes resulted differentially expressed, creating an environment similar to that of an infective parasite [43]. Upregulated genes were mostly categorized as cell-surface trypomastigote glycoproteins and downregulated as ribosomal and mitochondrial proteins (Fig. 3). This finding is similar to that of an earlier study, also using RNA sequencing, that compared two strains with opposing levels of virulence—gene expression of surface proteins is associated with parasite differentiation and level of infectivity [39].

Extracellular vesicles (EVs) serve as mediators in intercellular exchange of genetic material

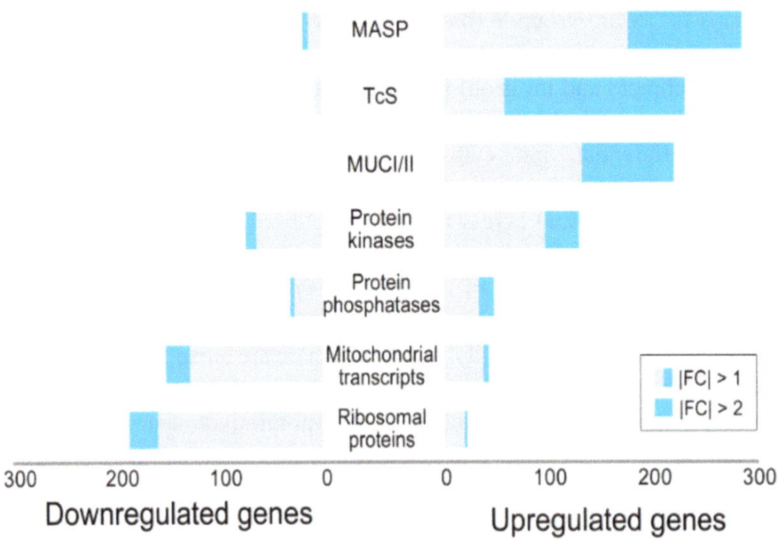

Fig. 3 Gene differential expression in parasites with overexpression of TcUNP1. Bars chart displaying the number of up- and downregulated genes within seven categories in the UBP1-OE transcriptome. Categories are MASP (mucin-associated surface proteins), TcS (trans-sialidase/trans-sialidase like), MUCI/II (mucin), protein kinases, protein phosphatases, mitochondrial transcripts, and ribosomal proteins. FC: fold change. Modified from Sabalette et al. [43]

in eukaryotes and are known to play roles in immune function such as antigen presentation. The enclosed RNA in EVs is also capable of inducing gene expression in recipient cells. EVs are also released by *T. cruzi* parasites, since these EVs and their RNA are not vulnerable to degradation in extracellular spaces, they are currently considered possible biomarkers for infection. Transcriptomic analysis of RNAs from EVs of three different *T. cruzi* life stages revealed over 200 single nucleotide polymorphisms (SNPs) associated with disease [44]. Some of these SNPs were stage-specific, and further research may uncover certain SNPs associated with gene expression and disease phenotype.

Although most of *T. cruzi* genome encodes for proteins of unknown function, the limited functional annotations available provide key functional insight on the parasite life cycle. As annotations are refined, transcriptomic approaches are poised to provide even greater depth to our understanding of these processes, and how they may vary among parasite strains, which may lead to new drug targets.

3 Mammalian Host Responses to *T. cruzi*

3.1 Host Genetic Susceptibility to *T. cruzi*

Almost any mammal with exposure to *T. cruzi* parasites may serve as a competent host, and the role of host–parasite relationship in sustaining infection is complex. In humans, as mentioned above, infection can lead to very diverse clinical outcomes; this led to the hypothesis that different host genetic backgrounds could be associated with differences in susceptibility to *T. cruzi* and/or in disease progression. Many pioneering studies were informed by work uncovering the immune mechanisms of parasite control and specifically targeted gene variants of the HLA, cytokines, and chemokines through genotyping. Thus, multiple SNPs modulating the immune response and its polarization have been described as associated with susceptibility/resistance to *T. cruzi* in diverse populations [45–47].

The development of NGS led these studies to move from genotyping a limited number of candidate SNPs to genome-wide association studies (GWAS)/whole exome sequencing (WES). The major advantage of these recent studies is to allow for untargeted/unbiased approaches for identifying host gene variants associated with Chagas disease [48, 49]. One of the largest studies included over 3400 patients from Colombia, Bolivia, Brazil, and Argentina and a GWAS was able to uncover new variants in H8, KLF4, and SAC3D1 genes that were associated with susceptibility to chronic Chagas cardiomyopathy [50].

Another study based on only 25 participants from six Brazilian families with multiple cases of Chagas disease was able to identify several mitochondrial and inflammation-related genes associated with chronic Chagas cardiomyopathy, based on WES [51], highlighting the power of this approach to identify potential rare and high-impact gene variants. A GWAS of 118 individuals from 19 native American populations identified PPP3CA variants associated with susceptibility to *T. cruzi*, with a significant positive selection on this gene in populations from the Amazon region, as opposed to a genetic drift effect, which suggested that selection may have favored a protective phenotype in Native Americans in the Amazon [52].

While these studies do point to a role of host genetics in the pathogenesis of Chagas disease, many of these associations still need to be replicated in broader populations, likely due to the complex nature of parasite–host relationship. Indeed, the presence of multiple genetic variants may work either in opposition or synergistically, making their effect more difficult to uncover. Additionally, factors such as infecting parasite strains or host environment have not been considered in these studies, so the relative contribution of host genetics to Chagas disease remains to be established.

3.2 *T. cruzi* Diversity in Hosts

Multiple efforts have been made over the years to characterize *T. cruzi* strains infecting hosts to investigate how parasite strain diversity may affect disease progression and assess potential associations of strains/DTUs to host species. However, initial methods relying on the isolation of strains in culture or direct genotyping by PCR may be biased toward strains better adapted to in vitro growth or more abundant in a sample, respectively. These limitations were in part solved by use of deep sequencing, which can discern the presence of multiple strains/DTUs. Thus, multiplicity of infection has been detected in naturally infected mammalian species such as nonhuman primates [53, 54], rodents [55], opossums [56], raccoons [57], dogs [58], and cats [59]. Infections in humans can also involve multiple strains/DTUs [60–62]. For example, a study in Mexico reported that nearly half of their human samples harbored multiple DTUs [63].

Such multiplicity of infection exemplifies the complexity of *T. cruzi* transmission cycles [57, 64], and more studies are needed to better understand transmission cycles at the level of DTUs and strains, which may differ according to vector and host species. Nonetheless, a major challenge of these studies remains the very low level of parasitemia in chronically infected mammalian hosts, which considerably complicates parasite genotyping [64]. As a result, parasite diversity may be underestimated when using single-copy genetic markers due a lack of sensitivity, while multicopy markers may be more difficult to interpret in terms of absolute multiplicity of parasite clones. The development of novel genotyping approaches may help addressing some of these issues [65].

In addition, a critical role of the multiplicity of infection in disease progression and pathogenesis is emerging. A longitudinal study in naturally infected Rhesus macaques showed that infections with a high parasite diversity were associated with a better control of the parasite and limited disease progression [53]. On the other hand, progression of parasitemia and cardiac disease over time was seen preferentially in macaques infected with low parasite diversity. This suggests that parasite strain assemblage is a key driver of disease progression, and multiplicity of infection may favor the host rather than the parasite, possibly due to some competition among strains within a host.

3.3 Host Transcriptomic Response to *T. cruzi* Infection

NGS has enabled a growing number of studies aimed at characterizing host response to infection through transcriptomic approaches, and RNA sequencing has substituted earlier microarray approaches [66–68]. The host cell response to *T. cruzi* infection has been investigated in various cell types, ranging from primary fibroblasts [35], stem cell-derived cardiomyocytes [69, 70], murine cardiomyocytes [71], mice heart [72, 73], and peripheral blood mononuclear cells (PBMCs) [74]. A key advantage of this approach is that host cell and parasite transcriptomes can be investigated simultaneously [35]. These studies documented extensive transcriptomic alterations in host cells, characterized by the modulation of immune response and cell cycle genes, together with changes in metabolic pathways, which may contribute to the provision of metabolites for *T. cruzi* amastigote differentiation and replication [35, 70]. Importantly, the transcriptomic responses appear to include a shared component, as well as a parasite strain-specific signature, both in vitro and in vivo [71, 72], although a greater diversity of strains still needs to be tested to fully assess these responses.

Interestingly, the transcriptomic response of stem cell-derived cardiomyocytes from asymptomatic (indeterminate) and symptomatic patients to in vitro infection with *T. cruzi* evidenced marked differences in gene expression profiles: gene pathways associated with antigen processing and presentation, as well as immunoregulatory genes, were downregulated in cells from indeterminate patients [74]. This was also associated with interferon alpha and gamma pathways.

On the other hand, cells from chronic Chagasic cardiomyopathy patients were associated with upregulated epithelial mesenchymal transition, mTORC1 signaling, and unfolded protein response pathways. These differences were interpreted as reflecting genetic differences among asymptomatic and symptomatic patients, although these were not directly assessed.

Another study of patient PBMCs focused on identifying transcriptomic changes involved in the early stages of disease progression toward cardiomyopathy, and identified that it was associated with reduced pathways for antigen presentation and immune cell activation, suggesting that early immune processes may mediate long-term clinical disease outcomes [74].

RNA sequencing also allowed the detection of microRNAs (miRNAs) in the plasma of patients, which could be specifically associated with Chagasic cardiomyopathy [75], and in infected mice, drug treatment altered the miRNA profile [73]. Nonetheless, a major limitation of these studies is that they do not incorporate host and parasite diversity, which also affect such transcriptomic responses, and more integrative studies are critically needed. The relative ease and standardization of RNA sequencing is poised to lead to an increasing number of studies in multiple model systems as well as in patients that should help disentangle host–parasite interactions in an integrative manner.

3.4 Congenital Transmission of *T. cruzi*

T. cruzi can be transmitted vertically to the fetus during pregnancy, and congenital transmission is a major cause of new infection in non-endemic countries [76, 77]. Transmission can occur in about 5–6% of pregnancies from an infected mother, and parasitemia level is one of the better established risk factors for transmission [78, 79]. However, the interactions of *T. cruzi* parasites with the placenta remain unclear. There is evidence to suggest that some strains may be better adapted for vertical transmission. Supporting this theory, a study in mice found that *T. cruzi* isolates from previous congenital cases resulted in vertical transmission at higher rates than moderate or high-virulence non-congenital strains [80]. Congenital strains modulated the expression of only 150 genes in the placenta compared to over 2500 genes modulated by virulent non-congenital strains, leading to anti-inflammatory responses and tolerance to congenital strains [80].

In humans, multiple differentially expressed genes have also been identified by comparing the placenta transcriptome of seropositive and seronegative mothers [81]. Again, several upregulated genes were associated with immune tolerance in pregnancy as well as with antiparasitic defense mechanisms. Downregulated genes were associated with chemotaxis and invasion inhibitors and human chorionic gonadotropin secretion. A specific gene of interest was the downregulation of *KISS1*, as low serum level of the kisspeptin protein is associated with preeclampsia, restricted intrauterine growth, and recurrent pregnancy loss. Furthermore, SNPs in placental genes *ADAM12* and *MMP2* have been found associated with congenital transmission [82].

NGS has also provided evidence for the transmission of multiple parasite genotypes between parent and infant, as well as for the presence of novel genotypes within the infant [83]. In fact, significantly higher levels of genetic diversity can be detected among infants than among their birthing parent, as well as several haplotypes that were exclusive to either parent or infant [84]. Paired haplotypes may represent strains that are more likely to result in congenital transmission, while parent-only haplotypes may represent the opposite. Infant-only haplotypes, on the other hand, may result in parasite strains restricted to the placental environment or be generated during transmission. As mentioned previously, some parasite strains or haplotypes may be more strongly associated with congenital transmission. Further, high-resolution parasite genotyping studies should help better understand how parasite diversity influences congenital transmission.

Importantly, fetal exposure to maternal infection with *T. cruzi*, even in the absence of congenital infection, may lead to alterations in newborns. Recently, epigenetic changes in DNA methylation, which may affect gene expression, have been detected in cord blood cells [85] and may help explain the increased risk of reduced fetal growth in these newborns [86].

3.5 Gastrointestinal Microbiome of the Host and *T. cruzi* Infection

The host and parasite genomes are not the only lens through which host–parasite interactions may be viewed. The expansion of NGS and metagenomic studies has also stimulated interest in the role of the intestinal microbiota. In Chagas disease patients with megaesophagus, intestinal microbiota showed relatively higher levels of bacterial concentration and variability compared to controls with a healthy esophagus, with a predominance of *Streptococcus* spp. and *Veillonella* [87]. Megacolon patients, alternatively, touted higher levels of anaerobes, which returned to normal levels after surgical treatment.

In mice, infection with *T. cruzi* decreased the relative abundance of *Bacteroides thetaiotaomicron*, *Faecalibaculum rodentium*, and *Lactobacillus johnsonii*, while *Akkermansia muciniphila* and *Staphylococcus xylosus* increased, providing important evidence that *T. cruzi* infection directly affects the gut microbiome of the host [88]. Expanding this work into more functional studies should help better understand the interactions of the host gut microbiome and the parasite, and may lead to new ways to mitigate infections.

4 Molecular Ecology of Triatomine Vectors

Triatomine bugs, a subfamily of *Reduviidae*, commonly referred to as "kissing bugs," spread *T. cruzi* through their feces [89]. About 130 species blood feed on vertebrates, and most often feed at night. All triatomine species are considered potential *T. cruzi* vectors, but the species of importance are those that tend to live in close proximity with humans; species of interest vary based on geography, but triatomines are abundant throughout the Americas [90]. NGS and metagenomic approaches have become key tool to assess the ecology of triatomines and establish *T. cruzi*

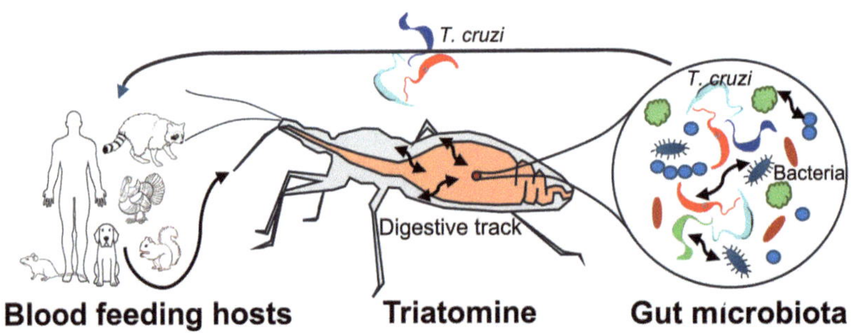

Fig. 4 One Health interactions in *T. cruzi* transmission cycles. Multiple vertebrates can serve as blood feeding sources for triatomine bugs, which become infected when feeding on an infected host. In their digestive track, *T. cruzi* parasites will multiply alongside bacteria from the microbiota. These bacteria have metabolic interactions among themselves and with the bugs, for example the provision of key metabolites such as vitamin B, as well as with *T. cruzi* when present, and these interactions can affect both bacteria and parasite growth. Parasites released in bug feces can infect new mammalian hosts from the bug feeding network to close the cycle

parasite transmission cycles. Both targeted and untargeted sequencing of DNA extracted from triatomines allows the identification of bug species, blood feeding sources, *T. cruzi* parasite diversity, and gut microbiota. Studies performed in an integrative context allow for a detailed analysis of the interactions among the vector–parasite–host triad, which will ultimately shape understanding of transmission cycles in a One Health framework (Fig. 4) [91–95]. However, each of these components can also be investigated individually in a more reductionist manner to gain a deeper appreciation for the intricacies of each field [87, 88, 96, 97].

4.1 Triatomine Blood Feeding Sources

One of the most straightforward uses of NGS in triatomines is to identify their blood feeding sources, which is essential to uncover parasite transmission cycles and infection risks for humans [92]. Although triatomine species are often grouped into sylvatic, synanthropic, or domestic cycles based on the habitat in which they are collected, blood meal characterization provides detailed information on the distribution/frequency of feeding hosts in each of these habitats and has revealed extensive overlaps and connections [97]. The advantage of NGS over conventional sequencing for blood meal analysis is its sensitivity and ability to detect blood from multiple hosts in a single bug, summarizing its recent feeding history [92, 96]. For example, *Mepraia parapatrica*, previously considered an exclusively sylvatic vector on the coast of Chile, was shown to feed on humans in nearby fishing communities [15, 98]. A similar feeding pattern was also detected in *Mepraia spinolai* and *Triatoma infestans* [97].

In vectors associated with primarily domestic cycles, blood meal characterization can provide critical insight into the contribution of humans and other host species to bug feeding and parasite transmission. For example, in Colombia, 83.3% of field-caught bugs were infected with *T. cruzi*, and humans were the predominant feeding source at 76.8% [99]. Analysis of *T. dimidiata* in Mexico also showed feeding habits that greatly centered around humans [94]. In addition, different genera of bugs seemed to prefer different nonhuman sources despite similar geographic location. Dogs, raccoons, and rodents were also identified as key players in the transmission cycle. Seven triatomine species in Texas predominantly feed on dogs and humans, and over half of bugs harbored a *T. cruzi* infection [96], evidencing a risk of infection for humans and pets.

One of the key questions underlying triatomine feeding behavior is whether blood meal sources are opportunistic and driven by host availability or selective for preferred host species, and many studies suggest that triatomines are mostly opportunistic feeders [93, 100]. Nonetheless, a recent study of *Mepraia spinolai* revealed differences in the relative availability of host species and blood meal frequencies, which suggests selective blood feeding in this triatomine [101]. This work highlights the need to better understand host abundance and availability to properly interpret blood feeding data. Extension of these studies will provide greater insight on triatomine feeding behavior and how it may be modulated by host availability, seasonality, or bug development, which in turn can better inform risk for human infection.

4.2 *T. cruzi* Infection in Triatomines

While triatomine infection rate with *T. cruzi* is a critical component of vectorial competence and risk of transmission, NGS can provide further information on parasite strains circulating in vectors. As mentioned above for mammalian hosts, NGS allowed the detection of multiple parasite haplotypes in single vectors, which may cover different DTUs. For example, up to 20 haplotypes have been detected within one *T. dimidiata* bug, with the ability to detect haplotypes that make up as little as 0.1% of the overall makeup [92]. A similar study showed that most field-collected bugs were infected with multiple DTUs, and all were infected with multiple haplotypes [94]. This trend is observable across multiple triatomine species in Ecuador, Brazil, and Texas, among others [102–104]. These observations concur with those in mammalian hosts and indicate that a large proportion natural infections through vectors involve multiple parasite strains that can cover multiple parasite DTUs. Such results open the possibility of further refining transmission cycles according to parasite DTUs, as well as to test for potential associations with hosts or habitats. In addition, it has been posited that parasite genetic mutation may take place in the midgut of the triatomine vector; it may also be where the genetic recombination/sexual reproduction described above occurs [105]. Therefore, high levels of parasite diversity within a single bug have significant implications for the continued diversification of *T. cruzi* parasites.

4.3 Gut Microbiota of Triatomines

The gut microbiota of triatomines has been an area of increased research as its role in bug development and vectorial competence is emerging, and NGS circumvents the limitations of earlier approaches based on bacterial culture [106]. The biological functions of triatomine gut microbiota are still not clear but are likely associated with the provision of key metabolites such as vitamin B by *Rhodococcus rhodnii* and other bacteria in *Rhodnius prolixus* [107, 108]. A rich diversity of bacteria is present in the gut of triatomines [92, 109–111], and uncovering their interactions and biological role in bug development and vectorial competence may lead to novel vector control interventions as suggested before [112, 113] (Fig. 4).

The microbiota does not appear to be transmitted vertically, but has to be acquired by early nymphal stages for their development as suggested for *Triatoma sanguisuga* [114] and several other triatomine species that present changes in microbiota composition between nymphal and adult stages [114]. Also, while the microbiota seems to vary according to triatomine species [89, 109, 110, 115], it is still unclear how its composition may be affected by habitat, blood feeding sources, or bug sex [92, 110], making integrative studies that incorporate these aspects critically needed. For example, comparing the microbiota in *T. sanguisuga* from Louisiana and Illinois showed major differences in bacterial composition, while the metabolic profile appeared highly conserved, suggesting that functional constraints drive microbiota composition [116].

On the other hand, once a triatomine becomes infected with *T. cruzi* parasites, the microbiota plays an integral role in parasite multiplication and establishment in the digestive tract [87, 91, 94, 111, 117–119]. Certain *T. cruzi* strains have antibacterial abilities, furthering their ability to colonize the vector midgut, and some bacteria have exhibited trypanolytic activity [120], highlighting the diversity of interactions that may be occurring in the microbiota community (Fig. 5).

For example, in *T. dimidiata* from Veracruz, Mexico, the presence of *Enterobacteriae* was correlated with the absence of *T. cruzi*, while *Nocardiaceae* and *Carnobacteriaceae* were correlated with the presence of *T. cruzi* [94]. Evidence also suggested that the bacterial diversity may be correlated with specific DTUs infecting the vector, with *Nocardiaceae* bacteria correlated with DTUs TcI, TcV, and TcVI. Conversely, *Carnobacteriaceae*, *Dietziaceae*, and *Paenibacillaceae* all correlated with the presence of TcI [94]. Comparable observations were made in *T. sanguisuga* in the Southeastern United States, indicating that interactions may be parasite DTU-specific [91].

Together, these studies support the idea that vector gut microbiota plays a key role in vector biology and in *T. cruzi* parasite proliferation. Further use of NGS could lead to a refined assessment of the interactions involved and lead to novel control strategies targeting some of these critical pathways. Furthermore, the recent identification of a complex core virome in *Rhodnius prolixus*, transmitted through a transovarial route at least in a laboratory colony, expands the diversity of organisms that may be interacting in the

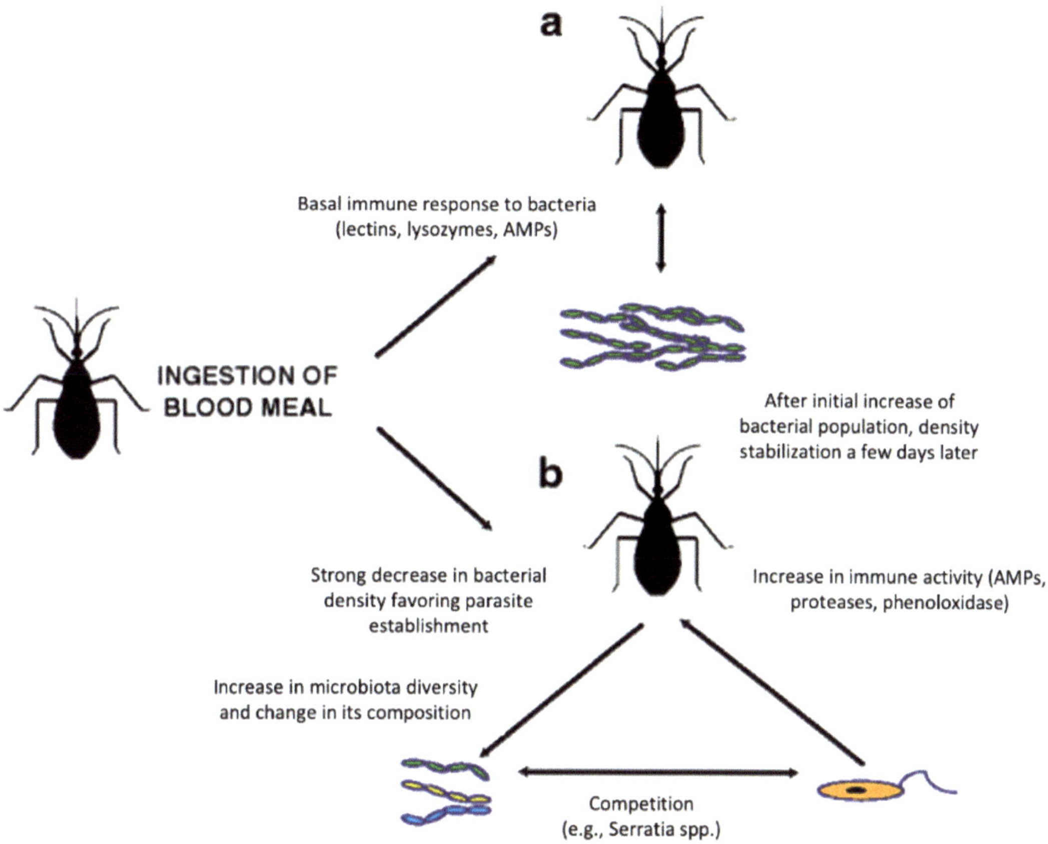

Fig. 5 Interaction between *T. cruzi* infection and vector microbiota. Model of triatomine host–microbiota–parasite interactions based on parasite immune modulation. (**a**) Triatomine homeostatic responses to blood meal ingestion that controls the density of bacteria in the gut. (**b**) Immune responses induced by *T. cruzi* infection control parasite development and decrease bacteria density; intense competition for nutrients between bacteria and the protozoan occurs in the gut. An overall increase in microbiota diversity concomitant to infection is observed. Modified from Diaz et al. [117]

bug digestive track and may have functional relevance [121].

4.4 Transcriptomic Responses of Triatomine Vectors

NGS-based transcriptomic approaches have also begun to be applied to triatomines to better understand their biology. For example, insight into response to parasite infection or insecticide exposure may lead to improved vector control. Most studies have focused on *Rhodnius prolixus*, given the availability of its genome sequence [122], allowing for reference-based transcript analysis. Transcriptomic responses of multiple tissues have been investigated, including the antennas [123, 124], the ovaries [125–128], the digestive tract [129], the fat body and the immune response [130], or the saliva [131], which provide unique perspectives on bug responses to feeding and host-seeking behavior. Some of these studies have been extended to the closely related *Rhodnius neglectus* [132]. The transcriptomic response of *T. infestans* to deltamethrin exposure has also been investigated to identify detoxifying mechanisms; this will be central to mitigating insecticide resistance [133], as it is of growing concern in this species and can impact the efficacy of control interventions [134, 135].

However, a major limitation of these transcriptomic studies is the lack of annotated triatomine genome sequences beyond that of *Rhodnius prolixus*, which makes the functional identification of differentially expressed genes and pathways challenging and likely incomplete given the high proportion of unknown transcripts [132]. The genome sequencing and annotation of other triatomine species should help solve this issue [136] and facilitate further transcriptomic studies. Indeed, comparative transcriptomic studies across triatomine species can reveal species specificities and genetic adaptions that can be relevant for parasite transmission and their vectorial capacity [137].

5 Conclusions

NGS has allowed for unprecedented insight into the *T. cruzi* parasite–vector–host triad and Chagas disease pathogenesis. The *T. cruzi* genome is complex and highly plastic, and further whole genome sequencing will provide additional information on parasite evolution and diversity.

Several contributors to Chagas disease pathogenesis have also been uncovered using NGS approaches, and expansion of RNA transcriptomics will likely bring more detailed understanding of the mechanisms and interactions among *T. cruzi*, hosts and vectors.

More integrative studies are needed to consider the complexity of Chagas disease more completely beyond individual components, and to fully uncover the interactions and relative contributions of host genetics, parasite strains, and host environment, vectors species, microbiota in parasite transmission and disease progression.

Thus, moving forward, NGS will continue to be an invaluable tool for answering questions that have persisted around Chagas disease for over 100 years. Several of the studies mentioned in this chapter were groundbreaking and foundational for further research in these fields of study. Undeniably, the use of NGS in Chagas disease will continue to grow in the coming years.

References

1. PAHO (2023) Chagas disease. https://www.paho.org/en/topics/chagas-disease
2. Lee BY, Bacon KM, Bottazzi ME, Hotez PJ (2013) Global economic burden of Chagas disease: a computational simulation model. Lancet Infect Dis 13(4):342–348. https://doi.org/10.1016/S1473-3099(13)70002-1
3. WHO (2023) Chagas disease (American trypanosomiasis). https://www.who.int/health-topics/chagas-disease#tab=tab_1
4. Rassi A Jr, Marin JAN, Rassi A (2017) Chronic Chagas cardiomyopathy: a review of the main pathogenic mechanisms and the efficacy of aetiological treatment following the BENznidazole evaluation for interrupting trypanosomiasis (BENEFIT) trial. Mem Inst Oswaldo Cruz 112(3):224–235. https://doi.org/10.1590/0074-02760160334
5. Chagas C (1909) Nova tripanozomiaze humana. Estudos sobre a morfologia e o ciclo evolutivo do Schizotrypanum cruzi n. gen., n. sp. Agente etiológico Agente etiológico de uma nova entidade mórbida para o homem. Mem Inst Oswaldo Cruz 1:159–218
6. Akintunde O, Tucker T, Carabetta VJ (2023) The evolution of next-generation sequencing technologies. arXiv. https://www.ncbi.nlm.nih.gov/pubmed/37292469
7. Matos GM, Lewis MD, Talavera-Lopez C, Yeo M, Grisard EC, Messenger LA, Andersson B (2022) Microevolution of *Trypanosoma cruzi* reveals hybridization and clonal mechanisms driving rapid genome diversification. Elife 11. https://doi.org/10.7554/eLife.75237
8. El-Sayed NM, Myler PJ, Blandin G, Berriman M, Crabtree J, Aggarwal G, Hall N (2005) Comparative genomics of trypanosomatid parasitic protozoa. Science 309(5733):404–409. https://doi.org/10.1126/science.1112181
9. Lewis MD, Llewellyn MS, Gaunt MW, Yeo M, Carrasco HJ, Miles MA (2009) Flow cytometric analysis and microsatellite genotyping reveal extensive DNA content variation in *Trypanosoma cruzi* populations and expose contrasts between natural and experimental hybrids. Int J Parasitol 39(12):1305–1317. https://doi.org/10.1016/j.ijpara.2009.04.001
10. Reis-Cunha JL, Coqueiro-Dos-Santos A, Pimenta-Carvalho SA, Marques LP, Rodrigues-Luiz GF, Baptista RP, Bartholomeu DC (2022) Accessing the variability of multicopy genes in complex genomes using unassembled next-generation sequencing reads: the case of *Trypanosoma cruzi* multigene fam-

ilies. mBio 13(6):e0231922. https://doi.org/10.1128/mbio.02319-22
11. Hoyos Sanchez MC, Ospina Zapata HS, Suarez BD, Ospina C, Barbosa HJ, Carranza Martinez JC, Duitama J (2024) A phased genome assembly of a Colombian *Trypanosoma cruzi* TcI strain and the evolution of gene families. Sci Rep 14(1):2054. https://doi.org/10.1038/s41598-024-52449-x
12. Wang W, Peng D, Baptista RP, Li Y, Kissinger JC, Tarleton RL (2021) Strain-specific genome evolution in *Trypanosoma cruzi*, the agent of Chagas disease. PLoS Pathog 17(1):e1009254. https://doi.org/10.1371/journal.ppat.1009254
13. NCBI (2023) Genome: Trypanosoma cruzi. https://www.ncbi.nlm.nih.gov/datasets/genome/?taxon=5693
14. Berna L, Rodriguez M, Chiribao ML, Parodi-Talice A, Pita S, Rijo G, Robello C (2018) Expanding an expanded genome: long-read sequencing of *Trypanosoma cruzi*. Microb Genom 4(5). https://doi.org/10.1099/mgen.0.000177
15. Breniere SF, Waleckx E, Barnabe C (2016) Over six thousand *Trypanosoma cruzi* strains classified into discrete typing units (DTUs): attempt at an inventory. PLoS Negl Trop Dis 10(8):e0004792. https://doi.org/10.1371/journal.pntd.0004792
16. Majeau A, Herrera C, Dumonteil E (2019) An improved approach to *Trypanosoma cruzi* molecular genotyping by next-generation sequencing of the mini-exon gene. Methods Mol Biol 1955:47–60. https://doi.org/10.1007/978-1-4939-9148-8_4
17. Messenger LA, Llewellyn MS, Bhattacharyya T, Franzen O, Lewis MD, Ramirez JD, Miles MA (2012) Multiple mitochondrial introgression events and heteroplasmy in *Trypanosoma cruzi* revealed by maxicircle MLST and next generation sequencing. PLoS Negl Trop Dis 6(4):e1584. https://doi.org/10.1371/journal.pntd.0001584
18. Lima-Cordon RA, Cahan SH, McCann C, Dorn PL, Justi SA, Rodas A, Stevens L (2021) Insights from a comprehensive study of *Trypanosoma cruzi*: a new mitochondrial clade restricted to north and Central America and genetic structure of TcI in the region. PLoS Negl Trop Dis 15(12):e0010043. https://doi.org/10.1371/journal.pntd.0010043
19. Souza TKM, Westphalen EVN, Westphalen SDR, Taniguchi HH, Elias CR, Motoie G, Tolezano JE (2022) Genetic diversity of *Trypanosoma cruzi* strains isolated from chronic Chagasic patients and non-human hosts in the state of Sao Paulo, Brazil. Mem Inst Oswaldo Cruz 117:e220125. https://doi.org/10.1590/0074-02760220125
20. Berna L, Greif G, Pita S, Faral-Tello P, Diaz-Viraque F, Souza RCM, Robello C (2021) Maxicircle architecture and evolutionary insights into *Trypanosoma cruzi* complex. PLoS Negl Trop Dis 15(8):e0009719. https://doi.org/10.1371/journal.pntd.0009719
21. Gerasimov ES, Zamyatnina KA, Matveeva NS, Rudenskaya YA, Kraeva N, Kolesnikov AA, Yurchenko V (2020) Common structural patterns in the Maxicircle divergent region of Trypanosomatidae. Pathogens 9(2). https://doi.org/10.3390/pathogens9020100
22. Cortez DR, Lima FM, Reis-Cunha JL, Bartholomeu DC, Villacis RAR, Rogatto SR, Marini MM (2022) Trypanosoma cruzi genomic variability: array comparative genomic hybridization analysis of clone and parental strain. Front Cell Infect Microbiol 12:760830. https://doi.org/10.3389/fcimb.2022.760830
23. Cruz-Saavedra L, Schwabl P, Vallejo GA, Carranza JC, Munoz M, Patino LH et al (2022) Genome plasticity driven by aneuploidy and loss of heterozygosity in *Trypanosoma cruzi*. Microb Genom 8(6). https://doi.org/10.1099/mgen.0.000843
24. Gaunt MW, Yeo M, Frame IA, Stothard JR, Carrasco HJ, Taylor MC, Miles MA (2003) Mechanism of genetic exchange in American trypanosomes. Nature 421(6926):936–939
25. Berry ASF, Salazar-Sanchez R, Castillo-Neyra R, Borrini-Mayori K, Chipana-Ramos C, Vargas-Maquera M, Chagas Disease Working Group in Arequipa (2019) Sexual reproduction in a natural *Trypanosoma cruzi* population. PLoS Negl Trop Dis 13(5):e0007392. https://doi.org/10.1371/journal.pntd.0007392
26. Schwabl P, Imamura H, Van den Broeck F, Costales JA, Maiguashca-Sanchez J, Miles MA, Llewellyn MS (2019) Meiotic sex in Chagas disease parasite *Trypanosoma cruzi*. Nat Commun 10(1):3972. https://doi.org/10.1038/s41467-019-11771-z
27. Tibayrenc M, Ayala FJ (2022) Microevolution and subspecific taxonomy of *Trypanosoma cruzi*. Infect Genet Evol 103:105344. https://doi.org/10.1016/j.meegid.2022.105344
28. Majeau A, Dumonteil E, Herrera C (2024) Identification of highly conserved *Trypanosoma cruzi* antigens for the development of a universal serological diagnostic assay. Emerg Microbes Infect 13(1):2315964
29. Majeau A, Murphy L, Herrera C, Dumonteil E (2021) Assessing *Trypanosoma cruzi* parasite diversity through comparative genomics: implications for disease epidemiology and diagnostics. Pathogens 10:212. https://doi.org/10.3390/pathogens10020212
30. Callejas-Hernandez F, Rastrojo A, Poveda C, Girones N, Fresno M (2018) Genomic assemblies of newly sequenced *Trypanosoma cruzi* strains reveal new genomic expansion and greater complexity. Sci Rep 8(1):14631. https://doi.org/10.1038/s41598-018-32877-2
31. Burle-Caldas GA, Dos Santos NSA, de Castro JT, Mugge FLB, Grazielle-Silva V, Oliveira AER, Teixeira SMR (2022) Disruption of active trans-sialidase genes impairs egress from mammalian host cells and generates highly attenuated Trypanosoma cruzi parasites. mBio 13(1):e0347821. https://doi.org/10.1128/mbio.03478-21
32. Santos VC, Oliveira AER, Campos ACB, Reis-Cunha JL, Bartholomeu DC, Teixeira SMR, Ferreira RS (2021) The gene repertoire of the main cys-

teine protease of *Trypanosoma cruzi*, cruzipain, reveals four sub-types with distinct active sites. Sci Rep 11(1):18231. https://doi.org/10.1038/s41598-021-97490-2
33. De Pablos LM, Osuna A (2012) Multigene families in *Trypanosoma cruzi* and their role in infectivity. Infect Immun 80(7):2258–2264. https://doi.org/10.1128/IAI.06225-11
34. Araujo PR, Teixeira SM (2011) Regulatory elements involved in the post-transcriptional control of stage-specific gene expression in *Trypanosoma cruzi*: a review. Mem Inst Oswaldo Cruz 106(3):257–266. https://doi.org/10.1590/s0074-02762011000300002
35. Li Y, Shah-Simpson S, Okrah K, Belew AT, Choi J, Caradonna KL, Burleigh BA (2016) Transcriptome remodeling in *Trypanosoma cruzi* and human cells during intracellular infection. PLoS Pathog 12(4):e1005511. https://doi.org/10.1371/journal.ppat.1005511
36. Minning TA, Weatherly DB, Atwood J 3rd, Orlando R, Tarleton RL (2009) The steady-state transcriptome of the four major life-cycle stages of *Trypanosoma cruzi*. BMC Genomics 10:370. https://doi.org/10.1186/1471-2164-10-370
37. Diaz-Viraque F, Chiribao ML, Libisch MG, Robello C (2023) Genome-wide chromatin interaction map for *Trypanosoma cruzi*. Nat Microbiol 8(11):2103–2114. https://doi.org/10.1038/s41564-023-01483-y
38. Queiroz RM, Charneau S, Mandacaru SC, Schwammle V, Lima BD, Roepstorff P, Ricart CA (2014) Quantitative proteomic and phosphoproteomic analysis of *Trypanosoma cruzi* amastigogenesis. Mol Cell Proteomics 13(12):3457–3472. https://doi.org/10.1074/mcp.M114.040329
39. Belew AT, Junqueira C, Rodrigues-Luiz GF, Valente BM, Oliveira AER, Polidoro RB, Teixeira SMR (2017) Comparative transcriptome profiling of virulent and non-virulent Trypanosoma cruzi underlines the role of surface proteins during infection. PLoS Pathog 13(12):e1006767. https://doi.org/10.1371/journal.ppat.1006767
40. Lima DA, Goncalves LO, Reis-Cunha JL, Guimaraes PAS, Ruiz JC, Liarte DB, Murta SMF (2023) Transcriptomic analysis of benznidazole-resistant and susceptible *Trypanosoma cruzi* populations. Parasit Vectors 16(1):167. https://doi.org/10.1186/s13071-023-05775-4
41. Sanchez-Valdez FJ, Padilla A, Wang W, Orr D, Tarleton RL (2018) Spontaneous dormancy protects *Trypanosoma cruzi* during extended drug exposure. Elife 7. https://doi.org/10.7554/eLife.34039
42. Desale H, Herrera C, Dumonteil E (2024) *Trypanosoma cruzi* amastigote transcriptome analysis reveals heterogenous populations with replicating and dormant parasites. Microbes Infect 26(1–2):105240. https://doi.org/10.1016/j.micinf.2023.105240
43. Sabalette KB, Sotelo-Silveira JR, Smircich P, De Gaudenzi JG (2023) RNA-Seq reveals that overexpression of TcUBP1 switches the gene expression pattern toward that of the infective form of *Trypanosoma cruzi*. J Biol Chem 299(5):104623. https://doi.org/10.1016/j.jbc.2023.104623
44. Gaur P, Chaturvedi A (2016) Mining SNPs in extracellular vesicular transcriptome of *Trypanosoma cruzi*: a step closer to early diagnosis of neglected Chagas disease. PeerJ 4:e2693. https://doi.org/10.7717/peerj.2693
45. Acosta-Herrera M, Strauss M, Casares-Marfil D, Martin J, Chagas Genetics CN (2019) Genomic medicine in Chagas disease. Acta Trop 197:105062. https://doi.org/10.1016/j.actatropica.2019.105062
46. Ayo CM, Bestetti RB, de Campos Junior E, Ronchi LS, Borim AA, Brandao CC, de Matttos LC (2021) MICA and KIR: immunogenetic factors influencing left ventricular systolic dysfunction and digestive clinical form of chronic Chagas disease. Front Immunol 12:714766. https://doi.org/10.3389/fimmu.2021.714766
47. Koh CC, Neves EGA, de Souza-Silva TG, Carvalho AC, Pinto CHR, Galdino AS, Dutra WO (2023) Cytokine networks as targets for preventing and controlling Chagas heart disease. Pathogens 12(2). https://doi.org/10.3390/pathogens12020171
48. Deng X, Sabino EC, Cunha-Neto E, Ribeiro AL, Ianni B, Mady C, Busch MP, Seielstad M, REDSII Chagas Study Group from the NHLBI Retrovirus Epidemiology Donor Study-II Component International (2013) Genome wide association study (GWAS) of Chagas cardiomyopathy in *Trypanosoma cruzi* seropositive subjects. PLoS One 8(11):e79629. https://doi.org/10.1371/journal.pone.0079629
49. Sabino EC, Franco LAM, Venturini G, Velho Rodrigues M, Marques E, Oliveira-da Silva LC, Pereira AC (2022) Genome-wide association study for Chagas cardiomyopathy identify a new risk locus on chromosome 18 associated with an immune-related protein and transcriptional signature. PLoS Negl Trop Dis 16(10):e0010725. https://doi.org/10.1371/journal.pntd.0010725
50. Casares-Marfil D, Strauss M, Bosch-Nicolau P, Lo Presti MS, Molina I, Chevillard C, Acosta-Herrera M (2021) A genome-wide association study identifies novel susceptibility loci in chronic Chagas cardiomyopathy. Clin Infect Dis 73(4):672–679. https://doi.org/10.1093/cid/ciab090
51. Ouarhache M, Marquet S, Frade AF, Ferreira AM, Ianni B, Almeida RR et al (2021) Rare pathogenic variants in mitochondrial and inflammation-associated genes may lead to inflammatory cardiomyopathy in Chagas disease. J Clin Immunol 41(5):1048–1063. https://doi.org/10.1007/s10875-021-01000-y
52. Couto-Silva CM, Nunes K, Venturini G, Araujo Castro ESM, Pereira LV, Comas D, Hunemeier T (2023) Indigenous people from Amazon show

genetic signatures of pathogen-driven selection. Sci Adv 9(10):eabo0234. https://doi.org/10.1126/sciadv.abo0234
53. Dumonteil E, Desale H, Tu W, Hernandez-Cuevas N, Shroyer M, Goff K et al (2023) Intra-host *Trypanosoma cruzi* strain dynamics shape disease progression: the missing link in Chagas disease pathogenesis. Microbiol Spectr 11(5):e0423622. https://doi.org/10.1128/spectrum.04236-22
54. Herrera C, Majeau A, Didier P, Falkenstein KP, Dumonteil E (2019) *Trypanosoma cruzi* diversity in naturally infected non-human primates in Louisiana assessed by deep sequencing of the mini-exon gene. Trans R Soc Trop Med Hyg 113(5):281–286. https://doi.org/10.1093/trstmh/try119
55. Pronovost H, Peterson AC, Ghersi Chavez B, Blum MJ, Dumonteil E, Herrera C (2020) Deep sequencing reveals multiclonality and new discrete typing units of *Trypanosoma cruzi* in rodents from the southern United States. J Microbiol Immunol Infect 53(4):622–633. https://doi.org/10.1016/j.jmii.2018.12.004
56. Llewellyn MS, Rivett-Carnac JB, Fitzpatrick S, Lewis MD, Yeo M, Gaunt MW, Miles MA (2011) Extraordinary *Trypanosoma cruzi* diversity within single mammalian reservoir hosts implies a mechanism of diversifying selection. Int J Parasitol 41(6):609–614. https://doi.org/10.1016/j.ijpara.2010.12.004
57. Majeau A, Cloherty E, Anderson AN, Straif-Bourgeois SC, Dumonteil E, Herrera C (2023) Genetic diversity of *Trypanosoma cruzi* infecting raccoons (Procyon lotor) in 2 metropolitan areas of southern Louisiana: implications for parasite transmission networks. Parasitology 150:1–8. https://doi.org/10.1017/S0031182023000070
58. Dumonteil E, Elmayan A, Majeau A, Tu W, Duhon B, Marx P, Herrera C (2020) Genetic diversity of *Trypanosoma cruzi* parasites infecting dogs in southern Louisiana sheds light on parasite transmission cycles and serological diagnostic performance. PLoS Negl Trop Dis 14(12):e0008932. https://doi.org/10.1371/journal.pntd.0008932
59. Dumonteil E, Desale H, Tu W, Duhon B, Wolfson W, Balsamo G, Herrera C (2021) Shelter cats host infections with multiple *Trypanosoma cruzi* discrete typing units in southern Louisiana. Vet Res 52(1):53. https://doi.org/10.1186/s13567-021-00923-z
60. Bowman NM, Balasubramanian S, Gilman RH, Parobek C, Calderon M, Waltmann A et al (2022) Deep sequencing to detect diversity of *Trypanosoma cruzi* infection in patients coinfected with human immunodeficiency virus and Chagas disease. J Infect Dis 225(2):243–247. https://doi.org/10.1093/infdis/jiab350
61. Ledezma AP, Blandon R, Schijman AG, Benatar A, Saldana A, Osuna A (2020) Mixed infections by different *Trypanosoma cruzi* discrete typing units among Chagas disease patients in an endemic community in Panama. PLoS One 15(11):e0241921. https://doi.org/10.1371/journal.pone.0241921
62. Murray KO, Saldana MA, Gunter SM, Manzanero R, Zielinski-Gutierrez E, Herrera C et al (2022) Diagnosis of acute Chagas disease in a Belizean child with evidence of a multiclonal *Trypanosoma cruzi* infection. Am J Trop Med Hyg 107(5):992–995. https://doi.org/10.4269/ajtmh.22-0338
63. Villanueva-Lizama L, Teh-Poot C, Majeau A, Herrera C, Dumonteil E (2019) Molecular genotyping of *Trypanosoma cruzi* by next-generation sequencing of the mini-exon gene reveals infections with multiple parasite discrete typing units in Chagasic patients from Yucatan, Mexico. J Infect Dis 219(12):1980–1988. https://doi.org/10.1093/infdis/jiz047
64. Izeta-Alberdi A, Pech-May A, Tun-Ku E, Mazariegos-Hidalgo CJ, Lopez-Cancino SA, Gutierrez S, Ramsey JM (2023) *Trypanosoma cruzi* in Mexican neotropical vectors and mammals: wildlife, livestock, pets, and human population. Salud Publica Mex, 65(2 mar–abr):114–126. https://doi.org/10.21149/13801
65. Schwabl P, Maiguashca Sanchez J, Costales JA, Ocana-Mayorga S, Segovia M, Carrasco HJ, Llewellyn MS (2020) Culture-free genome-wide locus sequence typing (GLST) provides new perspectives on *Trypanosoma cruzi* dispersal and infection complexity. PLoS Genet 16(12):e1009170. https://doi.org/10.1371/journal.pgen.1009170
66. Adesse D, Iacobas DA, Iacobas S, Garzoni LR, Meirelles Mde N, Tanowitz HB, Spray DC (2010) Transcriptomic signatures of alterations in a myoblast cell line infected with four distinct strains of *Trypanosoma cruzi*. Am J Trop Med Hyg 82(5):846–854. https://doi.org/10.4269/ajtmh.2010.09-0399
67. Cunha-Neto E, Dzau VJ, Allen PD, Stamatiou D, Benvenutti L, Higuchi ML, Liew CC (2005) Cardiac gene expression profiling provides evidence for cytokinopathy as a molecular mechanism in Chagas' disease cardiomyopathy. Am J Pathol 167(2):305–313. https://doi.org/10.1016/S0002-9440(10)62976-8
68. Goldenberg RC, Iacobas DA, Iacobas S, Rocha LL, da Silva de Azevedo Fortes F, Vairo L, Spray DC (2009) Transcriptomic alterations in *Trypanosoma cruzi*-infected cardiac myocytes. Microbes Infect 11(14–15):1140–1149. https://doi.org/10.1016/j.micinf.2009.08.009
69. Oliveira TGM, Venturini G, Alvim JM, Feijo LL, Dinardo CL, Sabino EC et al (2022) Different transcriptomic response to *T. cruzi* infection in hiPSC-derived cardiomyocytes from Chagas disease patients with and without chronic cardiomyopathy. Front Cell Infect Microbiol 12:904747. https://doi.org/10.3389/fcimb.2022.904747
70. Venturini G, Alvim JM, Padilha K, Toepfer CN, Gorham JM, Wasson LK, Seidman CE (2023) Cardiomyocyte infection by *Trypanosoma cruzi* promotes innate immune response and glycolysis

activation. Front Cell Infect Microbiol 13:1098457. https://doi.org/10.3389/fcimb.2023.1098457

71. Candray-Medina KS, Nakagama Y, Ito M, Nakagama S, Tshibangu-Kabamba E, Takeda N, Kido Y (2023) Differential cardiomyocyte transcriptomic remodeling during in vitro Trypanosoma cruzi infection using laboratory strains provides implications on pathogenic host responses. Trop Med Health 51(1):68. https://doi.org/10.1186/s41182-023-00552-6
72. de Castro TBR, Canesso MCC, Boroni M, Chame DF, Souza DL, de Toledo NE, Franco GR (2020) Differential modulation of mouse heart gene expression by infection with two Trypanosoma cruzi strains: a transcriptome analysis. Front Genet 11:1031. https://doi.org/10.3389/fgene.2020.01031
73. Silva Grijo Farani P, Iandra da Silva Ferreira B, Begum K, Vilar-Pereira G, Pereira IR, Fernandez-Figueroa EA, Moreira OC (2023) Treatment with benznidazole and pentoxifylline regulates microRNA transcriptomic profile in a murine model of Chagas chronic cardiomyopathy. PLoS Negl Trop Dis 17(3):e0011223. https://doi.org/10.1371/journal.pntd.0011223
74. Duque C, So J, Castro-Sesquen YE, DeToy K, Gutierrez Guarnizo SA, Jahanbakhsh F, Chagas Working G (2023) Immunologic changes are detectable in the peripheral blood transcriptome of clinically asymptomatic Chagas cardiomyopathy patients. bioRxiv. https://doi.org/10.1101/2023.10.03.560680
75. Villar SR, Herreros-Cabello A, Callejas-Hernandez F, Maza MC, Del Moral-Salmoral J, Gomez-Montes M, Girones N (2024) Discovery of circulating miRNAs as biomarkers of chronic Chagas heart disease via a small RNA-Seq approach. Sci Rep 14(1):1187. https://doi.org/10.1038/s41598-024-51487-9
76. Carlier Y, Altcheh J, Angheben A, Freilij H, Luquetti AO, Schijman AG et al (2019) Congenital Chagas disease: updated recommendations for prevention, diagnosis, treatment, and follow-up of newborns and siblings, girls, women of childbearing age, and pregnant women. PLoS Negl Trop Dis 13(10):e0007694. https://doi.org/10.1371/journal.pntd.0007694
77. Carlier Y, Sosa-Estani S, Luquetti AO, Buekens P (2015) Congenital Chagas disease: an update. Mem Inst Oswaldo Cruz 110(3):363–368. https://doi.org/10.1590/0074-02760140405
78. Buekens P, Cafferata ML, Alger J, Althabe F, Belizan JM, Bustamante N et al (2018) Congenital transmission of *Trypanosoma cruzi* in Argentina, Honduras, and Mexico: an observational prospective study. Am J Trop Med Hyg 98(2):478–485. https://doi.org/10.4269/ajtmh.17-0516
79. Howard EJ, Xiong X, Carlier Y, Sosa-Estani S, Buekens P (2014) Frequency of the congenital transmission of Trypanosoma cruzi: a systematic review and meta-analysis. BJOG 121(1):22–33. https://doi.org/10.1111/1471-0528.12396
80. Faral-Tello P, Greif G, Romero S, Cabrera A, Oviedo C, Gonzalez T, Robello C (2023) *Trypanosoma cruzi* isolates naturally adapted to congenital transmission display a unique strategy of transplacental passage. Microbiol Spectr 11(2):e0250422. https://doi.org/10.1128/spectrum.02504-22
81. Juiz NA, Torrejon I, Burgos M, Torres AMF, Duffy T, Cayo NM, Schijman AG (2018) Alterations in placental gene expression of pregnant women with chronic Chagas disease. Am J Pathol 188(6):1345–1353. https://doi.org/10.1016/j.ajpath.2018.02.011
82. Juiz NA, Cayo NM, Burgos M, Salvo ME, Nasser JR, Bua J, Schijman AG (2016) Human polymorphisms in placentally expressed genes and their association with susceptibility to congenital *Trypanosoma cruzi* infection. J Infect Dis 213(8):1299–1306. https://doi.org/10.1093/infdis/jiv561
83. Llewellyn MS, Messenger LA, Luquetti AO, Garcia L, Torrico F, Tavares SB et al (2015) Deep sequencing of the *Trypanosoma cruzi* GP63 surface proteases reveals diversity and diversifying selection among chronic and congenital Chagas disease patients. PLoS Negl Trop Dis 9(4):e0003458. https://doi.org/10.1371/journal.pntd.0003458
84. Hakim JMC, Waltmann A, Tinajeros F, Kharabora O, Machaca EM, Calderon M et al (2023) Amplicon sequencing reveals complex infection in infants congenitally infected with Trypanosoma Cruzi and informs the dynamics of parasite transmission. J Infect Dis 228(6):769–776. https://doi.org/10.1093/infdis/jiad125
85. Desale H, Buekens P, Alger J, Cafferata ML, Harville EW, Herrera C, Dumonteil E (2022) Epigenetic signature of exposure to maternal *Trypanosoma cruzi i*nfection in cord blood cells from uninfected newborns. Epigenomics 14(15):913–927. https://doi.org/10.2217/epi-2022-0153
86. Paternina-Caicedo A, Alger J, Cafferata ML, Carlier Y, Dumonteil E, Gibbons L et al (2024) *Trypanosoma cruzi* infection in pregnancies without congenital transmission is associated with reduced fetal growth: a cross-sectional study in Argentina, Honduras, and Mexico. Am J Trop Med Hyg 111:64. https://doi.org/10.4269/ajtmh.23-0493
87. Teotonio I, Dias N, Hagstrom-Bex L, Nitz N, Francisco AF, Hecht M (2019) Intestinal microbiota - a modulator of the *Trypanosoma cruzi*-vector-host triad. Microb Pathog 137:103711. https://doi.org/10.1016/j.micpath.2019.103711
88. Castaneda S, Munoz M, Hotez PJ, Bottazzi ME, Paniz-Mondolfi AE, Jones KM, Ramirez JD (2023) Microbiome alterations driven by *Trypanosoma cruzi* infection in two disjunctive murine models. Microbiol Spectr 11(3):e0019923. https://doi.org/10.1128/spectrum.00199-23
89. Brown JJ, Rodriguez-Ruano SM, Poosakkannu A, Batani G, Schmidt JO, Roachell W, Novakova E (2020) Ontogeny, species identity, and environment dominate microbiome dynamics in wild populations of kissing bugs (Triatominae). Microbiome 8(1):146. https://doi.org/10.1186/s40168-020-00921-x

90. Ceccarelli S, Balsalobre A, Vicente ME, Curtis-Robles R, Hamer SA, Ayala Landa JM et al (2022) American triatomine species occurrences: updates and novelties in the DataTri database. GigaByte 2022:gigabyte62. https://doi.org/10.46471/gigabyte.62
91. Dumonteil E, Pronovost H, Bierman EF, Sanford A, Majeau A, Moore R, Herrera C (2020) Interactions among *Triatoma sanguisuga* blood feeding sources, gut microbiota and *Trypanosoma cruzi* diversity in southern Louisiana. Mol Ecol 29(19):3747–3761. https://doi.org/10.1111/mec.15582
92. Dumonteil E, Ramirez-Sierra MJ, Pérez-Carrillo S, Teh-Poot C, Herrera C, Gourbière S, Waleckx E (2018) Detailed ecological associations of triatomines revealed by metabarcoding based on next-generation sequencing: linking triatomine behavioral ecology and *Trypanosoma cruzi* transmission cycles. Sci Rep 8(1):4140. https://doi.org/10.1038/s41598-018-22455-x
93. Murillo-Solano C, Lopez-Dominguez J, Gongora R, Rojas-Gulloso A, Usme-Ciro J, Perdomo-Balaguera E, Dumonteil E (2021) Diversity and interactions among triatomine bugs, their blood feeding sources, gut microbiota and *Trypanosoma cruzi* in the Sierra Nevada de Santa Marta in Colombia. Sci Rep 11(1):12306. https://doi.org/10.1038/s41598-021-91783-2
94. Murillo-Solano C, Ramos-Ligonio A, Lopez-Monteon A, Guzman-Gomez D, Torres-Montero J, Herrera C, Dumonteil E (2021) Diversity of *Trypanosoma cruzi* parasites infecting Triatoma dimidiata in Central Veracruz, Mexico, and their one health ecological interactions. Infect Genet Evol 95:105050. https://doi.org/10.1016/j.meegid.2021.105050
95. Polonio R, Lopez-Dominguez J, Herrera C, Dumonteil E (2021) Molecular ecology of Triatoma dimidiata in southern Belize reveals risk for human infection and the local differentiation of *Trypanosoma cruzi* parasites. Int J Infect Dis 108:320–329. https://doi.org/10.1016/j.ijid.2021.05.083
96. Balasubramanian S, Curtis-Robles R, Chirra B, Auckland LD, Mai A, Bocanegra-Garcia V, Hamer SA (2022) Characterization of triatomine bloodmeal sources using direct Sanger sequencing and amplicon deep sequencing methods. Sci Rep 12(1):10234. https://doi.org/10.1038/s41598-022-14208-8
97. Juan ES, Araya-Donoso R, Sierra-Rosales C, Correa JP, Quiroga N, Campos-Soto R, Botto-Mahan C (2023) Humans as blood-feeding sources in sylvatic triatomines of Chile unveiled by next-generation sequencing. Parasit Vectors 16(1):225. https://doi.org/10.1186/s13071-023-05841-x
98. Quiroga N, Correa JP, Campos-Soto R, San Juan E, Araya-Donoso R, Diaz-Campusano G, Botto-Mahan C (2022) Blood-meal sources and *Trypanosoma cruzi* infection in coastal and insular triatomine bugs from the Atacama Desert of Chile. Microorganisms 10(4). https://doi.org/10.3390/microorganisms10040785
99. Arias-Giraldo LM, Munoz M, Hernandez C, Herrera G, Velasquez-Ortiz N, Cantillo-Barraza O, Ramirez JD (2020) Identification of blood-feeding sources in Panstrongylus, Psammolestes, Rhodnius and Triatoma using amplicon-based next-generation sequencing. Parasit Vectors 13(1):434. https://doi.org/10.1186/s13071-020-04310-z
100. Hylton A, Fitzpatrick DM, Suepaul R, Dobson AP, Charles RA, Peterson JK (2020) Preliminary characterization of Triatomine bug blood meals on the Island of Trinidad reveals opportunistic feeding behavior on both human and animal hosts. Trop Med Infect Dis 5(4). https://doi.org/10.3390/tropicalmed5040166
101. De Bona S, Correa JP, San Juan E, Estay-Olea D, Quiroga N, Bacigalupo A et al (2023) Opportunistic or selective? Stage-dependent feeding behavior in a wild vector of Chagas disease. Int J Parasitol 53(1):55–64. https://doi.org/10.1016/j.ijpara.2022.10.003
102. Busselman RE, Curtis-Robles R, Meyers AC, Zecca IB, Auckland LD, Hodo CL et al (2024) Abundant triatomines in Texas dog kennel environments: triatomine collections, infection with Trypanosoma cruzi, and blood feeding hosts. Acta Trop 250:107087. https://doi.org/10.1016/j.actatropica.2023.107087
103. Honorato NRM, da Silva ANB, de Negreiros CCA, Aguiar LMA, Marliere NP, de Souza RCM, da Camara ACJ (2021) Triatomine and Trypanosoma cruzi discrete typing units distribution in a semi-arid area of northeastern Brazil. Acta Trop 220:105950. https://doi.org/10.1016/j.actatropica.2021.105950
104. Maiguashca Sanchez J, Sueto SOB, Schwabl P, Grijalva MJ, Llewellyn MS, Costales JA (2020) Remarkable genetic diversity of *Trypanosoma cruzi* and Trypanosoma rangeli in two localities of southern Ecuador identified via deep sequencing of mini-exon gene amplicons. Parasit Vectors 13(1):252. https://doi.org/10.1186/s13071-020-04079-1
105. Zingales B (2018) *Trypanosoma cruzi* genetic diversity: something new for something known about Chagas disease manifestations, serodiagnosis and drug sensitivity. Acta Trop 184:38–52. https://doi.org/10.1016/j.actatropica.2017.09.017
106. Gumiel, M., da Mota, F. F., Rizzo Vde, S., Sarquis, O., de Castro, D. P., Lima MM, .Azambuja, P. (2015). Characterization of the microbiota in the guts of *Triatoma brasiliensis* and *Triatoma pseudomaculata* infected by *Trypanosoma cruzi* in natural conditions using culture independent methods. Parasit Vectors, 8, 245. https://doi.org/10.1186/s13071-015-0836-z
107. Harington JS (1960) Synthesis of thiamine and folic acid by Nocardia rhodnii, the micro-symbiont of Rhodnius prolixus. Nature 188:1027–1028. https://doi.org/10.1038/1881027a0
108. Tobias NJ, Eberhard FE, Guarneri AA (2020) Enzymatic biosynthesis of B-complex vitamins is supplied by diverse microbiota in the *Rhodnius prolixus* anterior midgut following *Trypanosoma cruzi*

infection. Comput Struct Biotechnol J 18:3395–3401. https://doi.org/10.1016/j.csbj.2020.10.031
109. Mann AE, Mitchell EA, Zhang Y, Curtis-Robles R, Thapa S, Hamer SA, Allen MS (2020) Comparison of the bacterial gut microbiome of north American Triatoma spp. with and without *Trypanosoma cruzi*. Front Microbiol 11:364. https://doi.org/10.3389/fmicb.2020.00364
110. Rodriguez-Ruano SM, Skochova V, Rego ROM, Schmidt JO, Roachell W, Hypsa V, Novakova E (2018) Microbiomes of north American triatominae: the grounds for Chagas disease epidemiology. Front Microbiol 9:1167. https://doi.org/10.3389/fmicb.2018.01167
111. Waltmann A, Willcox AC, Balasubramanian S, Borrini Mayori K, Mendoza Guerrero S, Salazar Sanchez RS, Bowman NM (2019) Hindgut microbiota in laboratory-reared and wild *Triatoma infestans*. PLoS Negl Trop Dis 13(5):e0007383. https://doi.org/10.1371/journal.pntd.0007383
112. Beard CB, Durvasula RV, Richards FF (1998) Bacterial symbiosis in arthropods and the control of disease transmission. Emerg Infect Dis 4:581–591. https://doi.org/10.3201/eid0404.980408
113. Wang J, Gao L, Aksoy S (2023) Microbiota in disease-transmitting vectors. Nat Rev Microbiol 21(9):604–618. https://doi.org/10.1038/s41579-023-00901-6
114. Teal E, Herrera C, Dumonteil E (2023) Metabolomics of developmental changes in Triatoma sanguisuga gut microbiota. PLoS One 18(2):e0280868. https://doi.org/10.1371/journal.pone.0280868
115. Montoya-Porras LM, Omar TC, Alzate JF, Moreno-Herrera CX, Cadavid-Restrepo GE (2018) 16S rRNA gene amplicon sequencing reveals dominance of Actinobacteria in *Rhodnius pallescens* compared to *Triatoma maculata* midgut microbiota in natural populations of vector insects from Colombia. Acta Trop 178:327–332. https://doi.org/10.1016/j.actatropica.2017.11.004
116. Dumonteil E, Tu W, Jiménez FA, Herrera C (2024) Ecological interactions of *Triatoma sanguisuga* and risk for human infection with *Trypanosoma cruzi* in Illinois and Louisiana. J Med Entomol 19:tjae017. https://doi.org/10.1093/jme/tjae017
117. Diaz S, Villavicencio B, Correia N, Costa J, Haag KL (2016) Triatomine bugs, their microbiota and *Trypanosoma cruzi*: asymmetric responses of bacteria to an infected blood meal. Parasit Vectors 9(1):636. https://doi.org/10.1186/s13071-016-1926-2
118. Eberhard FE, Klimpel S, Guarneri AA, Tobias NJ (2022) Exposure to Trypanosoma parasites induces changes in the microbiome of the Chagas disease vector Rhodnius prolixus. Microbiome 10(1):45. https://doi.org/10.1186/s40168-022-01240-z
119. Orantes LC, Monroy C, Dorn PL, Stevens L, Rizzo DM, Morrissey L, Helms Cahan S (2018) Uncovering vector, parasite, blood meal and microbiome patterns from mixed-DNA specimens of the Chagas disease vector *Triatoma dimidiata*. PLoS Negl Trop Dis 12(10):e0006730. https://doi.org/10.1371/journal.pntd.0006730
120. Buarque DS, Gomes CM, Araújo RN, Pereira MH, Ferreira RC, Guarneri AA, Tanaka AS (2016) A new antimicrobial protein from the anterior midgut of *Triatoma infestans* mediates *Trypanosoma cruzi* establishment by controlling the microbiota. Biochimie 123:138–143. https://doi.org/10.1016/j.biochi.2016.02.009
121. Brito TF, Coelho VL, Cardoso MA, Brito IAA, Berni MA, Zenk FL, Pane A (2021) Transovarial transmission of a core virome in the Chagas disease vector Rhodnius prolixus. PLoS Pathog 17(8):e1009780. https://doi.org/10.1371/journal.ppat.1009780
122. Mesquita RD, Vionette-Amaral RJ, Lowenberger C, Rivera-Pomar R, Monteiro FA, Minx P, Oliveira PL (2015) Genome of *Rhodnius prolixus*, an insect vector of Chagas disease, reveals unique adaptations to hematophagy and parasite infection. Proc Natl Acad Sci USA 112(48):14936–14941. https://doi.org/10.1073/pnas.1506226112
123. Latorre-Estivalis JM, Grosse-Wilde E, da Rocha Fernandes G, Hansson BS, Lorenzo MG (2022) Changes in antennal gene expression underlying sensory system maturation in Rhodnius prolixus. Insect Biochem Mol Biol 140:103704. https://doi.org/10.1016/j.ibmb.2021.103704
124. Latorre-Estivalis JM, Sterkel M, Ons S, Lorenzo MG (2020) Transcriptomics supports local sensory regulation in the antenna of the kissing-bug *Rhodnius prolixus*. BMC Genomics 21(1):101. https://doi.org/10.1186/s12864-020-6514-3
125. Coelho VL, de Brito TF, de Abreu Brito IA, Cardoso MA, Berni MA, Araujo HMM, Pane A (2021) Analysis of ovarian transcriptomes reveals thousands of novel genes in the insect vector Rhodnius prolixus. Sci Rep 11(1):1918. https://doi.org/10.1038/s41598-021-81387-1
126. Leyria J, Orchard I, Lange AB (2020) Transcriptomic analysis of regulatory pathways involved in female reproductive physiology of *Rhodnius prolixus* under different nutritional states. Sci Rep 10(1):11431. https://doi.org/10.1038/s41598-020-67932-4
127. Leyria J, Orchard I, Lange AB (2020) What happens after a blood meal? A transcriptome analysis of the main tissues involved in egg production in *Rhodnius prolixus*, an insect vector of Chagas disease. PLoS Negl Trop Dis 14(10):e0008516. https://doi.org/10.1371/journal.pntd.0008516
128. Medeiros MN, Logullo R, Ramos IB, Sorgine MH, Paiva-Silva GO, Mesquita RD et al (2011) Transcriptome and gene expression profile of ovarian follicle tissue of the triatomine bug *Rhodnius prolixus*. Insect Biochem Mol Biol 41(10):823–831. https://doi.org/10.1016/j.ibmb.2011.06.004
129. Ribeiro JM, Genta FA, Sorgine MH, Logullo R, Mesquita RD, Paiva-Silva GO et al (2014) An insight into the transcriptome of the digestive tract of the bloodsucking bug, Rhodnius prolixus. PLoS

Negl Trop Dis 8(1):e2594. https://doi.org/10.1371/journal.pntd.0002594
130. Salcedo-Porras N, Oliveira PL, Guarneri AA, Lowenberger C (2022) A fat body transcriptome analysis of the immune responses of *Rhodnius prolixus* to artificial infections with bacteria. Parasit Vectors 15(1):269. https://doi.org/10.1186/s13071-022-05358-9
131. Santos DV, Gontijo NF, Pessoa GCD, Sant'Anna MRV, Araujo RN, Pereira MH, Koerich LB (2022) An updated catalog of lipocalins of the chagas disease vector *Rhodnius prolixus* (Hemiptera, Reduviidae). Insect Biochem Mol Biol 146:103797. https://doi.org/10.1016/j.ibmb.2022.103797
132. Carvalho-Costa TM, Tiveron RDR, Mendes MT, Barbosa CG, Nevoa JC, Roza GA, Oliveira CJF (2021) Salivary and intestinal transcriptomes reveal differential gene expression in starving, fed and *Trypanosoma cruzi*-infected Rhodnius neglectus. Front Cell Infect Microbiol 11:773357. https://doi.org/10.3389/fcimb.2021.773357
133. Traverso L, Latorre Estivalis JM, da Rocha Fernandes G, Fronza G, Lobbia P, Mougabure Cueto G, Ons S (2022) Transcriptomic modulation in response to an intoxication with deltamethrin in a population of *Triatoma infestans* with low resistance to pyrethroids. PLoS Negl Trop Dis 16(6):e0010060. https://doi.org/10.1371/journal.pntd.0010060
134. Gaspe MS, Cardinal MV, Fernandez MDP, Vassena CV, Santo-Orihuela PL, Enriquez GF, Gurtler RE (2021) Improved vector control of *Triatoma infestans* limited by emerging pyrethroid resistance across an urban-to-rural gradient in the Argentine Chaco. Parasit Vectors 14(1):437. https://doi.org/10.1186/s13071-021-04942-9
135. Marcet PL, Santo-Orihuela P, Messenger LA, Vassena CV (2021) Insights into the evolution and dispersion of pyrethroid resistance among sylvatic Andean *Triatoma infestans* from Bolivia. Infect Genet Evol 90:104759. https://doi.org/10.1016/j.meegid.2021.104759
136. Liu Q, Guo Y, Zhang Y, Hu W, Li Y, Zhu D, Zhou XN (2019) A chromosomal-level genome assembly for the insect vector for Chagas disease, Triatoma rubrofasciata. Gigascience 8(8). https://doi.org/10.1093/gigascience/giz089
137. Martinez-Barnetche J, Lavore A, Beliera M, Tellez-Sosa J, Zumaya-Estrada FA, Palacio V, Rodriguez MH (2018) Adaptations in energy metabolism and gene family expansions revealed by comparative transcriptomics of three Chagas disease triatomine vectors. BMC Genomics 19(1):296. https://doi.org/10.1186/s12864-018-4696-8

CRISPR-Cas in Parasites: Advances and Challenges in the Study of Neglected Tropical Diseases

Ana María Mejía-Jaramillo
and Omar Triana-Chávez

Abstract

Neglected tropical diseases (NTDs) mainly affect poor people living in rural areas of tropical and subtropical regions of the world. The study of gene function in the pathogens that cause these diseases has been delayed by inadequate genome editing techniques, lack of the interference system, genomic plasticity, or specific genomic features that hinder these processes. CRISPR-Cas offers a precise molecular application for targeted genome modification with various potential applications. Thus, CRISPR-Cas technology has opened new avenues for genome-wide studies in NTDs, providing insights into the biology of these parasites and enabling significant advances in genome editing. In this chapter, we show the relevance of CRISPR-Cas for studying agents causing tropical diseases, with emphasis mainly on kinetoplastids and apicomplexans. We will discuss how CRISPR-Cas has been used to edit genomes, investigate the molecular interactions between parasites and their hosts, develop high-throughput screening assays for identifying compounds with antiparasitic activity, or detect specific sequences for developing rapid and accurate diagnostic tests for parasitic infections. Additionally, this methodology has opened the possibility of performing genomic scale studies, which not only allow the study and functional characterization of genes but also open the possibility of searching for new treatments using their molecular targets and mechanisms, a challenge of great importance, considering that current treatments for all these diseases are inefficient and have the problem of drug resistance. Finally, we discuss how the CRISPR-Cas system could be used to study the gene expression regulation and epigenetic modifications involved in genomic functions.

Keywords

CRISPR-Cas · Gene editing · Base editing · High-throughput · Diagnostics · Neglected tropical diseases

1 Introduction

Infectious diseases are illnesses caused by pathogenic microorganisms such as bacteria, viruses, parasites, or fungi. These microorganisms can be spread directly or indirectly from one person to another, often through contact with body fluids,

A. M. Mejía-Jaramillo (✉) · O. Triana-Chávez
Biología y Control de Enfermedades Infecciosas-BCEI, Universidad de Antioquia, Medellín, Colombia
e-mail: maria.mejia3@udea.edu.co; omar.triana@udea.edu.co

J. D. Ramírez González (ed.), *Recent Advances in Parasitomics*,
https://doi.org/10.1007/978-3-031-70591-5_4

respiratory droplets, contaminated food or water, or vectors such as mosquitoes, bugs, ticks, or other insects. Among infectious diseases, vector-borne parasitic diseases such as leishmaniasis, malaria, and trypanosomiasis continue to be a major burden on human health worldwide. Many of these diseases are considered by the World Health Organization as neglected diseases because they mainly affect poor people living in rural areas of tropical and subtropical regions of the world. Despite their significant impact on global health, they historically receive less attention and research funding compared with other diseases such as HIV/AIDS and tuberculosis. Neglected tropical diseases (NTDs) remain relevant due to their profound impact on health, socioeconomic development, and global health security, emphasizing the importance of sustained efforts to control and eliminate these diseases [1].

While the genomes of many of the parasites causing these diseases were sequenced for the first time almost two decades ago [2–4], the study of gene function has been delayed by tedious or inadequate genome editing techniques, lack of the interference system in many of the parasite species, particular genomic features such as aneuploidy or the presence of multicopy gene families, genomic plasticity, presence of extra-chromosomal elements, which require two or more rounds of substitution with different resistance markers to modify the alleles of a gene. These processes are very time-consuming and are not always successful.

Early studies using CRISPR-Cas9 gene editing in eukaryotes provided new possibilities in parasitology (reviewed in [5]). Thus, CRISPR-Cas9 technology holds promise for addressing NTDs in several ways. CRISPR-Cas technology represents a transformative tool in molecular biology and biotechnology, offering unprecedented precision and versatility in genome editing and gene regulation [6, 7]. CRISPR-Cas9 technology has opened up new opportunities for genome-wide studies in NTDs, providing insights into the biology of parasites, potential options for intervention, and enabling significant advances in genome editing of *Leishmania* spp., *Trypanosoma cruzi*, *Trypanosoma brucei*, *Plasmodium* spp., and *Toxoplasma gondii*, among others [8–11].

In these parasites, the application of the CRISPR-Cas system has been successful, and it has been used for the functional study of single genes, multicopy genes, molecular diagnosis of parasites, and the implementation of high-throughput libraries for the functional characterization of genes during different stages of their life cycles and the development of drug resistance. However, it is essential to note that the application of CRISPR-Cas9 in the context of NTDs is still in its early stages, and several challenges must be addressed. These include ethical considerations, regulatory hurdles, off-target effects, and the need for delivery systems that can efficiently and safely deliver CRISPR components to target cells or organisms in the field. Despite these challenges, CRISPR-Cas holds great potential as a versatile tool for combating NTDs and improving global health outcomes. In this chapter, we aim to show the relevance of CRISPR-Cas for studying agents causing tropical diseases, with an emphasis on kinetoplastids and apicomplexans.

2 The CRISPR-Cas System in Bacteria

Clustered regularly interspaced short palindromic repeats (CRISPR) is a unique type of genomic element initially discovered in bacteria and archaea. It is an adaptive immunity mechanism to defend against phage or other foreign nucleic acid invasion. The CRISPR-Cas system comprises an array of short repeated DNA sequences called clustered regularly interspaced short palindromic repeats and CRISPR-associated (Cas) proteins expressed by the Cas genes [12].

Similar to other bacterial and archaeal defense mechanisms, the CRISPR-Cas system shows an immense diversity of Cas protein sequences, both in gene sequences and in the architecture of genomic loci. One of the most recent classifications of Cas proteins includes two classes, six types, and 33 subtypes. While in class 1, the CRISPR-Cas system requires multiple Cas proteins; in class 2, it requires only one large protein whose noticeable representatives are Cas9, Cas12, and Cas13, which belong to type II, V, and VI, respectively (reviewed in [13]).

The CRISPR-Cas-mediated immune response consists of three main steps: adaptation, expression, and interference. In the *adaptation phase*, complexes of proteins such as Cas1 and Cas2 are required to capture and catalytically insert the spacer, and in some systems, auxiliary proteins, such as Cas4, Csn2, Cas9, or non-Cas proteins for example host integration factor, or reverse transcriptase-like fusion proteins are needed [14]. In this phase, Cas proteins bind to target DNA, cut a segment of the DNA, duplicate it, and insert it into the array, converting it into a spacer. In the *expression phase*, an RNA is transcribed from the CRISPR array as a long pre-crRNA processed to produce the mature crRNA, which serves as a guide for the Cas effectors. In the *interference phase*, the crRNA forms a complex with the Cas proteins that recognizes a target sequence of an invading virus, plasmid, or any foreign nuclei acid, which is then cleaved by the Cas nuclease(s) [12]. The interference phase involves the formation of RNA loops when the crRNA guide region hybridizes to target DNA or base pairing between the crRNA guide region and target RNA, which will result in cleavage/degradation of the target (Fig. 1).

CRISPR-Cas systems must distinguish between self and invading DNA to prevent self-cleavage. Features that ensure cleavage of intruding genetic material are (1) the presence of a DNA motif flanking the complementary RNA–DNA region in types I, II, and V, known as PAM (protospacer adjacent motif); (2) the absence of RNA complementarity between the 5′ of the crRNA and the 3′ of the target RNA in type III; and (3) the presence of an RNA motif in RNA target in type VI known as protospacer flanking sequence (PFS) [15].

CRISPR offers a precise molecular tool for targeted genome modification with various potential applications. In this regard, the need for

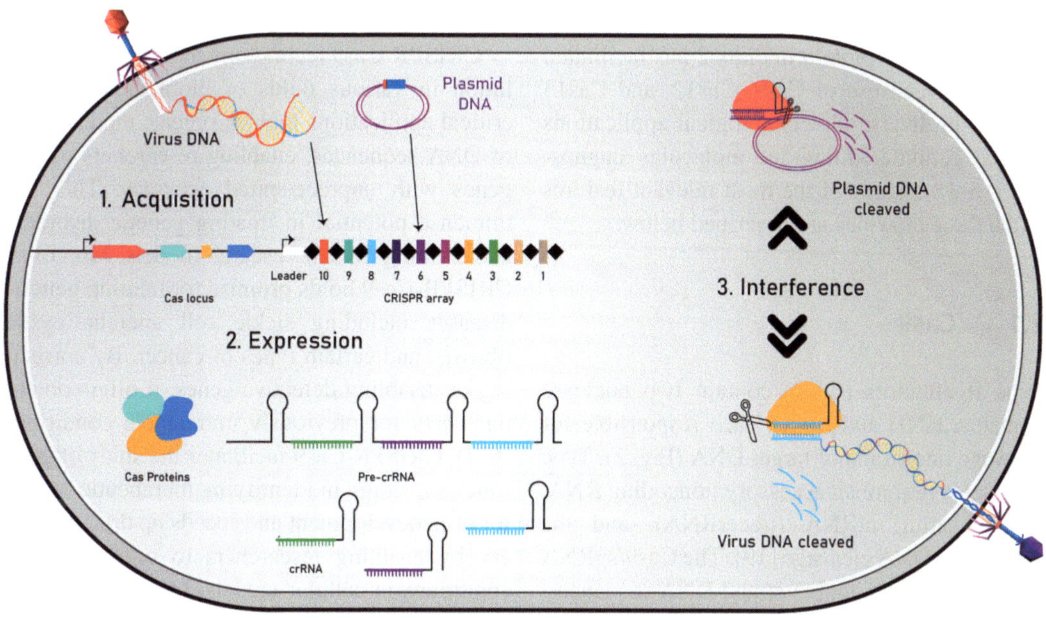

Fig. 1 The CRISPR-Cas system is involved in the adaptive immunity of bacteria and archaea. The immune response consists of three phases. First, the *adaptation phase*, in which the viruses (bacteriophages) or plasmids infect the bacteria, and the Cas proteins act as a self-programmable restriction enzyme, originating organized crRNAs inserted in a CRISPR array within the bacteria's genome. These crRNAs are then expressed, a step known as the *expression phase*. Finally, the *interference phase* controls the virus or plasmid DNA infection when a future infection occurs

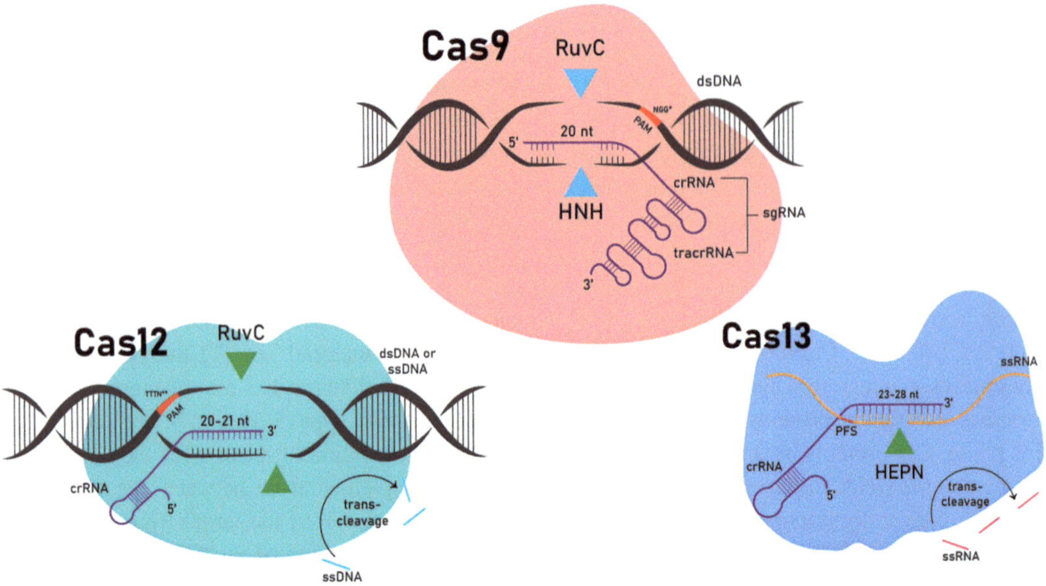

Fig. 2 CRISPR-Cas enzymes frequently used in editing and targeting of genes. The figure indicates the PAM region, cleavage site, and sgRNA features for Cas9, Cas12, and Cas13 enzymes. *RuvC*, RNase H domain; *HNH*, endonuclease domain; *PAM*, protospacer adjacent motif; *dsDNA*, double-strand DNA; *ssDNA*, single-strand DNA; *sgRNA*, single-guide RNA; *tracrRNA*, trans-activating crRNA; *crRNA*, CRISPR RNA; *PFS*, protospacer flanking sequence; *HEPN*, catalytic residues of the R-X4-H motifs within higher eukaryotes and prokaryotes nucleotide-binding (HEPN) domain; *ssRNA*, single-strand RNA

a single effector protein in class 2 has facilitated the widespread use of Cas9, Cas12, and Cas13 (Fig. 2) in diverse biotechnological applications such as genome editing and molecular diagnostics [16–18]. Some of the most relevant features of the three enzymes are described below.

2.1 Cas9

Type II effectors (Cas9) contain two nuclease domains, HNH and RuvC, each responsible for cleavage one strand of target DNA (Fig. 2). Type II systems require an accessory noncoding RNA, transactivating crRNA (tracerRNA), and the crRNA for DNA cleavage [19]. The Cas9/sgRNA complex binds to double-strand DNA sequences, which must be followed by an adjacent protospacer motif (PAM). Once the complex is bound, the two independent domains of the Cas9 nuclease will cleave each of the DNA strands three bases upstream of the PAM sequence, leaving a double-strand break (DSB).

CRISPR-Cas9 technology has sparked a revolution in various fields of biomedicine. Some critical applications include precise modification of DNA sequences, enabling researchers to edit genes with unprecedented accuracy. This has immense potential in treating genetic disorders and creating disease research models. Moreover, CRISPR-Cas9 holds promise for treating genetic diseases including sickle cell anemia, cystic fibrosis, and certain types of cancer. By correcting or disabling defective genes, it offers potential cures for previously untreatable conditions [6, 7]. CRISPR-Cas9 facilitates the study of gene function, aiding in identifying therapeutic targets for drug development and speeds up drug discovery by enabling researchers to mimic disease conditions in cellular models. In Biotechnology, the CRISPR-Cas9 system is used for various purposes, including the production of biofuels, industrial enzymes, and pharmaceuticals. It allows for the engineering of microorganisms with tailored metabolic pathways for specific applications [17, 20].

In the Parasitology field, CRISPR-Cas9 technology has shown promise in the study and control of parasites, offering new ways for understanding their biology and potentially developing novel control strategies. For instance, CRISPR-Cas9 enables precise manipulation of parasite genomes, allowing researchers to study gene function and identify potential drug targets [21, 22]. By disrupting or modifying specific genes, researchers can investigate their roles in parasite biology and pathogenesis. Parasites, such as *Plasmodium* spp., *Leishmania* spp., or *Trypanosoma cruzi*, often develop resistance to antiparasitic drugs [23–28]. CRISPR-Cas9 can be used to create drug-resistant parasite strains in the laboratory, providing insights into the mechanisms of drug resistance and helping to develop strategies to overcome it. On the other side, understanding the genes involved in parasite virulence and immune evasion is crucial for developing effective vaccines. CRISPR-Cas9 can aid in identifying and characterizing vaccine candidates by studying the effects of gene knockout or modification on parasite infectivity and immunogenicity. In addition, many parasites rely on arthropod vectors, such as mosquitoes or ticks, for transmission to their hosts. CRISPR-Cas9 can be used to modify genes in these vectors, potentially rendering them incapable of transmitting parasites. This approach, known as gene drive, holds promise for controlling vector populations and reducing the burden of vector-borne diseases [19].

Moreover, CRISPR-Cas9 can be used to investigate the molecular interactions between parasites and their hosts [29]. By disrupting parasite genes involved in host invasion, immune evasion, or virulence, researchers can elucidate the mechanisms underlying parasite–host interactions and identify targets for intervention. Finally, CRISPR-Cas9 technology can be used to develop high-throughput screening assays for identifying compounds with antiparasitic activity. By introducing CRISPR-mediated genetic reporters or markers into parasites, researchers can rapidly screen large compound libraries to identify potential drug candidates. Some of these features will be described below in more detail.

2.2 Cas12

Type V systems have Cas12 as the signature effector protein and cause stepwise, sequence-specific DNA cleavage using RuvC domain (Fig. 2). Subtype-specific variations of the target (DNA vs. RNA) and guide RNA requirements [crRNA or crRNA-traRNA or crRNA-scout (short complementarity untranslated)] have been observed in type V systems [30, 31]. CRISPR-Cas12, comparable to its counterpart Cas9, is a powerful genome editing tool derived from bacterial adaptive immune systems [32]. While Cas9 has been extensively studied and used for genome editing, Cas12 has unique features and applications. Its ability to recognize different protospacer adjacent motifs (PAM) sequences compared with Cas9 expands the range of targetable sequences, increasing the versatility of genome editing [33, 34].

Some Cas12 effectors can precisely bind and cut single-strand DNA (ssDNA) and double-strand DNA (dsDNA) at the site, but collateral excision remains only for ssDNA. Cas12 can be used to introduce targeted mutations or insertions/deletions (indels) in the DNA sequence of organisms. Cas12 has also been explored for epigenome editing, allowing researchers to modulate gene expression without altering the underlying DNA sequence. By fusing catalytically inactive Cas12 (dCas12) with effector domains capable of modifying chromatin, such as methyltransferases or demethylases, specific epigenetic modifications can be targeted to particular and desired genomic loci [33].

One of the most promising applications of Cas12 is in diagnostic applications. Cas12 can be programmed to detect specific DNA sequences. When it encounters the target sequence, it becomes activated and exhibits collateral cleavage activity, indiscriminately cleaving nearby nontargeted ssDNA molecules. This collateral activity can be harnessed to create highly sensitive and specific diagnostic assays to detect parasites, viruses, bacteria, and other pathogens [17, 35, 36]. Beyond diagnostics, Cas12-based platforms can detect nucleic acid in various contexts, including environmental monitoring, food safety, and forensic analysis. The high sensitivity, specificity, and simplicity of these assays make them attractive for point-of-care and field-based applications.

2.3 Cas13

Type VI effectors are unrelated to type II and V system effectors, contain two domains, and target genomic transcripts of invading DNAs, i.e., they cleave single-stranded RNA (ssRNA) (Fig. 2). The characteristic type VI nuclease, Cas13, binds crRNA and localizes a complementary RNA target, followed by RNase-processing activity via HEPN domains [13, 19, 37]. Cas13 proteins also show nonspecific collateral RNase activity triggered by target recognition and induce latency in virus-infected bacteria [13, 17].

CRISPR-Cas13, similar to its counterparts Cas9 and Cas12, offers promising applications in various fields, including parasitology. CRISPR-Cas13 can be adapted for highly sensitive and specific detection of parasite RNA [38, 39]. Comparable to CRISPR-Cas12-based assays, Cas13 can be programmed to detect specific sequences of parasitic RNA. This capability is valuable for developing rapid and accurate diagnostic tests for parasitic infections, such as malaria, leishmaniasis, and schistosomiasis. Thus, CRISPR-Cas13-based diagnostic tools offer significant potential for detecting parasitic infections, particularly in resource-limited settings where access to traditional diagnostic methods may be limited. These tools leverage the precise and sensitive nucleic acid detection capabilities of CRISPR-Cas13 to identify parasite RNA in clinical samples.

Moreover, CRISPR-Cas13 can be used for targeted gene knockdown or repression in parasitic organisms. By delivering Cas13 along with gRNAs targeting essential parasite genes, researchers can disrupt gene expression and investigate gene function [6, 40, 41]. This approach has the potential to uncover novel drug targets and mechanisms of parasite virulence or drug resistance. Some parasites, such as trypanosomatids and *Plasmodium* spp., are susceptible to viral infections that can impact their biology and pathogenicity. CRISPR-Cas13 systems can be engineered in parasites to provide antiviral defense against invading RNA viruses [29, 42]. By targeting and degrading viral RNA, Cas13 can help parasites combat viral infections and reduce their impact on host–parasite interactions. On the other hand, CRISPR-Cas13 can facilitate studies on drug resistance mechanisms in parasitic organisms. By targeting genes associated with drug resistance pathways, researchers can investigate how parasites develop resistance to antiparasitic drugs and identify potential targets for overcoming drug resistance. These applications will also be discussed further in the following sessions of this chapter.

3 Biotechnological Applications of CRISPR-Cas Systems on Parasites

3.1 Gene Editing

This approach has been used, because of the ability of Cas proteins to cleave specific sites from the genome, to induce a variety of modifications, including insertion/deletion (InDel) of bases or small fragments, gene knockin, introduction of mutations in both alleles of a target gene, complete knockout of genes, and C-terminal tagging (reviewed in [5, 43]), and has been used widely in different parasite species.

These modifications are mainly due to cell repair mechanisms when Cas enzymes generate double-strand breaks (DSBs). DSBs can be repaired primarily by homology direct repair (HDR) or by nonhomologous end joining pathway (NHEJ, i.e., *Toxoplasma gondii*). Homology repair is used mainly to introduce specific sequences at the cleavage site if a DNA donor is provided, which may include antibiotics, to facilitate selection of modified organisms (reviewed in [5]). NHEJ usually leads to a short insertion/deletion (indels) near the cut site. However, this type of repair is absent in *Plasmodium falciparum* and *P. yoelii* [44] as well as in in trypanosomatid parasites [45]. In these cases, microhomology-mediated end joining (MMEJ) is used which mainly generates deletions of different sizes and in some cases insertions [8, 11, 46].

As can be inferred from the above, two main components are required for these genomic editing applications: Cas proteins and single guides RNAs (sgRNAs). DNA donors may or may not be present. The most popular methodologies used in selected parasites to perform gene editing are described below, and some specific examples are mentioned.

3.1.1 Cas9

In parasites and eukaryotic organisms, the most widely used protein for this application has been Cas9 from *Streptococcus pyogenes* (SpCas9) and its high-fidelity variants. However, others, such as *Staphylococcus aureus* (SaCas9), are also widespread. In parasites, the most commonly used system to date has been the expression of the Cas9 protein in episomal or integrative plasmids (Fig. 3). These plasmids have undergone significant improvements, in terms of promoters, the amount and location of NLS sequences or the fusion with proteins such as luciferase [47] or GFP [48], which not only allow for the possibility of selecting parasites but also the generation of novel Cas9 proteins such as a bioluminescent/fluorescent version. For example, this Cas9 protein tagging has made it possible to follow the parasitic infection of *T. cruzi* in a murine model and to visualize sections of the infected tissues with fluorescent parasites [47]. The transfection of Cas9 in plasmids guarantees the expression of the enzyme permanently; however, in some cases, cytotoxicity has been reported [8] and off-target effects of Cas9 are always possible, which, although reduced with the eSpCas9 [49] and Cas9-HF [50] variants, may affect the outcome of the desired genotypes/phenotypes.

In this sense, the other popular strategy has been transfection with purified recombinant protein, which has the advantage of not generating off-target effects since its expression is transient but limits its use in the implementation of libraries. Thus, Soares-Medeiros et al. [51] showed that transfection of ribonucleoprotein (RNP) complexes composed of recombinant *S. aureus* Cas9 (SaCas9) and an in vitro transcribed sgRNA, resulted in rapid gene editing in kinetoplastids with frequencies close to 100%. This highly efficient genome editing, through SaCas9/sgRNA

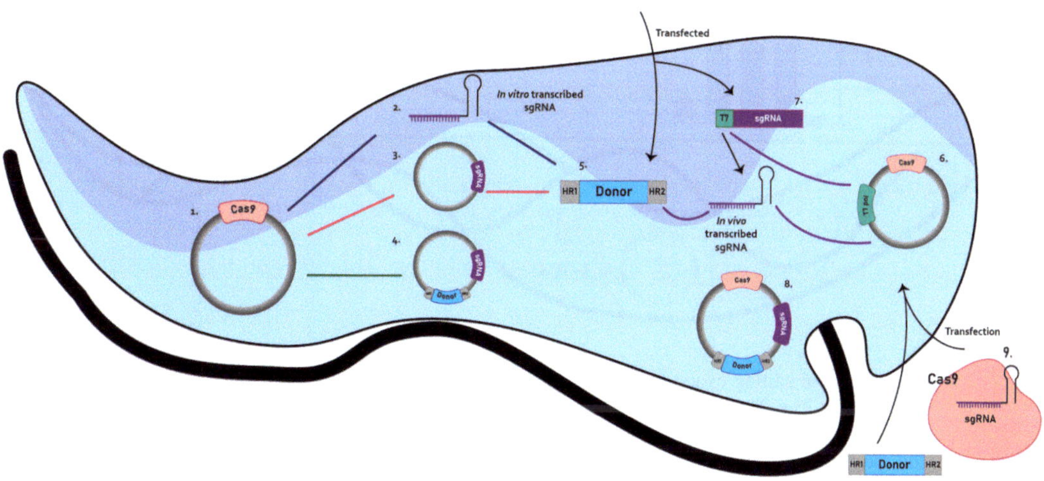

Fig. 3 Summary of the different strategies used for genomic editing in parasites. Plasmids containing CDS for the Cas9 enzyme (1) are transfected with in vitro transcribed sgRNA (2) or plasmid expressing the sgRNA (3). Both strategies could require a transfected DNA donor (5). In pathway (4), a plasmid containing sgRNA and DNA donor is transfected into parasites expressing Cas9. In pathway (6), plasmid containing Cas9 and T7 RNAP is used to obtain the in vivo transcribed sgRNA under the T7 promotor (7), which could be used with DNA donor. Plasmids with the Cas9, sgRNA, and DNA donor are employed (8). Finally, the RNP complex (Cas9 + sgRNA) could also be utilized (9)

RNP, was obtained for reporter and endogenous genes or even two simultaneous genes. Results were observed at different life cycle stages of *T. cruzi* strains and in *T. brucei* and *L. major* [51]. However, the authors reported that the size of the RPN, such as the one from *S. pyogenes*, affected knockout efficiency in these parasites. Recently, Asencio et al. [52] demonstrated that transfection of a ribonucleoprotein complex, composed of recombinant spCas9 and in vitro synthesized sgRNA, produces rapid and efficient genetic modifications for single genes and multigene families under selection marker-free conditions at various life cycle stages of *T. brucei*, *T. congolense*, and *Leishmania* spp. This approach was successfully developed to inactivate, knockdown, and mutate candidate genes, with great potential for inactivating multiple metabolic pathway genes; approaches that previously were restricted due to insufficient selection markers. However, although the authors did not perform the experiments in *T. cruzi*, they suggested that the enzyme they used could work in this model [52]. In our experience, using spCas9 for this parasite is efficient, even with a single NLS sequence, and it is possible to obtain double knockouts with a single DNA donor (Fig. 4).

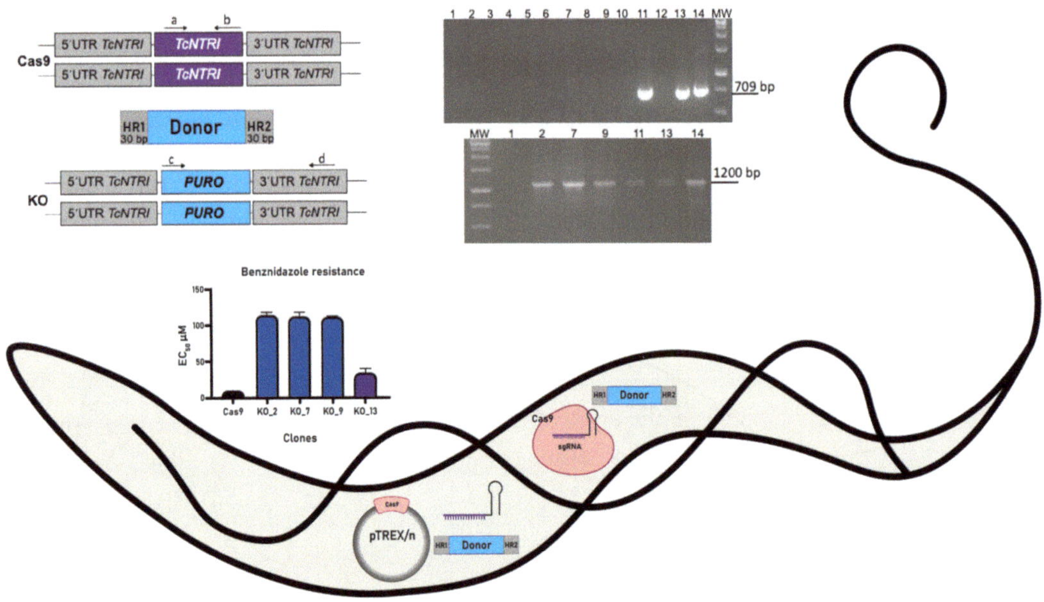

Fig. 4 CRISPR-Cas9 gene editing example in the parasite *Trypanosoma cruzi*. The NTRI (Nitroreductase I) gene was edited using in vitro transcribed gRNAs targeted to the CDS, and their role in resistance to benznidazole (Bz) drug was experimentally validated. The puromycin antibiotic gene was used as a DNA donor with 30 bp homology arms. The validation of editing was verified by PCR using specific primers against the NTRI CDS region (primers a and b), which amplify a region of 709 bp, or primers against the puromycin gene (primer c) and NTRI 3′-UTR region (primer d), that amplify a region of 1200 bp. The electrophoresis agarose gel shows no amplification of the NTRI gene in knockout parasites (lanes 2–10) and amplification of the puromycin gene (lanes 2, 7, and 9). Hemi-knockout parasites are also observed (lanes 11 and 13). Lane 1 corresponds to negative control in the PCR. Lane 14 corresponds to positive control (upper gel: Cas9 parasites; bottom gel: puromycin plasmid). The Bz susceptibility profile shows that knockout parasites gain resistance to this drug. In this case, two strategies were used: parasites expressing pTREX/n [48] Cas9 protein or parasites transfected directly with the RNP complex (Cas9 + sgRNA)

3.1.2 sgRNA

As mentioned above, sgRNA can be programmed to target a sequence of interest, which in the case of Cas9 proteins is approximately 20 bp, with a specificity that tolerates no more than four mismatches [53] and which has to be located immediately upstream of the PAM sequence (Fig. 2). This PAM sequence represents the only constraint on the molecular design of sgRNAs for Cas9 and will be dependent on the Cas9 [54]. Thus, in the case of SpCas9, an NGG sequence (where N is any nucleotide and G is the guanine base) is required for recognition, whereas for SaCas9 it is NNGRRT (where R is A or G). Currently, there are different tools for the design of sgRNAs from different parasite species, such as the one developed by Peng and Tarleton [55], known as the Eukaryotic Pathogen CRISPR-Cas9 technology guide RNA/DNA Design Tool [55], or the one developed by Beneke et al., LeishGEdit to retrieve primer sequences to target genes of interest for tagging and knockout [21].

Similar to Cas9, sgRNAs can be expressed using different methodologies, including plasmids and in vitro or in vivo synthesis using T7 promoters. Regarding plasmid expression for parasites, improvements have been made in the promoters used, the most popular of which have been the U6, ribosomal, and T7. Additionally, plasmids can have, in addition to the sgRNA, the Cas9 protein and DNA donor (Fig. 3).

3.1.3 DNA Donor

DNA donors will be needed depending on the type of repair required. As mentioned above, parasites can repair differently, but the MMEJ pathway is less efficient than the NHEJ repair pathway [11, 48, 56]. In this sense, to promote the desired editing type and repair via HDR, it is necessary to have a DNA donor with homology arms at both ends (Fig. 4). Depending on the parasite model, these have been reported to range from 30 nts in parasites such as *Leishmania* spp. [21] to 100 for *T. cruzi* [48]. However, in our experience, we have achieved efficient knockouts for this parasite with only 30 nts, using both the plasmid pTREX/Cas 9 [48] as well as the Cas9 RNP, and sgRNAs synthesized in vitro and simultaneously transfected with the DNA donors (Fig. 4).

For the synthesis of DNA donor, the most popular method has been PCR amplification and subsequent transfection into the parasites (Fig. 3). Currently, to reduce the complexity of their preparations, many plasmids have been designed that allow knockout by gene replacement with antibiotics and to tag the genes with particular labels [21]. However, in different models, DNA donors expressed on plasmids alone or with Cas9 and sgRNAs have also been used (Fig. 4) (reviewed in [57, 58]).

3.1.4 Successful Examples of Gene Editing Implemented in Parasites

Toxoplasma gondii was the first parasite in which the CRISPR-Cas9 system was implemented [59]. The strategies used in this parasite include a unique vector for Cas9 expression fused to a nuclear localization signal and other proteins (Cas9-NLS-GFP, Cas9-FLAG-NLS), and sgRNA transfected in the parasites with or without DNA donors to induce HDR and NHEJ-mediated DNA repair, respectively [59, 60]. In this parasite, the CRISPR-Cas9 system has been significantly improved, which can be applied without DNA donors or antibiotic selection, allowing the generation of genomic libraries, as discussed below [22, 61]. In other apicomplexans such as *P. falciparum*, the tools used have been similar; however, due to the lack of canonical NHEJ pathways, many of the strategies include transfection with vectors that simultaneously have sgRNA and DNA donor and even Cas9 [58].

For *T. cruzi*, the most widely used system is the one in which the CRISPR-Cas9 system was optimized by co-expressing the sgRNA and Cas9 in plasmids [48]. The strategies reported by these authors include one or two integrative vectors for constitutive expression of sgRNAs and Cas9 separately or one vector with both, which are co-transfected with a DNA donor to generate knockout cell lines for the genes of interest. Using a DNA repair template with a drug marker, the authors achieved highly efficient knockout and tagging of endogenous genes after 5 weeks of continuous drug pressure [62]. The vectors with a ribosomal promoter express Cas9 protein, fused to GFP, enabled the selection of ribonucleoprotein-expressing lines. This system has been widely used to generate knockouts of more than 20 genes, which accounts for the efficiency of the system (reviewed in [57]).

Additionally, CRISPR-Cas9 has been successfully implemented in species such as *L. mexicana*, *L. major*, *L. donovani*, *L. braziliensis*, *L. infantum*, and *T. brucei* [21, 63, 64]. Beneke et al. [21] developed a clone-free, PCR-based CRISPR-Cas9 method for rapid and accurate genome editing successfully applied to *L. mexicana*, *L. major*, *L. donovani*, and *T. brucei*. This system provides a toolkit with protocols for gene deletion, N- or C-terminal tagging, and a website for designing the primers needed to generate sgRNA and DNA repair cassettes [21]. In this system, transgenic parasites, stably expressing SpCas9 and T7 RNA polymerase (T7 RNAP), are transfected with two PCR fragments coding for sgRNA under the T7 promoter and a DNA repair template consisting of 30 bp homology arms flanking a drug selection marker. The authors reported that after 1 week of drug selection, 100% of the surviving parasites carried the desired genomic mutation, drastically reducing culture times compared with other methods. Another advantage of this method is that the PCR fragments are degraded within 48 h, which avoids further editing. This system has proven to be fast and reliable and has been used successfully to edit single genes [65]. Costa et al. [47] adapted the system developed by Beneke et al. [21] to *T. cruzi*, creating a cell line that stably expresses Cas9 and the T7 RNA polymerase [47], which has enabled the endogenous gene tagging and gene knockout in *T. cruzi* without cloning. Similar strategies with the T7-promoter-driven system have been implemented successfully in *P. falciparum* [28]. It is essential to highlight that the CRISPR-Cas9 methodology used for genomic editing in these parasites reduced the time to obtain genetically modified parasites from months to weeks (reviewed in [57, 58]).

3.2 Base Editing

One of the most promising applications of CRISPR-Cas for use in parasites without the need for a DNA donor is the Base Editor (BE). This system contains two primary components: a catalytically deficient Cas protein (dead Cas) or Cas9 nickase (nCas9), and a DNA-modifying enzyme, such as nucleoside deaminases, which can be classified into two types: cytosine base editors (CBE) and adenine base editors (ABE) [66].

These types of associations between modified Cas proteins and nucleoside deaminases allow for precise targeting to a site of interest, without the need for DSBs, HDR, or DNA donor templates, where deamination of cytidine generates uridine, which is read as thymidine by DNA polymerase. In contrast, the deamination of adenosine generates inosine, which is repaired as guanosine by the enzyme [67]. To make the system highly efficient, different modifications have been made such as using D10A nickases (nCas9) fused to monomers of a uracil glycosylase inhibitor (UGI) to prevent base excision repair during DNA replication, the insertion of a single-stranded (ss) non-sequence-specific DNA-binding domain (DBD) of the Rad51 protein to increase CBE editing rates, nuclear localization signals (NLS) modifications, the use of codons, among others, resulting in highly efficient base editing systems such as hyBE4max [68].

In this regard, Engstler and Beneke [69] implemented the CRISPR-Cas9 cytosine base editor (CBE) hyBE4max system in four species of *Leishmania* spp. to introduce stop codons. They created www.leishbaseedit.net to design CBE primers in kinetoplastids. These authors demonstrated how this tool can efficiently generate functional null mutants expressing only one sgRNA, achieving an editing rate of up to 100% in nonclonal populations [69].

Recently, necessary improvements to this base editing method were reported to minimize the effects on parasite growth and eliminate competition between deleterious and non-deleterious mutations. These improvements included the expression of not only T7 RNAP and the CBE hyBE4max but also of a highly efficient variant of a Cas12a nuclease derived from *Acidaminococcus* sp. that allows for the delivery and integration of CBE sgRNA expression cassettes at the 18S rRNA locus of *Leishmania* spp. The authors showed that these improvements achieved high editing rates quickly without significantly affecting parasite growth. In a final optimization step, the authors also improved the CBE sgRNA design and scoring tool www.leish-baseedit.net to select CBE sgRNAs that are less likely to generate edits and that do not result in a stop codon [34]. This system has great potential in parasites since it allows functional genetic screening without the need for DNA donors.

3.3 High-Throughput Use of CRISPR-Cas System

Although one of the main applications of the CRISPR-Cas system is undoubtedly the editing of single genes, in recent years, the system has been scalable to the whole genome so that the function of multiple genes can be studied quickly and efficiently. The high-throughput CRISPR-Cas system is based on the transfection of sgRNA library specific for each protein-coding gene into parasites constitutively expressing the Cas9 enzyme. Upon DSB by Cas9 and with or without a repair cassette, the repair pathway generates a loss of gene function. From the surviving knockout parasites, plasmids are recovered and sequenced to identify the specific sgRNAs of each gene (Fig. 5) [11, 21, 61, 64].

The system has been applied to the functional study of different parasites, such as *Leishmania* spp. [64, 69–71] and *T. brucei* [21] and some apicomplexa such as *T. gondii* [61] and *P. falciparum* [25]. One advantage of this method is that parasites transfected with the sgRNA libraries can be used to study various biological processes, such as drug resistance, cell differentiation, differences in growth, and metabolic needs. This section will discuss some of the applications and methodologies used in these parasites.

3.3.1 Construction of sgRNA Libraries, Transfection, and Identification

The first step in using the CRISPR-Cas system on a large scale is the design of the sgRNA library. In the studies reported in the literature, at least 6–10 sgRNAs for each protein-coding gene must be synthesized. These sgRNAs can align to the 5′-UTR, 3′-UTR, or in the CDS, or have a combination of guides covering the entire gene sequence. The sgRNAs are cloned into a plasmid and used for the transfection into parasites stably expressing the Cas9 protein. In most studies, the identification of the edited genes is based on the recovery of the sgRNAs and subsequent sequencing, as explained above (Fig. 5).

3.3.2 CRISPR-Cas9 Screening

Screening is based on identifying genes whose loss is negatively or positively selected under an experimental condition of interest (e.g., selection with drugs). In this sense, experience with different biological models shows that after selection during a series of passages (usually 3), some sgRNA will be eliminated from the population, indicating genes whose loss generates a biological disadvantage under the experimental condition, or gene essentiality, and others will be enriched, showing that the loss of the gene provides an advantage to the parasites under the experimental condition (Fig. 5).

3.3.3 CRISPR-Cas High-Throughput Applications in Medically Important Parasites

Recently, a high-throughput strategy using CRISPR-Cas9 was developed in *Leishmania infantum* to identify the mechanisms of action and drug resistance. For this purpose, a cloning vector compatible with Illumina sequencing was built, where sgRNAs were expressed under the control of a ribosomal RNA (rRNA) promoter. In this vector, 49,754 sgRNAs corresponding to six sgRNAs per parasite CDS were cloned and subsequently expressed in *L. infantum* Cas9 parasites. These parasites were either pressured or not with different leishmanicidal drugs. This strategy allowed the identification of new genes involved

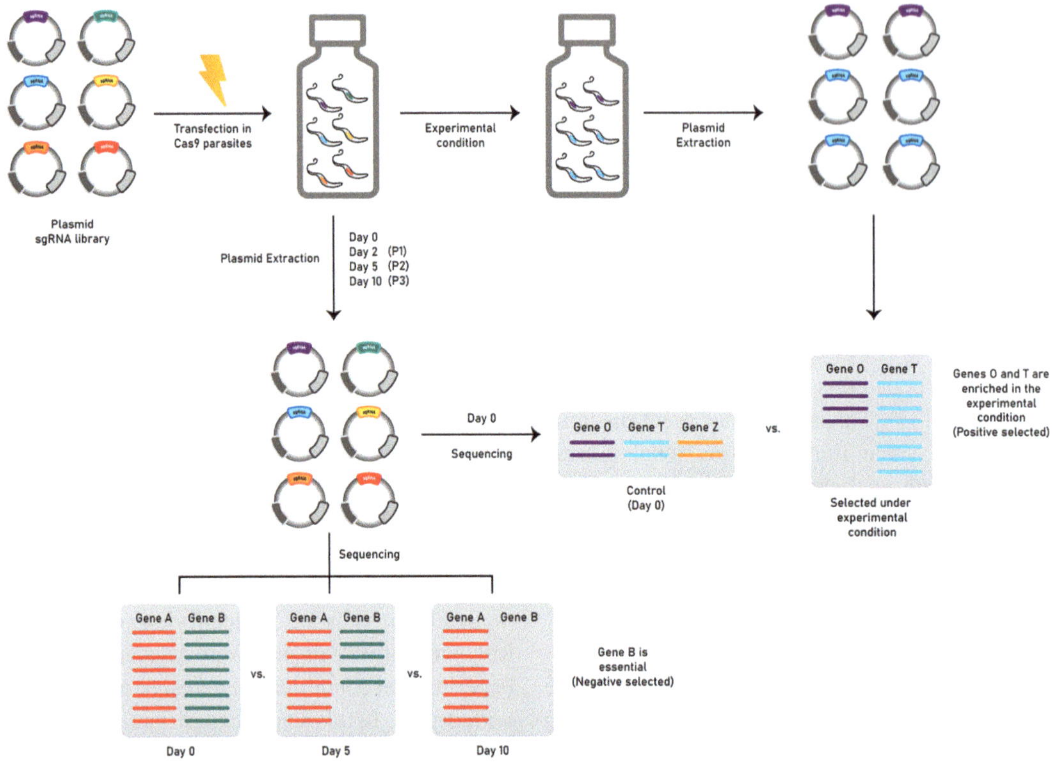

Fig. 5 Overview of sgRNA library construction, sequencing, and gene selection in high-throughput CRISPR-Cas experiments. At least 6–10 gRNAs for each gene are cloned in plasmids, and then the entire library is transfected in the parasite. After selection with a particular experimental condition (e.g., drug selection), the sgRNAs that are negatively or positively selected are identified by sequencing

in the resistance to MTF and AMB, which were validated by complete gene knockout using the CRISPR-Cas9 strategy [64].

In other parasites, such as *T. gondii*, the CRISPR-Cas9 system was implemented on a high-throughput study to evaluate the contribution of each gene during different cellular processes. *Toxoplasma gondii* is the only apicomplexan that exhibits NHEJ repair, so Cas9-induced DBSs are repaired without needing a homologous DNA template. The use of the RH strain expressing Cas9 and transfection with a library with 10 sgRNA for each of the 8158 parasite proteins has been successful in targeting genes involved in different parasite functions, such as drug resistance and essential genes in its life cycle [34]. On the other hand, in the case of infection of human fibroblasts, more than 200 specific genes were identified that revealed essential functions during infection of human cells, with some of them being key in the process of cell invasion, as is the case of the gene encoding the CLAMP protein (Claudin-like apicomplexan microneme protein) [72].

3.4 Variations in CRISPR-Cas Systems: Applications and New Challenges

The applications of the CRISPR-Cas system for the study of gene function are tremendous. However, the system has its limitations and disadvantages, including the irreversible change in the DNA sequence and that DNA repair occurs in an imprecise manner, sometimes making it challenging to associate gene loss with phenotypic changes. An alternative to modulate gene expression without knockout is to modify epigenetic signals and thus control gene expression without changing the DNA sequence. The catalytically inactive Cas9 enzyme (dCas9) can be used as an activator (CRIPSRa) or inhibitor of transcription (CRISPRi). The dCas9 binds to the target sequence but does not cleave it, facilitating access to other gene regions of interest for studying the function and regulation of gene expression. The system is based on the constitutive expression of proteins fused to dCas9, which are responsible for maintaining transcriptional control. The CRISPRi system has been widely used in *P. falciparum* and *P. yoelli*, allowing for the knockdown of genes related to artemisinin and chloroquine resistance and under nutritional and heat stress conditions [25, 73]. On the other hand, combining chromatin immunoprecipitation studies with ChIP-seq and the CRISPRi/a system will allow us to identify the regions of relaxed chromatin to which the gRNAs of interest will be targeted.

More recently, two systems based on protein methylases and demethylases fused to the enzyme dCas, responsible for writing and deleting DNA epigenetic marks, respectively, were designed to control gene expression in insects and humans. The CRISPRoff and CRISPRon system acts reversibly and can be used to study gene regulation in different organisms [74]. Likewise, histone-modifying proteins, such as some acetyltransferases or deacetylases, can be fused to the dCas9 enzyme, making it possible to identify gene regulation due to histone modifications. The challenge is to adapt these systems to study the control of gene regulation in parasites, where regulation occurs mainly at the posttranscriptional level. In *Plasmodium* spp., for example, the CRISPRi system was used to study the function of essential genes [75].

Another novel system to study gene function in *Plasmodium* spp. was recently designed where a single-stranded binding protein (dCas9-SSAP) is fused to dCas9. SSAPs are DNA-binding proteins that promote recombination with homologous DNA donor sequences. This system has been used to edit genomes of mammalian cell lines and was validated for the study of gene editing and tagging in *P. yoelii* [73].

3.5 CRISPR-Cas Systems in the Molecular Diagnosis of Parasites

As mentioned above, the CRISPR-Cas system offers a precise molecular tool to cut genetic material since as nucleases they recognize specific target sequences, making them potent candidates for pathogen detection and diagnosis. These systems rely on Cas type II enzymes that physically bind to target sequences or type V and VI enzymes that bind to and specifically and collaterally cleave a DNA or RNA sequence, respectively, cleavage that allows the release of a probe that can be detected on different platforms [16–18].

Currently reported CRISPR-Cas detection systems can be classified into three groups according to the different Cas effectors: Cas9-based, Cas12-based, and Cas13-based [17], with the latter two systems being of particular diagnostic importance due to the collateral cleavage they present (Fig. 2). This intrinsic activity allows the use of labeled probes that facilitate the visualization of the reaction (Fig. 6). Among the platforms that use these systems are the SHERLOCK platform [38], HOLMES [35], DETECTR [76], and CRISPRDx [77], among others, which rapidly detect target nucleic acids with high sensitivity, specificity, simplicity, and low cost. Diagnostics using CRISPR-Cas technology have been applied successfully in pathogen detection, genotyping, single nucleotide polymorphism (SNP) discrimination, and cancer mutation detection [38].

The SHERLOCK (specific high sensitivity enzymatic reporter unlocking) system combines isothermal amplification of target sequences, followed by target recognition via the Cas13a protein of *Leptotrichia wade* (LwaCas13a) and collateral excision of an ssRNA probe to report the presence of pathogen genetic material [38]. This system has undergone significant optimization since its first development in 2017, with improvement of (1) sensitivity, by including an accessory protein to amplify the signal or replacing RPA with LAMP (loop-mediated isothermal

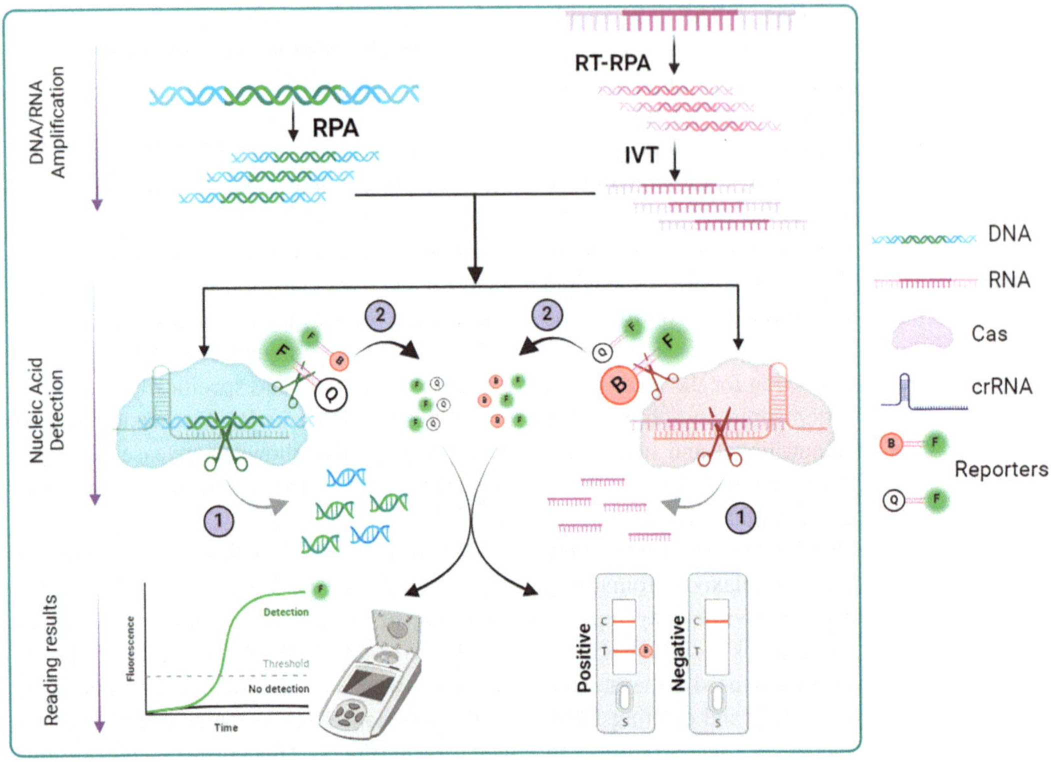

Fig. 6 CRISPR-Cas proteins for diagnostic pathogen applications. The DNA or RNA is amplified using recombinase polymerase amplification (RPA) or reverse-transcriptase RPA. The DNA obtained by RT-PCR is also amplified using in vitro transcription (IVT). For the detection of nucleic acids, the collateral activity of Cas12 and Cas13 is taken advantage of using biotinylated or fluorescent probes. The results could be visualized by flow lateral strips or using fluorescence reading, respectively. Created with BioRender.com

amplification), (2) specificity, by optimizing the primer and sgRNA, (3) throughput, by multiplex detection using additional enzymes (including a cocktail of LwaCas13a, PsmCas13b (*Prevotella* sp. MA2016), CcaCas13b (*Capnocytophaga canimorsus* Cc5), and AsCas12a (*Acidaminococcus* sp. BV3L6)), and (4) validation as point-of-care diagnostics using lateral flow and ultrafast RNA extraction methods [39]. This detection system is capable of detecting DNA as target RNA. It has been successfully applied to detecting Zika virus, all serotypes of dengue virus [39], *Plasmodium* spp. [78], *T. brucei* [79], bacterial isolates, antibiotic resistance genes, human DNA genotypes, and mutations in cancer [38].

Recently, a new detection method combining multiple isothermal detection enzymes (MIRA) and the Cas13 enzyme was designed to diagnose *Trichomonas vaginalis*. This system, known as MIRA-CRISPR-Cas13, uses lateral flow strips to detect the parasite's genomic DNA and shows high sensitivity and specificity in the molecular diagnosis of this parasite [80]. These advances show that diagnosis is increasingly being improved by renewing amplification systems or visualization technologies.

The first report on using Cas12a or Cpf1 was related to developing HOLMES (a one-hour, low-cost, multipurpose, highly efficient system). HOLMES can detect DNA/RNA viruses and distinguish virus genotypes and human SNPs from cell lines and clinical specimens within 1 h [35]. Similarly, DETECTOR has been used to detect viruses and discriminate among human papillomavirus (HPV) genotypes in virus-infected human cell lines or clinical patient samples [76].

Class 2 V-B-type CRISPR-Cas12b also demonstrated the same trans DNA excision activity as

Cas12a. Recently, Li and colleagues employed Cas12b to create an improved version of HOLMES (called HOLMESv2). Having a wide reaction temperature range for trans excision, Cas12b can be combined with various isothermal amplification methods, such as cLAMP. HOLMESv2 is a nucleic acid detection system that supports the quantitative determination of target DNA. In HOLMESv2, a DNA polymerase Bst 3.0 containing 5′-3′ polymerase activity for both DNA or RNA templates was used to directly amplify RNA targets for a simplified RNA detection protocol, being the most straightforward CRISPR biosensor system for the detection of this type of nucleic acids so far [81].

Today, many variations and improvements have been made to the diagnostic system using CRISPR-Cas12 technology. Currently, several reports show its application in the diagnosis of *P. falciparum*, *P. vivax*, *T. gondii*, *Schistosoma haematobium*, and *Cryptosporidium parvum*, among others (reviewed in [82]). Among the advantages over other molecular diagnostic tests, the Cas12-based design allows diagnosis directly at the point of patient care. It is low-cost, shows high specificity, and low dependence on robust equipment, as it will enable visualization by lateral flow strips or fluorescence. Additionally, reaction times are increasingly shorter, and the reaction is done in a single step, facilitating use in the point-of-care (POC). Finally, assays performed with these parasites indicate that CRISPR-Cas12 has specificity and sensitivity comparable to traditional assays and sometimes better [82].

Another interesting aspect to highlight is using CRISPR-Cas12 to diagnose drug-resistant parasite strains. In this regard, a combination of the LAMP and Cas12 technique, known as CRISPR-LAMP, was recently designed and used to diagnose some polymorphisms in the pfcrt, PfMDR1, and PfKelch13 genes, which facilitated the search for sensitivity and resistance to different antimalarial drugs [36].

In addition to Cas12 and Cas13, the Cas9 enzyme has also been used to diagnose *T. gondii*. This system combines RPA with CRISPR technology, known as RPA-CRISPR-Cas9, and helped identify *T. gondii* in clinical samples from cats and other animals. Thus, the Cas9 enzyme is an alternative to conventional methods used to diagnose these parasites [83].

Although none of these systems are commercially available at present, the development of these systems in the diagnosis of parasites of medical importance is vital because, in addition to increasing specificity and sensitivity concerning gold-standard tests, they can be performed at the POC, which is a necessary and fundamental issue for the control of these neglected diseases.

4 Conclusions and Perspectives

The CRISPR-Cas system is highly versatile. In recent years, the variations using both modified Cas enzymes and domains of other regulatory proteins of gene expression opened new possibilities to be tested in parasites of medical importance. Using these new technologies in parasites where genetic modification has been a limitation raises new challenges to studying gene function in these medically essential microorganisms. Additionally, this methodology has opened the possibility of performing high-throughput studies, which not only allow the study and functional characterization of genes but also opens the possibility of searching for new treatments with their molecular targets and mechanisms of action at the genomic level, a challenge of great importance, considering that current treatments for all these diseases are inefficient and have the confounding problem of drug resistance. Also, an inducible CRISPR-Cas9 system would allow researchers to generate conditional knockouts that, except in *T. gondii* and *T. brucei*, have yet to be reported in most parasite models described here, which is fundamental for the study of essential genes. Finally, the CRISPR-Cas system presents new alternatives for identifying and diagnosing these infectious agents. Hence, the design of rapid, easy-to-use, and inexpensive tests at the point of patient care is essential to respond to this need in low-resource countries.

Acknowledgments The authors are supported by Universidad de Antioquia, SGR Grant BPIN 2020000100479, and Grant 1115-918-91933 Minciencias-Colombia. We are very grateful to Professor Carl Lowenberger for his valuable comments and English revision and to Hader Ospina for the design of Fig. 6.

References

1. Satapathy P, Rustagi S, Kumar P et al (2024) Understanding noma: WHO's recognition and the path forward in global health. Trans R Soc Trop Med Hyg 6:trae031. https://doi.org/10.1093/trstmh/trae031
2. Berriman M, Ghedin E, Hertz-Fowler C et al (2005) The genome of the African trypanosome *Trypanosoma brucei*. Science 309:416–422. https://doi.org/10.1126/science.1112642
3. El-Sayed NM, Myler PJ, Bartholomeu DC et al (2005) The genome sequence of *Trypanosoma cruzi*, etiologic agent of Chagas disease. Science 309:409–415. https://doi.org/10.1126/science.1112631
4. Ivens AC, Peacock CS, Worthey EA et al (2005) The genome of the kinetoplastid parasite, *Leishmania major*. Science 309:436–442. https://doi.org/10.1126/science.1112680
5. Bryant JM, Baumgarten S, Glover L et al (2019) CRISPR in parasitology: not exactly cut and dried! Trends Parasitol 35(6):409–422. https://doi.org/10.1016/j.pt.2019.03.004
6. Ebrahimi S, Khosravi MA, Raz A et al (2023) CRISPR-Cas technology as a revolutionary genome editing tool: mechanisms and biomedical applications. IBJ 27:219–246. https://doi.org/10.61186/ibj.27.5.219
7. Wang M, Chen M, Wu X et al (2023) CRISPR applications in cancer diagnosis and treatment. Cell Mol Biol Lett 28:73. https://doi.org/10.1186/s11658-023-00483-4
8. Peng D, Kurup SP, Yao PY et al (2015) CRISPR-Cas9-mediated single-gene and gene family disruption in *Trypanosoma cruzi*. mBio 6(1):e02097-14. https://doi.org/10.1128/mBio.02097-14
9. Sollelis L, Ghorbal M, Macpherson CR et al (2015) First efficient CRISPR-Cas9-mediated genome editing in *Leishmania* parasites. Cell Microbiol 17:1405–1412. https://doi.org/10.1111/cmi.12456
10. Zhang W-W, Lypaczewski P, Matlashewski G (2017) Optimized CRISPR-Cas9 genome editing for *Leishmania* and its use to target a multigene family, induce chromosomal translocation, and study DNA break repair mechanisms. mSphere 2. https://doi.org/10.1128/msphere.00340-16
11. Zhang W-W, Matlashewski G (2015) CRISPR-Cas9-mediated genome editing in *Leishmania donovani*. mBio 6:1–14. https://doi.org/10.1128/mBio.00861-15
12. Mojica FJM, Díez-Villaseñor C, Soria E, Juez G (2000) Biological significance of a family of regularly spaced repeats in the genomes of Archaea, bacteria and mitochondria. Mol Microbiol 36(1):244–246. https://doi.org/10.1046/j.1365-2958.2000.01838.x
13. Makarova KS, Wolf YI, Iranzo J et al (2020) Evolutionary classification of CRISPR–Cas systems: a burst of class 2 and derived variants. Nat Rev Microbiol 18:67–83. https://doi.org/10.1038/s41579-019-0299-x
14. Silas S, Mohr G, Sidote DJ et al (2016) Direct CRISPR spacer acquisition from RNA by a natural reverse transcriptase–Cas1 fusion protein. Science 351:aad4234. https://doi.org/10.1126/science.aad4234
15. Gleditzsch D, Pausch P, Müller-Esparza H et al (2019) PAM identification by CRISPR-Cas effector complexes: diversified mechanisms and structures. RNA Biol 16:504–517. https://doi.org/10.1080/15476286.2018.1504546
16. Lau A, Ren C, Lee LP (2021) Critical review on where CRISPR meets molecular diagnostics progress in biomedical engineering. Prog Biomed Eng 3:1–12. https://doi.org/10.1088/2516-1091/abbf5e
17. Li Y, Li S, Wang J, Liu G (2019) CRISPR/Cas systems towards next-generation biosensing. Trends Biotechnol 37:730–743. https://doi.org/10.1016/j.tibtech.2018.12.005
18. Mustafa MI, Makhawi AM (2021) SHERLOCK and DETECTR: CRISPR-Cas systems as potential rapid diagnostic tools for emerging infectious diseases. J Clin Microbiol 59. https://doi.org/10.1128/jcm.00745-20
19. Grissa I, Vergnaud G, Pourcel C et al (2015) An updated evolutionary classification of CRISPR-Cas systems. Nat Rev Microbiol 13:722–736. https://doi.org/10.1038/nrmicro3569
20. Bisht D, Salave S, Desai N et al (2024) Genome editing and its role in vaccine, diagnosis, and therapeutic advancement. Int J Biol Macromol 269:131802. https://doi.org/10.1016/j.ijbiomac.2024.131802
21. Beneke T, Madden R, Makin L et al (2017) A CRISPR Cas9 high-throughput genome editing toolkit for kinetoplastids. R Soc Open Sci 4:170095. https://doi.org/10.1098/rsos.170095
22. Sidik SM, Huet D, Lourido S (2018) CRISPR-Cas9-based genome-wide screening of *Toxoplasma gondii*. Nat Protoc 13:307–323. https://doi.org/10.1038/nprot.2017.131
23. Bello AR, Nare B, Freedman D et al (1994) PTR1: a reductase mediating salvage of oxidized pteridines and methotrexate resistance in the protozoan parasite *Leishmania major*. Proc Natl Acad Sci USA 91:11442–11446. https://pubmed.ncbi.nlm.nih.gov/7972081/
24. do Monte-Neto RL, Coelho AC, Raymond F et al (2011) Gene expression profiling and molecular characterization of antimony resistance in *Leishmania amazonensis*. PLoS Negl Trop Dis 5:e1167. https://doi.org/10.1371/journal.pntd.0001167
25. Lucky AB, Wang C, Li X et al (2024) Transforming the CRISPR/dCas9-based gene regulation technique into a forward screening tool in *Plasmodium falci-*

parum. iScience 27:109602. https://doi.org/10.1016/j.isci.2024.109602
26. Mejía AM, Hall BS, Taylor MC et al (2012) Benznidazole-resistance in *Trypanosoma cruzi* is a readily acquired trait that can arise independently in a single population. J Infect Dis 206:220–228. https://doi.org/10.1093/infdis/jis331
27. Mejía-Jaramillo AM, Fernández GJ, Montilla M et al (2012) *Trypanosoma cruzi* strains resistant to benznidazole occurring in Colombia. Biomedica 32(2):196–205. https://doi.org/10.7705/biomedica.v32i2.458
28. Wagner JC, Platt RJ, Goldfless SJ et al (2014) Efficient CRISPR-Cas9-mediated genome editing in *Plasmodium falciparum*. Nat Methods 11:915–918. https://doi.org/10.1038/nmeth.3063
29. Kupczok A, Bollback JP (2014) Motif depletion in bacteriophages infecting hosts with CRISPR systems. BMC Genomics 15:663. https://doi.org/10.1186/1471-2164-15-663
30. Harrington LB, Ma E, Chen JS et al (2020) A scoutRNA is required for some type V CRISPR-Cas systems. Mol Cell 79:416–424.e5. https://doi.org/10.1016/j.molcel.2020.06.022
31. Yan WX, Hunnewell P, Alfonse LE et al (2019) Functionally diverse type V CRISPR-Cas systems. Science 363:88–91. https://doi.org/10.1126/science.aav7271
32. Newsom S, Parameshwaran HP, Martin L, Rajan R (2021) The CRISPR-Cas mechanism for adaptive immunity and alternate bacterial functions fuels diverse biotechnologies. Front Cell Infect Microbiol 10:619763. https://doi.org/10.3389/fcimb.2020.619763
33. Senthilnathan R, Ilangovan I, Kunale M et al (2023) An update on CRISPR-Cas12 as a versatile tool in genome editing. Mol Biol Rep 50:2865–2881. https://doi.org/10.1007/s11033-023-08239-1
34. May NH, Schmid A, Meiser E, Beneke T (2024) Co-expression of the AsCas12a ultra variant, a T7 RNA polymerase and a cytosine base editor greatly increases transfection and editing rates in *Leishmania* species. bioRxiv. https://doi.org/10.1101/2024.02.28.582587
35. Li L, Li S, Wang J (2018) CRISPR-Cas12b-assisted nucleic acid detection platform. bioRxiv. https://doi.org/10.1101/362889
36. Parikesit AA, Hermantara R, Kevin G, Sidhartha E (2023) Designing hybrid CRISPR-Cas12 and LAMP detection systems for treatment-resistant *Plasmodium falciparum* with in silico method. Narra J 3:e301. https://doi.org/10.52225/narra.v3i3.301
37. O'Connell MR (2019) Molecular mechanisms of RNA targeting by Cas13-containing type VI CRISPR–Cas systems. J Mol Biol 431:66–87. https://doi.org/10.1016/j.jmb.2018.06.029
38. Gootenberg JS, Abudayyeh OO, Lee JW et al (2017) Nucleic acid detection with CRISPR-Cas13a/C2c2. Science 356:438–442. https://doi.org/10.1126/science.aam9321
39. Myhrvold C, Freije CA, Gootenberg JS et al (2018) Field-deployable viral diagnostics using CRISPR-Cas13. Science 360:444–448. https://doi.org/10.1126/science.aas8836
40. Granados-Riveron JT, Aquino-Jarquin G (2018) CRISPR–Cas13 precision transcriptome engineering in cancer. Cancer Res 78:4107–4113. https://doi.org/10.1158/0008-5472.CAN-18-0785
41. Yu J, Shin J, Yu J et al (2024) Programmable RNA base editing with photoactivatable CRISPR-Cas13. Nat Commun 15:673. https://doi.org/10.1038/s41467-024-44867-2
42. Zakharova A, Albanaz ATS, Opperdoes FR et al (2022) *Leishmania guyanensis* M4147 as a new LRV1-bearing model parasite: phosphatidate phosphatase 2-like protein controls cell cycle progression and intracellular lipid content. PLoS Negl Trop Dis 16:e0010510. https://doi.org/10.1371/journal.pntd.0010510
43. Wu X, Kriz AJ, Sharp PA (2014) Target specificity of the CRISPR-Cas9 system. Quan Biol 2(2):59–70. https://doi.org/10.1007/s40484-014-0030-x
44. Kirkman LA, Lawrence EA, Deitsch KW (2014) Malaria parasites utilize both homologous recombination and alternative end joining pathways to maintain genome integrity. Nucleic Acids Res 42:370–379. https://doi.org/10.1093/nar/gkt881
45. Passos-Silva DG, Rajão MA, Nascimento De Aguiar PH et al (2010) Overview of DNA repair in *Trypanosoma cruzi*, *Trypanosoma brucei*, and *Leishmania major*. J Nucleic Acids 2010:1–14. https://doi.org/10.4061/2010/840768
46. Glover L, Jun J, Horn D (2011) Microhomology-mediated deletion and gene conversion in African trypanosomes. Nucleic Acids Res 39:1372–1380. https://doi.org/10.1093/nar/gkq981
47. Costa FC, Francisco AF, Jayawardhana S et al (2018) Expanding the toolbox for *Trypanosoma cruzi:* a parasite line incorporating a bioluminescence-fluorescence dual reporter and streamlined CRISPR/Cas9 functionality for rapid *in vivo* localisation and phenotyping. PLoS Negl Trop Dis 12(4):e0006388. https://doi.org/10.1371/journal.pntd.0006388
48. Lander N, Li ZH, Niyogi S, Docampo R (2015) CRISPR/Cas9-induced disruption of paraflagellar rod protein 1 and 2 genes in *Trypanosoma cruzi* reveals their role in flagellar attachment. mBio 6:e01012–e01015. https://doi.org/10.1128/mBio.01012-15
49. Slaymaker IM, Gao L, Zetsche B et al (2016) Rationally engineered Cas9 nucleases with improved specificity. Science 351:84–88. https://doi.org/10.1126/science.aad5227
50. Kleinstiver BP, Pattanayak V, Prew MS et al (2016) High-fidelity CRISPR-Cas9 nucleases with no detectable genome-wide off-target effects. Nature 529:490–495. https://doi.org/10.1038/nature16526
51. Soares-Medeiros LC, South L, Peng D et al (2017) Rapid, selection-free, high-efficiency genome editing in protozoan parasites using CRISPR-cas9 ribo-

nucleoproteins. mBio 8(6):e01788-17. https://doi.org/10.1128/mBio.01788-17
52. Asencio C, Hervé P, Morand P et al (2024) *Streptococcus pyogenes* Cas9 ribonucleoprotein delivery for efficient, rapid and marker-free gene editing in *Trypanosoma* and *Leishmania*. Mol Microbiol 121:1079. https://doi.org/10.1111/mmi.15256
53. Hsu PD, Scott DA, Weinstein JA et al (2013) DNA targeting specificity of RNA-guided Cas9 nucleases. Nat Biotechnol 31:827–832. https://doi.org/10.1038/nbt.2647
54. Grenier F, Lucier J-F, Rodrigue S (2015) Selection and validation of spacer sequences for CRISPR-Cas9 genome editing and transcription regulation in bacteria. In: Leblanc BP, Rodrigue S (eds) DNA-protein interactions. Springer, New York, pp 233–244
55. Peng D, Tarleton R (2015) EuPaGDT: a web tool tailored to design CRISPR guide RNAs for eukaryotic pathogens. Microb Genom 1:e000033. https://doi.org/10.1099/mgen.0.000033
56. Glover L, McCulloch R, Horn D (2008) Sequence homology and microhomology dominate chromosomal double-strand break repair in African trypanosomes. Nucleic Acids Res 36:2608–2618. https://doi.org/10.1093/nar/gkn104
57. Lander N, Chiurillo MA (2019) State-of-the-art CRISPR/Cas9 technology for genome editing in trypanosomatids. J Eukaryot Microbiol 66:981–991. https://doi.org/10.1111/jeu.12747
58. Lander N, Chiurillo MA, Docampo R (2016) Genome editing by CRISPR /Cas9: a game change in the genetic manipulation of protists. J Eukaryot Microbiol 63:679–690. https://doi.org/10.1111/jeu.12338
59. Shen B, Brown KM, Lee TD, Sibley LD (2014) Efficient gene disruption in diverse strains of *Toxoplasma gondii* using CRISPR/CAS9. mBio 5:e01114. https://doi.org/10.1128/mBio.01114-14
60. Sidik SM, Hackett CG, Tran F et al (2014) Efficient genome engineering of *Toxoplasma gondii* using CRISPR/Cas9. PLoS One 9:e100450. https://doi.org/10.1371/journal.pone.0100450
61. Sidik SM, Huet D, Ganesan SM et al (2016) A genome-wide CRISPR screen in *Toxoplasma* identifies essential apicomplexan genes. Cell 166:1423–1435.e12. https://doi.org/10.1016/j.cell.2016.08.019
62. Lander N, Chiurillo MA, Storey M et al (2016) CRISPR/Cas9-mediated endogenous C-terminal tagging of *Trypanosoma cruzi* genes reveals the acidocalcisome localization of the inositol 1,4,5-trisphosphate receptor. J Biol Chem 291(49):25505–25515. https://doi.org/10.1074/jbc.M116.749655
63. Espada CR, Quilles JC, Albuquerque-Wendt A et al (2021) Effective genome editing in *Leishmania (Viannia) braziliensis* stably expressing Cas9 and T7 RNA polymerase. Front Cell Infect Microbiol 11:772311. https://doi.org/10.3389/fcimb.2021.772311
64. Queffeulou M, Leprohon P, Fernandez-Prada C et al (2024) CRISPR-Cas9 high-throughput screening to study drug resistance in *Leishmania infantum*. mBio 15(7):e0047724. https://doi.org/10.1128/mbio.00477-24
65. Martel D, Beneke T, Gluenz E et al (2017) Characterisation of casein kinase 1.1 in *Leishmania donovani* using the CRISPR Cas9 toolkit. Biomed Res Int 2017:4635605. https://doi.org/10.1155/2017/4635605
66. Yang L, Chen J (2020) A tale of two moieties: rapidly evolving CRISPR/Cas-based genome editing. Trends Biochem Sci 45:874–888. https://doi.org/10.1016/j.tibs.2020.06.003
67. Xiang J, Xu W, Wu J et al (2023) Nucleoside deaminases: the key players in base editing toolkit. Biophys Rep 9:325. https://doi.org/10.52601/bpr.2023.230029
68. Zhang X, Chen L, Zhu B et al (2020) Increasing the efficiency and targeting range of cytidine base editors through fusion of a single-stranded DNA-binding protein domain. Nat Cell Biol 22:740–750. https://doi.org/10.1038/s41556-020-0518-8
69. Engstler M, Beneke T (2023) Gene editing and scalable functional genomic screening in *Leishmania* species using the CRISPR/Cas9 cytosine base editor toolbox LeishBASEedit. eLife 12:e85605. https://doi.org/10.7554/eLife.85605
70. Baker N, Catta-Preta CMC, Neish R et al (2021) Systematic functional analysis of *Leishmania* protein kinases identifies regulators of differentiation or survival. Nat Commun 12:1244. https://doi.org/10.1038/s41467-021-21360-8
71. Beneke T, Gluenz E (2020) Bar-seq strategies for the LeishGEdit toolbox. Mol Biochem Parasitol 239:111295. https://doi.org/10.1016/j.molbiopara.2020.111295
72. Wu S-Z, Wei H-X, Jiang D et al (2020) Genome-wide CRISPR screen identifies host factors required by *Toxoplasma gondii* infection. Front Cell Infect Microbiol 9:460. https://doi.org/10.3389/fcimb.2019.00460
73. Zhang C, Yang S, Quansah E et al (2024) The dCas9-based genome editing in *Plasmodium yoelii*. mSphere 9:e0009524. https://doi.org/10.1128/msphere.00095-24
74. Nuñez JK, Chen J, Pommier GC et al (2021) Genome-wide programmable transcriptional memory by CRISPR-based epigenome editing. Cell 184:2503–2519.e17. https://doi.org/10.1016/j.cell.2021.03.025
75. Toenhake CG, Fraschka SAK, Vijayabaskar MS et al (2018) Chromatin accessibility-based characterization of the gene regulatory network underlying *Plasmodium falciparum* blood-stage development. Cell Host Microbe 23:557–569.e9. https://doi.org/10.1016/j.chom.2018.03.007
76. Chen JS, Ma E, Harrington LB et al (2018) CRISPR-Cas12a target binding unleashes indiscriminate single-stranded DNase activity. Science 360:436–439. https://doi.org/10.1126/science.aar6245
77. Brogan DJ, Chaverra-Rodriguez D, Lin CP et al (2021) Development of a rapid and sensitive CasRx-based diagnostic assay for SARS-CoV-2. ACS

Sens 6(11):3957–3966. https://doi.org/10.1021/acssensors.1c01088
78. Lee RA, De Puig H, Nguyen PQ et al (2020) Ultrasensitive CRISPR-based diagnostic for field-applicable detection of *Plasmodium* species in symptomatic and asymptomatic malaria. Proc Natl Acad Sci USA 117:25722–25731. https://doi.org/10.1073/pnas.2010196117
79. Sima N, Dujeancourt-Henry A, Perlaza BL et al (2022) SHERLOCK4HAT: a CRISPR-based tool kit for diagnosis of human African trypanosomiasis. EBioMedicine 85:104308. https://doi.org/10.1016/j.ebiom.2022.104308
80. Yang Z, Wang J, Qi Y et al (2024) A novel detection method based on MIRA-CRISPR/Cas13a-LFD targeting the repeated DNA sequence of *Trichomonas vaginalis*. Parasit Vectors 17:14. https://doi.org/10.1186/s13071-023-06106-3
81. Li L, Li S, Wu N et al (2019) HOLMESv2: a CRISPR-Cas12b-assisted platform for nucleic acid detection and DNA methylation quantitation. ACS Synth Biol 8(10):2228–2237. https://doi.org/10.1021/acssynbio.9b00209
82. Li X, Dang Z, Tang W et al (2023) Detection of parasites in the field: the ever-innovating CRISPR-Cas12a. Biosensors 14(3):145. https://doi.org/10.3390/bios14030145
83. Wu M, Wu H, Chen X et al (2024) RPA-CRISPR/Cas9-based method for the detection of *Toxoplasma gondii*: a proof of concept. Vet Parasitol 327:110115. https://doi.org/10.1016/j.vetpar.2024.110115

Part II

Intestinal Parasites

Blastocystomics

Daisy Shaw and Anastasios D. Tsaousis

Abstract

Blastocystis sp. is a eukaryotic gut microorganism with a questionable pathogenicity, which often makes it the subject of debate whether it should be considered a parasite or a commensal. This stramenopile is thought to be carried in the gastrointestinal tract of several billion people worldwide and is associated with symptoms in only a few of these cases. At present, there are at least 46 subtypes (ST) colonizing mammalian hosts within the *Blastocystis* genus, with others that have been found to colonize a variety of other vertebrates and invertebrates. It is via this subtyping system that mammalian isolates are named, rather than the standard binomial nomenclature of a genus and species name. Most of the genomes of these 46 subtypes are yet to be characterized, with only ST1–ST4 and ST6–ST9 were annotated, or draft genomes assembled. Metabolomics, on the other hand, can be used to identify the metabolites produced by *Blastocystis*, which can be applied as a tool to understand the processes that this microbe is undertaking under certain conditions. Another area of interest is the associations of *Blastocystis* with other inhabitants of the gut microbiome, and how it could be affecting the composition of these populations. Generally, it is found that *Blastocystis* occurs alongside a high diversity of other gut microbiota, although some studies disagree, saying it could be correlated to dysbiotic conditions such as those associated with irritable bowel syndrome (IBS). In this chapter, we evaluate the current status quo on the different omics techniques currently used to characterize *Blastocystis* biology and suggest further avenues to be explored in the future. These will help us better understand its elusive role in the ecosystem that is the gut of humans and other animals.

Keywords

Blastocystis · Gut microbiome · Omics techniques · *Blastocystis* genomics · *Blastocystis* transcriptomics · *Blastocystis* metabolomics · Microbiomics

1 Introduction

Blastocystis sp. is a controversial parasitic protist, first observed and published by A. Alexeieff [1] and Émile Brumpt [2], who named it

D. Shaw · A. D. Tsaousis (✉)
Laboratory of Molecular and Evolutionary Parasitology, RAPID Group, School of Biosciences, University of Kent, Canterbury, Kent, UK
e-mail: a.tsaousis@kent.ac.uk

J. D. Ramírez González (ed.), *Recent Advances in Parasitomics*,
https://doi.org/10.1007/978-3-031-70591-5_5

'*Blastocystis enterocola*' and '*Blastocystis hominis*', respectively. They identified the microorganism as an intestinal yeast, which it was recognized as until later being declared a protozoan [3], after failing to thrive on fungal growth media [4]. *Blastocystis* sp. was determined as a stramenopile by Silberman in 1996 after successfully sequencing the small-subunit ribosomal RNA (SSU-rRNA) and generating a relevant phylogeny [5]. This came as a surprise, since it does not display the typical characteristics of other organisms within this clade. Still today, it remains the only known stramenopile to infect humans [5, 6]. Its closest relative is *Proteromonas* sp., which has been found to colonize the intestines of cold-blooded reptiles [7]. The current taxonomic system used for *Blastocystis* sp. colonizing humans and other animals is slightly different from the standard nomenclature used for naming other microorganisms. In 2007, Stensvold and colleagues proposed a standard naming system where all mammalian and avian isolates of the protist are denoted as *Blastocystis* sp., followed by a subtype number to identify the different variants. At the time, only subtypes 1–9 had been identified [8], but at present, more than 44 subtypes have been recognized [9]. *Blastocystis* sp. (shown in Fig. 1) is highly polymorphic, with at least four distinct forms being described in the literature, namely cyst, vacuolar, granular and amoeboid forms, with multivacuolar and avacuolar also being suggested [10]. The cyst form, discovered more recently than its morphological counterparts [11], is understood to be the transmissible form of the protist [12].

One key characteristic of *Blastocystis* subcellular structure that commonly raises questions is its unconventional mitochondrion-related organelles (MROs) [13]. While they appear as typical mitochondria with cristae, they do lack some of the pathways and electron transport chain enzymes that are common in aerobic mitochondria. An incomplete array of enzymes of the TCA cycle has been discovered in *Blastocystis* [13], and although hydrogenase has been detected, production of hydrogen has not been demonstrated [13, 14].

Although known to be one of the most common human intestinal protists, estimated to infect between 1 and 2 billion people worldwide [15], there is still much debate about the role of *Blastocystis* sp. in the gastrointestinal tract, specifically about whether it should be considered a commensal or a pathogen [16]. Clinical symptoms that could be associated with *Blastocystis* sp. are those generally evident with other self-limiting gastrointestinal infections, such as diarrhoea, abdominal discomfort and nausea [17]. It, however, remains unclear if the presence of *Blastocystis* sp. in these cases is the primary cause of these clinical presentations, or has merely only been associated with disease as it is sometimes the only potentially pathogenic agent positively detected in stools of symptomatic patients [18]. Many studies suggest that the correlation between *Blastocystis* positive patients and symptomatic status is not significant, suggesting that the numbers of symptomatic carriers are no different from asymptomatic carriers. So, what is the true role of this controversial parasite in the human gastrointestinal tract?

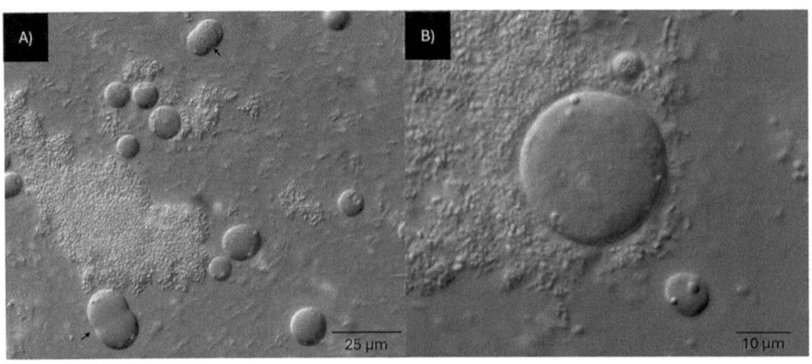

Fig. 1 (**a**) Overview of *Blastocystis* in xenic culture, with dividing cells indicated by arrowheads. (**b**) Single *Blastocystis* cell. Microscopy pictures provided by Dr. Vasana Jinatham (Mae Fah Lunag University, Thailand)

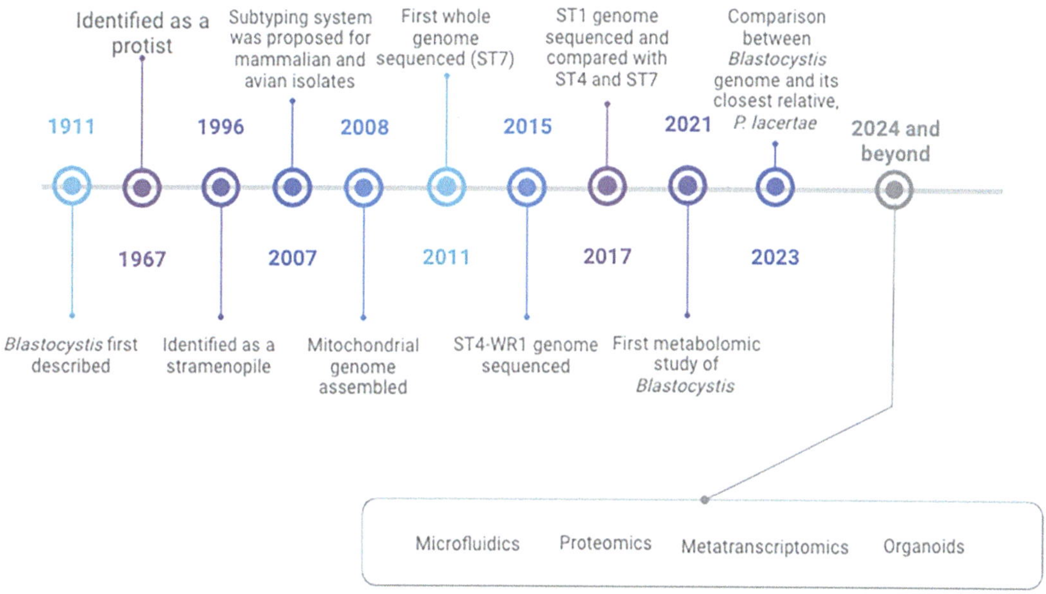

Fig. 2 Timeline of *Blastocystis* discoveries and directions for the future

The aim of this chapter is to provide the status quo on the various technologies that have currently been used (Fig. 2) to understand the biology of *Blastocystis*: from genomics and transcriptomics to metagenomics and metabolomics.

2 Genomics/Transcriptomics

Genomics, the study of a (micro)organism's complete set of DNA, allows insight into how the organisms have evolved and diversified and can be a useful tool to predict future outbreaks of pathogens [19]. Early genomic investigations of *Blastocystis* (Fig. 2) focused on deciphering the function of the MROs that the microorganism harbours, which are thought to function as they do today largely due to lateral gene transfer (LGT) from prokaryotes and other eukaryotes [20]. In 2008, three studies looked into the mitochondrial genome of *Blastocystis*. One of these was by Stechmann et al., who generated 12,767 expressed sequence tags (ESTs) and identified 115 clusters encoding potential mitochondrial proteins [13]. In the same year, Wawrzyniak et al. used shotgun sequencing data to describe the 29,270 bp circular DNA of *Blastocystis*, which has many genes in common with mitochondrial genes, but is lacking those coding for complexes III, IV and V of the electron transport chain [21]. To add to the confusion of the identity of these organelles, both of these studies also found genes encoding the hydrogenosomal enzymes iron hydrogenase and pyruvate ferredoxin oxidoreductase [13, 21], therefore it remained unclear if this organelle is a mitochondrion, mitosome or hydrogenosome [14]. Comparison of the ST1 (NandII) and ST4 (DMP/02-328) mitochondrial genomes showed a divergence between subtypes that exceeds 50% in some genes, which is more than typically seen between the mitochondrial genomes of different species of the same genus [14]. When human isolates of ST4 were compared with ST3 via multilocus sequence typing (MLST), this subtype showed far less intra-subtype variation, with this homogeneity suggesting a far more recent colonization of humans by ST4 than by ST3 [22].

The first whole *Blastocystis* genome reported was that of ST7, by Denoeud et al., and with a size of 18.8 Mbp, it was the smallest stramenopile genome to have been sequenced at the time [6], with evidence to suggest LGT from bacteria, followed by secondary LGT between eukaryotes [6]. LGT was further confirmed by Eme et al. in 2017, whereby through genomic and transcriptomic analysis, it was shown that 2.5% of the protein-coding genes in ST1 have been recently laterally acquired, with this percentage varying between different subtypes. Most of these genes are thought to be of prokaryotic origin, particularly from Proteobacteria, Bacteroidetes and Firmicutes. The majority of these genes are thought to have contributed to the ability of *Blastocystis* to colonize the gastrointestinal tract, with identification of five LGT genes involved in the oxidative stress response, seven in immune evasion and virulence, 17 in carbohydrate metabolism and others involved in anaerobic metabolism and nitrogen metabolism [23]. Further analysis of the ST7 genome several years later showed that roughly 15% of *Blastocystis* nuclear genes exhibit termination codons created via polyadenylation, and these genes can be linked to those encoding enzymes, although the significance of this is not well understood [24].

In 2015, Wawrzyniak et al. published the draft genome sequence of the WR1 isolate of ST4 using the Illumina-HiSeq 2000 system [25]. ST4-WR1 has an even smaller genome (12.91 Mbp) than ST7, but interestingly both subtypes have the same number (45) of MRO genes [13, 25]. After further analysis of the ST4 genome, genes of 20 secreted proteases were identified, which may be integral to the successful colonization of this subtype in the host [25].

In 2017, the genome of *Blastocystis* ST1 was first published by Gentekaki et al. in a comparison with ST4 and ST7, showing an assembled genome size of 16.5 Mbp across 580 scaffolds. It was stated in this study that differences seen between the genomes of these three first sequenced subtypes of *Blastocystis* could be impacted by the different methods and procedures used for the gene assemblies, particularly as the ST7 and ST4-WR1 genomes were annotated before the discovery of polyadenylated termination codons [26]. In this chapter, it was shown via comparative genomics that the genetic diversity between compared subtypes was far higher than the genetic diversity between separate species of other parasites. For example, 20% of ST7 genes are not found in ST1, whilst there is only a 1% difference in protein-coding genes between *Leishmania major* and *Leishmania infantum*, which are recognized as two separate species. More recently, draft genomes of ST2, ST3, ST6, ST8 and ST9 were assembled [27] and these are included in Table 1.

More recently, in 2023, the genome and transcriptomes of *Proteromonas lacertae* were published. *P. lacerate* is the closest canonical relative of *Blastocystis* and has the typical morphological features of the stramenopila that *Blastocystis* lacks. Comparison between the two genomes

Table 1 Summary of current genome information available for different subtypes of *Blastocystis* and its closest canonical relative *Proteromonas lacertae* for comparison

ST	Isolate	Genome size (Mbp)	G+C content (%)	Number of scaffolds	Scaffold N50 (kb)	Scaffold L50	Number of contigs	Genome coverage	GenBank accession no.
ST1	NandII	16.5	54.6	580	79.4	68	1092	80×	LXWW00000000
ST2	Flemming	12.7	54.0	969	20.1	180	1014	100×	JZRJ00000000
ST3	ZGR	11.7	52.0	917	20.8	169	971	100×	JZRK00000000
ST4	WR1	12.91	39.6	1301	29.9	128	1355	300×	JPUL02000000
ST6	SSI:754	15.4	43	879	33.6	119	899	100×	JZRM00000000
ST7	B	18.8	45.2	54	900.6	7	155	12.4×	CABX01000000
ST8	Dmp/08-128	12.2	39.5	947	22.9	160	967	100×	JZRN00000000
ST9	F5323	11.7	43.0	871	23.3	154	897	100×	JZRO00000000
	Proteromonas lacertae	52.25	27.1	1449	92,586	125	1449	54.3	PRJNA386230

ST, subtype; *Mbp*, mega base pairs; *kb*, kilobases

showed that *Blastocystis* has a much lower gene number than other stramenopiles [28]. This comparative study provided a hypothesis for the evolution and adaptations of *Blastocystis*. It is thought that the reduced genome size of *Blastocystis*, in comparison to other stramenopiles, is due to a progressive genome-wide streamlining, with particular loss of organellar genes, such as those for flagella and peroxisomes, organelles that *Blastocystis* lacks [28].

3 Proteomics

The field of proteomics is one that is relatively untouched in the protists field. To date, there are no investigations into the proteome of *Blastocystis* sp., so this remains an area of research open for future investigations. Organisms that have been characterized include Trypanosomes [29], *Giardia* [30], *Entamoeba* [31] and apicomplexan parasites such as *Plasmodium* [32] and *Toxoplasma* [33]. These proteomic investigations have either focused on characterizing specific organelles (e.g. mitochondria) or investigated the pathogenic factors of these microbial parasites.

In 2020, Barylyuk et al. published a new methodology to investigate the proteome of the tachyzoite form of the apicomplexa *Toxoplasma gondii* using hyperplexed localization of organelle proteins by isotope tagging (hyperLOPIT). This is a method that can be utilized to capture the subcellular location of proteins after cellular and organellar separation via varying methods of density-gradient fractionation [34]. The proteome of another apicomplexan parasite, *Cryptosporidium parvum*, has been investigated using the WoLF PSORT II program to predict subcellular locations of proteins identified via mass spectrometry (MS; [35, 36]) and analysis of protein–protein interactions were analysed using the STRING database [37]. A similar method could be undertaken to produce a draft proteome of *Blastocystis*, although this is yet to be performed. These investigations will allow a deeper understanding of the proteins present in the microorganism at specific times or under specific conditions [38], which could uncover more information about how *Blastocystis* is colonizing the host, and mechanisms that could be attributed to pathogenicity, commensalism or symbiosis.

4 Microbiomics

The association of *Blastocystis* sp. with other microbial inhabitants of the gastrointestinal tract is an area of great interest, and studying such trends can be a useful tool in elucidating the true role of *Blastocystis* sp. in the human host. This branch of 'omics', which involves identification and analysis of the general population of an ecosystem, can be referred to as metagenomics or microbiomics [39]. Certain gastrointestinal disorders, such as irritable bowel syndrome (IBS) and types of inflammatory bowel disease (IBD), such as Crohn's disease and ulcerative colitis, are often correlated with intestinal dysbiosis [40, 41], or an imbalance between canonical beneficial and harmful members of the microbiota. Several studies report a positive association between *Blastocystis* colonization and IBS patients [42, 43], whilst others show *Blastocystis* colonization is mostly associated with gut eubiosis [44]. This evidence is conflicting, further giving reason for the debate around its clinical status.

Initial studies focusing on the association between *Blastocystis* and gut pathology in humans did show some contradicting results. Some of the earlier studies found that symptoms cleared after eradication of *Blastocystis*; however, there were also cases where *Blastocystis* infection cleared without intervention, but symptoms still persisted [45]. Other studies suggested that the abnormal gut environment of those with intestinal disorders may provide conditions that aid *Blastocystis* proliferation, but it was equally possible that it was the cause of reported symptoms [42].

Many of the more recent studies into the microbiota diversity of individuals with and without *Blastocystis* colonization show similar findings, in that *Blastocystis* presence is often associated with high bacterial diversity and eubiosis of the gut [44]. Bacteria of the families Ruminococcaceae, Prevotellaceae [46],

Muribaculaceae, Rikenellaceae and Acidaminococcaceae [47] are commonly found to colonize guts of those who are positive for *Blastocystis*. On the other hand, the families Enterobacteriaceae [46], Enterococcae, Streptoccaceae and Lactobacillaceae [44] are found in reduced quantities in those colonized with *Blastocystis*. Despite many showing the converse, there are still some studies that suggest *Blastocystis* is an indicator of dysbiosis, with one study finding carriage was associated with a decrease in the anti-inflammatory bacteria *Faecalibacterium prausnitzii* [48]. In contrast, other publications report seeing an increase in *F. prausnitzii* with *Blastocystis* carriage [46, 49]. It remains unclear whether the associations just mentioned are due to *Blastocystis* creating a favourable/unfavourable environment for the microorganisms it is positively/negatively associated with and essentially being an engineer of its own ecosystem, or whether it is the converse; particular compositions of the gut microbiota provide optimal/inadequate growth conditions for this protist.

A recent investigation by Jeffery et al. into the difference in *Blastocystis* prevalence between populations treated and not treated with commercially available antibiotics found that *Blastocystis* was significantly less prevalent in the stool samples of those who had been treated with antibiotics within the last month than those who had not [50]. The antibiotics used were those belonging to groups that specifically target prokaryotes, with typical antibacterial modes of action such as inhibition of the synthesis of cell walls, nucleic acids and proteins, making it unlikely that antibiotic treatment is directly affecting any enteric eukaryotic microbes, which includes *Blastocystis*. This suggests that reductions in *Blastocystis* prevalence in these cases may be caused by secondary extinction, a term usually used within the biodiversity field to describe the decline seen in certain populations when other populations that they rely on, or co-exist with in their microenvironment, are made extinct [51]. This suggests a dependency on certain bacterial species for the colonization and persistence of *Blastocystis* in the intestinal environment [50]. This study also revealed that co-colonization with different *Blastocystis* subtypes was found several times in the non-treated population, whilst only once in those treated with the antibiotics [50], which could be a result of unfavourable conditions causing *Blastocystis* subtypes to compete with one another.

Linked to the findings that *Blastocystis* growth and occurrence is influenced by the composition of the surrounding microbial population, an investigation by Rajamanikam et al., using ST3, suggests that the surrounding environment could be influencing the pathogenic status of *Blastocystis* [47]. They showed, using in vitro growth profiling, specific protease activity and assessing the ability to proliferate cancer cells, that inoculating ST3 from a symptomatic patient with the bacterial microbiota of an asymptomatic patient leads to a reduction in activity associated with pathogenic infection. It is possible that protein expression is altered, as isolated antigens from asymptomatic individuals inoculated with the bacterial microbiota of symptomatic individuals produced greater colonic cancer cell proliferation than antigens of *Blastocystis* from asymptomatic individuals cultured in their original bacterial microbiota.

The majority of investigations into *Blastocystis* are focused on the human perspective; however, subtype specific in vivo studies have shown that *Blastocystis* ST4 is able to alter the bacterial composition in murine subjects. It is able to do so in healthy subjects, but also in those with DSS-induced colitis, as colonization showed a reduction in the gastroenteritis causing *Bacteroides* and *Escherichia-Shigella* [52], which suggests the potential to reduce colitis-associated symptoms. Whilst this study suggests ST4 could have possible benefits to its host, sequencing data from humans colonized with ST7 suggest that this subtype, which has previously been noted to occur often with those presenting gastrointestinal symptoms [53], is associated with a decrease in bacterial alpha diversity of the gut microbiota [54]. In 2019, Yason et al. published an in vitro study investigating the effect of two different isolates of ST7 (ST7-B and ST7-H) on representative gut microbial bacteria. They showed that *Blastocystis* had a positive effect on the growth of some of the gut bacteria, but the opposite was true for *Bifidobacterium longum* and

Lactobacillus brevis, which are both generally considered as healthy enteric inhabitants [55, 56]. Further investigations suggested that *Blastocystis* affects *B. longum* growth via oxidative stress [57]. The above findings suggest that we could be looking at the potential for *Blastocystis* pathogenicity status being subtype specific.

The main caveat to the aforementioned ST4 murine study is that this was an investigation into *Blastocystis* in a controlled laboratory environment, infecting the mice with a human-derived subtype of the protist at concentrations that may not occur naturally, and therefore may not be fully representative of the gastrointestinal effects it has in a naturally occurring in vivo setting. A study by Betts et al., published in 2021, investigated in a wild population of European water voles (*Arvicola amphibius*), the microbiota associated with *Blastocystis* carriers and non-carriers. They found that, unlike in humans, in the water-vole population, *Blastocystis* colonization had no associations with specific bacterial compositions of the gut [58]. The next year, Deng, Chen et al. published an investigation into the changes in composition of the gut microbiome in the presence of *Blastocystis* in captive forest musk deer (*Moschus berezovskii*). This was the first study of *Blastocystis* in this species and a mixture of ST5 and ST10 was found to be colonizing these artiodactyls. ST5 colonization, when compared to non-carrier controls, was seen to have no significant difference in bacterial composition in the gut. ST10, however, was associated with higher beta diversity and higher bacterial richness than the non-carrier controls. Specifically, the phyla bacteroidota were enriched in this group [59], which has also been reported in humans [47, 49] (Table 2).

5 Metabolomics

Metabolomics is the study of the low-molecular weight molecules found intra- and extracellularly as substrates and by-products of biochemical processes within microorganisms [62]. The current uncertainty around *Blastocystis* pathogenicity makes the study of its metabolome an important tool in understanding its role in the gastrointestinal tract, not only in the effects it has on host cells but also on the neighbouring gut flora. The best-established methods for detecting and identifying metabolites are mass spectrometry (MS) and nuclear magnetic resonance (NMR) [63]. MS has a high sensitivity, making it a valuable technique for characterization of small molecules; however, its long and laborious sample preparation process can lead to the loss of sample material and compromises the integrity of the metabolites, making reproducibility low [64]. Whilst MS detects at the molecular level, NMR detects at the atomic level [63], which provides a lower sensitivity, therefore requiring a larger sample size, however, is far more reproducible and acts as a better quantitative tool [64].

Very few studies have been carried out on the faecal metabolome of *Blastocystis* in humans, including studies comparing it to the faecal metabolome of those not carrying *Blastocystis*. In 2021, Betts et al. published the first in vivo study with a focus on *Blastocystis* and the gut metabolome. They used a previously established one-dimensional proton NMR technique as a metabolite detection method [64], as has been used for exploring other host–parasite interactions in recent years [65]. On the amino acid level, decreases in alanine, glycine, histidine, isoleucine, methionine, threonine, tryptophan and valine were observed in *Blastocystis* colonized samples [65]. An increase of these particular amino acids is generally thought to be associated with inflammation [66], suggesting *Blastocystis* colonization has a possible anti-inflammatory effect. Of the metabolites seen to decrease, histidine is of particular interest. Both histidine and its metabolic product histamine are often found in higher abundance in patients with gastrointestinal conditions such as IBS, and many of these patients report a worsening of symptoms after ingesting histamine-releasing food items [67], further evidencing its probable role in the symptomatic gut. The decrease in levels seen with *Blastocystis* positive individuals could be due to direct anti-inflammatory abilities of the protist, but equally could be due to a positive association

Table 2 Bacterial taxa positively associated with colonization of the gastrointestinal tract, identified via analysis from human stool samples

Positively associated with *Blastocystis* colonization				
Phyla	Classes	Orders	Families	Genera
Firmicutes [47, 49, 60]	Clostridia [46, 47, 49, 60]	Clostridiales [46, 49, 60]	Rumincoccaceae [46, 47, 49, 60]	Acetanaerobacterium [46]
Bacteroidota [47, 49, 59]	Mollicutes [46]	Bacteroidales [47, 49]	Prevotellaceae [46, 47, 49, 61]	Acetivibrio [46]
Actinobacteria [47]	Bacteroidia [47, 49]	Aeromonadales [49, 60]	Lachnospiraceae [49, 60, 61]	Coprococcus [46]
Elusimicrobiota [60]	Coriobacteriia [47]	Oscillospirales [47]	Clostridiaceae [49, 60, 61]	Hespellia [46]
Lentisphaerota [60]	Negativicutes [47]	Christensenellales [47]	Succinivibrionaceae [49]	Oscillibacter [46, 49]
Euryarchaeota [60]	Bacilli [47]	Coriobacteriales [47]	Christensellaceae [47]	Papillibacter [46]
	Elusimicrobia [60]	Veillonellales [47]	Eggerthellaceae [47]	Sporobacter [46]
	Lentisphaeria [60]	Selenomonadales [47]	Veillonellaceae [47]	Ruminococcus [46, 61]
	Metanobacteria [60]	Erysipeotrichales [47]	Butyricicoccaceae [47]	Prevotella [46, 49, 61]
	Deltaproteobacteria [60]	Victivallales [60]	Erysipelatoclostridiaceae [47]	Roseburia [46]
		Methanobacteriales [60]	Barnesiellaceae [47]	Faecalibacterium [46, 49]
		Elusimicrobiales [60]	Methanobacteriaceae [61]	Flavonifracter [49]
		Acidaminococcales [60]	Erysipelotrichaceae [61]	Clostridium [49, 60]
		Desulfovibrionales [60]	Pasteurellaceae [61]	Succinivibrio [49]
				Slackia [47]
				Megasphaera [47]
				Allisonella [47]
				Asteroleplasma [47]
				Coprobacter [47]
				Methanobrevibacter [60, 61]

of *Blastocystis* with bacterial communities that degrade histamine, or even a negative association with bacterial communities that produce histamine [65].

On the contrary, a metabolite produced by *Blastocystis* has been seen to trigger the host immune system. A study by Wojciech et al., using both in vitro and in vivo techniques, showed that the metabolite indole-3-acetaldehyde (I3AA), produced by ST7, is responsible for pro-inflammatory expansion by triggering CD4+ T cells, specifically Th17 cells [68], which are known to be involved in a number of immune-mediated diseases [69]. By affecting CD4+ T cell recognition of self-flora antigens, I3AA is also responsible for reducing the amount of anti-inflammatory regulatory T cells [68]. These findings support previous findings that ST7 is associated with diarrhoeal patients [53].

6 Discussion

In this chapter we discussed the use of various 'omics' techniques to study *Blastocystis* sp. (Fig. 3), a relatively neglected eukaryotic micro-organism. Many questions remain about *Blastocystis*, with the most predominant being

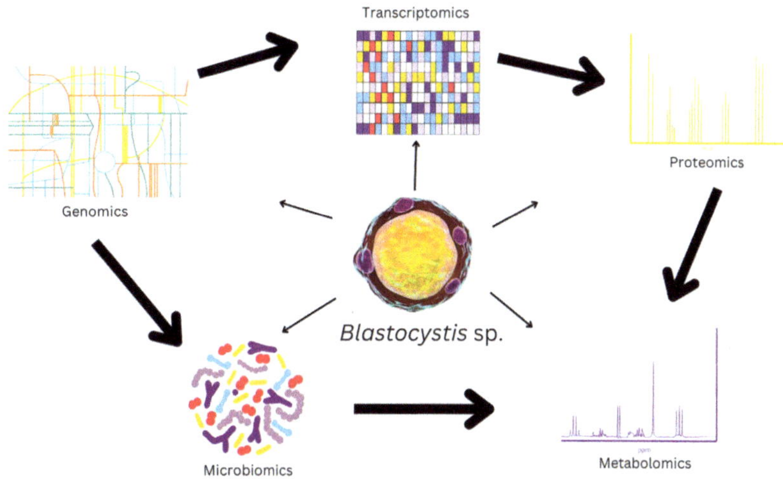

Fig. 3 A range of 'omics' methods can be used to better understand *Blastocystis*, including genomics, transcriptomics, proteomics, metabolomics and microbiomics. All of these interlink to help create a broader picture of *Blastocystis* in health and disease

about its clinical and pathogenic status, as there has been much dispute about whether colonization is correlated to symptomatology. This could be tackled using a combination of omics techniques to expand our understanding of *Blastocystis* at the subtype level.

The proteome of *Blastocystis* has had a lack of attention. We do not have a good understanding about this branch of *Blastocystis*, but we can try to adopt methodologies based on those successfully used for other protists, such as the hyperLOPIT method mentioned previously that was used to study the proteome of *Toxoplasma gondii* [34]. This could be complimented with genetic manipulation studies as well. The first successful transfection of *Blastocystis* was performed by Li et al. in 2019, where ectopic proteins were inserted into *Blastocystis* ST7 via a nanoluciferase (Nluc) reporter system [70]. The genetic manipulation of the *Blastocystis* genome opens doors for further investigations into the roles of certain genes found to be present and lacking in particular subtypes of *Blastocystis*, while studying the overall adaptations of this peculiar organism as well. Especially, further investigation into the genes acquired through LGT could also give us an idea about the other species *Blastocystis* was commonly in contact with in the environment, and whether this aligns with current metagenomic studies, or whether colonization and composition has evolved over time.

One obstacle for investigating *Blastocystis* biology is the difficulty of maintaining some of the subtypes in culture. While most of them can be maintained for a few weeks under xenic conditions (with other microorganisms) and with specialized media, currently only ST1 NUH9 [71], ST4-WR1, ST4-S1 [72], ST7H and ST7B [6] can be maintained axenically [73]. Successful axenization of further subtypes would allow more investigations into the role of *Blastocystis* in the gut environment.

To further our understanding in the field of *Blastocystis*, it is also necessary to consider novel methodologies to study its interactions with both the host and other members of the enteric microbial community. Due to the nature of the gastrointestinal tract, it is difficult to accurately recapitulate the conditions that *Blastocystis* would be found under in a human host, therefore making it difficult to have confidence that data obtained from in vitro studies translates directly to what is happening in vivo. Moving away from two-dimensional culturing to more three-dimensional approaches will undoubtedly enable more characteristics of the in vivo environment to be maintained. The cultivation of in vitro enteroids, the form of organoids made from intestinal cells, has been successfully maintained for up to 2 months [74]. Organoids have been successfully produced from adult stem cells, pluripotent stem cells, induced pluripotent stem cells and embryonic stem cells, with differentiation allowing the growth of different cell types of the gastrointestinal tract [75].

Another avenue that shows promise is the use of microfluidic technologies to develop gut-on-a-chip devices, which can be used to sustain contact between host cells with chosen microbial communities in an environment that can be manipulated to replicate certain conditions of the in vivo gut [76]. One of the main challenges of replicating an enteric environment in vitro is the differing conditions observed in each part of the gastrointestinal tract. For example, the oxygen levels can fluctuate from 5 to 0.5% from the duodenum to the colon [77]. Organ-on-a-chip devices that are able to overcome this problem could be revolutionary for the three-dimensional in vitro culturing of obligate anaerobic microorganisms such as *Blastocystis* [4].

Despite a century of study (Fig. 2), we remain ignorant of the basic biology, adaptations and synergistic/antagonistic roles of *Blastocystis* in the gut. What is such a common gut resident doing there? What are the cellular and biochemical adaptations that make *Blastocystis* so successful? A multidisciplinary approach is required to answer these fundamental questions by integrating new advances in medical, veterinary, engineering and multi-omics research (Fig. 3). This avenue of research will enable us to provide an answer to this 100-year-old question that has troubled the scientific community regarding the role of *Blastocystis* in health and disease.

Acknowledgments The authors acknowledge the COST Action CA21105-Blastocystis under One Health, which is supported by COST (European Cooperation in Science and Technology).

References

1. Alexeieff A (1911) Sur la nature des formations dites kystes de *Trichomonas intestinalis*. C R Soc Biol 71:296–298
2. Brumpt E (1912) *Blastocystis hominis* N. sp et formes voisines. Bull Exotic Pathol Soc 5:725–730
3. Zierdt CH, Rude WS, Bull BS (1967) Protozoan characteristics of *Blastocystis hominis*. Am J Clin Pathol 48:495–501. https://doi.org/10.1093/ajcp/48.5.495
4. Jeremiah S, Parija S (2013) *Blastocystis*: taxonomy, biology and virulence. Trop Parasitol 3:17. https://doi.org/10.4103/2229-5070.113894
5. Silberman JD, Sogin ML, Leipe DD, Clark CG (1996) Human parasite finds taxonomic home. Nature 380:398–398. https://doi.org/10.1038/380398a0
6. Denoeud F, Roussel M, Noel B et al (2011) Genome sequence of the stramenopile *Blastocystis*, a human anaerobic parasite. Genome Biol 12:R29. https://doi.org/10.1186/gb-2011-12-3-r29
7. Kostka M (2017) Opalinata. In: Handbook of the Protists. Springer, Cham, pp 543–565
8. Stensvold CR, Suresh GK, Tan KSW et al (2007) Terminology for *Blastocystis* subtypes—a consensus. Trends Parasitol 23:93–96. https://doi.org/10.1016/j.pt.2007.01.004
9. Santin M, Figueiredo A, Molokin A et al (2023) Division of *Blastocystis* ST10 into three new subtypes: ST42-ST44. J Eukaryot Microbiol. https://doi.org/10.1111/jeu.12998
10. Tan KSW, Singh M, Yap EH (2002) Recent advances in *Blastocystis hominis* research: hot spots in terra incognita. Int J Parasitol 32:789–804. https://doi.org/10.1016/S0020-7519(02)00005-X
11. Stenzel DJ, Boreham PFL (1991) A cyst-like stage of *Blastocystis hominis*. Int J Parasitol 21:613–615. https://doi.org/10.1016/0020-7519(91)90069-J
12. Puthia MK, Tan KS (2011) Blastocystosis. Oxford University Press
13. Stechmann A, Hamblin K, Pérez-Brocal V et al (2008) Organelles in *Blastocystis* that blur the distinction between mitochondria and hydrogenosomes. Curr Biol 18:580–585. https://doi.org/10.1016/j.cub.2008.03.037
14. Perez-Brocal V, Clark CG (2008) Analysis of two genomes from the mitochondrion-like organelle of the intestinal parasite *Blastocystis*: complete sequences, gene content, and genome organization. Mol Biol Evol 25:2475–2482. https://doi.org/10.1093/molbev/msn193
15. Scanlan PD, Stensvold CR (2013) *Blastocystis*: getting to grips with our guileful guest. Trends Parasitol 29:523–529. https://doi.org/10.1016/j.pt.2013.08.006
16. Roberts T, Stark D, Harkness J, Ellis J (2014) Update on the pathogenic potential and treatment options for *Blastocystis* sp. Gut Pathog 6:17. https://doi.org/10.1186/1757-4749-6-17
17. Tan KSW (2008) New insights on classification, identification, and clinical relevance of *Blastocystis* spp. Clin Microbiol Rev 21:639–665. https://doi.org/10.1128/CMR.00022-08
18. Stensvold CR, Lewis HC, Hammerum AM et al (2009) *Blastocystis*: unravelling potential risk factors and clinical significance of a common but neglected parasite. Epidemiol Infect 137:1655–1663. https://doi.org/10.1017/S0950268809002672
19. (2019) A genomic approach to microbiology. Nat Rev Genet 20:311–311. https://doi.org/10.1038/s41576-019-0131-5

20. Tsaousis AD, Yarlett N, Tan KSW (2019) The mitochondrion-related organelles of *Blastocystis*. In: Hydrogenosomes and mitosomes: mitochondria of anaerobic eukaryotes. Springer, pp 267–286
21. Wawrzyniak I, Roussel M, Diogon M et al (2008) Complete circular DNA in the mitochondria-like organelles of *Blastocystis hominis*. Int J Parasitol 38:1377–1382. https://doi.org/10.1016/j.ijpara.2008.06.001
22. Stensvold CR, Alfellani M, Clark CG (2012) Levels of genetic diversity vary dramatically between *Blastocystis* subtypes. Infect Genet Evol 12:263–273. https://doi.org/10.1016/j.meegid.2011.11.002
23. Eme L, Gentekaki E, Curtis B et al (2017) Lateral gene transfer in the adaptation of the anaerobic parasite *Blastocystis* to the gut. Curr Biol 27:807–820. https://doi.org/10.1016/j.cub.2017.02.003
24. Klimeš V, Gentekaki E, Roger AJ, Eliáš M (2014) A large number of nuclear genes in the human parasite *Blastocystis* require mRNA polyadenylation to create functional termination codons. Genome Biol Evol 6:1956–1961. https://doi.org/10.1093/gbe/evu146
25. Wawrzyniak I, Courtine D, Osman M et al (2015) Draft genome sequence of the intestinal parasite *Blastocystis* subtype 4-isolate WR1. Genom Data 4:22–23. https://doi.org/10.1016/j.gdata.2015.01.009
26. Gentekaki E, Curtis BA, Stairs CW et al (2017) Extreme genome diversity in the hyper-prevalent parasitic eukaryote *Blastocystis*. PLoS Biol 15:e2003769. https://doi.org/10.1371/journal.pbio.2003769
27. Andersen LO, Bonde I, Nielsen HB, Stensvold CR (2015) A retrospective metagenomics approach to studying *Blastocystis*. FEMS Microbiol Ecol 91:1–9. https://doi.org/10.1093/femsec/fiv072
28. Záhonová K, Low RS, Warren CJ et al (2023) Evolutionary analysis of cellular reduction and anaerobicity in the hyper-prevalent gut microbe *Blastocystis*. Curr Biol 33:2449–2464.e8. https://doi.org/10.1016/j.cub.2023.05.025
29. Panigrahi AK, Ogata Y, Zíková A et al (2009) A comprehensive analysis of *Trypanosoma brucei* mitochondrial proteome. Proteomics 9:434–450. https://doi.org/10.1002/pmic.200800477
30. Kim J, Bae S-S, Sung M-H et al (2009) Comparative proteomic analysis of trophozoites versus cysts of *Giardia lamblia*. Parasitol Res 104:475–479. https://doi.org/10.1007/s00436-008-1223-x
31. Biller L, Matthiesen J, Kühne V et al (2014) The cell surface proteome of *Entamoeba histolytica*. Mol Cell Proteomics 13:132–144. https://doi.org/10.1074/mcp.M113.031393
32. Johnson JR, Yates JR (2014) The *Plasmodium* proteome. In: Molecular approaches to malaria. ASM Press, Washington, DC, pp 85–94
33. Xia D, Sanderson SJ, Jones AR et al (2008) The proteome of *Toxoplasma gondii*: integration with the genome provides novel insights into gene expression and annotation. Genome Biol 9:R116. https://doi.org/10.1186/gb-2008-9-7-r116
34. Barylyuk K, Koreny L, Ke H et al (2020) A comprehensive subcellular atlas of the *toxoplasma* proteome via hyperLOPIT provides spatial context for protein functions. Cell Host Microbe 28:752–766.e9. https://doi.org/10.1016/j.chom.2020.09.011
35. Li T, Liu H, Jiang N et al (2021) Comparative proteomics reveals *Cryptosporidium parvum* manipulation of the host cell molecular expression and immune response. PLoS Negl Trop Dis 15:e0009949. https://doi.org/10.1371/journal.pntd.0009949
36. Wang L, Cui Z, Li N et al (2023) Comparative proteomics reveals *Cryptosporidium parvum* infection disrupts cellular barriers. J Proteome 287:104969. https://doi.org/10.1016/j.jprot.2023.104969
37. Szklarczyk D, Gable AL, Lyon D et al (2019) STRING v11: protein–protein association networks with increased coverage, supporting functional discovery in genome-wide experimental datasets. Nucleic Acids Res 47:D607–D613. https://doi.org/10.1093/nar/gky1131
38. Aslam B, Basit M, Nisar MA et al (2017) Proteomics: technologies and their applications. J Chromatogr Sci 55:182–196. https://doi.org/10.1093/chromsci/bmw167
39. Bokulich NA, Ziemski M, Robeson MS, Kaehler BD (2020) Measuring the microbiome: best practices for developing and benchmarking microbiomics methods. Comput Struct Biotechnol J 18:4048–4062. https://doi.org/10.1016/j.csbj.2020.11.049
40. Kaur N, Chen C-C, Luther J, Kao JY (2011) Intestinal dysbiosis in inflammatory bowel disease. Gut Microbes 2:211–216. https://doi.org/10.4161/gmic.2.4.17863
41. Menees S, Chey W (2018) The gut microbiome and irritable bowel syndrome. F1000Res 7:1029. https://doi.org/10.12688/f1000research.14592.1
42. Giacometti A, Cirioni O, Fiorentini A et al (1999) Irritable bowel syndrome in patients with *Blastocystis hominis* infection. Eur J Clin Microbiol Infect Dis 18:436–439. https://doi.org/10.1007/s100960050314
43. Rostami A, Riahi SM, Haghighi A et al (2017) The role of *Blastocystis* sp. and *Dientamoeba fragilis* in irritable bowel syndrome: a systematic review and meta-analysis. Parasitol Res 116:2361–2371. https://doi.org/10.1007/s00436-017-5535-6
44. Deng L, Wojciech L, Gascoigne NRJ et al (2021) New insights into the interactions between *Blastocystis*, the gut microbiota, and host immunity. PLoS Pathog 17:e1009253. https://doi.org/10.1371/journal.ppat.1009253
45. Doyle PW, Helgason MM, Mathias RG, Proctor EM (1990) Epidemiology and pathogenicity of *Blastocystis hominis*. J Clin Microbiol 28:116–121. https://doi.org/10.1128/jcm.28.1.116-121.1990
46. Audebert C, Even G, Cian A et al (2016) Colonization with the enteric protozoa *Blastocystis* is associated with increased diversity of human gut bacterial microbiota. Sci Rep 6:25255. https://doi.org/10.1038/srep25255

47. Rajamanikam A, Isa MNM, Samudi C et al (2023) Gut bacteria influence *Blastocystis* sp. phenotypes and may trigger pathogenicity. PLoS Negl Trop Dis 17:e0011170. https://doi.org/10.1371/journal.pntd.0011170
48. Nourrisson C, Scanzi J, Pereira B et al (2014) Blastocystis is associated with decrease of fecal microbiota protective bacteria: comparative analysis between patients with irritable bowel syndrome and control subjects. PLoS One 9:e111868. https://doi.org/10.1371/journal.pone.0111868
49. Stensvold CR, Sørland BA, Berg RPKD et al (2022) Stool microbiota diversity analysis of *Blastocystis*-positive and *Blastocystis*-negative individuals. Microorganisms 10:326. https://doi.org/10.3390/microorganisms10020326
50. Jeffery IB, Cotter PD, Scanlan PD (2022) Collateral damage in the human gut microbiome - *Blastocystis* is significantly less prevalent in an antibiotic-treated adult population compared to non-antibiotic treated controls. Front Cell Infect Microbiol 12. https://doi.org/10.3389/fcimb.2022.822475
51. Brodie JF, Aslan CE, Rogers HS et al (2014) Secondary extinctions of biodiversity. Trends Ecol Evol 29:664–672. https://doi.org/10.1016/j.tree.2014.09.012
52. Deng L, Wojciech L, Png CW et al (2022) Experimental colonization with *Blastocystis* ST4 is associated with protective immune responses and modulation of gut microbiome in a DSS-induced colitis mouse model. Cell Mol Life Sci 79:245. https://doi.org/10.1007/s00018-022-04271-9
53. Deng L, Tay H, Peng G et al (2021) Prevalence and molecular subtyping of *Blastocystis* in patients with Clostridium difficile infection, Singapore. Parasit Vectors 14:277. https://doi.org/10.1186/s13071-021-04749-8
54. Deng L, Lee JWJ, Tan KSW (2022) Infection with pathogenic *Blastocystis* ST7 is associated with decreased bacterial diversity and altered gut microbiome profiles in diarrheal patients. Parasit Vectors 15:312. https://doi.org/10.1186/s13071-022-05435-z
55. Malinen E, Rinttilä T, Kajander K et al (2005) Analysis of the fecal microbiota of irritable bowel syndrome patients and healthy controls with real-time PCR. Am J Gastroenterol 200:373–382
56. Rajilić-Stojanović M, Biagi E, Heilig HGHJ et al (2011) Global and deep molecular analysis of microbiota signatures in fecal samples from patients with irritable bowel syndrome. Gastroenterology 141:1792–1801. https://doi.org/10.1053/j.gastro.2011.07.043
57. Yason JA, Liang YR, Png CW et al (2019) Interactions between a pathogenic *Blastocystis* subtype and gut microbiota: in vitro and in vivo studies. Microbiome 7:30. https://doi.org/10.1186/s40168-019-0644-3
58. Betts EL, Hoque S, Torbe L et al (2021) Parasites, drugs and captivity: *Blastocystis*-microbiome associations in captive water voles. Biology (Basel) 10:457. https://doi.org/10.3390/biology10060457
59. Deng L, Chen S, Meng W et al (2022) Changes in gut microbiota composition associated with the presence of enteric Protist *Blastocystis* in captive forest musk deer (*Moschus Berezovskii*). Microbiol Spectr 10. https://doi.org/10.1128/spectrum.02269-21
60. Kodio A, Coulibaly D, Koné AK et al (2019) *Blastocystis* colonization is associated with increased diversity and altered gut bacterial communities in healthy Malian children. Microorganisms 7:649. https://doi.org/10.3390/microorganisms7120649
61. Gabrielli S, Furzi F, Fontanelli Sulekova L et al (2020) Occurrence of *Blastocystis*-subtypes in patients from Italy revealed association of ST3 with a healthy gut microbiota. Parasite Epidemiol Control 9:e00134. https://doi.org/10.1016/j.parepi.2020.e00134
62. Doerr A (2017) Global metabolomics. Nat Methods 14:32–32. https://doi.org/10.1038/nmeth.4112
63. Alseekh S, Fernie AR (2018) Metabolomics 20 years on: what have we learned and what hurdles remain? Plant J 94:933–942. https://doi.org/10.1111/tpj.13950
64. Newton JM, Betts EL, Yiangou L et al (2021) Establishing a metabolite extraction method to study the metabolome of *Blastocystis* using NMR. Molecules 26:3285. https://doi.org/10.3390/molecules26113285
65. Betts EL, Newton JM, Thompson GS et al (2021) Metabolic fluctuations in the human stool obtained from *Blastocystis* carriers and non-carriers. Metabolites 11:883. https://doi.org/10.3390/metabo11120883
66. Marchesi JR, Holmes E, Khan F et al (2007) Rapid and noninvasive metabonomic characterization of inflammatory bowel disease. J Proteome Res 6:546–551. https://doi.org/10.1021/pr060470d
67. Fabisiak A, Włodarczyk J, Fabisiak N et al (2017) Targeting histamine receptors in irritable bowel syndrome: a critical appraisal. J Neurogastroenterol Motil 23:341–348. https://doi.org/10.5056/jnm16203
68. Wojciech L, Png CW, Koh EY et al (2023) A tryptophan metabolite made by a gut microbiome eukaryote induces pro-inflammatory T cells. EMBO J 42:10.15252/embj.2022112963
69. Tesmer LA, Lundy SK, Sarkar S, Fox DA (2008) Th17 cells in human disease. Immunol Rev 223:87–113. https://doi.org/10.1111/j.1600-065X.2008.00628.x
70. Li F-J, Tsaousis AD, Purton T et al (2019) Successful genetic transfection of the colonic Protistan parasite *Blastocystis* for reliable expression of ectopic genes. Sci Rep 9:3159. https://doi.org/10.1038/s41598-019-39094-5

71. Wong KHS, Ng GC, Lin RTP et al (2008) Predominance of subtype 3 among *Blastocystis* isolates from a major hospital in Singapore. Parasitol Res 102:663–670. https://doi.org/10.1007/s00436-007-0808-0
72. Chen XQ, Singh M, Ho LC et al (1997) A survey of *Blastocystis* sp. in rodents. Lab Anim Sci 47:91–94
73. Ho LC, Singh M, Suresh G et al (1993) Axenic culture of *Blastocystis hominis* in Iscove's modified Dulbecco's medium. Parasitol Res 79:614–616. https://doi.org/10.1007/BF00932249
74. Hamilton CA, Young R, Jayaraman S et al (2018) Development of in vitro enteroids derived from bovine small intestinal crypts. Vet Res 49:54. https://doi.org/10.1186/s13567-018-0547-5
75. Puschhof J, Pleguezuelos-Manzano C, Clevers H (2021) Organoids and organs-on-chips: insights into human gut-microbe interactions. Cell Host Microbe 29:867–878. https://doi.org/10.1016/j.chom.2021.04.002
76. Ashammakhi N, Nasiri R, de Barros NR et al (2020) Gut-on-a-chip: current progress and future opportunities. Biomaterials 255:120196. https://doi.org/10.1016/j.biomaterials.2020.120196
77. Jalili-Firoozinezhad S, Gazzaniga FS, Calamari EL et al (2019) A complex human gut microbiome cultured in an anaerobic intestine-on-a-chip. Nat Biomed Eng 3:520–531. https://doi.org/10.1038/s41551-019-0397-0

Deep Sequencing Applications for Intestinal Parasite Detection and Differentiation

Christen Rune Stensvold and Richard S. Bradbury

Abstract

The detection and differentiation of intestinal parasites has previously relied mainly or primarily on microscopy, culture, and, to some extent, targeted PCR methods. Here, we give examples of how next-generation sequencing methods have been useful for improving parasite diagnostics and opening up new avenues for the generation of DNA data pertaining not only to the parasites but also to other organisms, enabling analysis of trans-kingdom relationships.

Keywords

Deep sequencing · Parasitology · Ecology · Taxonomy · Public Health Microbiology · Diagnostics

C. R. Stensvold (✉)
Laboratory of Parasitology, Department of Bacteria, Parasites, and Fungi, Statens Serum Institut, Copenhagen, Denmark
e-mail: run@ssi.dk

R. S. Bradbury
School of Public Health and Tropical Medicine, James Cook University, Townsville, QLD, Australia
e-mail: richard.bradbury@jcu.edu.au

1 Introduction

Eukaryotic organisms residing in the intestines that rely on a host to complete their life cycles, commonly known as "parasites," encompass both helminths and protists. Fifteen protistan genera belonging to very diverse groups of taxa (Amoebozoa, flagellates, ciliates, Apicomplexa, and Stramenopiles) and over 50 helminth genera (mainly trematodes, nematodes, and cestodes) infect or colonize the intestinal tract of humans [1]. Other eukaryotic organisms that might infect or colonize the gut include microsporidia and yeasts, respectively, some of which might be considered part of the *eukaryome*, a term that was coined to indicate the collection of eukaryotic organisms populating the gut [2].

In high-income nations, there has been a significant decline in the prevalence of individuals carrying metazoan helminth infections over the past century [3, 4]. This trend has also been observed in selected low-income nations, likely due to a combination of control programs and improved hygiene infrastructure. Despite this progress, helminthic disease remains a significant public health problem globally. Approximately 909 million people are infected with soil-transmitted helminths (*Ascaris*,

J. D. Ramírez González (ed.), *Recent Advances in Parasitomics*,
https://doi.org/10.1007/978-3-031-70591-5_6

Trichuris, and the hookworms) [5], 614 million have strongyloidiasis [6], and 142 million are affected by schistosomiasis [7], while 82 million have foodborne trematode infections [7]. As an exception, pinworm infections (i.e., infections with *Enterobius vermicularis*) continue to be relatively common even in highly developed countries [8].

Single-celled parasitic protists continue to readily infect or colonize humans. Genera belonging to the protozoa infect the epithelial lining, and among these, *Cryptosporidium* causes cryptosporidiosis, which can cause even large outbreaks of diarrhea [9, 10] and may be fatal in immunocompromised hosts [11]. *Giardia duodenalis* is another common pathogenic flagellate that can cause chronic gastrointestinal symptoms and growth retardation in children [12, 13]. Among the amoebas, mainly one species has often been associated with human disease, namely *Entamoeba histolytica* [14]. Some coccidian parasites, including *Cyclospora* and *Cystoisospora*, both of which are genetically related to *Cryptosporidium*, may also produce disease, with diarrhea as the most prominent feature.

2 Metabarcoding

Essential to investigating the impact of intestinal parasites on health and disease are accurate diagnostic tools, including DNA-based technology such as PCR and sequencing, plus accurate and exhaustive reference databases. PCR-based assays were introduced into diagnostic laboratories some 25 years ago [4] with trailblazing impact due to high sensitivity, accuracy, and room for automation and standardization. However, a major drawback of PCR-based diagnostic approaches is the fact that molecular assays such as conventional PCR or real-time PCR only target one (singleplex) or a few (multiplex) targets at a time. An increase in targets typically leads to reduced diagnostic sensitivity.

Metabarcoding, the targeted amplicon-based deep sequencing of housekeeping genes to detect and taxonomically differentiate microorganisms was introduced in the 2000s and initially widely used, typically known as "16S PCR" [15]. This enabled not only a one-fits-all approach to detecting bacteria but also simulations of bacterial communities in, e.g., complex samples, such as slurry and fecal samples, and the methods are still widely used in more modern laboratories as of today. Metabarcoding has revolutionized our understanding of diversity and ecology of microorganisms in different habitats [16] and of trans-kingdom relationships between parasite and bacterial communities. Metabarcoding involves large-scale taxonomic identification of taxa in biological samples through the analysis of short DNA fragments of one or more genes ("DNA barcodes"). A number of metabarcoding methods have been developed to detect and differentiate intestinal parasites in mammalian hosts, including humans and in other complex samples, such as sewage material [17–38].

The housekeeping gene targets are genes that are required for the survival of the cell, any deleterious mutation of which results in immediate cell death. For this reason, such genes have very slow mutation rates and therefore present excellent metabarcoding targets, as they have highly conserved regions, while the variable regions of these genes are generally conserved intraspecies while being highly variable interspecies. The housekeeping genes most commonly targeted for metabarcoding are within the mitochondrial genome and the ribosomal genes in the nuclear genome, which code for ribosomal RNA (Table 1). Small subunit (SSU) ribosomal RNA (rRNA) genes consistently present in both pro- and eukaryotic organisms are today avidly used as taxonomic markers and are probably the genes most commonly represented in DNA reference databases [39].

At the time of writing, Illumina high-throughput sequencing (HTS) technology is most widely used for metabarcoding. Although Oxford Nanopore long-read sequencing is increasingly employed, the higher error rates compared to Illumina HTS have meant that where low-abundance organisms are targeted, Illumina HTS has some advantages. Illumina MiSeq and HiSeq enable the sequencing of short ~300 bp (up to

Table 1 Housekeeping genes of parasitic species and genotypes colonizing or infecting in the intestinal tract of mammalian hosts commonly targeted in metabarcoding (amplicon-based deep sequencing) assays

Gene target	Abbreviation	Nuclear/mitochondrial	Frequency of usage
Cytochrome c oxidase subunit 1	*cox1*	Nuclear	Commonly employed—helminths
Cytochrome c oxidase subunit 3	*Cox3*	Nuclear	Rarely employed—helminths
12 Svedberg ribosomal RNA	*18S rRNA*	Nuclear	Moderately employed—protozoa and helminths
16 Svedberg ribosomal RNA	*16S rRNA*	Nuclear	Moderately employed—protozoa and helminths
5.8 Svedberg ribosomal RNA	*5.8S rRNA*	Nuclear	Rarely employed—protozoa and helminths
18 Svedberg ribosomal RNA	*18S rRNA*	Nuclear	Commonly employed—protozoa and helminths
28 Sverdberg ribosomal RNA	*12S rRNA*	Nuclear	Rarely employed—protozoa and helminths
Internal transcribed spacer region 1	ITS 1	Nuclear	Rarely employed—helminths
Internal transcribed spacer region 2	ITS 2	Nuclear	Moderately employed—helminths

~600 bp by pair-ended sequencing) amplicons. It should be mentioned in this respect that the development and utilization of pan-eukaryotic *18S rRNA* diagnostic primer sets for Illumina sequencing has been hampered by the diversity in the nuclear 18S rRNA gene, which makes it a valuable taxonomic marker for many groups of parasites, but which means that it has been challenging to identify a single primer set that will detect all genera of intestinal parasites [40, 41], given the very wide diversity of parasitic phyla infecting human intestinal tract. Any given 300–600-bp region of *18S rRNA* might be sufficient to differentiate some organisms at species level but not others. This is why no less than three eukaryotic primer pairs were used in the 16S + 18S method introduced a few years ago [42], and this one-fits-most method has been widely used to detect and differentiate the eukaryotic fauna in complex samples and also enabled analysis of trans-kingdom associations between, e.g., parasites and bacteria (see Sect. 6.5).

Factors that have been found to influence the efficiency of a metabarcoding assay include but may not be limited to PCR priming locations, GC content in target template, DNA template concentration, gene copy number, relative abundance, thermocycling conditions, and nonspecific binding of primers [20]. Also, as can be understood from the paragraph above, the choice of primers may play a role. Choi and Park found 1413 and 915 operational taxonomic units (OTUs) when using primer sets targeting the V9 and V4 regions, respectively [43]. It has been argued, however, that most of the variability observed at the barcode level might be caused by sequencing errors, and application of a denoising tool, such as *DADA2*, appears useful to increase the likelihood of limiting nonfactual diversity [16].

Conclusively, amplicon-based HTS targeting ribosomal genes can in principle:

- Obviate the need for cloning.
- Be a "one-fits-all," or, at least, a "one-fits-most" approach to detecting organisms, provided the absence of preferential amplification, sufficient primer availability, and a minimal amount of sequencing error.

2.1 Metabarcoding for the Detection and Differentiation of Intestinal Parasitic Protists

The role of most parasitic single-celled species found in stool samples remains elusive. These parasites mostly comprise organisms that have been referred to as "common luminal intestinal parasitic protists" (CLIPPs) [44], most of which have been subject to controversy among clinicians, biologists, and public health microbiologists over the years.

Challenges to pinpointing the clinical and public health significance of CLIPPs include but are not limited to the following:

1. Lack of data on parasite infection/colonization in the healthy population
2. Lack of knowledge on mixed colonization/infection by genetic variants (species, subtypes, etc.)
3. Lack of molecular data and knowledge on cryptic genetic variation within species and lack of information on variability in virulence among morphologically identical organisms
4. Limitations of diagnostic tools, tools used for surveillance, and reference databases
5. Lack of understanding about the influence parasites have on gut ecology.

Although little is known about the role of host-associated micro-eukaryotes outside the field of parasitology, we know that CLIPPs exhibit a high degree of cryptic genetic diversity, which means that morphologically identical organisms can exhibit an even high degree of genetic diversity (see below). We moreover know that some CLIPPs are shared between human and nonhuman hosts and that some parasites in animals that are genetically related to CLIPPs found in humans assist host animals in the metabolism of nutrients and vitamin production among other activities [35, 39], and hence the role of CLIPPs in human metabolism should also be clarified for that purpose. The relevance of this purpose is emphasized by the fact that we already know that some CLIPPs are related to certain gut microbiota profiles (see below). Similar challenges are faced with some helminth infections of humans, for which cryptic species may exist [45] and sequences in databases may be incomplete or incorrect [46].

Metabarcoding has been widely applied for the detection and differentiation of parasitic protozoa species and subtypes in fecal samples. For some of the CLIPPs, such as *Entamoeba* and *Blastocystis*, PCR and Sanger sequencing have readily been used to generate thousands of ribosomal DNA sequences that have populated the NCBI nucleotide databases. It could therefore seem remarkable that other CLIPPs, such as *Endolimax nana*, which has been estimated to colonize almost 1 billion people worldwide [47], are barely represented by DNA sequences in databases such as the NCBI databases; the number of sequences is 36 at the current stage of writing. A similar situation exists for *Iodamoeba*, a parasite that can be found at a rate of up to 14% in humans living in areas in the Philippines and Indonesia [48, 49]; for this genus, only 79 DNA sequences are currently available in the NCBI database.

However, acquiring SSU rDNA sequences for *Endolimax* and *Iodamoeba* is challenging due to several factors. First, there is often a lack of cultured material for these parasites and therefore, the acquisition of specific DNA sequences relies on amplification from DNA extracted from either isolated parasite cysts or directly from stool containing the cysts. Second, the SSU rDNA sequences of these organisms are notably lengthy, around 2.5 kbp (kilobase pairs) [50, 51]. When attempting PCR using low-specificity eukaryotic primers, there is a preference for amplifying shorter and more abundant SSU rDNA from co-infecting or co-colonizing organisms that may be present in the intestinal environment. This phenomenon is frequently observed with *Blastocystis* sp., which is often found in samples positive for *Endolimax* and *Iodamoeba*. In fact, it is very rare to see, for instance, *Iodamoeba* in a solitary parasite infection. The SSU rDNA sequences of *Blastocystis* are approximately 700 base pairs shorter than those of *Endolimax* and *Iodamoeba* and therefore more readily amplified by general eukaryotic primers such as the RD5/RD3 primer pair [52].

Even when amplification of DNA specific to *Iodamoeba* or *Endolimax* is achieved, for instance by using diagnostic primers, Sanger sequencing of the PCR product frequently results in sequences of low quality. These low-quality sequences are characterized by double peaks and patterns that lack synchronization, mainly due to high intra-genome variation among ribosomal gene copies, including variations in homopolymer length [50]. Consequently, when exclusively relying on Sanger sequencing, a preliminary cloning step has proven necessary for sequencing the SSU rDNA of *Iodamoeba* and *Endolimax* [50, 51]. These are only a few examples of common intestinal parasites for which reference DNA data is remarkably limited, and we have here mentioned some of the reasons why this is so. This serves to exemplify the fact that even with the availability of deep sequencing methods and highly automated workflows, diagnostics and research are still severely limited by the lack of reference data.

Much work has been focused on protozoan amoebae of the genus *Entamoeba*. Perhaps the first deep sequencing-based method ever to be developed for intestinal parasites was the one used for detecting and differentiating species of *Entamoeba* in fecal samples was a method that used a genus-specific primer pair in an amplicon-based pyrosequencing assay [53]. Vlčková and colleagues applied *18S rRNA* metabarcoding to analysis of the diversity of *Entamoeba* spp. in humans and great apes from Africa [54]. The nested PCR primers employed in this study amplified a 270-bp region of the *Entamoeba* genus 18S rRNA gene, specifically. As with all metabarcoding primer design, this required the design of primers binding in a conserved region of the gene flanked by a sufficiently discriminatory region that all species could be detected and distinguished by subsequent Illumina MiSeq HTS. This method was highly effective when applied to nonhuman primate (NHP) fecal samples, detecting and identifying 36 unique 18S rRNA *Entamoeba* haplotypes within six monophyletic clusters (representing the species *Entamoeba coli*, *Entamoeba dispar*, *Entamoeba hartmanni*, *E. histolytica*, *Entamoeba invadens*, and *Entamoeba nuttalli*). Several human individuals residing close to the ape colonies were passing DNA of an *Entamoeba* spp. traditionally associated with NHPs (*E. nuttalli*), but previously shown to be capable of infecting humans [55]. Several humans and great apes were also passing DNA of a novel *Entamoeba* OTU 19, with ~93% sequence homology at the target site to the reptile amoeba, *E. invadens* [54]. These findings demonstrate the efficacy of deep sequencing in the detection of previously undetected species and genotypes. The depth of MiSeq HTS also allowed the detection of mixed infections in 63% of the 132 human and 58 NHP samples tested, with up to seven haplotypes being found in a single sample (mean 2.27 OTUs per sample) [54].

An incidental finding of the above work was that the default algorithm for chimaera detection in the bioinformatic program *DADA2* yielded a large number of false positives (sequences incorrectly identified as chimaera) and an alternative method for filtering spurious haplotypes and detecting chimeric sequences needed to be applied [54].

Indeed, amplicon-based HTS has proven useful for detecting and differentiating novel ribosomal lineages or conditional lineages of *Entamoeba*, even when only a few hundred bp are sequenced [56].

Lane and colleagues recently applied a universal parasite diagnostic test (UPDx) previously developed for the detection and differentiation of protozoa and helminths in blood to a limited number of cloacal swabs and caecum tissue samples from various bird species [40]. This nested PCR metabarcoding assay was developed with a restriction enzyme digestion step prior to each PCR to reduce host DNA amplification when using universal *18S rRNA* primers. The primers used were ineffective at amplification of entamoebae and trichomonads but did amplify some undescribed Apicomplexa and alveolates [40]. The further trialing of this approach in mammalian feces containing protozoa and helminths is indicated.

2.2 Metabarcoding for the Detection and Differentiation of Intestinal Helminths

Metabarcoding for the detection and characterization of intestinal helminths was adopted later than for protozoa. Among challenges encountered specific to helminth applications is the often much lower numbers of helminth eggs or larvae shed compared with the often high numbers of protozoan trophozoites and cysts. Finding gene targets universal to cestodes, nematodes, trematodes, and Acanthocephala, which do not also amplify host DNA has also proved impossible. Therefore, most helminth metabarcoding assays are limited to specific families, classes, or phyla of helminths. Nevertheless, metabarcoding has been successfully applied to detect and differentiate intestinal helminth infections in human, NHP, and other mammalian fecal samples [23, 57].

An early attempt to detect and differentiate helminths using metabarcoding was performed on wild rat feces in Japan. A universal *18S rRNA* target was employed, followed by Illumina MiSeq HTS [17]. The results of this metabarcoding approach were compared with data on helminths detected in the host intestinal tract during necropsy and identified by a combination of morphology and *18S rRNA* PCR with Sanger sequencing [17]. The metabarcoding assay identified large amounts of DNA from Alveolata, Apicomplexa, fungi, and plants at phylum level in many individual fecal samples. Metazoan DNA sequences were far rarer (likely reflecting the relatively limited distribution of metazoan eggs and larvae in host feces). When analyzed to subfamily level, specific helminth and Amoebozoa groups became evident in some cases [17]. Further analysis and comparison with necropsy samples identified likely reads of *Strongyloides* spp., *Nippostrongylus brasiliensis*, and *Hymenolepis* sp. DNA, as well as DNA from various other nematodes [17], including bird parasites, which likely represent fecal passage of DNA from the rat's diets. Ultimately, this was a groundbreaking study; still, limitations in sequence database completeness, read length, and difficulty in discrimination to species level limited the efficacy of the approach.

Later, the same group improved their methodology by in silico screening a wide array of *18S rRNA* and *28S rRNA* primers to find the optimal set for the identification of parasites in feces [58]. Application of these to rat and bovine specimens, followed by MiSeq sequencing and QIIME bioinformatic analysis revealed that nonspecific amplification of host DNA was insignificant for all primer sets tested; however, no primer set tested could detect all parasite genera. Furthermore, the short sequence length continued to hamper taxonomic resolution to species level [58].

In some cases, helminth targets chosen have allowed differentiation to haplotype level. However, such discriminatory targets detect only a limited range of parasites, generally restricted to a single genus or genetically similar group, such as Strongylidae and Ancylostomatidae [23, 57]. Barratt and colleagues developed sets of PCR primers to amplify a discriminatory region of *cox1* and two regions of *Strongyloides 18S rRNA* (hyper-variable region (HVR) I and HVR IV) for Illumina MiSeq HTS genotyping of *Strongyloides* spp. directly on DNA from stool [57]. Very usefully, the *cox1* PCR primers also allowed simultaneous amplification and discrimination of strongyle nematodes within the specimens. This approach has been applied to numerous humans, dog, and NHP fecal samples from various parts of the world and was effective in identifying and discriminating clades and haplotypes of *S. stercoralis* and *S. fuelleborni* but also detected *cox1* DNA sequences corresponding to *Necator americanus*, *Ancylostoma caninum*, *Ancylostoma ceylanicum*, and *Oesophagostomum* sp. in some samples [57, 59]. The additional findings of *Metastrongylus salmi* (pig lungworm) and *Strongyloides ratti* (rat threadworm) in some dog fecal samples [59] almost certainly represented spurious passage of DNA after the dogs had ingested infected hosts or host feces, raising the issue of confounding data from coprophagia and ingestion of other hosts, a phenomenon discussed further elsewhere in this chapter (Sect. 2).

Challenges associated with low helminth DNA abundance and the volume of other eukaryotic DNA encountered when sequencing of DNA extracted directly from feces have led some researchers to apply metabarcoding to populations of cultured nematode larvae. This has been trialed most successfully as the "nemabiome" approach developed and validated in ruminants by Avramenko within the Gilleard group at McGill University in Canada [19, 21, 22]. This approach applied modified NC1 and NC2 primers, amplifying the ITS-2 region to metabarcode nematode larvae cultured from ruminant feces via Illumina MiSeq [19, 21, 22]. This method was applied to characterizing populations of gastrointestinal nematodes (GIN) in large-scale surveys of cattle [21] and sheep farms [25] and for the monitoring and detection of anthelmintic therapy nonresponse in cattle [21] and sheep herds [26]. Apart from these practical applications for agricultural diagnostics, the nemabiome assay has also allowed better characterization of seasonal [27, 60] and geographical [34] variations in ruminant GIN population structures, informing epidemiology and intervention.

A Strongyle ITS-2 metabarcoding approach has also been applied to direct fecal and cultured Strongylid larval DNA from populations of NHPs in Central Africa [23]. This method was highly effective in separating and characterizing populations of Strongylids, specifically within the genera *Necator* and *Oesophagostomum* [23]. It also allowed comparison of this method when applied directly to duplicate fecal DNA extracts and duplicate cultured nematode larval population DNA extracts [23]. There was greater intra-duplicate correlation in fecal DNA extract results than for larval culture results. While cultured larvae and fecal results from the same individual corelated, there was greater haplotype diversity in the fecal DNA sequences than those from cultures of the same fecal sample [23]. This raises a limitation of the nemabiome-like approach involving metabarcoding of cultured nematode larvae, in that we do not know whether all haplotypes of target Strongyles are equally capable of being cultured in vitro, nor if they will culture at the same rate or outcompete other species or genotypes in culture conditions.

Metabarcoding may also be used for the detection of parasitic helminth life stages in the environment. The NC1 and NC2 primers, previously adapted for the nemabiome assay, have been applied directly to both feces and soil for the detection and characterization of parasitic nematode populations [31]. An approach for the detection of cestodes, trematodes, and nematodes targeting regions in the platyhelminth mitochondrial 12S rRNA gene, the nematode 12S rRNA gene and a platyhelminth + nematode mitochondrial *16S rRNA* target on mock communities in fecal, soil, tissue, and pond water matrices has been trialed and shows promise for future application [61].

2.3 Metabarcoding for Mapping Eukaryotic Communities

In spite of groundbreaking discoveries regarding the diversity of microeukaryotes in the distal human gut dating back to 2008 [62], deep sequencing technologies employed in gut microbiome research have predominantly focused on bacterial species. While bacteria in the gut significantly outnumber other components of the gut microbiome, including archaea, fungi, and non-fungal eukaryotes such as CLIPPs, these latter members, with their considerably larger genomes, may have roles that would go unnoticed if research on the role of gut microbiome in health and disease solely relied on the analysis of bacterial taxa [2].

Mapping of the eukaryome in mammals, including humans, is still in its infancy. A recent review of gut microbiome studies in livestock only included one single mentioning of profiling of the eukaryotic fauna [63].

One advantageous aspect of employing metaribosomalomics is the ability to compare bacterial communities within the digestive tract of individuals with and without parasites. This allows for the examination of significant differences in bacterial composition and offers a basis for hypothesizing the extent to which such disparities may be influenced by the colonization or infection of parasites (see below).

Studying the distribution of parasitic genera common in humans in nonhuman hosts and the relationship between such parasites and bacterial communities in nonhuman hosts can help us hypothesize on the clinical and public health significance of these parasites in humans.

The 16S + 18S rRNA gene assay already mentioned was evaluated for use on human fecal samples that had already been amply tested by a variety of state-of-the-art methods, including microscopy and species-specific PCRs for a variety of pathogenic protozoa and CLIPPS [33]. The assay was found to provide a good and cost-effective alternative to these methods; however, the lack of sensitivity for *Giardia* and *Dientamoeba* was highlighted. This concern was corroborated in other studies where the method was used to map parasitic communities in pigs [29] and in samples from Swedish wastewater treatment plants [64]. In the wastewater sample DNAs, the assay enabled detection and differentiation of *Blastocystis* subtypes and a variety of the Amoebozoa commonly infecting/colonizing humans, but the study did not report on any findings of *Giardia* and *Dientamoeba*. The issue with low sensitivity for *Giardia* was also pinpointed in the study of protozoa and CLIPPs in pigs [29]. Here, the 16S + 18S assay efficiently detected *Balantioides coli*, *Cryptosporidium* (*suis* and *scrofarum*), *Blastocystis* (subtypes 1, 3, 5, and 15), *Entamoeba* (*polecki* and *hartmanni*), and *Iodamoeba buetschlii*. The study also employed a real-time PCR for *Giardia* and *Enterocytozoon bieneusi*, which were found in 86/259 (33%) and 34/37 (92%) of tested samples, respectively. These were not detected by the metabarcoding assay.

A relatively high GC content in target template may impede successful amplification [65]. Indeed, the GC content of *Giardia* SSU rDNA sequences amounts to about 72%. In comparison, similar calculations for *Blastocystis* and genera belonging to the Amoebozoa yield GC contents between 30% and 50%. *Dientamoeba* has a relatively low GC content, only about 39%, and even in the sections amplified by the primers, the GC content is at a similar level. Hence, the GC content alone cannot be used to predict the ability of an metabarcoding assay to amplify any given organism. Another team also failed to find *Giardia* (and *Cryptosporidium*) using a metabarcoding method based on ITS2 and 18S primers when profiling gut eukaryotic communities in pigs [66]. The remaining eukaryome of the pigs was very reminiscent of the one identified by Stensvold and colleagues [29], so there was little reason to believe that the pigs should not, to a large extent, be possible for both *Cryptosporidium* and *Giardia*.

Although an assay such as the 16S + 18S assay may have its disadvantages relating to its lack of sensitivity for a couple of parasitic genera, it does seem to provide a fairly good approach to surveying communities of pathogenic and nonpathogen protozoa in fecal samples or sludge material. What is more, it allows for detection at species level and enable detection of mixed infection/colonization by multiple species within a genus. For example, in the study of pig samples [29], mixed infection with *C. suis* and *C. scrofarum* was observed in eight cases, a finding that might very well not have been clear if conventional PCR with Sanger sequencing had been employed, unless species-specific primers were used.

In the study of the wastewater samples, *Blastocystis* subtypes 1–4 were detected in almost all, apart from only a couple in which one of these subtypes was missing. This distribution of subtypes in sewage material was unsurprising given the subtype distribution in humans in Europe, which is mainly ST1–ST4 [67]. However, one more subtype was detected by the assay in quite a few of the samples, namely ST8, which is uncommon in human fecal samples. This finding let the authors speculate that feces from other hosts than humans could be present in the sewage sampled, and since rats may spend time in sewers, the authors speculated that rats could contribute to the sewage material and that rats could be natural hosts of ST8 since it was seen in relatively many samples. Hence, the metabarcoding of sewage samples was leading to the generation of a new hypothesis in the field of genetic diversity and host specificity.

As a final example, in the pig study, the assay used enabled the detection of no less than six different genetic variants of *Blastocystis* sp. ST5 [29]. Given the fact that ST5 is a common (if not the most common) subtype in pigs, being able to differentiate up to at least six different genetic variants of the subtype in a population might make it a useful tool in studies aiming to map the epidemiology and host impact of ST5.

Chana and colleagues explored the utility of 12S and 16S mitochondrial ribosomal targets for diversity mapping in helminths species via Sanger sequencing [68]. The 12S rRNA target trialed performed well in correctly categorizing the phylogenetic relationships of various helminth species, while the *16S rRNA* target performed poorly. The group specifically chose mitochondrial targets for a number of reasons. These were that the whole mitochondrial genome of many helminths has now been sequenced, allowing easy sequencing comparison and that the low rates of mutation in the mitochondrial ribosomal genes result in far less indels than are seen in nuclear gene targets such as ITS-1, ITS-2, 18S rRNA, and 28S rRNA [68]. The points might be applied to diversity mapping via HTS-based amplicon sequencing of the 12S rRNA gene in the future.

Conclusively, a metabarcoding method such as the one reviewed here may be considered a "most-fits-most" tool to map eukaryotic diversity even in complex samples such as feces and sewage and with high discriminatory power. However, as a diagnostic tool it has little value in terms of ruling out infection/colonization. Moreover, Illumina-based sequencing provides relatively short reads that for some parasites, such as *Blastocystis*, may enable differentiation even at genotype/allele/haplotype level, while for others (e.g., *Eimeria* and some helminths), identification is possible only at genus level.

Short-read HTS, i.e., Illumina, has been widely used for mapping of microbial communities, including eukaryotic communities. However, recently, long-read sequencing by the use of Nanopore long-read sequencing technology has been used in various studies of *Entamoeba* and, particularly, *Blastocystis* [56, 69–73]. While short reads of *Blastocystis* are often sufficient to identify an organism to subtype level, near-complete sequences are required in studies aiming to assess whether a sequence represents a new or already existing subtype [74].

2.4 Metabarcoding for Deciphering Cryptic Genetic Diversity in Parasitic Protists

The finding that quadrinucleate *Entamoeba* cysts of equal size and similar morphological features could have phenotypically very different properties encouraged investigations into the genetic make-up of these cysts. It became clear that there was ~1.5% genetic variation across the SSU rRNA gene, and today, we think of *E. dispar* as the "apathogenic variant" of *E. histolytica.*

This example of cryptic genetic diversity has prompted similar research questions and investigations into the genetic diversity of morphologically similar parasites, including those referred to as CLIPPs.

In 2012, Stensvold and colleagues used PCR, cloning, and Sanger sequencing to produce the first DNA sequences reflecting *Iodamoeba* SSU rRNA genes ever documented [75]. Two different ribosomal lineages were identified (RL1 and RL2, respectively), which were shown to differ by no less than 31% across the SSU rRNA genes.

Ten years later, a similar study was published on *Endolimax*, the sister taxon of *Iodamoeba.* Here, PCR and PacBio long-read sequencing technology helped the team sequence the SSU rRNA genes of *Endolimax*, which were shown to differ by up to at least 16%. As for *Iodamoeba*, again, two ribosomal lineages could be identified.

In 2023, amplicon-based HTS was employed by Stensvold and colleagues [76] to screen for genetic diversity within species of *Entamoeba*, namely *E. coli* and *E. hartmanni.* Both are parasites of probably cosmopolitan distribution. When the study was carried out, it was already known that *E. coli*, a parasite that was found to colonize at least 37% of Sulawesi villagers [49], encompassed two ribosomal lineages, *E. coli* ST1 and *E. coli* ST2 [77]. In the new study, data not only confirmed the existence of two lineages, but suggested the existence of three ribosomal lineages of *E. coli*, differing by up to 11% across the SSU rRNA genes [76]. Similarly, for *E. hartmanni*, we also now have evidence of at least three ribosomal lineages, subtypes 1–3,

The findings mentioned for these Amoebozoa have two major implications. First, the existence of such degrees of genetic diversity in morphologically similar organisms challenged current species concepts. Genetic boundaries between species are typically in the vicinity of maximum 3–5% [78], and so "species" such as *E. coli* and the others mentioned in this section could be viewed as species complexes rather than species. The second implication pertains to the resulting incentive to study the ribosomal lineages in a public health context, where the aim is to identify whether the lineages differ in terms of geographic distribution and overall epidemiology, including host specificity, clinical significance, and so on.

Importantly, with the increased use of HTS technology, it is not unusual to obtain DNA data in the absence of morphological data. In such situations, difficulty may arise in identifying OTUs as specific parasite species and in describing new species detected by sequencing only. For instance, in a recent study of *Strongyloides* spp. in carnivores, sequences of a feline *Strongyloides* were amplified via Sanger sequencing from cultured larvae [79]. The authors assigned these as likely belonging to *S. felis*, first described by Chandler as a parasite of cats from India [80], and only reported since in Northern Queensland, Australia by Speare [81]. While this was a likely identification, as no sequences are available for *S. felis*, a definitive assignment of the sequences produced to this species cannot be made. The widespread loss of morphological skills throughout the parasitology community internationally will only compound issues such as this [46]. Similar issues arise when probably novel species of helminths and protists are detected by sequencing only, as the International Commission on Zoonotic Nomenclature still requires morphological descriptions to accompany any description of a new species. It may be necessary in the future to allow the naming of species based on sequence data only, but with a specific *caveat* in that name (for example, a postscript such as "S.O." to denote "sequence only"), designating that this species has only been described based on DNA sequences and no correlated morphologic descriptions are available.

2.5 Metabarcoding for Identifying Trans-kingdom Taxonomic Relationships

The role of the gut microbiota in health and disease is a topic of intense investigation these years. Although data are conflicting, some clinical phenotypes/diagnoses have been negatively associated with the presence of microeukaryotes in stool, including inflammatory bowel disease (ulcerative colitis and, especially, Crohn's disease) [82, 83] and irritable bowel syndrome [84].

An important strategy to delineating the role of intestinal parasites in human and animal health and disease could be to study relationships between gut microbial taxa across taxonomic kingdoms. In addition to direct parasite–host interaction (e.g., immunological process), parasites may exert indirect effects on the host by selecting for specific types of bacteria that maintain various functions of benefit or harm to the host. It could also be speculated that bacteria select for the presence of certain microeukaroytes, so that these may only establish under conditions that are favored by selected bacteria [85]; in the latter case, we could learn to study parasites as biomarkers [44].

Studies of gut microbiota diversity typically include analyses of diversity and differences in taxonomic composition. It will be beyond the aim of this book chapter to provide a full account of studies investigating gut microbiota trans-kingdom relationships in human and other mammalian hosts. Therefore, in the following, only some examples from the recent scientific literature will be given, focusing on a parasite of unknown clinical and public health significance (*Blastocystis*) and a parasite of well-known clinical significance (*Giardia*).

Blastocystis has been found to be associated with higher bacterial richness [86, 87], higher evenness [86], and higher bacterial diversity [88–92].

A remarkable feature of *Blastocystis* is the fact that it has been strongly associated with increased bacterial richness and higher abundance of several anaerobes [93] and negatively associated with, e.g., Proteobacteria [88, 92]. Bacterial taxa that have been positively associated with *Blastocystis* particularly include *Prevotella* [83, 88, 89, 94–97], but also *Akkermansia muciniphila* [98, 99], *Faecalibacterium* [87, 88, 92, 96, 100], *Lactobacillus* [101], *Roseburia* [92],

Ruminococcus [83, 89, 96, 97], *Methanobrevibacter* [89, 97], *Bifidobacterium* [102], *Flavonifracter* [88], *Clostridium* [88], *Succinivibrio* [88], and *Oscillibacter* [88].

Similarly, taxa that have been negatively associated with *Blastocystis* include primarily *Bacteroides* [83, 88–91, 95, 102], but also *Akkermansia* [94, 101, 103], *Enterococci* [96], Clostridial cluster XIVa [95], *Bifidobacterium* [101], *Escherichia* spp. [88, 90, 100, 101], *Faecalibacterium* [97, 104], *Klebsiella* [88], and *Pseudomonas* [88].

As seen, some bacteria have been identified as both positively and negatively associated with *Blastocystis* colonization; however, specific features to the cohorts studied (age, gender, diet, underlying conditions, subtype of *Blastocystis*, methods used, etc.) may account for differences in results.

Of course, the idea of using selected parasites as probiotics and/or immunomodulators has been introduced. It was already shown that rodents are suitable hosts of *Blastocystis* ST1, one of the most common subtypes of *Blastocystis* in humans [105]. Recently, a study using an experimental mouse model that colonization with *Blastocystis* ST1 exerts beneficial effects on host health through modulating the gut microbiota and adaptive immune responses [99].

Reminiscent trends have been seen for other CLIPPs [106], although genera such as *Entamoeba* and *Dientamoeba* have not yet been studied nearly to the same extent as *Blastocystis* in a gut microbiota context.

Meanwhile, infection with *G. duodenalis*, an acknowledged pathogen, has been positively associated with dysbiotic-characteristic taxa such as *Escherichia* [100, 107], and a range of other taxa; for an extensive summary, see the work by Fekete and colleagues [107]. McGregor and colleagues observed that study individuals with asymptomatic and symptomatic giardiasis represented distinctive microbial diversity profiles, with both being distinguishable from the profiles of healthy volunteers [108]. In mice, it was shown that depletion of the intestinal flora by antibiotic treatment had a severe effect on the infection dynamics of *Giardia*, with not only duration of infection being affected but also parasite burden, which increased significantly [109].

Incidentally, a recent study revealed that *Giardia*-positive individuals may in fact exhibit features of eubiosis, but possibly only in the presence of *Blastocystis* (refs).

Studies of trans-kingdom relationships in nonhuman mammals include studies of calves [110, 111].

Intestinal helminth infection also alters the intestinal microbiota in humans. In the Philippines, those with intestinal nematodes or *Schistosoma japonicum* demonstrated significant differences in gut microbiome compared to noninfected persons. Indeed, the microbiota of those with nematodes alone was closer to those with *S. japonicum* alone than to noninfected individuals. Helminth infected persons showed greater richness and diversity of intestinal bacterial genera, and particularly increased abundance of *Lactobacillaceae* and *Faecalibacterium* [112]. Meanwhile, in Cameroon, significant differences in the composition of the intestinal microbiome for those infected with intestinal nematodes compared to uninfected people were also observed [113]. *Bacteroidales*, *Treponema*, *Prevotella stercorea*, *Anaerovibrio*, and *Succinivibrio* were all present in greater abundance in nematode-infected people in the Cameroonian cohort. These changes were so distinct that when random forest classifier machine learning was applied to determine if a person was likely to be infected with nematodes based solely on their bacterial microbiome composition, this could be predicted with ~90% accuracy [113].

A meta-analysis of studies found that helminths increased alpha diversity and promoted changes in the predominant species within the microbiome [114]. Specifically, *Firmicutes* and *Clostridia* dominated in those who were helminth infected, and there was a collapse of a gradient of species in the class *Bacteroidia* that is seen in the helminth-uninfected gut and residing in the lower intestine (specifically *T. trichiura* and *E. vermicularis*) most strongly influenced these changes in microbiota composition [114]. While the effects of these changes remain to be elucidated, there are suggestions that helminth-induced changes to the gut microbiome may alter immunity in affected hosts [113]. Furthermore, the shift toward *Clostridia* may benefit transmission of at least some helminth genera, noting that increased abundance of the Peptostreptococcaceae family (within the class *Clostridia*) has been associated with accelerated *Trichuris* egg hatching within the gut of hosts [115]. All of these findings must be interpreted in the context of significant initial microbiota variation, even by village. In Malaysia, village of origin was more strongly associated with microbiome composition of the gut than helminth infection [116]. The effects of anthelmintics on the gut microbiota must also be considered when interpreting data on the microbiome of helminth infected vs uninfected persons when conducting studies in highly helminth-endemic settings [116]. One final factor in trans-kingdom relationships, which is only beginning to be explored, is the intrinsic microbiota of the helminths themselves, and how this might interact with and potentially alter the host gut microbiome [117].

3 Metagenomics

While metabarcoding is the most prevalent application of omics for intestinal parasites, other applications such as shotgun sequencing are increasingly being applied to fecal parasite investigations.

Srivathsan [118, 119] used an Illumina HiSeq shotgun metagenomic sequencing approach to determine the diet and intestinal parasites of leaf-feeding monkeys (*Pygathrix nemaeus*) *cox1* and ribosomal gene sequences from various parasites were found in the fecal DNA. These included the protozoa *E. histolytica*, *E. invadens*, *E. histolytica*, *Entamoeba* sp., and various subtypes of *Blastocystis* sp., as well as the intestinal helminths *Strongyloides fuelleborni*, additional *Strongyloides* sp. DNA, and *Oesophagostomum* sp. [118, 119]. Arthropod sequences were also recovered in some samples [118].

The finding of arthropod sequences in the aforementioned study [118] raises a major challenge when performing HTS directly on stool for parasite detection: when DNA sequences of a taxon not expected in a test host are recovered, does this represent true intestinal colonization with that OTU, or simple fecal passage following environmental ingestion with food or water? This is particularly important in the context of dogs, which may perform coprophagia on human feces and subsequently pass DNA of helminths and protozoa present in those human samples [120]. Such passage may confound the true source and zoonotic potential of specific parasites when their DNA is subsequently identified as being passed in dog feces by sequencing. To this end it should also be mentioned that microbiome DNA data from fecal samples from predators may be confounded by data stemming from pray animals devoured by the predator in question.

The report of *E. invadens* from an NHP from the above study [118] highlights another challenge for fecal HTS diagnostics regarding species discrimination. When performing metagenomic sequencing, particularly when using Illumina HTS, short lengths of sequence may match species OTUs are common, but the longer lengths allowing greater taxonomic differentiation may not be present. As *E. invadens* is a reptile parasite, it was unexpected in a monkey. Perhaps this OTU actually represented a genetically similar, but distinct, *Entamoeba* sp., such as *Entamoeba* OTU 19 found by Vlčková [54] in African great apes and humans DNA. It may also have represented innocuous passage of true *E. invadens* cysts by the monkey following ingestion in water or leaves that had been contaminated by reptile feces.

Illumina MiSeq metagenomics sequencing of human fecal samples from Cote d'Ivoire were aligned with 36 pathogen sequences, including 12 helminths and ten protozoa. This approach did not perform well for the detection and diagnosis of any infecting parasites. Commensal *E. coli* infections were misidentified as pathogenic *E. histolytica* in two samples, while microscopically identified *Chilomastix mesnili*, *G. duodenalis*, and *Blastocystis* spp. were not detected, despite being included in the reference sequences used for pathogen detection.

Superior results were found when using a very similar approach to direct metagenomic sequencing of helminths and protists in the feces of people from Cameroon [113]. When helminth detection via metagenomics was compared to qPCR results, significant positive correlations were observed, with the best correlation (Spearman's $p = 0.74$) seen for *Ascaris lumbricoides*, a helminth that sheds an abundance of eggs into the feces daily, whereas *N. americanus* and *T. trichiura* diagnosis correlations were far lower ($p = -0.63$ and -0.36, respectively). While protists were not included in comparisons of metagenomics to qPCR, several *Entamoeba* spp. *Cryptoporidium* spp., and *G. duodenalis* were detected by metagenomic sequencing in these same samples [113].

4 Whole Genome Sequencing

Although full genome data are extremely useful to make available reference DNA data and predict the properties of any given organism, including its clinical potential in host–parasite relationships, in contrast to other areas within the field of microbiology, whole genome sequencing (WGS) is currently not a practical approach for characterizing parasites. Only a small number of parasites can be readily cultured, and other types of isolation and enrichment procedures are generally challenging. Adjuvant identification data, such as morphology data, may be difficult to acquire, further complicating the confirmation of taxa at, e.g., species level. Additionally, parasites typically have much larger genomes, around 20 megabases (Mb), in comparison to the relatively shorter genomes of viruses and bacteria (e.g., approximately 30 kb for severe acute respiratory syndrome coronavirus 2). This results in WGS being both costly and time-consuming for parasites. Moreover, the difficulty in isolating parasites from other organisms often leads to issues with competing templates during amplification and sequencing procedures.

Furthermore, the reference data pipeline for WGS in the context of parasites is still in its early stages of development. Therefore, even if WGS data were available, the analysis of genome data would be a complex and time-consuming task, which makes the whole situation a circular problem. Nevertheless, considering the acknowledged diversity within CLIPPs, it is reasonable to anticipate extensive genomic diversity. This diversity could, in turn, have diverse effects on the host and, particularly, the host's gut microbiota among different CLIPPs. Hence, prioritizing the mapping of their genomes is essential.

5 Concluding Remarks

The implementation of harmonized multiplex DNA target panels designed for detecting human gastrointestinal infections increasingly seen in clinical microbiology laboratories has improved the ability to identify common intestinal protozoa, bacterial, and viral pathogens in fecal samples. While these innovative diagnostic tools offer numerous advantages, their increasing use, which do not rely on traditional microscopy-based techniques, has resulted in a gradual and widespread decline in expertise related to parasite morphology for identification. This decline in microscopy and morphology skills may or may not have adverse effects on patient care, public health, and epidemiology. The lack of proficiency in morphology could lead to missed and incorrect diagnoses, as well as inaccurate descriptions of newly discovered human parasitic diseases [46, 121].

HTS-based methods have proven their use when it comes to the simultaneous detection and differentiation of intestinal parasites. Although approaches such as metaribosomalomics (metabarcoding) and shot-gun sequencing may not yet have been developed sufficiently to replace state-of-the-art methods in the clinical microbiology laboratory, the methods can provide useful data that can be used to inform research projects and populate databases.

Major limitations specific to short-read amplicon-based HTS methods currently include lack of resolution and insufficient databases when it comes to mapping pro- and eukaryotic fecal communities of host species that have not been sampled to any larger extent. Efforts continue to combine morphological observations with HTS-based DNA data to better populate publicly available DNA sequence databases. Moreover, in such amplicon-based approaches, the extent to which preferential amplification of some species over others will typically remain uncertain, as it may depend on (1) whether the DNA extraction used is more efficient for some organisms, and (2) the composition of each individual fecal sample, including the distribution and diversity of bacteria, parasites, and fungi.

References

1. Hamad I, Raoult D, Bittar F (2016) Repertory of eukaryotes (eukaryome) in the human gastrointestinal tract: taxonomy and detection methods. Parasite Immunol 38:12–36
2. Lukeš J, Stensvold CR, Jirků-Pomajbíková K, Wegener PL (2015) Are human intestinal eukaryotes beneficial or commensals? PLoS Pathog 11:e1005039
3. Verweij JJ, van Lieshout L (2011) Intestinal parasitic infections in an industrialized country; a new focus on children with better DNA-based diagnostics. Parasitology 138:1492–1498
4. Verweij JJ, Stensvold CR (2014) Molecular testing for clinical diagnosis and epidemiological investigations of intestinal parasitic infections. Clin Microbiol Rev 27:371–418
5. Collaborators GDaI (2020) Intestinal nematode infections—level 3 cause. Lancet 396:S29
6. Buonfrate D, Bisanzio D, Giorli G, Odermatt P, Fürst T, Greenaway C, French M, Reithinger R, Gobbi F, Montresor A, Bisoffi Z (2020) The global prevalence of Strongyloides stercoralis infection. Pathogens 9
7. Anonymous. (2018) Global, regional, and national incidence, prevalence, and years lived with disability for 354 diseases and injuries for 195 countries and territories, 1990-2017: a systematic analysis for the Global Burden of Disease Study 2017. Lancet 392:1789–1858
8. Röser D, Simonsen J, Nielsen HV, Stensvold CR, Mølbak K (2013) Dientamoeba fragilis in Denmark: epidemiological experience derived from four years of routine real-time PCR. Eur J Clin Microbiol Infect Dis 32:1303. https://doi.org/10.1007/s10096-013-1880-2
9. Widerström M, Schönning C, Lilja M, Lebbad M, Ljung T, Allestam G, Ferm M, Björkholm B, Hansen A, Hiltula J, Långmark J, Löfdahl M, Omberg M, Reuterwall C, Samuelsson E, Widgren K, Wallensten A, Lindh J (2014) Large outbreak of Cryptosporidium hominis infection transmitted through the public water supply, Sweden. Emerg Infect Dis 20:581–589
10. Mac Kenzie WR, Hoxie NJ, Proctor ME, Gradus MS, Blair KA, Peterson DE, Kazmierczak JJ, Addiss DG, Fox KR, Rose JB (1994) A massive outbreak in Milwaukee of cryptosporidium infection transmitted through the public water supply. N Engl J Med 331:161–167
11. Gerace E, Lo Presti VDM, Biondo C (2019) Cryptosporidium infection: epidemiology, pathogenesis, and differential diagnosis. Eur J Microbiol Immunol (Bp) 9:119–123
12. Dougherty M, Bartelt LA (2022) Giardia and growth impairment in children in high-prevalence settings: consequence or co-incidence? Curr Opin Infect Dis 35:417–423
13. Rogawski ET, Bartelt LA, Platts-Mills JA, Seidman JC, Samie A, Havt A, Babji S, Trigoso DR, Qureshi S, Shakoor S, Haque R, Mduma E, Bajracharya S, Gaffar SMA, Lima AAM, Kang G, Kosek MN, Ahmed T, Svensen E, Mason C, Bhutta ZA, Lang DR, Gottlieb M, Guerrant RL, Houpt ER, Bessong PO, Investigators M-EN (2017) Determinants and impact of giardia infection in the first 2 years of life in the MAL-ED birth cohort. J Pediatric Infect Dis Soc 6:153–160
14. Stanley SL (2003) Amoebiasis. Lancet 361:1025–1034
15. Tringe SG, Hugenholtz P (2008) A renaissance for the pioneering 16S rRNA gene. Curr Opin Microbiol 11:442–446
16. Flegontova O, Lukes J, Horak A (2023) Intragenomic diversity of the V9 hypervariable domain in eukaryotes has little effect on metabarcoding. iScience 26:107291
17. Tanaka R, Hino A, Tsai IJ, Palomares-Rius JE, Yoshida A, Ogura Y, Hayashi T, Maruyama H, Kikuchi T (2014) Assessment of helminth biodiversity in wild rats using 18S rDNA based metagenomics. PLoS One 9:e110769
18. Aivelo T, Medlar A, Löytynoja A, Laakkonen J, Jernvall J (2015) Tracking year-to-year changes in intestinal nematode communities of rufous mouse lemurs (Microcebus rufus). Parasitology 142:1095–1107
19. Avramenko RW, Redman EM, Lewis R, Yazwinski TA, Wasmuth JD, Gilleard JS (2015) Exploring the gastrointestinal "Nemabiome": deep amplicon sequencing to quantify the species composition of parasitic nematode communities. PLoS One 10:e0143559
20. Bradley IM, Pinto AJ, Guest JS (2016) Design and evaluation of Illumina MiSeq-compatible, 18S rRNA gene-specific primers for improved characterization of mixed phototrophic communities. Appl Environ Microbiol 82:5878–5891
21. Avramenko RW, Redman EM, Lewis R, Bichuette MA, Palmeira BM, Yazwinski TA, Gilleard JS (2017) The use of nemabiome metabarcoding to explore gastro-intestinal nematode species diver-

sity and anthelmintic treatment effectiveness in beef calves. Int J Parasitol 47:893–902
22. Avramenko RW, Bras A, Redman EM, Woodbury MR, Wagner B, Shury T, Liccioli S, Windeyer MC, Gilleard JS (2018) High species diversity of trichostrongyle parasite communities within and between Western Canadian commercial and conservation bison herds revealed by nemabiome metabarcoding. Parasit Vectors 11:299
23. Pafčo B, Čížková D, Kreisinger J, Hasegawa H, Vallo P, Shutt K, Todd A, Petrželková KJ, Modrý D (2018) Metabarcoding analysis of strongylid nematode diversity in two sympatric primate species. Sci Rep 8:5933
24. Wylezich C, Belka A, Hanke D, Beer M, Blome S, Höper D (2019) Metagenomics for broad and improved parasite detection: a proof-of-concept study using swine faecal samples. Int J Parasitol 49:769–777
25. Redman E, Queiroz C, Bartley DJ, Levy M, Avramenko RW, Gilleard JS (2019) Validation of ITS-2 rDNA nemabiome sequencing for ovine gastrointestinal nematodes and its application to a large scale survey of UK sheep farms. Vet Parasitol 275:108933
26. Queiroz C, Levy M, Avramenko R, Redman E, Kearns K, Swain L, Silas H, Uehlinger F, Gilleard JS (2020) The use of ITS-2 rDNA nemabiome metabarcoding to enhance anthelmintic resistance diagnosis and surveillance of ovine gastrointestinal nematodes. Int J Parasitol Drugs Drug Resist 14:105–117
27. Wang T, Avramenko RW, Redman EM, Wit J, Gilleard JS, Colwell DD (2020) High levels of third-stage larvae (L3) overwinter survival for multiple cattle gastrointestinal nematode species on western Canadian pastures as revealed by ITS2 rDNA metabarcoding. Parasit Vectors 13:458
28. Gogarten JF, Calvignac-Spencer S, Nunn CL, Ulrich M, Saiepour N, Nielsen HV, Deschner T, Fichtel C, Kappeler PM, Knauf S, Müller-Klein N, Ostner J, Robbins MM, Sangmaneedet S, Schülke O, Surbeck M, Wittig RM, Sliwa A, Strube C, Leendertz FH, Roos C, Noll A (2020) Metabarcoding of eukaryotic parasite communities describes diverse parasite assemblages spanning the primate phylogeny. Mol Ecol Resour 20:204–215
29. Stensvold CR, Jirků-Pomajbíková K, Tams KW, Jokelainen P, Berg RPKD, Marving E, Petersen RF, Andersen LO, Angen Ø, Nielsen HV (2021) Parasitic intestinal protists of zoonotic relevance detected in pigs by metabarcoding and real-time PCR. Microorganisms 9
30. Poissant J, Gavriliuc S, Bellaw J, Redman EM, Avramenko RW, Robinson D, Workentine ML, Shury TK, Jenkins EJ, McLoughlin PD, Nielsen MK, Gilleard JS (2021) A repeatable and quantitative DNA metabarcoding assay to characterize mixed strongyle infections in horses. Int J Parasitol 51:183–192
31. Davey ML, Utaaker KS, Fossøy F (2021) Characterizing parasitic nematode faunas in faeces and soil using DNA metabarcoding. Parasit Vectors 14:422
32. Beaumelle C, Redman EM, de Rijke J, Wit J, Benabed S, Debias F, Duhayer J, Pardonnet S, Poirel MT, Capron G, Chabot S, Rey B, Yannic G, Gilleard JS, Bourgoin G (2021) Metabarcoding in two isolated populations of wild roe deer (Capreolus capreolus) reveals variation in gastrointestinal nematode community composition between regions and among age classes. Parasit Vectors 14:594
33. Chihi A, O'Brien Andersen L, Aoun K, Bouratbine A, Stensvold CR (2022) Amplicon-based next-generation sequencing of eukaryotic nuclear ribosomal genes (metabarcoding) for the detection of single-celled parasites in human faecal samples. Parasite Epidemiol Control 17:e00242
34. De Seram EL, Redman EM, Wills FK, de Queiroz C, Campbell JR, Waldner CL, Parker SE, Avramenko RW, Gilleard JS, Uehlinger FD (2022) Regional heterogeneity and unexpectedly high abundance of Cooperia punctata in beef cattle at a northern latitude revealed by ITS-2 rDNA nemabiome metabarcoding. Parasit Vectors 15:17
35. Vonaesch P, Billy V, Mann AE, Morien E, Habib A, Collard JM, Dede M, Kapel N, Sansonetti PJ, Parfrey LW, Afribiota I (2023) The eukaryome of African children is influenced by geographic location, gut biogeography, and nutritional status. Microlife 4:uqad033
36. Donohue ME, Hert ZL, Karrick CE, Rowe AK, Wright PC, Randriamanandaza LJ, Zakamanana F, Nomenjanahary ES, Everson KM, Weisrock DW (2023) Lemur gut microeukaryotic community variation is not associated with host phylogeny, diet, or habitat. Microb Ecol 86:2149–2160
37. Garcia-Montoya GM, Galvan-Diaz AL, Alzate JF (2023) Metataxomics reveals blastocystis subtypes mixed infections in Colombian children. Infect Genet Evol 113:105478
38. Rozo-Montoya N, Bedoya-Urrego K, Alzate JF (2023) Monitoring potentially pathogenic protists in sewage sludge using Metataxonomics. Food Waterborne Parasitol 33:e00210
39. del Campo J, Bass D, Keeling PJ (2019) The eukaryome: diversity and role of microeukaryotic organisms associated with animal hosts. Funct Ecol 34:2045–2054
40. Lane M, Kashani M, Barratt JL, Qvarnstrom Y, Yabsley MJ, Garrett KB, Bradbury RS (2023) Application of a universal parasite diagnostic test to biological specimens collected from animals. Int J Parasitol Parasites Wildl 20:20–30
41. Owens LA, Friant S, Martorelli Di Genova B, Knoll LJ, Contreras M, Noya-Alarcon O, Dominguez-Bello MG, Goldberg TL (2024) VESPA: an optimized protocol for accurate metabarcoding-based characterization of vertebrate eukaryotic

endosymbiont and parasite assemblages. Nat Commun 15:402
42. Ring HC, Thorsen J, Saunte DM, Lilje B, Bay L, Riis PT, Larsen N, Andersen LO, Nielsen HV, Miller IM, Bjarnsholt T, Fuursted K, Jemec GB (2017) The follicular skin microbiome in patients with Hidradenitis Suppurativa and healthy controls. JAMA Dermatol 153:897–905
43. Choi J, Park JS (2020) Comparative analyses of the V4 and V9 regions of 18S rDNA for the extant eukaryotic community using the Illumina platform. Sci Rep 10:6519
44. Stensvold CR (2019) Pinning down the role of common luminal intestinal parasitic protists in human health and disease - status and challenges. Parasitology 146:695–701
45. Cháves-González LE, Morales-Calvo F, Mora J, Solano-Barquero A, Verocai GG, Rojas A (2022) What lies behind the curtain: cryptic diversity in helminth parasites of human and veterinary importance. Curr Res Parasitol Vector Borne Dis 2:100094
46. Bradbury RS, Sapp SGH, Potters I, Mathison BA, Frean J, Mewara A, Sheorey H, Tamarozzi F, Couturier MR, Chiodini P, Pritt B (2022) Where have all the diagnostic morphological parasitologists gone? J Clin Microbiol 60:e0098622
47. Poulsen CS, Stensvold CR (2016) Systematic review on Endolimax nana: a less well studied intestinal ameba. Trop Parasitol 6:8–29
48. Lee KJ, Ahn YK, Yong TS (2000) A small-scale survey of intestinal parasite infections among children and adolescents in Legaspi city, The Philippines. Korean J Parasitol 38:183–185
49. Cross JH, Clarke MD, Durfee PT, Irving GS, Taylor J, Partono F, Joesoef A, Hudojo OS (1975) Parasitology survey and seroepidemiology of amoebiasis in South Kalimantan (Borneo), Indonesia. Southeast Asian J Trop Med Public Health 6:52–60
50. Stensvold CR, Lebbad M, Clark CG (2011) Last of the human protists: the phylogeny and genetic diversity of Iodamoeba. Mol Biol Evol 29:39. https://doi.org/10.1093/molbev/msr238
51. Hocke EF, Jamy M, Burki F, Clark CG, Stensvold CR (2022) Unravelling the phylogeny of a common intestinal Protist: intrageneric diversity of Endolimax. Protist 173:125908
52. Clark CG (1997) Extensive genetic diversity in Blastocystis hominis. Mol Biochem Parasitol 87:79–83
53. Stensvold CR, Lebbad M, Verweij JJ, Jespersgaard C, von Samson-Himmelstjerna G, Nielsen SS, Nielsen HV (2010) Identification and delineation of members of the Entamoeba complex by pyrosequencing. Mol Cell Probes 24:403–406
54. Vlčková K, Kreisinger J, Pafčo B, Čížková D, Tagg N, Hehl AB, Modrý D (2018) Diversity of Entamoeba spp. in African great apes and humans: an insight from Illumina MiSeq high-throughput sequencing. Int J Parasitol 48:519–530
55. Levecke B, Dorny P, Vercammen F, Visser LG, Van Esbroeck M, Vercruysse J, Verweij JJ (2015) Transmission of Entamoeba nuttalli and Trichuris trichiura from nonhuman primates to humans. Emerg Infect Dis 21:1871–1872
56. Stensvold CR, Berg R, Maloney JG, Molokin A, Santin M (2023) Molecular characterization of Blastocystis and Entamoeba of muskoxen and sheep in Greenland. Int J Parasitol 53:673–685
57. Barratt JLN, Lane M, Talundzic E, Richins T, Robertson G, Formenti F, Pritt B, Verocai G, Nascimento de Souza J, Mato Soares N, Traub R, Buonfrate D, Bradbury RS (2019) A global genotyping survey of Strongyloides stercoralis and Strongyloides fuelleborni using deep amplicon sequencing. PLoS Negl Trop Dis 13:e0007609
58. Kounosu A, Murase K, Yoshida A, Maruyama H, Kikuchi T (2019) Improved 18S and 28S rDNA primer sets for NGS-based parasite detection. Sci Rep 9:15789
59. Beknazarova M, Barratt JLN, Bradbury RS, Lane M, Whiley H, Ross K (2019) Detection of classic and cryptic Strongyloides genotypes by deep amplicon sequencing: a preliminary survey of dog and human specimens collected from remote Australian communities. PLoS Negl Trop Dis 13:e0007241
60. Wang T, Redman EM, Morosetti A, Chen R, Kulle S, Morden N, McFarland C, Vineer HR, Colwell DD, Morgan ER, Gilleard JS (2021) Seasonal epidemiology of gastrointestinal nematodes of cattle in the northern continental climate zone of western Canada as revealed by internal transcribed spacer-2 ribosomal DNA nemabiome barcoding. Parasit Vectors 14:604
61. Chan AHE, Saralamba N, Saralamba S, Ruangsittichai J, Chaisiri K, Limpanont Y, Charoennitiwat V, Thaenkham U (2022) Sensitive and accurate DNA metabarcoding of parasitic helminth mock communities using the mitochondrial rRNA genes. Sci Rep 12:9947
62. Scanlan PD, Marchesi JR (2008) Micro-eukaryotic diversity of the human distal gut microbiota: qualitative assessment using culture-dependent and -independent analysis of faeces. ISME J 2:1183–1193
63. Forcina G, Perez-Pardal L, Carvalheira J, Beja-Pereira A (2022) Gut microbiome studies in livestock: achievements, challenges, and perspectives. Animals (Basel) 12
64. Stensvold CR, Lebbad M, Hansen A, Beser J, Belkessa S, O'Brien Andersen L, Clark CG (2020) Differentiation of Blastocystis and parasitic archamoebids encountered in untreated wastewater samples by amplicon-based next-generation sequencing. Parasite Epidemiol Control 9:e00131
65. Benita Y, Oosting RS, Lok MC, Wise MJ, Humphery-Smith I (2003) Regionalized GC content of template DNA as a predictor of PCR success. Nucleic Acids Res 31:e99
66. Ramayo-Caldas Y, Prenafeta-Boldu F, Zingaretti LM, Gonzalez-Rodriguez O, Dalmau A, Quintanilla

R, Ballester M (2020) Gut eukaryotic communities in pigs: diversity, composition and host genetics contribution. Anim Microbiome 2:18
67. Alfellani MA, Stensvold CR, Vidal-Lapiedra A, Onuoha ES, Fagbenro-Beyioku AF, Clark CG (2013) Variable geographic distribution of Blastocystis subtypes and its potential implications. Acta Trop 126:11–18
68. Chan AHE, Chaisiri K, Morand S, Saralamba N, Thaenkham U (2020) Evaluation and utility of mitochondrial ribosomal genes for molecular systematics of parasitic nematodes. Parasit Vectors 13:364
69. Maloney JG, Molokin A, Santin M (2020) Use of Oxford Nanopore MinION to generate full-length sequences of the Blastocystis small subunit (SSU) rRNA gene. Parasit Vectors 13:595
70. Maloney JG, Santin M (2021) Mind the gap: new full-length sequences of Blastocystis subtypes generated via Oxford Nanopore Minion sequencing allow for comparisons between full-length and partial sequences of the small subunit of the ribosomal RNA gene. Microorganisms 9
71. Maloney JG, Jang Y, Molokin A, George NS, Santin M (2021) Wide genetic diversity of Blastocystis in white-tailed deer (Odocoileus virginianus) from Maryland, USA. Microorganisms 9
72. Maloney JG, Molokin A, Segui R, Maravilla P, Martinez-Hernandez F, Villalobos G, Tsaousis AD, Gentekaki E, Munoz-Antoli C, Klisiowicz DR, Oishi CY, Toledo R, Esteban JG, Koster PC, de Lucio A, Dashti A, Bailo B, Calero-Bernal R, Gonzalez-Barrio D, Carmena D, Santin M (2022) Identification and molecular characterization of four new Blastocystis subtypes designated ST35-ST38. Microorganisms 11
73. Baek S, Maloney JG, Molokin A, George NS, Cortes Vecino JA, Santin M (2022) Diversity of Blastocystis subtypes in horses in Colombia and identification of two new subtypes. Microorganisms 10
74. Stensvold CR, Clark CG (2020) Pre-empting Pandora's box: blastocystis subtypes revisited. Trends Parasitol 36:229–232
75. Stensvold CR, Lebbad M, Clark CG (2012) Last of the human Protists: the phylogeny and genetic diversity of Iodamoeba. Mol Biol Evol 29:39–42
76. Stensvold CR, Ascuña-Durand K, Chihi A, Belkessa S, Kurt Ö, El-Badry A, van der Giezen M, Clark CG (2022) Further insight into the genetic diversity of Entamoeba coli and Entamoeba hartmanni. J Eukaryot Microbiol. https://doi.org/10.1111/jeu.12949:e12949
77. Stensvold CR, Lebbad M, Victory EL, Verweij JJ, Tannich E, Alfellani M, Legarraga P, Clark CG (2011) Increased sampling reveals novel lineages of Entamoeba: consequences of genetic diversity and host specificity for taxonomy and molecular detection. Protist 162:525–541
78. Zhao F, Filker S, Xu K, Li J, Zhou T, Huang P (2019) Effects of intragenomic polymorphism in the SSU rRNA gene on estimating marine microeukaryotic diversity: a test for ciliates using single-cell high-throughput DNA sequencing. Limnol Oceanogr Methods 17:533–543
79. Ko PP, Suzuki K, Canales-Ramos M, Aung M, Htike WW, Yoshida A, Montes M, Morishita K, Gotuzzo E, Maruyama H, Nagayasu E (2020) Phylogenetic relationships of Strongyloides species in carnivore hosts. Parasitol Int 78:102151
80. Chandler AC (1925) The species of Strongyloides (Nematoda). Parasitology 17:426–433
81. Speare R, Tinsley D (1986) Strongyloides felis: an "old" worm rediscovered in Australian cats. Aust Vet Pract 16:10–18
82. Petersen AM, Stensvold CR, Mirsepasi H, Engberg J, Friis-Møller A, Porsbo LJ, Hammerum AM, Nordgaard-Lassen I, Nielsen HV, Krogfelt KA (2013) Active ulcerative colitis is associated with low prevalence of Blastocystis and Dientamoeba fragilis infection. Scand J Gastroenterol 48:638
83. Andersen LO, Bonde I, Nielsen HB, Stensvold CR (2015) A retrospective metagenomics approach to studying Blastocystis. FEMS Microbiol Ecol 91:1
84. Krogsgaard LR, Engsbro AL, Stensvold CR, Nielsen HV, Bytzer P (2015) The prevalence of intestinal parasites is not greater among individuals with irritable bowel syndrome: a population-based case-control study. Clin Gastroenterol Hepatol 13:507–513.e2
85. Stensvold CR, van der Giezen M (2018) Associations between gut microbiota and common luminal intestinal parasites. Trends Parasitol 34:369–377
86. Even G, Lokmer A, Rodrigues J, Audebert C, Viscogliosi E, Segurel L, Chabe M (2021) Changes in the human gut microbiota associated with colonization by Blastocystis sp. and Entamoeba spp. in non-industrialized populations. Front Cell Infect Microbiol 11:533528
87. Castañeda S, Muñoz M, Villamizar X, Hernández PC, Vásquez LR, Tito RY, Ramírez JD (2020) Microbiota characterization in Blastocystis-colonized and Blastocystis-free school-age children from Colombia. Parasit Vectors 13:521
88. Stensvold CR, Sorland BA, Berg R, Andersen LO, van der Giezen M, Bowtell JL, El-Badry AA, Belkessa S, Kurt O, Nielsen HV (2022) Stool microbiota diversity analysis of Blastocystis-positive and Blastocystis-negative individuals. Microorganisms 10
89. Gabrielli S, Furzi F, Fontanelli Sulekova L, Taliani G, Mattiucci S (2020) Occurrence of Blastocystis-subtypes in patients from Italy revealed association of ST3 with a healthy gut microbiota. Parasite Epidemiol Control 9:e00134
90. Lappan R, Classon C, Kumar S, Singh OP, de Almeida RV, Chakravarty J, Kumari P, Kansal S, Sundar S, Blackwell JM (2019) Meta-taxonomic analysis of prokaryotic and eukaryotic gut flora in stool samples from visceral leishmaniasis cases and endemic controls in Bihar State India. PLoS Negl Trop Dis 13:e0007444

91. Krogsgaard LR, Andersen LO, Johannesen TB, Engsbro AL, Stensvold CR, Nielsen HV, Bytzer P (2018) Characteristics of the bacterial microbiome in association with common intestinal parasites in irritable bowel syndrome. Clin Transl Gastroenterol 9:161
92. Kodio A, Coulibaly D, Koné AK, Konaté S, Doumbo S, Guindo A, Bittar F, Gouriet F, Raoult D, Thera MA, Ranque S (2019) Colonization is associated with increased diversity and altered gut bacterial communities in healthy Malian children. Microorganisms 7
93. Han N, Peng X, Qiang Y, Zhang T, Li X, Zhang W (2023) Genetic characteristics of Blastocystis sp. ST3 at the genome level in the Chinese population. Parasitol Res 122:2719–2727
94. Alzate JF, Toro-Londono M, Cabarcas F, Garcia-Montoya G, Galvan-Diaz A (2020) Contrasting microbiota profiles observed in children carrying either Blastocystis spp. or the commensal amoebas Entamoeba coli or Endolimax nana. Sci Rep 10:15354
95. O'Brien Andersen L, Karim AB, Roager HM, Vigsnæs LK, Krogfelt KA, Licht TR, Stensvold CR (2016) Associations between common intestinal parasites and bacteria in humans as revealed by qPCR. Eur J Clin Microbiol Infect Dis 35:1427–1431
96. Kim MJ, Lee YJ, Kim TJ, Won EJ (2021) Gut microbiome profiles in colonizations with the enteric protozoa Blastocystis in Korean populations. Microorganisms 10
97. Renelies-Hamilton J, Noguera-Julian M, Parera M, Paredes R, Pacheco L, Dacal E, Saugar JM, Rubio JM, Poulsen M, Koster PC, Carmena D (2019) Exploring interactions between Blastocystis sp., Strongyloides spp. and the gut microbiomes of wild chimpanzees in Senegal. Infect Genet Evol 74:104010
98. Huang LS, Yeh YM, Chiu SF, Huang PJ, Chu LJ, Huang CY, Cheng FW, Chen LC, Lin HC, Shih YW, Lin WN, Huang KY (2023) Intestinal microbiota analysis of different Blastocystis subtypes and Blastocystis-negative individuals in Taiwan. Biomed J. https://doi.org/10.1016/j.bj.2023.100661:100661
99. Deng L, Wojciech L, Png CW, Kioh YQD, Ng GC, Chan ECY, Zhang Y, Gascoigne NRJ, Tan KSW (2023) Colonization with ubiquitous protist Blastocystis ST1 ameliorates DSS-induced colitis and promotes beneficial microbiota and immune outcomes. NPJ Biofilms Microbiomes 9:22
100. Iebba V, Santangelo F, Totino V, Pantanella F, Monsia A, Di Cristanziano V, Di Cave D, Schippa S, Berrilli F, D'Alfonso R (2016) Gut microbiota related to Giardia duodenalis, Entamoeba spp. and Blastocystis hominis infections in humans from Côte d'Ivoire. J Infect Dev Ctries 10:1035–1041
101. Caudet J, Trelis M, Cifre S, Soriano JM, Rico H, Merino-Torres JF (2022) Interplay between intestinal bacterial communities and unicellular parasites in a morbidly obese population: a neglected trinomial. Nutrients 14
102. Behboud S, Solhjoo K, Erfanian S, Pirestani M, Abdoli A (2022) Alteration of gut bacteria composition among individuals with asymptomatic Blastocystis infection: a case-control study. Microb Pathog 169:105639
103. Tito RY, Chaffron S, Caenepeel C, Lima-Mendez G, Wang J, Vieira-Silva S, Falony G, Hildebrand F, Darzi Y, Rymenans L, Verspecht C, Bork P, Vermeire S, Joossens M, Raes J (2019) Population-level analysis of Blastocystis subtype prevalence and variation in the human gut microbiota. Gut 68:1180–1189
104. Nourrisson C, Scanzi J, Pereira B, NkoudMongo C, Wawrzyniak I, Cian A, Viscogliosi E, Livrelli V, Delbac F, Dapoigny M, Poirier P (2014) Blastocystis is associated with decrease of fecal microbiota protective bacteria: comparative analysis between patients with irritable bowel syndrome and control subjects. PLoS One 9:e111868
105. Růžková J, Květoňová D, Jirků M, Lhotská Z, Stensvold CR, Parfrey LW, Pomajbíková KJ (2018) Evaluating rodent experimental models for studies of Blastocystis ST1. Exp Parasitol 191:55–61
106. Gotfred-Rasmussen H, Stensvold CR, Ingham AC, Johannesen TB, Andersen LO, Röser D, Nielsen HV (2021) Impact of metronidazole treatment and Dientamoeba Fragilis colonization on gut microbiota diversity. J Pediatr Gastroenterol Nutr 73:23–29
107. Fekete E, Allain T, Siddiq A, Sosnowski O, Buret AG (2020) Giardia spp. and the gut microbiota: dangerous liaisons. Front Microbiol 11:618106
108. McGregor BA, Razmjou E, Hooshyar H, Seeger DR, Golovko SA, Golovko MY, Singer SM, Hur J, Solaymani-Mohammadi S (2023) A shotgun metagenomic analysis of the fecal microbiome in humans infected with Giardia duodenalis. Parasit Vectors 16:239
109. Maertens B, Gagnaire A, Paerewijck O, De Bosscher K, Geldhof P (2021) Regulatory role of the intestinal microbiota in the immune response against Giardia. Sci Rep 11:10601
110. Fu Y, Zhang K, Yang M, Li X, Chen Y, Li J, Xu H, Dhakal P, Zhang L (2023) Metagenomic analysis reveals the relationship between intestinal protozoan parasites and the intestinal microecological balance in calves. Parasit Vectors 16:257
111. Hares MF, Griffiths BE, Johnson F, Nelson C, Haldenby S, Stewart CJ, Duncan JS, Oikonomou G, Coombes JL (2023) Specific pathway abundances in the neonatal calf faecal microbiome are associated with susceptibility to Cryptosporidium parvum infection: a metagenomic analysis. Anim Microbiome 5:43
112. Gordon CA, Krause L, McManus DP, Morrison M, Weerakoon KG, Connor MC, Olveda RM, Ross AG, Gobert GN (2020) Helminths, polyparasitism, and the gut microbiome in the Philippines. Int J Parasitol 50:217–225

113. Rubel MA, Abbas A, Taylor LJ, Connell A, Tanes C, Bittinger K, Ndze VN, Fonsah JY, Ngwang E, Essiane A, Fokunang C, Njamnshi AK, Bushman FD, Tishkoff SA (2020) Lifestyle and the presence of helminths is associated with gut microbiome composition in Cameroonians. Genome Biol 21:122
114. Kupritz J, Angelova A, Nutman TB, Gazzinelli-Guimaraes PH (2021) Helminth-induced human gastrointestinal dysbiosis: a systematic review and meta-analysis reveals insights into altered taxon diversity and microbial gradient collapse. mBio 12:e0289021
115. Sargsian S, Chen Z, Lee SC, Robertson A, Thur RS, Sproch J, Devlin JC, Tee MZ, Er YX, Copin R, Heguy A, Pironti A, Torres VJ, Ruggles KV, Lim YAL, Bethony J, Loke P, Cadwell K (2022) Clostridia isolated from helminth-colonized humans promote the life cycle of Trichuris species. Cell Rep 41:111725
116. Tee MZ, Er YX, Easton AV, Yap NJ, Lee IL, Devlin J, Chen Z, Ng KS, Subramanian P, Angelova A, Oyesola O, Sargsian S, Ngui R, Beiting DP, Boey CCM, Chua KH, Cadwell K, Lim YAL, Loke P, Lee SC (2022) Gut microbiome of helminth-infected indigenous Malaysians is context dependent. Microbiome 10:214
117. Jenkins TP, Brindley PJ, Gasser RB, Cantacessi C (2019) Helminth microbiomes - a hidden treasure trove? Trends Parasitol 35:13–22
118. Srivathsan A, Sha JC, Vogler AP, Meier R (2015) Comparing the effectiveness of metagenomics and metabarcoding for diet analysis of a leaf-feeding monkey (Pygathrix nemaeus). Mol Ecol Resour 15:250–261
119. Srivathsan A, Ang A, Vogler AP, Meier R (2016) Fecal metagenomics for the simultaneous assessment of diet, parasites, and population genetics of an understudied primate. Front Zool 13:17
120. Bradbury RS, Pafčo B, Nosková E, Hasegawa H (2021) Strongyloides genotyping: a review of methods and application in public health and population genetics. Int J Parasitol 51:1153–1166
121. Mewara A, Gile GH, Mathison B, Zhao H, Pritt B, Bradbury RS (2023) Lophomonas as a respiratory pathogen-jumping the gun. J Clin Microbiol. https://doi.org/10.1128/jcm.00845-23:e0084523

Part III

Apicomplexans

Where Is *Cryptosporidium* Population Genomics Going Next: The Challenge and the Promise

Randi Turner, Rodrigo de Paula Baptista, Benjamin M. Rosenthal, Jessica C. Kissinger, and Asis Khan

Abstract

Forty years have passed since species of *Cryptosporidium*, single-cellular protozoan parasites were first recognized as significant human pathogens (as opportunistic pathogens in AIDS patients), and more than 10 years since these enteric food and waterborne parasites were recognized as major sources of debilitating infant and childhood diarrhea. In that time, our appreciation of the diversity of this species complex has grown, with a dawning appreciation of complexities in their geography and distribution among hosts. However, the discovery of myriad species and genotypes in the genus chastens researchers hoping to understand the particularities of a given infection or outbreak and underscores the limitations inherent to available diagnostic assays. Here, we briefly review foundational aspects of *Cryptosporidium* biology and introduce recent methodological advances that herald a newfound opportunity to characterize parasite diversity at global, regional, community, and clinical levels and call for renewed commitment to sampling these parasites, widely, in order to finally understand the processes driving parasite evolution and the practical consequences of those processes for clinical management and public health control.

Keywords

Cryptosporidium · Cryptosporidiosis · Population genetics · Life cycle · Genome

R. Turner
Animal Parasitic Disease Laboratory, Beltsville Agricultural Research Center, Agricultural Research Service, United States Department of Agriculture, Beltsville, MD, USA

Center for Tropical and Emerging Global Diseases, University of Georgia, Athens, GA, USA

R. de Paula Baptista
Division of Infectious Diseases and Department of Medicine, Houston Methodist Hospital, Houston, TX, USA

B. M. Rosenthal · A. Khan (✉)
Animal Parasitic Disease Laboratory, Beltsville Agricultural Research Center, Agricultural Research Service, United States Department of Agriculture, Beltsville, MD, USA
e-mail: asis.khan@usda.gov

J. C. Kissinger
Center for Tropical and Emerging Global Diseases, University of Georgia, Athens, GA, USA

Department of Genetics, University of Georgia, Athens, GA, USA

Institute of Bioinformatics, University of Georgia, Athens, GA, USA

J. D. Ramírez González (ed.), *Recent Advances in Parasitomics*,
https://doi.org/10.1007/978-3-031-70591-5_7

1 History

Cryptosporidium is a protozoan parasite prominent across the globe. It was first discovered by Edward Tyzzer in 1907, though for many years its zoonotic potential was not appreciated, having been found primarily in asymptomatic vertebrate animals [1]. In 1976, the first human infections of *Cryptosporidium* were reported in a 3-year-old child and a 39-year-old immunosuppressed patient. Both patients experienced protracted bouts of watery diarrhea [2, 3]. This marked the beginning of the recognition of *Cryptosporidium* as a significant threat to human health. It became a defining feature of the AIDS epidemic in the early 1980s, causing illness and death by inducing persistent, dehydrating diarrhea; effective treatment options were lacking [4–7]. In 1982, the CDC published a report of 21 cases of *Cryptosporidium* infections in AIDS patients since December 1979 [8]. Just a year later, more than 50 cases had been reported [5]. Although antiretroviral therapy has significantly decreased the rate of infection among individuals with HIV/AIDS in developed countries, cryptosporidiosis remains a serious complication of HIV/AIDS and malnourished children in developing countries.

Contaminated drinking and recreational waters account for most of the reported *Cryptosporidium* exposures in developed countries. Waterborne outbreaks are reported each year in Europe and the United States [9]. Two significant incidents drew attention to the link between cryptosporidiosis and water supplies. The first occurred in 1989 when approximately 5000 individuals in Swindon and Oxfordshire were affected by an outbreak of waterborne cryptosporidiosis, prompting the UK government to establish the Expert Group on *Cryptosporidium* in Water Supplies [10]. The second event took place in 1993 when over 400,000 people were infected by a major outbreak of cryptosporidiosis in Milwaukee, WI, USA [11]. In November 2010, a waterborne outbreak of cryptosporidiosis involved approximately 27,000 residents of Östersund, Sweden, with a rapid onset and high attack rate, particularly among young and middle-aged individuals [12].

Cryptosporidium continues to pose a major health threat across the globe. According to a Global Enteric Multicenter Study (GEMS) that ran from 2007 to 2011, cryptosporidiosis is a major concern for children under 5 years of age in Africa and Asia. The study, which involved more than 22,500 children, revealed that *Cryptosporidium* is the second most significant cause of diarrhea in this age group, after rotavirus [13–15]. In 2015, it was estimated to be responsible for more than 60,000 deaths in children under 5 [14, 16]. As a result, there has been a surge in cases of *Cryptosporidium* worldwide. Global prevalence of *Cryptosporidium* infection is estimated at around 7.6%. The prevalence varies greatly by geography; prevalence in low- and middle-income countries averages 10.4%, but some countries report prevalence as high as 69% [17]. High-income countries report an average prevalence of only 4.3%, but chlorine resistance explains notable outbreaks even there, where water receives thorough treatment. Between 2001 and 2010, *Cryptosporidium* was the most prevalent cause of waterborne-acquired disease in the United States [13]. In August 2023, the United Kingdom experienced a surge of cases, with more than 2400 cases reported in a matter of weeks [18].

2 Cryptosporidiosis

Infection causes protracted bouts of watery diarrhea, known as cryptosporidiosis [19, 20]. Symptoms are usually self-limiting but can be life-threatening to children and immunocompromised individuals. Infections in children can also result in malnutrition and stunted growth, even in the absence of active diarrhea [21], perhaps because infection compromises the integrity of the intestinal–epithelial barrier and may disrupt nutrient absorption [20, 22].

Treatment options for cryptosporidiosis are extremely limited. No vaccine is available, and the sole FDA-approved drug, nitazoxanide, is ineffective in immunocompromised patients [4, 23, 24]. Recent developments have identified calcium-dependent protein kinases (CDPKs) as novel targets for therapies. Because CPDKs are required for parasite motility and secretion [25], CPDK inhibitors inhibit the growth of *Cryptosporidium* in silico and in vitro [26–28].

Transmission typically occurs by the fecal–oral route through contaminated food and water. Oocysts are resistant to chlorine-based disinfectants like those used in the treatment of drinking water and recreational swimming pools [29, 30]. Respiratory infections and inhalation transmission have also been reported [31]. Recently, it has been documented that respiratory *Cryptosporidium* infection commonly occurs in immunocompetent children in cities with poor sanitation, supporting the concept of person-to-person transmission of *Cryptosporidium* through aerosols [32, 33]. A recent study also showed evidence of direct anthroponotic transmission through providing care of an infected child in the United Kingdom [34]. In low- and middle-income countries, cryptosporidiosis is often endemic, with poor sanitation and hygiene as the primary risk factors. Studies have also noted overcrowding, young age, contact with animals, and malnutrition as other major risk factors [35].

3 Life Cycle

Cryptosporidium undergoes its entire life cycle within one host [36]. Infection is initiated by the host-ingesting oocysts, each containing four infectious sporozoites (Fig. 1) [24, 38, 39].

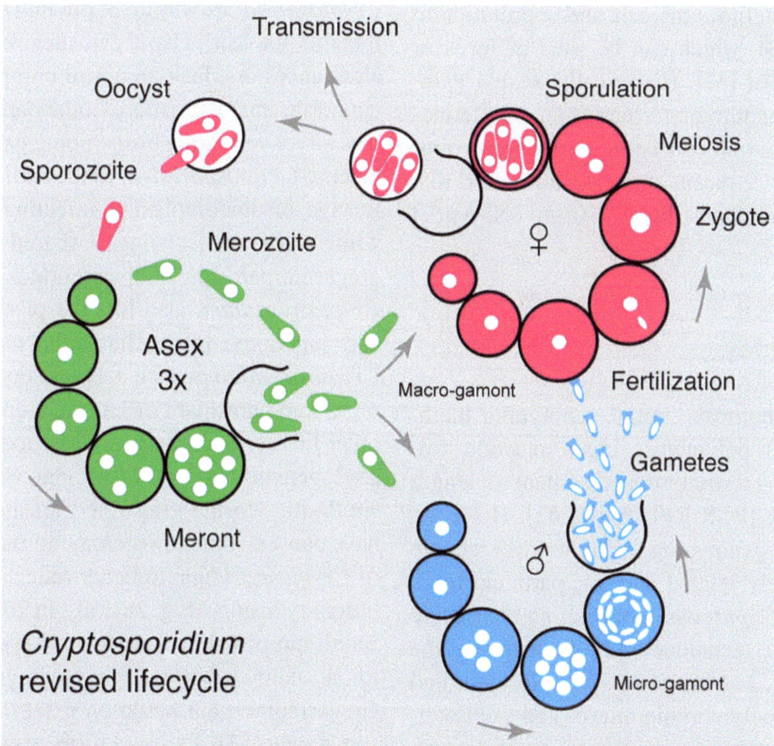

Fig. 1 Life cycle of *Cryptosporidium parvum*. Infection is initiated by ingestion of oocyst containing four infectious sporozoites. Sporozoites invade epithelial cells and asexually replicate by synchronous schizogony for three cycles, forming a meront with eight merozoites (green). Following the third cycle, reinvading merozoites become sexually committed. Macro-gamonts (pink) become cell-cycle arrested, increasing in mass and accumulating essential proteins, lipids, and carbohydrates for future oocyst formation. Micro-gamont (blue) undergo four rounds of mitotic cell division to form 16 gametes before rupturing and fertilizing female gametes. The resulting zygote undergoes meiosis and sporulation to form an oocyst that can be shed or reinfect the host. Modified from English et al. [37] under Creative Commons license

Following excystation, the sporozoites attach to the apical intestinal wall and invade the host cell, forming a parasitophorous vacuole (PV). Within the PV, the sporozoites undergo asexual replication by synchronous schizogony to form a meront with eight merozoites. Upon rupture, merozoites then reinvade and perform asexual reproduction for two additional cycles. Following the final round of asexual reproduction, reinvading merozoites become sexually committed, forming macro- and micro-gamonts. Female gametes, which are derived from macro-gamonts, are cell-cycle arrested and remain haploid; they grow in mass and stockpile proteins, lipids, and carbohydrates for the future oocyst. On the other hand, micro-gamonts undergo four rounds of mitotic cell division to form 16 male gametes before rupturing and fertilizing female gametes. The resulting zygote undergoes meiosis and sporulation to form an oocyst, which can be shed in feces or reinfect the host [37]. Historically, sexual gametes were thought to be formed from an intermediate type II meront consisting of 4N merozoites [24]. However, a recent study demonstrated that gametes could come directly from 8N type I meronts [37].

4 Species

Molecular Diagnosis Initial genotyping methods employed polymerase chain reaction followed by restriction fragment length polymorphism (PCR-RFLP) [40, 41]. However, RFLP-based genotyping cannot differentiate between closely related species, particularly *C. hominis* and *C. parvum*, requiring an alternative subgenotyping technique [42]. Sequencing the genomes of *C. parvum* and *C. hominis* identified a few highly polymorphic micro- and minisatellite loci flanked by conserved sequences amenable to amplification. These sequences were utilized to develop microsatellite and minisatellite locus-specific PCR assays capable of identifying intra-species differentiation. The small ribosomal subunit (SSU or 18S rRNA) gene is the most widely used genetic target to differentiate species infecting humans from those infecting other animal hosts; this molecule exhibits limited intraspecific variation [42]. Other loci, such as actin, heat shock protein 70 (*hsp70*), *Cryptosporidium* cell outer wall protein (COWP), and thrombospondin-related adhesive protein (TRAP-C2) have also been used for species identification [43–45]. However, accurate identification can be difficult because, at any one such locus, distinct species may bear identical or nearly identical sequences. Multi-loci sequencing has been proposed to help address this issue, but it can be costly and time-consuming [46]. Current methods cannot easily detect infection with multiple species of *Cryptosporidium*, and genetic introgression can confound their application [47, 48].

Taxonomy The precise placement of *Cryptosporidium* within the phylum Apicomplexa remains unclear. These parasites, evolved from algal ancestors, lack an apical complex or mitochondria, characteristic of other apicomplexans that infect vertebrate hosts. Some data argued for placing *Cryptosporidium* as the earliest diverging lineage of apicomplexans infecting vertebrates, while other studies suggest shared origins with gregarine parasites of invertebrates (which, like *Cryptosporidium*, also lost the plastid organelle and functions) [49]. The initial assignment of *Cryptosporidium* to the subclass coccidia, based on shared parasitism of intestinal epithelium and other life cycle similarities [50], does not receive phylogenetic support. Molecular and biological similarities to the gregarine parasites of insects have caused some to recommend reclassification of *Cryptosporidium* to better reflect evident evolutionary history (Fig. 2a) [52]. In 2014, Cavalier-Smith proposed reclassifying *Cryptosporidium* to a new subclass, Cryptogregaria, within Gregarinomorphia based on extensive 18S ribosomal data [53]. This was further revised in 2019 by Adl et al., proposing a subclass Cryptogregarinorida within the Gregarinasina class [54]. More recent transcriptomic analyses of a wider array of apicomplexans have suggested that *Cryptosporidium* is distinct from gregarine parasites, though the exact relationship remains unclear [49, 51, 52]. Regardless, this

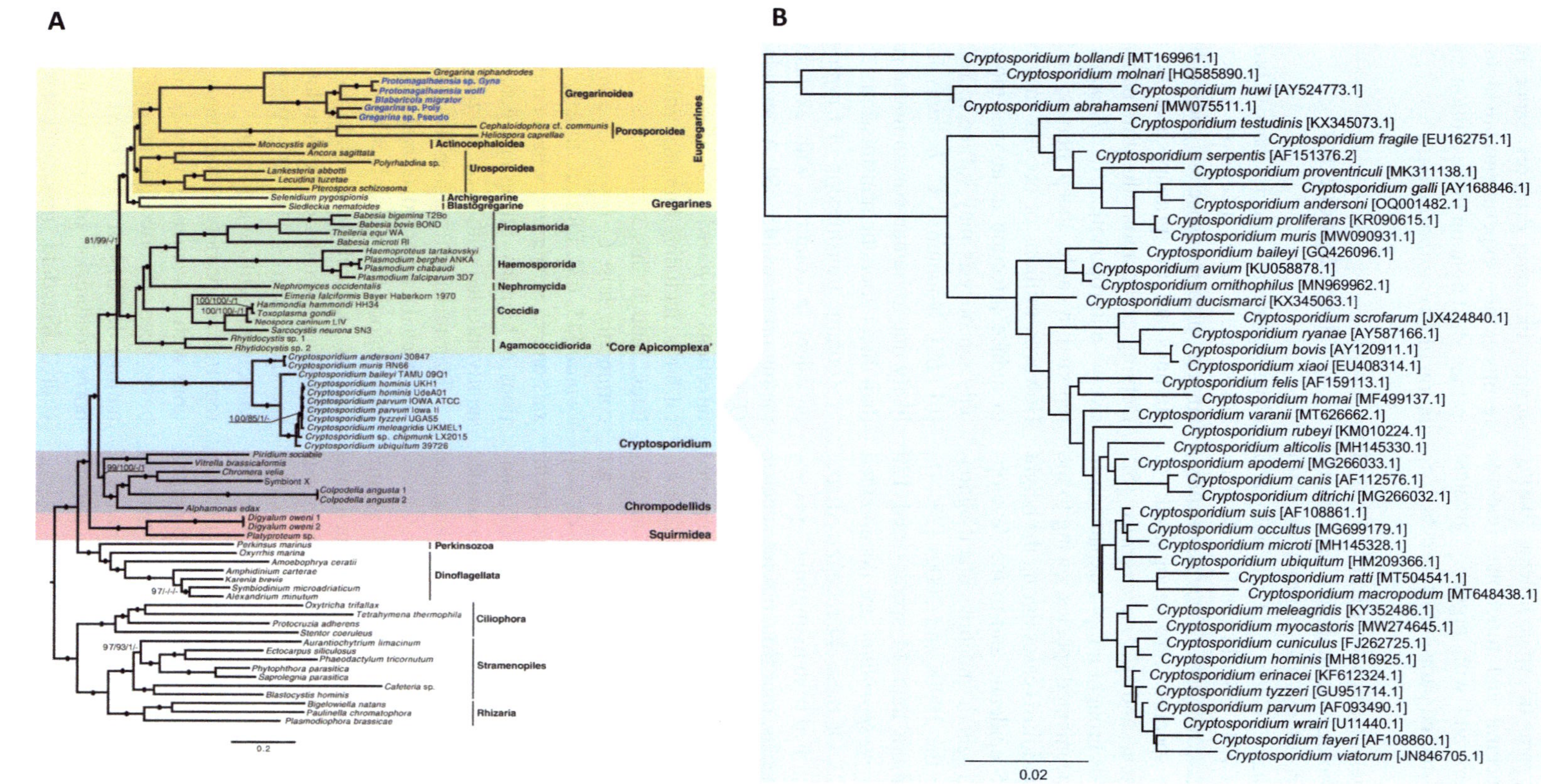

Fig. 2 Phylogeny of the Apicomplexa. (**a**) Consensus maximum likelihood tree of the Apicomplexa. Accession numbers in brackets. Reproduced from Salomaki et al. [51] under Creative Commons license. (**b**) Neighbor-joining phylogenetic tree of recognized species of *Cryptosporidium* based on 18S sequences. Accession number in brackets. Tree was constructed using Geneious Prime software (Biomatters Ltd., Auckland, New Zealand) with global alignment with free end gaps and a cost matrix of 93%

genus deserves recognition as related to, but distinct from, other parasitic Apicomplexa, sharing only certain structural and functional features common to others and with arguably greatest affinity to lesser-studied parasites of insects.

Cryptosporidium species are identified based on oocyst morphology, natural host specificity, genetic characterizations, and compliance with the International Code of Zoological Nomenclature (ICZN) [55]. In the first decades following the discovery of *Cryptosporidium*, only three such species were recognized: *C. muris*, *C. parvum*, and *C. meleagridis* [52, 56–58]. The 1970s brought a rapid expansion in the number of documented species, owing to the surge in cryptosporidiosis cases during the height of the HIV/AIDS epidemic and to the development of advanced genetic and molecular diagnostic techniques in the 1990s and 2000s [38, 45, 52]. Currently, more than 44 species and 120 genotypes of *Cryptosporidium* are documented (Fig. 2b) [38, 52]. Of these 44 recognized species, 29 were identified from mammalian species, six from avian species, four from piscine species, and one from amphibian species [52]. Although 19 *Cryptosporidium* species have been reported to cause human infection, *C. parvum* and *C. hominis* (previously known as *Cryptosporidium parvum* human genotype, or genotype 1), are responsible for ~95% of human cases [45, 52]. Global distribution of these species varies, with *C. hominis* more prevalent in industrialized countries and *C. parvum* more prevalent in countries where a preponderance of people remain engaged in farming [59].

Many species constitute organismal groups that can and do freely interbreed, but genetic and phenotypic distinctions can arise among microbes that vary in their amenability to genetic exchange. The degree of reproductive isolation among species of *Cryptosporidium* remains unclear. Interestingly, a recent experiment documented genetic exchange between *C. parvum*, a parasite of cattle and humans, and *C. tyzzeri*, a parasite of mice; these species differ 5% in their genetic makeup [60]. The offspring of the genetic exchange between *C. parvum* and *C. tyzzeri* showed combined features from both parental strains, including drug-resistance markers. The results of these experiments are highly significant as they shed light on the evolution and host adaptation of *Cryptosporidium*.

5 Host Specificity

Cryptosporidium species have a broad range of hosts; their host specificity varies drastically among parasite species and subtypes [45]. Humans are the primary host of *C. hominis*. Defying expectations that this parasite is exclusively anthroponotic, *C. hominis* has been reported in cattle, sheep, horses, and goats [24]. However, millions of oocysts may be necessary to induce infection in animal models, whereas only ten oocysts may suffice to infect a person [61, 62]. Nonhuman *C. hominis* infections are likely due to human-contaminated run-off rather than direct zoonotic transmission.

Zoonotic transmission of *Cryptosporidium* is prevalent in countries with extensive livestock raising practices [45, 59]. Alongside *C. hominis*, *C. parvum* is the most common cause of human infections. *C. parvum* has a broad range of natural hosts, including humans, wildlife, and ungulates, though this varies by strain. Cattle also commonly host *C. bovis*, *C. andersoni*, and *C. ryanae* [45, 63]. Pigs are mostly infected by *C. suis* and *C. scrofarum*, which rarely cause infections in humans. Sheep and goats are predominantly infected with *C. ubiquitium* and *C. xiaoi*.

Transmission from non-livestock animals has also been documented. *C. meleagridis* is the third most reported human-infecting *Cryptosporidium* species, particularly in children and immunocompromised individuals from low- and middle-income countries [16, 56]. While it is mostly associated with avian hosts, it has also been reported in a variety of mammals. Human infections with *C. cuniculus*, usually found in rabbits, have been reported in several industrialized nations. In 2014, it was linked to several waterborne outbreaks of cryptosporidiosis in the United Kingdom, making it the only species, aside from *C. parvum* and *C. hominis*, to do so

[64, 65]. Rodents are also an important *Cryptosporidium* reservoir, hosting six human-infecting species, including *C. muris*, *C. viatorum*, and *Cryptosporidium* chipmunk genotype I (recently proposed as *C. mortiferum*) [45]. Infections of *C. canis* and *C. felis*, acquired from household pets, have also been documented; they are more prevalent in low- and middle-income countries [66, 67].

One outstanding question is what drives host-specificity. For example, one cluster of genetically similar species of *Cryptosporidium* (including *C. parvum*, *C. hominis*, *C. cuniculus*, *C. tyzzeri*, and *C. meleagridis*) demonstrate vastly different host preferences.

6 Genome

Significant scrutiny of *Cryptosporidium* genomes started with two benchmark studies: The first *C. parvum* genome (IOWA isolate) was assembled in 2004, using modified HAPPY mapping techniques, to identify eight chromosomes (ranging from 0.88 to 1.34 Mb) and a total size of around 9.1 Mb [38, 68]. While this approach yielded a remarkably chromosomal-level assembly, representing all eight chromosomes, five gaps remained and some assembled regions were poorly supported. That same year, the first *C. hominis* genome was sequenced using similar techniques [68]. It sequenced the laboratory maintained TU502 isolate, originating from a Ugandan child. This assembly was significantly more fragmented, consisting of more than 1400 contigs and 246 gaps. These genomes showed an extreme bias toward AT content, with a GC content of approximately 31%. This contrasts with other apicomplexan parasites such as *Plasmodium* (19.4%) and *Toxoplasma* (52%). These studies also identified very compact *Cryptosporidium* genomes: ~77% of the genome was predicted to be coding regions, containing ~3900 proteins [26]. Only 5% of predicted protein-coding genes harbor introns, although the gene prediction (including introns) was mainly in silica since there was a lack of evidence to be used for proper annotation, such as proteomic data and expression data covering all life cycles. Striking contrasts to other apicomplexan parasites, both organisms lack genomes for either the apicoplast or mitochondrion, as well as genes for synthesizing simple sugars, de novo biosynthesis of amino acids, and nucleotides, despite the presence of genes related to glycolysis-based metabolism. The lack of these features underscores the dependency of *Cryptosporidium* on their host for nutrient acquisition.

In 2015, the original 2004 *C. parvum* assembly was reannotated using new RNAseq evidence, improving many of the ambiguities in the gene model [69]. This annotation became widely adopted as the reference genome (*Cp*IRef) and stood as the only complete assembly, until recently.

In 2021, Baptista et al. generated a gapless assembly of *C. parvum* using a combination of PacBio, Oxford Nanopore (ONT), and Illumina sequencing [26]. The isolate they sequenced (*C. parvum* IOWA-ATCC, *Cp*IA) was from the same lineage as the original *Cp*IRef isolate. Despite a 14-year propagation window between the two isolates, the two genome sequences showed extremely high pair-wise sequence identity. The new *Cp*IA assembly suggested a revised orientation, with inversions and relocations identified on chromosomes 2, 4, and 5. Boundaries for these regions were located within ambiguous regions or gaps in the *Cp*IRef assembly. Synteny comparisons with closely related *C. hominis* (*Ch*30976) and *C. tyzzeri* (*Ct*UGA55) genomes supported the revised arrangement (Fig. 3a).

Excitingly, they also discovered evidence of recombination and duplication between telomeric regions (Fig. 3b). One subtelomeric region in chromosome 1 showed significant compression in the *Cp*IA assembly, suggesting that there may be multiple copies within the genome. Comparative analysis showed that this region contained 13 genes that vary in copy number among species of *Cryptosporidium*, including rRNAs (18S, 28S, and 5S) and a cluster of MEDLE genes. In addition, reads from the subtelomeric region showed sequence diversity, indicating there may be recombination or conversion occurring during sexual reproduction.

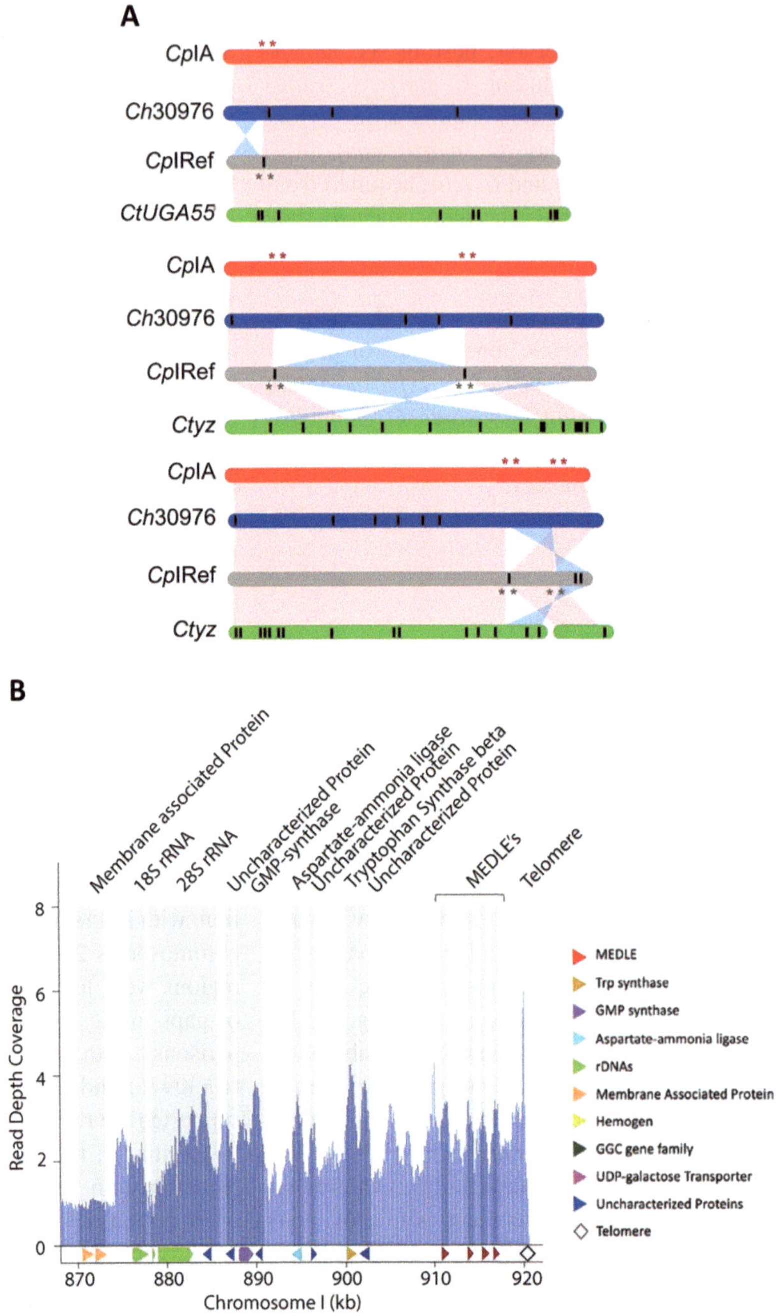

Fig. 3 Genetic structure of newly sequenced *C. parvum* IOWA-ATCC (*Cp*IA). (**a**) Synteny map of chromosomes 2, 4, and 5 comparing *Cp*IA with *C. parvum* IOAWII (*Cp*IRef), *C. hominis* (*Ch30976*), and *C. tyzzeri* (*Ct*UGA55). Vertical black lines represent known assembly gaps. (**b**) Duplication of subtelomeric regions in Illumina reads were mapped to the Chr1 region of the long-read assembly of *Cp*IA to identify read compression. Vertical gray bars correspond to annotated genes (represented below bar chart). Reproduced from Baptista et al. [26] under Creative Commons license

Subtelomeric duplication has been noted in *Plasmodium* species, where antigenic *var* genes are located [70–72]. Additionally, remodeling to the heterochromatin within the subtelomeric regions has been linked to host-cell evasion [73, 74]. In *Trypanosoma brucei*, subtelomeric recombination plays a role in VSG switching, a surface-level antigen that contributes to host-cell evasion [75, 76].

Recently, Baptista et al. generated the first telomere-to-telomere assembly of *C. parvum* using a hybrid approach of ONT and Illumina sequencing [77, 78]. The new assembly confirms the presence of duplicated subtelomeric regions of chromosomes 1, 7, and 8 and addresses ambiguous orientation and structures seen on chromosomes 2, 4, and 5. Importantly, the assembly has been extensively annotated with support of PacBio Isoseq and ONT direct RNAseq data to include UTRs, long non-coding RNAs, and antisense RNAs with the same gene names used in the current reference genome (*Cp*IRef). This achievement provides a critical resource for the community.

Advances in sequencing have led to a recent expansion in available genome assemblies. However, of the 44 recognized species, only 15 have assembled genomes, and only 8 of those are annotated (Fig. 4). Aside from *C. parvum*, all available assemblies remain fragmented [26, 47]. Additionally, several genomes, including *C. tyzzeri* UGA55 and several *C. hominis* genomes like ChUDEA, seem to be at the chromosomal level and of very good quality. However, they used *C. parvum* information to construct chromosomal scaffolds, which may introduce some bias in terms of chromosomal structure. Therefore, although they appear to be almost complete, they are not species-specific and may not be entirely evidence-based. Hence, there is a dire need to generate and annotate more complete genome sequences to better understand genetic relationships within and between species. This is particularly true when looking at closely related species such as *C. parvum*, *C. hominis*, and *C. cuniculus*, where species show greatly disparate host-specificity despite high genetic identity.

One continuing challenge for generating and annotating genome sequences is the availability of genetic material. Whole genome sequencing typically requires hundreds of nanograms or up to a microgram of high-quality DNA, depending on the sequencing method. For *Cryptosporidium*, vast numbers of oocysts are required to achieve this, as each oocyst only contains 40 fg of DNA. Achieving requisite template has been difficult because most species lacked methods for culture or reproduction in lab animal models. Recent advances have shown success in developing in vitro culturing methods. Hollow-fiber tubes, as well as silk-scaffolded cell culture mimicking an intestinal structure, have been successful in cultivating species of *Cryptosporidium* [85, 86]. Excitingly, intestinal organoid cultures have been used to sustain and propagate *C. parvum* for 2–3 weeks with full completion of the parasite life cycle [85, 87].

Additionally, multiple displacement amplification (MDA) provides a promising alternative to conventional PCR for whole genome amplification, when combined with long-read sequencing for low-input samples. Agyabeng-Dadzie et al. successfully generated almost complete genome sequences for *Cryptosporidium meleagridis*, *Staphylococcus aureus*, *Enterococcus faecium*, and *Escherichia coli* with only 0.025 ng of DNA

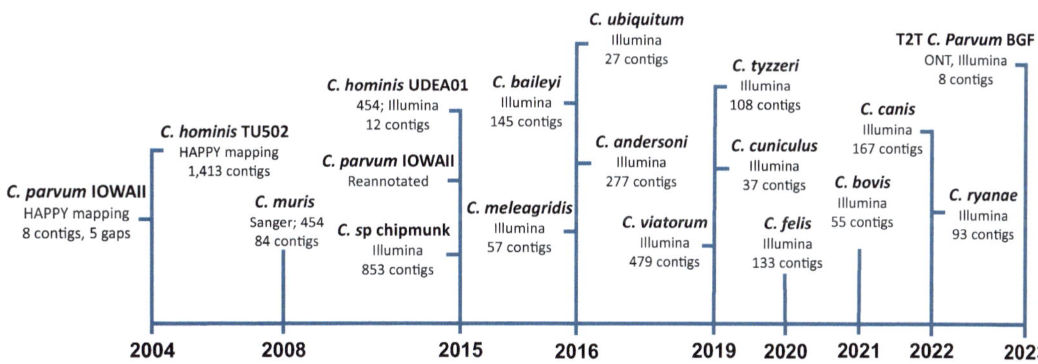

Fig. 4 Timeline of whole genome sequencing of *Cryptosporidium* species. Data from NCBI Genome Browser. GenBank accessions (left to right): GCA_000165345.1, GCA_000006425.1, GCA_000006515.1, GCA_000831705.1, GCA_002223825.1, GCA_001593445.1, GCA_001593455.1, GCA_001835705.1, GCA_001865355.1, GCA_004337795.1, GCA_004337835.1, GCA_007210665.1, GCA_014529505.1, GCA_009768925.2, GCA_027243985.1 [68, 69, 79–84]

[88, 89]. They also developed CADECT, a pipeline to remove concatemeric sequences produced by MDA, to improve de novo alignment significantly [88]. These approaches aid the study of organisms that are difficult to culture, and hold promise for addressing clinical and produce industry needs concerning scarce but consequential parasite contamination.

Single-cell genomics techniques also provide a potential way to overcome material limitations. Recent studies have shown success in combining whole genome amplification and next-generation sequencing from single oocysts [60, 90]. Sequencing from ten oocysts at 20× depth using Illumina sequencing was sufficient to provide near complete genome coverage (99.8%) [90]. However, generating a reference genome, using either this approach or long-read whole genome sequencing, would require highly purified oocysts from a relatively homogenous population [88].

7 Population Genetics

Widespread genetic diversity exists within species. Genotyping is determined using the highly polymorphic 60 kDa glycoprotein (*gp60*) involved in cell invasion. This gene contains several regions exhibiting high mutation rates, including a microsatellite region near the 5′ end with tandem-repeating serine coding sequences [47, 65, 91]. Subtypes are named as a Roman numeral followed by a letter for each species and genotype (i.e., *C. hominis* Ia or *C. parvum* IIa) [24, 45, 47, 92]. These can be further divided based on the number and type of repeated serine-coding sequences (with A representing TCA, G representing TCG, and T representing TCT). Some isolates share the same serine-coding repeats, but differ in their copies of other repetitive sequences elsewhere in the gene (represented by R) [65].

Due to the high sequence variability, primers for *gp60* are not broadly applicable. Primers designed for *C. parvum* and *C. hominis* will not amplify more divergent *Cryptosporidium* species, and the existence of typing tools are often limited by whole genome sequence availability [65]. Recent genome sequencing efforts have drastically expanded the number of species with *gp60* typing systems, including *C. felis*, *C. canis*, *C. bovis*, and *C. ryanae* [35, 67, 93–95].

Single marker sequencing provides a low-resolution understanding of genetic diversity. *gp60* undergoes strong selective pressure, and therefore may not fairly represent genome-wide variation [96–99]. Further confounding this, recombination is widely documented in *Cryptosporidium* populations, including a recent report of recombination between closely related species [60, 97, 100].

Despite its shortcomings, *gp60* remains a useful marker for understanding global population structures. Subtypes have been linked to phenotypic differences within a species such as geographic distribution, host-preference, and even transmissibility, though the underlying causes remain unclear [47, 63, 100–103].

C. parvum encompasses abundant *gp60* diversity, with more than 20 documented subtypes [47]. Three are dominant in humans: IIa, IId, and IIc. These vary in geographic distribution and host preference. *C. parvum* IIc has been found almost exclusively in humans, while IIa and IId occur more often in other animals [59, 102]. A recent genomic study argued that these deserve recognition as genetically distinct subspecies (Fig. 5) [102]. Phylogenetic analysis of whole genome sequences showed that while all *C. parvum* isolates clustered together, distinct separation of the anthroponotic IIc isolates warranted their recognition as *C. parvum anthroponosum* (*C. p. anthroponosum*); other subtypes, proposed as *C. parvum parvum* (*C. p. parvum*). Geographic distributional differences were also observed. *C. p. anthroponosum*, along with *C. hominis*, was more prevalent in low- and middle-income countries, while *C. p. parvum* isolates were dominant in high-income countries. Despite the genetic and geographic difference, authors found evidence of frequent genetic exchanges between the subspecies.

A similar phenomenon has been observed in *C. hominis*. Recent studies of global *C. hominis* isolates have shown two distinct populations

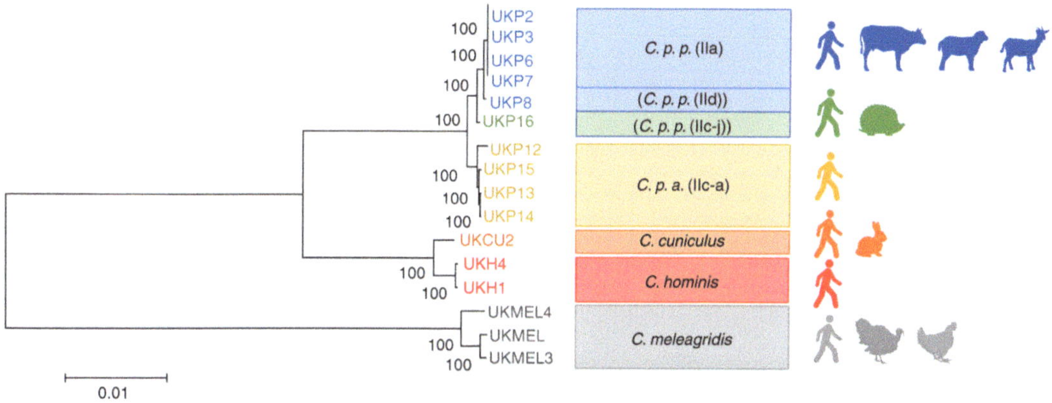

Fig. 5 Phylogenetic separation of anthroponotic *C. parvum anthroponosum* (*C.p.a*) and zoonotic *C. parvum parvum* (*C.p.p*). Maximum likelihood phylogeny based on 16 human infective *Cryptosporidium* isolates using 61 loci. Reproduced with permission from Nader et al. [102]

(Fig. 6a) [100, 103, 104]. One population (clade 2/population 1, proposed as *C. hominis aquapotentis*, *C. h. aquapotentis*) was entirely comprised of IbA10G1 isolates, most originating in Europe. The other population (proposed as *C. hominis hominis*, *C. h. hominis*) showed a much more diverse origin and subtypes, including nearly all the African and Asian isolates. The populations were estimated to have diverged around 500 years ago, experiencing very little genetic exchange since the divergence. One group noted that the estimated time of the divergence coincided with an improvement in water quality in European countries, hypothesizing that may have driven selective pressure for *C. h. aquapotentis* to become more directly transmissible; another suggested lower transmission intensity as responsible [100, 103].

One study noted that isolates within *C. h. hominis* clustered in 13 distinct populations (Fig. 6b) [100]. Isolates from Africa and Asia are largely clustered by geographic origin irrespective of their *gp60* subtypes. Asian isolates formed two sister clades, whereas African isolates clustered by country of origin. Isolates from the United States, on the other hand, are clustered primarily by their *gp60* subtype. The most dominant subtype, IfA12G1R5, showed three distinct population clusters (populations 6, 13, and 14). They proposed that IfA12G1R5 was first brought to the United States from two different origins: one from East Africa (population 6) and one from Europe (population 14; Fig. 6c). As these isolates circulated, and they recombined to form a new strain (population 14). Population 14 isolates first appeared in 2014 and rapidly became the dominant circulating isolate [100].

These studies also highlighted mucin-like proteins as potential virulence factors. Mucins have a role in mediating cell–cell interactions and infectivity of *Cryptosporidium* [105, 106]. Glycosylation of mucin proteins mediates host interactions for a variety of pathogens [107]. In *C. parvum*, MUC5 is highly expressed in early infection and constitutes a determinant of host adaptation [106]. Among the most divergent genes between *C. h. hominis* and *C. h. aquapotentis* were a cluster of mucin-like proteins (CHUDEA2_430, CHUDEA2_440, and CHUDEA2_450) [103]. CHUDEA2_430, a *muc5* homolog, was estimated to have diverged around the time the two *C. hominis* subspecies diverged. Importantly, eight novel glycosylation sites in CUDEA2_430 are unique to *C. h. aquapotentis* haplotypes. Similarly, *C. h. hominis* population 14, noted to be hyper-transmissible, went through selective genetic sweeps that resulted in the acquisition of two invasion-associated mucin-like glycoproteins (cgd6_40 and cdg8_700) [100].

A continuing limitation for population analysis is the lack of global sampling of isolates for whole genome sequences [44, 47, 100]. Most

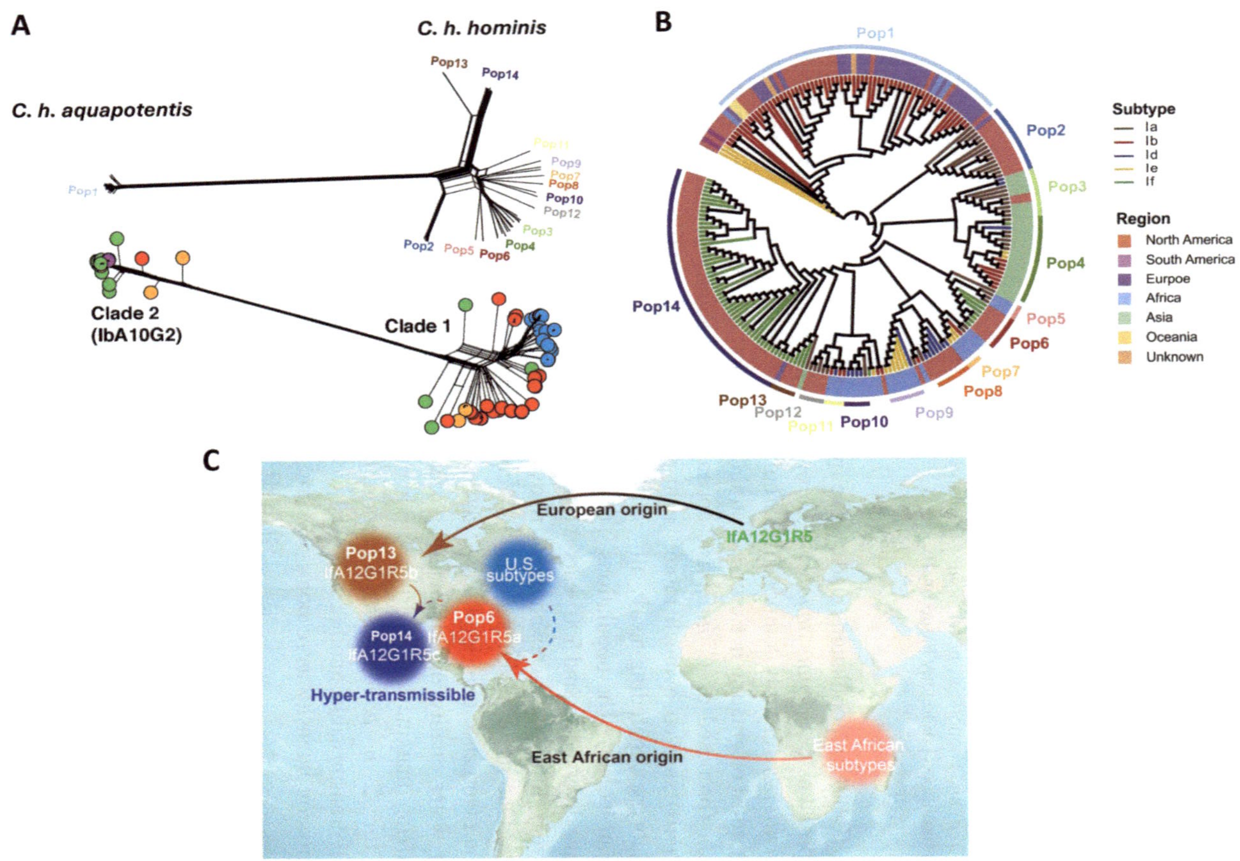

Fig. 6 Global population structure of *Cryptosporidium hominis*. (**a**) Splitstree analysis from two separate studies showing two distinct populations: *C. hominis aquapotentis* (*C. h. aquapotentis*) associated with *gp60* subtype IbA10G2 and *C. hominis hominis* (*C. h. hominis*). *Top*: Analysis of 199 *C. hominis* isolates based on 9394 whole genome SNPs. Reproduced with permission from Huang et al. [100]. *Bottom*: Analysis of 114 *C. hominis* isolates based on 5618 whole genome SNPs. Reproduced with permission from Tichkule et al. [103]. (**b**) Circular tree inferred by maximum likelihood analysis of 199 isolates using 12,736 whole genome SNPs. Branch color denotes *gp60* subtype; background color shows geographic origin and outer circle notes the population cluster. Reproduced with permission from Huang et al. [100]. (**c**) Proposed model for the origin and genetic exchanges of the three IfA12G1R5 populations within the United States. IfA12G1R5 was first brought in from two separate origins: East Africa (population 6) and Europe (population 13). While circulating, these population recombined to form a third population (population 14). Reproduced with permission from Huang et al. [100]

publicly available sequences, including those generated in the above studies, originate from isolates derived from the United States, the United Kingdom, and China. Asian, African, and South American countries are particularly underrepresented. Species representation is also limited, with *C. parvum* accounting for 65% of available sequences and *C. hominis* 31%. Increasing representation will be critical for developing an understanding of global population structures.

8 The Challenge and Promise

Recent advances have broadened our understanding of the genetic structure and diversity of *Cryptosporidium*. However, the community still lacks access to unbiased, global samples of isolates thoroughly characterized and differentiated by genetic markers providing requisite discriminating power. Challenges in developing whole-genome sequencing (WGS) and high-resolution genotyping markers for medically important eukaryotic parasites, such as *Cryptosporidium*, include:

(a) A lack of adequate in vitro propagation techniques.
(b) Few animal models to expand protozoan parasites in animals.
(c) The requirement for fresh, unfrozen samples in order to make use of available immunomagnetic separation (IMS) procedures that enhance downstream sequencing.

Capture enrichment sequencing (CES) may help overcome these challenges. This approach has been applied successfully to concentrate *Wolbachia* DNA from whole insect DNA extracts [108], to enrich *Yersinia pestis* DNA from Black Death victims [109], and to detect and characterize felid species for veterinary diagnosis and discovery [110]. CES, using SureSelect technology, was used to characterize parasite genomes directly from clinical samples, validated on bone marrow and splenic aspirates of visceral leishmaniasis patients in Nepal [111]. Additionally, RNA enrichment methods for high-resolution quantitative transcriptional analysis, using SureSelect CES, have enriched fungal transcripts up to 1600-fold [112]. This approach uses biotinylated, single-stranded RNA probes designed to hybridize with specific DNA regions of interest. These hybridized fragments can be isolated from complex mixtures, such as fecal DNA extracts, and sequenced. Probe sets can be designed to be highly diverse, capturing targets across the genome and between species. This approach has been used in other apicomplexan parasites such as *Plasmodium*. A pilot study in *Cryptosporidium* showed that this approach can generate nearly full genome coverage from as little as 5 ng of input DNA [113–115]. A larger study is underway to develop a diverse probe set capable of capturing the global genetic diversity of *Cryptosporidium* species [116–118].

A further hindrance to understanding global population structure is mixed infections. Co-infections with multiple species or strains have been widely reported in animals and humans [101, 119–121]. In an endemic region, individuals with cryptosporidiosis are often infected with multiple species of *Cryptosporidium* with varying levels of relatedness [122–124]. Hence, estimation of the rate and relatedness structure of such mixed infections is a critical factor in correctly constructing the population genetic structure of *Cryptosporidium*, particularly in highly endemic regions. Current identification methods have difficulty resolving the multiplicity of infections and mixed genetic backgrounds [43, 44, 47]. Multi-locus sequencing has been proposed as a way to address the issue, though it can be prohibitively time-consuming and costly [46, 96, 98]. High-resolution WGS data will provide opportunities to estimate the number of strains and species, and their relative proportions, within a mixed sample [125, 126].

Genetic recombination is also documented in *Cryptosporidium* both between strains and closely related species [100, 102, 103, 127]. However, we are in the infancy of understanding the role of recombination in fixing the population structure of *Cryptosporidium* and its impact on host adaptation and spatial structure on parasite dynamics and evolution. Recent research clearly underscores how protozoan parasites employ secreted pathogenesis determents (SPDs) to infect and cause disease in their hosts [128]. The introgression of these SPDs from virulent pathogens can induce virulence in previously avirulent pathogens [128]. Interestingly, it has been proposed that adaptive introgression might have influenced the recent emergence of the anthroponotic subspecies of *Cryptosporidium* 55–164 years ago [102].

Single-cell sequencing may offer a path to understanding mixed infections and genetic recombination. A recent study was not only able to detect recombination between species but also resolve the placement of the recombination events [60]. This approach could be used in combination with isolation strategies to study clinical stool samples and gain a greater understanding of global population variance [48]. This technology could also be vital in generating new genomes. Generally, millions of *Cryptosporidium* oocysts are required to provide the genetic material needed for whole genome sequencing. Studies using single-cell techniques have shown success in generating whole genome coverage from as few as ten oocysts [60, 90]. In combination with next-generation long-read sequencing, such an approach could greatly expand the number of available reference genomes available for study. To fairly characterize and understand global diversity, to understand the genetic and epidemiological processes that drive diversification, and to understand the epidemiological and immunological consequences of such diversity, more reference genomes from more regions are required. Promising new tools offer the means to capture genetic diversity, the composition and relatedness of multiple species infections, and the role of genetic introgression, which all influence local population structure and the factors controlling the host adaptation of *Cryptosporidium*.

Acknowledgments We acknowledge CryptoDB (https://cryptodb.org/cryptodb/app) [129, 130] for providing a publicly available repository for *Cryptosporidium* genomic data. CryptoDB is part of the Eukaryotic Pathogen, Vector, and Host Informatics Resources (VEuPathDB), which is funded by the National Institute of Allergy and Infectious Diseases (NIH/DHHS) under Contract No. NIH HHS 75N930C00077. VEuPathDB also receives funding from the Wellcome Trust (UK) to support informatics efforts focusing on kinetoplastida and fungal organisms with special emphasis on improving functional annotation of genomes. Additional computing resources were provided by the SciNet HPC Consortium (https://scinet.usda.gov/).

Author Contributions R.T., R.P.B., B.M.R., J.C.K., and A.K. wrote the paper.

Competing Interest The authors declare no competing interests.

Funding Information This work was financially supported by USDA CRIS Project 8042-32420-007-00D to R.T., B.M.R., and A.K.; National Institute for Allergy and Infectious Diseases (NIAID), R01 AI148667 to Travis Glenn and J.C.K.; and Houston Methodist Research Institute personal startup funds to R.P.B.

References

1. Levine ND (1980) Some corrections of coccidian (Apicomplexa: protozoa) nomenclature. J Parasitol 66:830–834
2. Meisel JL, Perera DR, Meligro C, Rubin CE (1976) Overwhelming watery diarrhea associated with a cryptosporidium in an immunosuppressed patient. Gastroenterology 70:1156
3. Nime AF, Burek DJ, Page LD, Holscher AM, Yardley HJ (1976) Acute enterocolitis in a human being infected with the protozoan cryptosporidium. Gastroenterology 70:592–598
4. Amadi B, Mwiya M, Musuku J, Watuka A, Sianongo S, Ayoub A, Kelly P (2002) Effect of nitazoxanide on morbidity and mortality in Zambian children with cryptosporidiosis: a randomised controlled trial. Lancet 360:1375–1380
5. Ma P, Soave R (1983) Three-step stool examination for cryptosporidiosis in 10 homosexual men with protracted watery diarrhea. J Infect Dis 147:824–828
6. Navin RT, Juranek DD (1984) Cryptosporidiosis: clinical, epidemiologic, and parasitologic review. Clin Infect Dis 6:313–327
7. Tzipori S, Widmer G (2008) A hundred-year retrospective on cryptosporidiosis. Trends Parasitol 24:184–189
8. Centers for Disease Control (CDC) (1982) Cryptosporidiosis: assessment of chemotherapy of males with acquired immune deficiency syndrome (AIDS). Morb Mortal Wkly Rep 31:589–592

9. European Centre for Disease Prevention and Control (2019) Cryptosporidiosis—annual epidemiological report for 2017
10. Richardson AJ, Frankenberg RA, Buck AC, Selkon JB, Colbourne JS, Parsons JW, Mayon-White RT (1991) An outbreak of waterborne cryptosporidiosis in Swindon and Oxfordshire. Epidemiol Infect 107:485–495
11. Corso SP, Kramer HM, Blair AK, Addiss GD, Davis PJ, Haddix CA (2003) Costs of illness in the 1993 waterborne cryptosporidium outbreak, Milwaukee, Wisconsin. Emerg Infect Dis 9:426–431
12. Widerstrom M, Schonning C, Lilja M, Lebbad M, Ljung T, Allestam G, Ferm M, Bjorkholm B, Hansen A, Hiltula J, Langmark J, Lofdahl M, Omberg M, Reuterwall C, Samuelsson E, Widgren K, Wallensten A, Lindh J (2014) Large outbreak of Cryptosporidium hominis infection transmitted through the public water supply, Sweden. Emerg Infect Dis 20:581–589
13. Delahoy JM, Omore R, Ayers LT, Schilling AK, Blackstock JA, Ochieng BJ, Moke F, Jaron P, Awuor A, Okonji C, Juma J, Farag HT, Nasrin D, Panchalingam S, Nataro PJ, Kotloff LK, Levine MM, Oundo J, Roellig MD, Xiao L, Parsons BM, Laserson K, Mintz DE, Breiman FR, O'Reilly EC (2018) Clinical, environmental, and behavioral characteristics associated with Cryptosporidium infection among children with moderate-to-severe diarrhea in rural western Kenya, 2008–2012: the Global Enteric Multicenter Study (GEMS). PLoS Negl Trop Dis 12:e0006640
14. Kotloff LK, Nataro PJ, Blackwelder CW, Nasrin D, Farag HT, Panchalingam S, Wu Y, Sow OS, Sur D, Breiman FR, Faruque SA, Zaidi KA, Saha D, Alonso LP, Tamboura B, Sanogo D, Onwuchekwa U, Manna B, Ramamurthy T, Kanungo S, Ochieng BJ, Omore R, Oundo OJ, Hossain A, Das KS, Ahmed S, Qureshi S, Quadri F, Adegbola AR, Antonio M, Hossain JM, Akinsola A, Mandomando I, Nhampossa T, Acácio S, Biswas K, O'Reilly EC, Mintz DE, Berkeley YL, Muhsen K, Sommerfelt H, Robins-Browne MR, Levine MM (2013) Burden and aetiology of diarrhoeal disease in infants and young children in developing countries (the Global Enteric Multicenter Study, GEMS): a prospective, case-control study. Lancet 382:209–222
15. Sow OS, Muhsen K, Nasrin D, Blackwelder CW, Wu Y, Farag HT, Panchalingam S, Sur D, Zaidi MKA, Faruque GSA, Saha D, Adegbola R, Alonso LP, Breiman FR, Bassat Q, Tamboura B, Sanogo D, Onwuchekwa U, Manna B, Ramamurthy T, Kanungo S, Ahmed S, Qureshi S, Quadri F, Hossain A, Das KS, Antonio M, Hossain JM, Mandomando I, Nhampossa T, Acácio S, Omore R, Oundo OJ, Ochieng BJ, Mintz DE, O'Reilly EC, Berkeley YL, Livio S, Tennant MS, Sommerfelt H, Nataro PJ, Ziv-Baran T, Robins-Browne MR, Mishcherkin V, Zhang J, Liu J, Houpt RE, Kotloff LK, Levine MM (2016) The burden of cryptosporidium diarrheal disease among children < 24 months of age in moderate/high mortality regions of sub-Saharan Africa and South Asia, utilizing data from the Global Enteric Multicenter Study (GEMS). PLoS Negl Trop Dis 10:e0004729
16. Troeger C, Forouzanfar M, Rao CP, Khalil I, Brown A, Reiner CR, Fullman N, Thompson LR, Abajobir A, Ahmed M, Alemayohu AM, Alvis-Guzman N, Amare TA, Antonio AC, Asayesh H, Avokpaho E, Awasthi A, Bacha U, Barac A, Betsue DB, Beyene SA, Boneya JD, Malta CD, Dandona L, Dandona R, Dubey M, Eshrati B, Fitchett ARJ, Gebrehiwot TT, Hailu BG, Horino M, Hotez JP, Jibat T, Jonas BJ, Kasaeian A, Kissoon N, Kotloff K, Koyanagi A, Kumar AG, Rai KR, Lal A, Razek EAMH, Mengistie AM, Moe C, Patton G, Platts-Mills AJ, Qorbani M, Ram U, Roba SH, Sanabria J, Sartorius B, Sawhney M, Shigematsu M, Sreeramareddy C, Swaminathan S, Tedla AB, Jagiellonian T-MR, Ukwaja K, Werdecker A, Widdowson M-A, Yonemoto N, Zaki SEM, Lim SS, Naghavi M, Vos T, Hay IS, Murray LJC, Mokdad HA (2017) Estimates of global, regional, and national morbidity, mortality, and aetiologies of diarrhoeal diseases: a systematic analysis for the Global Burden of Disease Study 2015. Lancet Infect Dis 17:909–948
17. Dong S, Yang Y, Wang Y, Yang D, Yang Y, Shi Y, Li C, Li L, Chen Y, Jiang Q, Zhou Y (2020) Prevalence of cryptosporidium infection in the global population: a systematic review and meta-analysis. Acta Parasitol 65:882–889
18. Peake L, Inns T, Jarvis C, King G, Rabie H, Henderson J, Wensley A, Jarratt R, Roberts C, Williams C, Orife O, Browning L, Neilson M, Mccarthy C, Millar P, Love N, Elwin K, Robinson G, Mannes T, Young N, Chalmers R, Elson R, Vivancos R (2023) Preliminary investigation of a significant national cryptosporidium exceedance in the United Kingdom, August 2023 and ongoing. Eur Secur 28:2300538
19. Guérin A, Striepen B (2020) The biology of the intestinal intracellular parasite Cryptosporidium. Cell Host Microbe 28:509–515
20. Lamisere H, Bhalchandra S, Kane VA, Zeng X-L, Mo D, Adams W, Estes KM, Ward DH (2022) Differential response to the course of Cryptosporidium parvum infection and its impact on epithelial integrity in differentiated versus undifferentiated human intestinal enteroids. Infect Immun 90:e0039722
21. Yang X, Guo Y, Xiao L, Feng Y (2021) Molecular epidemiology of human cryptosporidiosis in low- and middle-income countries. Clin Microbiol Rev 34:e00087-19
22. Kumar A, Chatterjee I, Anbazhagan AN, Jayawardena D, Priyamvada S, Alrefai WA, Sun J, Borthakur A, Dudeja PK (2018) Cryptosporidium parvum disrupts intestinal epithelial barrier function via altering expression of key tight junction and adherens junction proteins. Cell Microbiol 20:e12830

23. Amadi B, Mwiya M, Sianongo S, Payne L, Watuka A, Katubulushi M, Kelly P (2009) High dose prolonged treatment with nitazoxanide is not effective for cryptosporidiosis in HIV positive Zambian children: a randomised controlled trial. BMC Infect Dis 9:195
24. Dhal KA, Panda C, Yun S-I, Mahapatra KR (2022) An update on cryptosporidium biology and therapeutic avenues. J Parasit Dis 46:923–939
25. Lourido S, Moreno SN (2015) The calcium signaling toolkit of the apicomplexan parasites Toxoplasma gondii and Plasmodium spp. Cell Calcium 57:186–193
26. Baptista PR, Li Y, Sateriale A, Sanders JM, Brooks LK, Tracey A, Ansell REB, Jex RA, Cooper WG, Smith DE, Xiao R, Dumaine EJ, Georgeson P, Pope JB, Berriman M, Striepen B, Cotton AJ, Kissinger CJ (2022) Long-read assembly and comparative evidence-based reanalysis of cryptosporidium genome sequences reveal expanded transporter repertoire and duplication of entire chromosome ends including subtelomeric regions. Genome Res 32:203–213
27. Castellanos-Gonzalez A, White CA, Ojo KK, Vidadala RSR, Zhang Z, Reid CM, Fox WMA, Keyloun RK, Rivas K, Irani A, Dann MS, Fan E, Maly JD, Voorhis VCW (2013) A novel calcium-dependent protein kinase inhibitor as a lead compound for treating cryptosporidiosis. J Infect Dis 208:1342–1348
28. Hulverson AM, Choi R, Arnold LMS, Schaefer AD, Hemphill A, Mccloskey CM, Betzer PD, Müller J, Vidadala SRR, Whitman RG, Rivas LK, Barrett KL, Hackman CR, Love SM, Mcnamara WC, Shaughnessy KT, Kondratiuk A, Kurnick M, Banfor NP, Lynch JJ, Freiberg MG, Kempf JD, Maly JD, Riggs WM, Ojo KK, Voorhis VCW (2017) Advances in bumped kinase inhibitors for human and animal therapy for cryptosporidiosis. Int J Parasitol 47:753–763
29. Hlavsa CM, Cikesh LB, Roberts AV, Kahler MA, Vigar M, Hilborn DE, Wade JT, Roellig MD, Murphy LJ, Xiao L, Yates MK, Kunz MJ, Arduino JM, Reddy CS, Fullerton EK, Cooley AL, Beach JM, Hill RV, Yoder SJ (2018) Outbreaks associated with treated recreational water — United States, 2000–2014. MMWR Morb Mortal Wkly Rep 67:547–551
30. Shields MJ, Hill RV, Arrowood JM, Beach JM (2008) Inactivation of Cryptosporidium parvum under chlorinated recreational water conditions. J Water Health 6:513–520
31. Sponseller KJ, Griffiths KJ, Tzipori S (2014) The evolution of respiratory cryptosporidiosis: evidence for transmission by inhalation. Clin Microbiol Rev 27:575–586
32. Ginn O, Rocha-Melogno L, Bivins A, Lowry S, Cardelino M, Nichols D, Tripathi SN, Soria F, Andrade M, Bergin M, Deshusses MA, Brown J (2021) Detection and quantification of enteric pathogens in aerosols near open wastewater canals in cities with poor sanitation. Environ Sci Technol 55:14758–14771
33. Mor SM, Tumwine JK, Ndeezi G, Srinivasan MG, Kaddu-Mulindwa DH, Tzipori S, Griffiths JK (2010) Respiratory cryptosporidiosis in HIV-seronegative children in Uganda: potential for respiratory transmission. Clin Infect Dis 50:1366–1372
34. Mckerr C, Chalmers MR, Elwin K, Ayres H, Vivancos R, O'Brien JS, Christley MR (2022) Cross-sectional household transmission study of cryptosporidium shows that C. hominis infections are a key risk factor for spread. BMC Infect Dis 22:114
35. Wang W, Wan M, Yang F, Li N, Xiao L, Feng Y, Guo Y (2021) Development and application of a gp60-based subtyping tool for Cryptosporidium bovis. Microorganisms 9:2067
36. Bouzid M, Kintz E, Hunter PR (2018) Risk factors for Cryptosporidium infection in low and middle income countries: a systematic review and meta-analysis. PLoS Negl Trop Dis 12:e0006553
37. English DE, Guérin A, Tandel J, Striepen B (2022) Live imaging of the Cryptosporidium parvum life cycle reveals direct development of male and female gametes from type I meronts. PLoS Biol 20:e3001604
38. Khan A, Shaik SJ, Grigg EM (2018) Genomics and molecular epidemiology of cryptosporidium species. Acta Trop 184:1–14
39. Tzipori S, Ward H (2002) Cryptosporidiosis: biology, pathogenesis and disease. Microbes Infect 4:1047–1058
40. Awad-El-Kariem FM, Warhurst DC, Mcdonald V (1994) Detection and species identification of cryptosporidium oocysts using a system based on PCR and endonuclease restriction. Parasitology 109(Pt 1):19–22
41. Leng X, Mosier DA, Oberst RD (1996) Differentiation of Cryptosporidium parvum, C. muris, and C. baileyi by PCR-RFLP analysis of the 18S rRNA gene. Vet Parasitol 62:1–7
42. Roellig DM, Xiao L (2020) Cryptosporidium genotyping for epidemiology tracking. Methods Mol Biol 2052:103–116
43. Elwin K, Robinson G, Pérez-Cordón G, Chalmers MR (2022) Development and evaluation of a real-time PCR for genotyping of cryptosporidium spp. from water monitoring slides. Exp Parasitol 242:108366
44. Hassan ME, Örmeci B, Derosa CM, Dixon RB, Sattar AS, Iqbal A (2021) A review of cryptosporidium spp. and their detection in water. Water Sci Technol 83:1–25
45. Ryan U, Zahedi A, Feng Y, Xiao L (2021) An update on zoonotic cryptosporidium species and genotypes in humans. Animals 11:3307
46. Robinson G, Pérez-Cordón G, Hamilton C, Katzer F, Connelly L, Alexander LC, Chalmers MR (2022) Validation of a multilocus genotyping scheme for subtyping Cryptosporidium parvum for epidemiological purposes. Food Waterborne Parasitol 27:e00151

47. Baptista PR, Cooper WG, Kissinger CJ (2021) Challenges for cryptosporidium population studies. Genes 12:894
48. Morris A, Robinson G, Swain TM, Chalmers MR (2019) Direct sequencing of Cryptosporidium in stool samples for public health. Front Public Health 7:360
49. Mathur V, Kolísko M, Hehenberger E, Irwin ATN, Leander SB, Kristmundsson Á, Freeman AM, Keeling JP (2019) Multiple independent origins of apicomplexan-like parasites. Curr Biol 29:2936–2941.e5
50. Pinto P, Ribeiro AC, Kváč M, Tsaousis DA (2022) Cryptosporidium. In: Lifecycles of pathogenic protists in humans. Springer, pp 331–389
51. Salomaki DE, Terpis XK, Rueckert S, Kotyk M, Varadínová KZ, Čepička I, Lane EC, Kolisko M (2021) Gregarine single-cell transcriptomics reveals differential mitochondrial remodeling and adaptation in apicomplexans. BMC Biol 19:77
52. Ryan UM, Feng Y, Fayer R, Xiao L (2021) Taxonomy and molecular epidemiology of Cryptosporidium and giardia - a 50 year perspective (1971-2021). Int J Parasitol 51:1099–1119
53. Cavalier-Smith T (2014) Gregarine site-heterogeneous 18S rDNA trees, revision of gregarine higher classification, and the evolutionary diversification of Sporozoa. Eur J Protistol 50:472–495
54. Adl MS, Bass D, Lane EC, Lukeš J, Schoch LC, Smirnov A, Agatha S, Berney C, Brown WM, Burki F, Cárdenas P, Čepička I, Chistyakova L, Campo DJ, Dunthorn M, Edvardsen B, Eglit Y, Guillou L, Hampl V, Heiss AA, Hoppenrath M, James YT, Karnkowska A, Karpov S, Kim E, Kolisko M, Kudryavtsev A, Lahr JGD, Lara E, Gall LL, Lynn HD, Mann GD, Massana R, Mitchell ADE, Morrow C, Park SJ, Pawlowski WJ, Powell JM, Richter JD, Rueckert S, Shadwick L, Shimano S, Spiegel WF, Torruella G, Youssef N, Zlatogursky V, Zhang Q (2019) Revisions to the classification, nomenclature, and diversity of eukaryotes. J Eukaryot Microbiol 66:4–119
55. Xiao L, Fayer R, Ryan U, Upton SJ (2004) Cryptosporidium taxonomy: recent advances and implications for public health. Clin Microbiol Rev 17:72–97
56. Slavin D (1955) Cryptosporidium meleagridis (sp. nov.). J Comp Pathol Therap 65:262–266
57. Tyzzer EE (1907) A sporozoan found in the peptic glands of the common mouse. Exp Biol Med 5:12–13
58. Tyzzer EE (1912) Cryptosporidium parvum (sp. nov.), a Coccidium. Arch Protistenkunde 26
59. King P, Tyler MK, Hunter RP (2019) Anthroponotic transmission of Cryptosporidium parvum predominates in countries with poorer sanitation: a systematic review and meta-analysis. Parasit Vectors 12:16
60. Shaw S, Cohn SI, Baptista PR, Xia G, Melillo B, Agyabeng-Dadzie F, Kissinger CJ, Striepen B (2024) Genetic crosses within and between species of cryptosporidium. Proc Natl Acad Sci 121:e2313210120
61. Chappell LC, Tzipori S, Akiyoshi ED, Okhuysen CP, Tanriverdi S, Langer-Curry R, Widmer G (2006) Cryptosporidium hominis: experimental challenge of healthy adults. Am J Trop Med Hyg 75:851–857
62. Widmer G, Köster CP, Carmena D (2020) Cryptosporidium hominis infections in non-human animal species: revisiting the concept of host specificity. Int J Parasitol 50:253–262
63. Chalmers MR, Robinson G, Elwin K, Elson R (2019) Analysis of the Cryptosporidium spp. and gp60 subtypes linked to human outbreaks of cryptosporidiosis in England and Wales, 2009 to 2017. Parasit Vectors 12:95
64. Puleston LR, Mallaghan MC, Modha ED, Hunter RP, Nguyen-Van-Tam SJ, Regan MC, Nichols LG, Chalmers MR (2014) The first recorded outbreak of cryptosporidiosis due to Cryptosporidium cuniculus (formerly rabbit genotype), following a water quality incident. J Water Health 12:41–50
65. Xiao L, Feng Y (2017) Molecular epidemiologic tools for waterborne pathogens Cryptosporidium spp. and Giardia duodenalis. Food Waterborne Parasitol 8-9:14–32
66. Esch JK, Petersen AC (2013) Transmission and epidemiology of zoonotic protozoal diseases of companion animals. Clin Microbiol Rev 26:58–85
67. Jiang W, Roellig MD, Lebbad M, Beser J, Troell K, Guo Y, Li N, Xiao L, Feng Y (2020) Subtype distribution of zoonotic pathogen Cryptosporidium felis in humans and animals in several countries. Emerg Microbes Infect 9:2446–2454
68. Abrahamsen SM, Templeton JT, Enomoto S, Abrahante EJ, Zhu G, Lancto AC, Deng M, Liu C, Widmer G, Tzipori S, Buck AG, Xu P, Bankier TA, Dear HP, Konfortov AB, Spriggs FH, Iyer L, Anantharaman V, Aravind L, Kapur V (2004) Complete genome sequence of the apicomplexan, Cryptosporidium parvum. Science 304:441–445
69. Isaza PJ, Galván LA, Polanco V, Huang B, Matveyev VA, Serrano GM, Manque P, Buck AG, Alzate FJ (2015) Revisiting the reference genomes of human pathogenic Cryptosporidium species: reannotation of C. parvum Iowa and a new C. hominis reference. Sci Rep 5:16324
70. Guizetti J, Barcons-Simon A, Scherf A (2016) Trans-acting GC-rich non-coding RNA at var expression site modulates gene counting in malaria parasite. Nucleic Acids Res 44:9710–9718
71. Mok WB, Ribacke U, Sherwood E, Wahlgren M (2008) A highly conserved segmental duplication in the subtelomeres of Plasmodium falciparum chromosomes varies in copy number. Malar J 7:46
72. Zhang X, Alexander N, Leonardi I, Mason C, Kirkman AL, Deitsch WK (2019) Rapid antigen diversification through mitotic recombination in the human malaria parasite Plasmodium falciparum. PLoS Biol 17:e3000271

73. Barcons-Simon A, Carrington M, Siegel NT (2023) Decoding the impact of nuclear organization on antigenic variation in parasites. Nat Microbiol 8:1408–1418
74. Bunnik ME, Cook BK, Varoquaux N, Batugedara G, Prudhomme J, Cort A, Shi L, Andolina C, Ross SL, Brady D, Fidock AD, Nosten F, Tewari R, Sinnis P, Ay F, Vert J-P, Noble SW, Roch LGK (2018) Changes in genome organization of parasite-specific gene families during the Plasmodium transmission stages. Nat Commun 9:1910
75. Cestari I, Stuart K (2018) Transcriptional regulation of telomeric expression sites and antigenic variation in trypanosomes. Curr Genomics 19:119–132
76. Saha A, Nanavaty PV, Li B (2020) Telomere and subtelomere R-loops and antigenic variation in trypanosomes. J Mol Biol 432:4167–4185
77. Baptista RP, Xiao R, Li Y, Glenn TC, Kissinger JC (2023) New T2T assembly of Cryptosporidium parvum IOWA annotated with reference genome gene identifiers. bioRxiv. https://doi.org/10.1101/2023.06.13.544219
78. Menon KV, Okhuysen CP, Chappell LC, Mahmoud M, Mahmoud M, Meng Q, Doddapaneni H, Vee V, Han Y, Salvi S, Bhamidipati S, Kottapalli K, Weissenberger G, Shen H, Ross CM, Hoffman LK, Cregeen JS, Muzny MD, Metcalf AG, Gibbs AR, Petrosino FJ, Sedlazeck JF (2022) Fully resolved assembly of Cryptosporidium parvum. Gigascience 11:giac010
79. Guo Y, Cebelinski E, Matusevich C, Alderisio AK, Lebbad M, Mcevoy J, Roellig MD, Yang C, Feng Y, Xiao L (2015) Subtyping novel zoonotic pathogen cryptosporidium chipmunk genotype I. J Clin Microbiol 53:1648–1654
80. Ifeonu OO, Chibucos CM, Orvis J, Su Q, Elwin K, Guo F, Zhang H, Xiao L, Sun M, Chalmers MR, Fraser MC, Zhu G, Kissinger CJ, Widmer G, Silva CJ (2016) Annotated draft genome sequences of three species of Cryptosporidium: Cryptosporidium meleagridis isolate UKMEL1, C. baileyi isolate TAMU-09Q1 and C. hominis isolates TU502_2012 and UKH1. Pathog Dis 74:ftw080
81. Liu S, Roellig MD, Guo Y, Li N, Frace AM, Tang K, Zhang L, Feng Y, Xiao L (2016) Evolution of mitosome metabolism and invasion-related proteins in Cryptosporidium. BMC Genomics 17:1006
82. Sateriale A, Šlapeta J, Baptista R, Engiles BJ, Gullicksrud AJ, Herbert TG, Brooks FC, Kugler ME, Kissinger CJ, Hunter AC, Striepen B (2019) A genetically tractable, natural mouse model of cryptosporidiosis offers insights into host protective immunity. Cell Host Microbe 26:135–146.e5
83. Xu P, Widmer G, Wang Y, Ozaki SL, Alves MJ, Serrano GM, Puiu D, Manque P, Akiyoshi D, Mackey JA, Pearson RW, Dear HP, Bankier TA, Peterson LD, Abrahamsen SM, Kapur V, Tzipori S, Buck AG (2004) The genome of Cryptosporidium hominis. Nature 431:1107–1112
84. Xu Z, Li N, Guo Y, Feng Y, Xiao L (2020) Comparative genomic analysis of three intestinal species reveals reductions in secreted pathogenesis determinants in bovine-specific and non-pathogenic Cryptosporidium species. Microb Genom 6:e000379
85. Cardenas D, Bhalchandra S, Lamisere H, Chen Y, Zeng X-L, Ramani S, Karandikar CU, Kaplan LD, Estes KM, Ward DH (2020) Two- and three-dimensional bioengineered human intestinal tissue models for cryptosporidium. Methods Mol Biol 2052:373–402
86. Morada M, Lee S, Gunther-Cummins L, Weiss ML, Widmer G, Tzipori S, Yarlett N (2016) Continuous culture of Cryptosporidium parvum using hollow fiber technology. Int J Parasitol 46:21–29
87. Korwin-Mihavics BR, Dews EA, Genova BMD, Huston CD (2023) Organoid-based in vitro systems to model Cryptosporidium parvum infection in 2D and 3D. bioRxiv. https://doi.org/10.1101/2023.09.29.560165
88. Dadzie FA, Beaudry MS, Deyanov A, Slanis H, Duong MQ, Turner R, Khan A, Arias CA, Kissinger JC, Glenn TC, Baptista RDP (2024) Evaluating the benefits and limits of multiple displacement amplification with whole-genome oxford nanopore sequencing. bioRxiv 12:2024.02.09.579537
89. Penumarthi LR, Baptista RP, Beaudry MS, Glenn TC, Kissinger JC (2024) A new chromosome-level genome assembly and annotation of Cryptosporidium meleagridis. bioRxiv. https://doi.org/10.1101/2024.02.16.580748
90. Troell K, Hallström B, Divne A-M, Alsmark C, Arrighi R, Huss M, Beser J, Bertilsson S (2016) Cryptosporidium as a testbed for single cell genome characterization of unicellular eukaryotes. BMC Genomics 17:471
91. Strong BW, Gut J, Nelson GR (2000) Cloning and sequence analysis of a highly polymorphic Cryptosporidium parvum gene encoding a 60-kilodalton glycoprotein and characterization of its 15- and 45-kilodalton Zoite surface antigen products. Infect Immun 68:4117–4134
92. Xiao L (2010) Molecular epidemiology of cryptosporidiosis: an update. Exp Parasitol 124:80–89
93. Fan Y, Huang X, Guo S, Yang F, Yang X, Guo Y, Feng Y, Xiao L, Li N (2021) Subtyping Cryptosporidium xiaoi, a common pathogen in sheep and goats. Pathogens 10:800
94. Rojas-Lopez L, Elwin K, Chalmers MR, Enemark LH, Beser J, Troell K (2020) Development of a gp60-subtyping method for Cryptosporidium felis. Parasit Vectors 13:39
95. Yang X, Huang N, Jiang W, Wang X, Li N, Guo Y, Kváč M, Feng Y, Xiao L (2020) Subtyping Cryptosporidium ryanae: a common pathogen in bovine animals. Microorganisms 8:1107
96. Abal-Fabeiro LJ, Maside X, Bello X, Llovo J, Bartolomé C (2013) Multilocus patterns of genetic variation across Cryptosporidium species suggest

balancing selection at the gp60 locus. Mol Ecol 22:4723–4732
97. Garcia-R CJ, Hayman STD (2017) Evolutionary processes in populations of Cryptosporidium inferred from gp60 sequence data. Parasitol Res 116:1855–1861
98. Widmer G, Lee Y (2010) Comparison of single- and multilocus genetic diversity in the protozoan parasites Cryptosporidium parvum and C. hominis. Appl Environ Microbiol 76:6639–6644
99. Widmer G, Lee Y, Hunt P, Martinelli A, Tolkoff M, Bodi K (2012) Comparative genome analysis of two Cryptosporidium parvum isolates with different host range. Infect Genet Evol 12:1213–1221
100. Huang W, Guo Y, Lysen C, Wang Y, Tang K, Seabolt MH, Yang F, Cebelinski E, Gonzalez-Moreno O, Hou T, Chen C, Chen M, Wan M, Li N, Hlavsa MC, Roellig DM, Feng Y, Xiao L (2023) Multiple introductions and recombination events underlie the emergence of a hyper-transmissible Cryptosporidium hominis subtype in the USA. Cell Host Microbe 31:112–123.e4
101. Dettwiler I, Troell K, Robinson G, Chalmers MR, Basso W, Rentería-Solís MZ, Daugschies A, Mühlethaler K, Dale IM, Raghavendra BJ, Ruf M-T, Poppert S, Meylan M, Olias P (2022) TIDE analysis of cryptosporidium infections by gp60 typing reveals obscured mixed infections. J Infect Dis 225:686–695
102. Nader LJ, Mathers CT, Ward JB, Pachebat AJ, Swain TM, Robinson G, Chalmers MR, Hunter RP, Oosterhout VC, Tyler MK (2019) Evolutionary genomics of anthroponosis in Cryptosporidium. Nat Microbiol 4:826–836
103. Tichkule S, Cacciò MS, Robinson G, Chalmers MR, Mueller I, Emery-Corbin JS, Eibach D, Tyler MK, Oosterhout VC, Jex RA (2022) Global population genomics of two subspecies of Cryptosporidium hominis during 500 years of evolution. Mol Biol Evol 39:msac056
104. Cabarcas F, Galvan-Diaz LA, Arias-Agudelo ML, García-Montoya MG, Daza MJ, Alzate FJ (2021) Cryptosporidium hominis phylogenomic analysis reveals separate lineages with continental segregation. Front Genet 12:740940
105. Lippuner C, Ramakrishnan C, Basso UW, Schmid WM, Okoniewski M, Smith CN, Hässig M, Deplazes P, Hehl BA (2018) RNA-Seq analysis during the life cycle of Cryptosporidium parvum reveals significant differential gene expression between proliferating stages in the intestine and infectious sporozoites. Int J Parasitol 48:413–422
106. O'Connor MR, Burns BP, Ha-Ngoc T, Scarpato K, Khan W, Kang G, Ward H (2009) Polymorphic mucin antigens CpMuc4 and CpMuc5 are integral to Cryptosporidium parvum infection in vitro. Eukaryot Cell 8:461–469
107. Lin B, Qing X, Liao J, Zhuo K (2020) Role of protein glycosylation in host-pathogen interaction. Cells 9:1022
108. Kent NB, Salichos L, Gibbons GJ, Rokas A, Newton GLI, Clark EM, Bordenstein RS (2011) Complete bacteriophage transfer in a bacterial endosymbiont (Wolbachia) determined by targeted genome capture. Genome Biol Evol 3:209–218
109. Schuenemann JV, Bos K, Dewitte S, Schmedes S, Jamieson J, Mittnik A, Forrest S, Coombes KB, Wood WJ, Earn DJD, White W, Krause J, Poinar NH (2011) Targeted enrichment of ancient pathogens yielding the pPCP1 plasmid of Yersinia pestis from victims of the Black Death. Proc Natl Acad Sci 108:E746–E752
110. Lee SJ, Mackie SR, Harrison T, Shariat B, Kind T, Kehl T, Löchelt M, Boucher C, Vandewoude S (2017) Targeted enrichment for pathogen detection and characterization in three felid species. J Clin Microbiol 55:1658–1670
111. Domagalska AM, Imamura H, Sanders M, Broeck DVF, Bhattarai RN, Vanaerschot M, Maes I, D'Haenens E, Rai K, Rijal S, Berriman M, Cotton AJ, Dujardin J-C (2019) Genomes of Leishmania parasites directly sequenced from patients with visceral leishmaniasis in the Indian subcontinent. PLoS Negl Trop Dis 13:e0007900
112. Cai W, Nunziata S, Rascoe J, Stulberg JM (2019) SureSelect targeted enrichment, a new cost effective method for the whole genome sequencing of Candidatus Liberibacter asiaticus. Sci Rep 9:18962
113. Khan A, Ferreira ECA, Grigg ME (2019) Development of SureSelect target enrichment for whole genome sequencing of cryptosporidium directly from stool samples. In: VIIth International Giardia and Cryptosporidium Conference, p 99
114. Widmer G, Carmena D, Kváč M, Chalmers MR, Kissinger CJ, Xiao L, Sateriale A, Striepen B, Laurent F, Lacroix-Lamandé S, Gargala G, Favennec L (2020) Update on Cryptosporidium spp.: highlights from the seventh international giardia and Cryptosporidium conference. Parasite 27:14
115. Khan A, Alves-Ferreira EVC, Vogel H, Botchie S, Ayi I, Pawlowic MC, Robinson G, Chalmers RM, Lorenzi H, Grigg ME (2024) Phylogenomic reconstruction of Cryptosporidium spp. captured directly from clinical samples reveals extensive genetic diversity. bioRxiv. https://doi.org/10.1101/2024.04.17.589752
116. Kissinger JC, Glenn TC Capturing the genomic variation present in Cryptosporidium and Cryptosporidiosis (1R01AI148667-01A1)
117. CryptoCapture.org: a large community effort to survey the population genetic structure of human-infecting cryptosporidia [Online]. CryptoCapture.org. Accessed 9 January 2024
118. Bayona-Vásquez NJ, Sullivan AH, Beaudry MS, Khan A, Baptista RP, Petersen KN, Bhuiyan M, Brunelle B, Robinson G, Chalmers RM, Alves-Ferreira E, Grigg ME, Kissinger JC, Glenn TC (2024) Whole genome targeted enrichment and sequencing of human-infecting Cryptosporidium spp. bioRxiv. https://doi.org/10.1101/2024.03.29.586458

119. Kader EAMN, Blanco M-A, Ali-Tammam M, Ghaffar EABREA, Osman A, Sheikh EN, Rubio MJ, Fuentes DI (2012) Detection of Cryptosporidium parvum and Cryptosporidium hominis in human patients in Cairo, Egypt. Parasitol Res 110:161–166
120. Paparini A, Yang R, Chen L, Tong K, Gibson-Kueh S, Lymbery A, Ryan MU (2017) Cryptosporidium in fish: alternative sequencing approaches and analyses at multiple loci to resolve mixed infections. Parasitology 144:1811–1820
121. Zintl A, Neville D, Maguire D, Fanning S, Mulcahy G, Smith VH, Waal DT (2007) Prevalence of Cryptosporidium species in intensively farmed pigs in Ireland. Parasitology 134:1575–1582
122. Cama V, Gilman HR, Vivar A, Ticona E, Ortega Y, Bern C, Xiao L (2006) Mixed cryptosporidium infections and HIV. Emerg Infect Dis 12:1025–1028
123. Korpe SP, Gilchrist C, Burkey C, Taniuchi M, Ahmed E, Madan V, Castillo R, Ahmed S, Arju T, Alam M, Kabir M, Ahmed T, Petri AW, Haque R, Faruque GSA, Duggal P (2019) Case-Control study of cryptosporidium transmission in Bangladeshi households. Clin Infect Dis 68:1073–1079
124. Sannella RA, Suputtamongkol Y, Wongsawat E, Cacciò MS (2019) A retrospective molecular study of cryptosporidium species and genotypes in HIV-infected patients from Thailand. Parasit Vectors 12:91
125. Chang H-H, Worby JC, Yeka A, Nankabirwa J, Kamya RM, Staedke GS, Dorsey G, Murphy M, Neafsey ED, Jeffreys EA, Hubbart C, Rockett AK, Amato R, Kwiatkowski PD, Buckee OC, Greenhouse B (2017) THE REAL McCOIL: A method for the concurrent estimation of the complexity of infection and SNP allele frequency for malaria parasites. PLoS Comput Biol 13:e1005348
126. O'Brien DJ, Amenga-Etego L, Li R (2016) Approaches to estimating inbreeding coefficients in clinical isolates of Plasmodium falciparum from genomic sequence data. Malar J 15:473
127. Corsi IG, Tichkule S, Sannella RA, Vatta P, Asnicar F, Segata N, Jex RA, Oosterhout VC, Cacciò MS (2023) Recent genetic exchanges and admixture shape the genome and population structure of the zoonotic pathogen Cryptosporidium parvum. Mol Ecol 32:2633–2645
128. Lorenzi H, Khan A, Behnke MS, Namasivayam S, Swapna LS, Hadjithomas M, Karamycheva S, Pinney D, Brunk BP, Ajioka JW, Ajzenberg D, Boothroyd JC, Boyle JP, Darde ML, Diaz-Miranda MA, Dubey JP, Fritz HM, Gennari SM, Gregory BD, Kim K, Saeij JP, Su C, White MW, Zhu XQ, Howe DK, Rosenthal BM, Grigg ME, Parkinson J, Liu L, Kissinger JC, Roos DS, Sibley LD (2016) Local admixture of amplified and diversified secreted pathogenesis determinants shapes mosaic Toxoplasma gondii genomes. Nat Commun 7:10147
129. Warrenfeltz S, Kissinger CJ (2020) Accessing cryptosporidium omic and isolate data via CryptoDB.org. Methods Mol Biol:139–192
130. Warrenfeltz S, Kissinger JC, EupathDB Team (2020) Correction to: accessing cryptosporidium omic and isolate data via CryptoDB.org. Methods Mol Biol 2052:C1

Advances in *Toxoplasma* Functional Genomics

Hisham S. Alrubaye and Jon P. Boyle

Abstract

Toxoplasma gondii is a globally prevalent pathogen of humans and most warm-blooded animals, and also a genetically tractable model system for the study of intracellular parasitism. The modern field has been characterized by consistent progress in the application of cutting-edge genetic manipulation and functional genomics. Here, we highlight the most recent advances in the field and discuss results from representative manuscripts from the constantly evolving field of *Toxoplasma* functional genomics.

Keywords

Toxoplasma gondii · Transcriptomics · Single-cell RNA sequencing · Functional genomics · CRISPR screening · Next-generation sequencing

H. S. Alrubaye · J. P. Boyle (✉)
Department of Biological Sciences, Dietrich School of Arts and Sciences, University of Pittsburgh, Pittsburgh, PA, USA
e-mail: boylej@pitt.edu

1 Introduction

Toxoplasma gondii is the causative agent of toxoplasmosis, a parasite belonging to the phylum Apicomplexa. Like all of its relatives, it has an intracellular parasitic lifestyle, but *T. gondii* is unique in having the ability to infect, and be transmitted by, a comparatively wide array of intermediate hosts. Moreover, with few exceptions (such as syncytiotrophoblasts of the placenta [1]), *T. gondii* can infect virtually any nucleated cell in nearly all warm-blooded animals. Although local incidence rates vary tremendously by region, it is estimated that one-third of the world population is infected with this parasite [2, 3]. In immunocompetent individuals, *T. gondii* infection rarely results in disease in the host, and this feature allows it to replicate, persist, and disseminate to new hosts.

The intracellular lifestyle of *T. gondii* is initiated by invasion into the host cell and formation of a membrane-bound compartment called the parasitophorous vacuole (PV). PV formation is mediated by parasite-secreted proteins that are introduced into the host cell where it localizes to the plasma membrane and functions as a receptor for a surface-localized parasite protein [4]. Once

J. D. Ramírez González (ed.), *Recent Advances in Parasitomics*,
https://doi.org/10.1007/978-3-031-70591-5_8

the parasite establishes anchorage it forms a moving junction (MJ), a tight connection between the parasite and the host cell's plasma membrane that propels the parasite into the PV [5]. The parasite resides in the PV which is completely delineated from the host cytoplasm and is extensively modified by the parasite to enable its growth and propagation.

T. gondii can persist within its host environment using an extensive arsenal of effector proteins. Notably, *T. gondii* injects these effectors into host cells irrespective of invasion suggesting remodeling of host cells can occur both in invaded cells and bystanders that are not invaded by are directly impacted by parasite effector secretion [6, 7]. Presumably, the injection of effectors into targeted cells changes the internal environment of the host cells making it (a) permissive for infection, replication, and encystation and (b) suitable to escape immune responses employed by host innate and adaptive immune cells.

In this chapter, we focus on functional genomics in *T. gondii*, including recent work using single-cell and long-read transcriptomics in the parasite and recent technical advances for genetic manipulation and how they have impacted our understanding of *Toxoplasma* development and impact on the host cell. These advancements have revolutionized our ability to dissect the intricacies of *T. gondii's* genetic makeup. From classical approaches like forward and reverse genetics to cutting-edge technologies such as CRISPR (Clustered Regularly Interspersed Short Palindromic Repeats)-Cas9, the toolkit available for studying this parasite's genome has never been more diverse or powerful. Moreover, we will explore how multi-omics has provided a comprehensive view of *T. gondii's* molecular repertoire such as the intricate regulatory networks determining virulence, cell cycle progression, and development into diverse life stages.

2 Functional Genomics Advances: Transcriptomics

The *T. gondii* transcriptome has been extensively queried over the years, including early studies using spotted [8–12]and *in situ* synthesized [13–16] microarray technologies. With the emergence of RNAseq as a routine approach, the number of transcriptional studies in *T. gondii* has increased dramatically, and based on a crude search at the NCBI short read archive at the time of this writing, there are 99 Bioprojects deposited with *T. gondii* listed as the primary organism. Fortunately, many of these microarray and RNAseq datasets have been curated and uploaded onto the main functional genomics repository for *T. gondii*, www.toxodb.org [17]. Advances in transcriptomics are rapid and vast and have been extensively reviewed over the years [18–23]; therefore, our aim here will be to focus on *Toxoplasma* transcriptomic data emerging from two major technical advances in recent years: direct, single molecule long read RNA sequencing, and single-cell RNA sequencing.

2.1 Single-Molecule, Long-Read RNA Sequencing: A Direct View into the Extent and Nature of Splice Variation in *Toxoplasma gondii*

Single-molecule sequencing obtained via Oxford Nanopore or PacBio RS technologies has a distinct advantage over its predecessors and antecedents: exceptionally long read length. Early EST and genomic sequencing projects relied on Sanger sequencing, which could average as long as 800–900 bp, while high throughput pyrosequencing-based approaches (such as those used by 454 and Illumina) were smaller (ranging from 20–36 bp in early days to ~150 bp currently). Single-molecule reads theoretically can be as big as the pieces of DNA being sequenced, and single reads as long as 200 kb are routine [24], while reads larger than a megabase are a reality, especially when care is taken to preserve DNA integrity and minimize shearing [25]. As a testament to its utility, long-read genome sequencing on *T. gondii* reversed a long-held belief that the *T. gondii* genome had 14 chromosomes when, in reality, it had only 13 (first described in [26] and then further confirmed in [27]). As the Nanopore platform is amenable to all types of nucleic acid, this approach can also be applied directly to RNA of any size isolated from any organism, allowing for individual RNA species to be sequenced directly permitting the nature and extent of isoform variation to be determined with minimal sample manipulation. The first such published study in *T. gondii* obtained ~300,000 individual reads covering over 60% of the *T. gondii* transcriptome, and >1000 of the ~8000 *T. gondii* genes had reads mapping to them that were at least 95% as long as the predicted full-length transcript [28]. By precise mapping of splice sites and characterizing splice variants by integrating long read data with short read data and the current version of the *T. gondii* genome assembly, the authors could identify splice variants across thousands of transcripts and then determine whether they would encode functional proteins or not. It was found that the bulk of transcript variants (including both splice variants as well as those with "retained" introns) were "nonproductive" [28], meaning that they would produce a truncated protein based on splice variation introducing a premature stop codon. This not only improves genome annotation (as some predicted introns were never found to be removed from mature transcripts) but also allows for a much more thorough identification of the protein-coding gene splice variants that would produce functional protein than previously available using short-read sequencing (where novel, functional splice sites can be identified but not in their context within the entire transcript). While the majority of retained introns (70%) lead to truncated protein products, there are likely to be some alternately spliced transcripts that do produce functional proteins that may be important in the function of specific gene products. Fortunately, these data can be easily visualized on the public repository, www.toxodb.org. It will be interesting to see if there is more extensive and functional splice variation when comparing different *T. gondii* life stages. However, this exposes a limitation of the single-molecule sequencing approaches, which require higher quality and more highly concentrated nucleic acid compared with second-generation Illumina sequencing as there is no amplification step and larger molecules are more prone to shearing and nuclease activity. However, as these technologies become more sensitive, these approaches will become more feasible.

2.2 Single-Cell RNAseq in *Toxoplasma gondii*: Unraveling the Complexity of the Cell Cycle and Developmental Transitions

With the advent of multiple approaches for transcriptionally profiling single cells using microfluidics (e.g., "DropSeq" [29] and derived techniques) or combinatorial barcoding and/or flow cytometry [30], it has become possible to identify subpopulations of *T. gondii* parasites within what had been previously analyzed as homogenous populations. The first such studies in *T. gondii* used flow cytometric sorting into the individual reaction well sand Smart-Seq library construction [31, 32]. Subsequent studies have used microfluidics given the proliferation of the 10X genomics devices in most sequencing core facilities. Examination of the single-cell transcriptomes of *T. gondii* tachyzoites from a highly lab-adapted strain (RH) allowed for the unsynchronized parasites to be placed along the entire cell cycle using unsupervised clustering [31]. This cell cycle projection was retained among presumably dividing parasites along with others that were undergoing developmental change when the parasite strain PRU was chemically induced to transition into a slow-growing bradyzoite form. Consistent with prior work [33] using bulk RNAseq data to study the tachyzoite to bradyzoite transition, these data showed that certain phases of the cell cycle were critical checkpoints in developmental trajectories [31]. However, the value of single-cell RNAseq is its greatest when examining rare cell populations, and in this study, the authors identified a single *T. gondii* tachyzoite that did not have detectable expression of the transcript encoding SAG1, a dominant tachyzoite surface antigen. When comparing the thousands of detected transcripts in this single cell to other SAG1+ cells, they found a transcriptional profile that was similar to what had been previously shown using bulk RNA sequencing as being similar to *T. gondii* merozoites, life stages that drive sexual recombination and that are restricted to development in the cat-gut. One of the genes distinguishing this single SAG1$^+$ parasite from others was an Apetela-2-like transcription factor called AP2IX-1. Remarkably ectopic overexpression of this factor in otherwise wild-type *T. gondii* parasites caused a significant decrease in SAG1 transcript abundance and surface expression [31]. These data not only revealed a potentially unappreciated role for AP2IX-1 in sexual development (or repression of asexual development) but also highlighted that, in a given population of presumably homogeneous parasites, there may be those expressing surface antigens and transcription factors relevant to a completely different life stage. This is the greatest immediate benefit of the proliferation of single-cell sequencing reagents, protocols, and equipment: the ability to obtain transcriptional profiles from very rare populations of cells and learn new biology, even if there are only a few of these rare cells.

Heterogeneous populations of parasites that have undergone distinct recent passage histories and differ in their ability to form cysts *in vitro* and *in vivo* were investigated using scRNAseq in an effort to understand how *T. gondii* adapts to *in vitro* culture [34]. A frustrating observation in many laboratories is that *T. gondii* parasites rapidly adapt to *in vitro* cultivation, becoming impaired in their developmental capacity in the process. Parasites that were continuously passed in mice and had high cyst-forming capacity were grown *in vitro* in astrocytes or fibroblasts for up to 7 days after being inoculated with bradyzoites isolated directly from *in vivo* cysts. All populations demonstrated a subset of parasites that were in the process of converting to bradyzoite-like developmental stages in the absence of any external stressors (like pH), a phenomenon that is difficult to detect without the use of scRNAseq or other single-cell technologies. However, what was particularly remarkable was that parasites grown in astrocytes for 7 days had a larger cluster of parasites that were in the process of developing into bradyzoites than those grown in HFFs, and this was consistent with their ability to form a large number of cysts in vivo (performed in separate experiments). The rich scRNAseq dataset can now be mined for genes and/or transcriptional networks that might undergo relatively rapid changes in cyst-formation capability depending on the host cell type. In this case, scRNAseq gives a much more precise and quantitative view of the developmental state of these individual parasites, allowing for new hypotheses to be developed about the root cause(s) [34].

Single-cell RNA sequencing was performed in another study in the context of the tachyzoite-to-bradyzoite transition in the context of the genetic ablation of a critical gene that regulates the transition of *T. gondii* from tachyzoite to bradyzoite called BFD1 [32]. Wild-type tachyzoites showed a similar cell cycle profile as in other studies, and those exposed to stress induce bradyzoite formation to express bradyzoite-specific genes and exit the cell cycle-driven cluster [32]. When BFD1 mutants were analyzed in the same fashion, differences with their wild-type counterparts only emerged under stress conditions, where some of the BFD1 mutants still clustered with the replicating tachyzoites or were found in completely distinct clusters, owing to their poor expression of bradyzoite-specific genes [32]. While bulk RNAseq could provide some of the same information about the impact of BFD1 deletion on parasite developmental responses to stress, the use of single-cell RNAseq allowed for a more precise evaluation of precisely when during the tachyzoite to bradyzoite transition that BFD1 was important.

2.3 Predicting the Future: Emerging Single-Cell Technologies and Their Potential Utility in *T. gondii*

As single-cell RNAseq analyses become more routine, we anticipate that less tractable life stages such as those that undergo sexual development in the cat will be a focus of these technologies. While the biology of sexual development is starting to be deciphered [35–37], the heterogeneity [38] of the parasite population in the cat intestinal epithelium is the perfect setting for single-cell transcriptomics to shine. Other uses for single-cell transcriptomics could include analysis of *T. gondii* mutants that are incapable of forming cysts yet still persist in the mouse brain (i.e., ROCY1/BFD2 knockouts [39]). By comparing mutant parasites isolated from mouse brains to wild type, one could determine if there is a subset of *T. gondii* wild-type parasites that have a transcriptional profile that is similar to ROCY1 mutants, indicating the existence of a persisting, but non-cyst-like, population in the mouse brain. While the focus of this chapter is on the use of new functional genomics technologies on *Toxoplasma* itself, in terms of the host cell dual RNAseq on both the parasite and host simultaneously is now relatively routine, and very amenable to single-cell transcriptomics. In the next section, we discuss the use of PertubSeq to examine the impact of gene deletions on *T. gondii* on the host transcriptome at the single-cell level, and this approach should also be useful for querying the impact of the ablation of host or parasite genes on host or parasite transcript abundance at the single cell level. Finally, we also expect that emerging single-cell proteomics techniques [40–43] will soon be applied to *T. gondii*, as these techniques are becoming more and more sensitive meaning that simultaneous quantification of tens to hundreds of parasite proteins will be possible in individual cells.

3 Genetic Manipulation Techniques

Over the past decade, there has been rapid progress in the enhancement of high-throughput genetic approaches in eukaryotes, spanning comprehensive genome-wide functional analyses to sophisticated single-cell studies. This period has undergone an unprecedented advancement in our capacity to delve deeply into the function of individual genes of *T. gondii*. Traditional methods, such as chemically induced mutagenesis, insertion mutagenesis, and genetic crosses, have played pivotal roles in identifying genetic loci associated with traits including virulence, differentiation, cell cycle regulation, and drug resistance within T. *gondii* [9, 44–48].

However, despite their importance, these conventional approaches are not without their challenges. Chemically induced and insertion mutagenesis techniques necessitate large parasite populations per experiment. Additionally, these approaches are highly nonspecific making it challenging to parse mutations that are relevant to the phenotype and those that are inconsequential. Similarly, genetic crosses, while providing valuable insights, require the completion of the parasite's sexual life cycle, a process that to date can only occur within the *T. gondii* definitive feline host.

Nevertheless, these traditional approaches have been instrumental in advancing our understanding of the complex biological processes underlying *T. gondii's* biology and survival strategies. They have provided insights into the genetic determinants governing various functions, shedding light on the parasite's adaptation mechanisms and its interactions with host cells. These approaches have unraveled key genetic pathways and regulatory networks, paving the way for subsequent advancements in high-throughput genomic techniques. For example, chemically induced mutagenesis was used in *T. gondii* to generate temperature-sensitive mutants and drug-resistance mutants including Arabinosyl nucleosides-resistant mutants [49], FUDR-resistant mutants [50], sulfonamides-resistant mutants [51], and 6-thioxanthine-resistant mutants [45]. These studies were fundamental to performing forward and reverse genetics with selectable markers. This approach generates mutation across the genome that allows studying essential genes such as those responsible for cell cycle [46, 52] and egress [48, 53]. Insertion mutagenesis led to the identification of BSR4 as a bradyzoite-specific gene [54] and ROP4 [55].

Classical genetics involving genetic crosses between multiple strains of *T. gondii* that vary in their virulence phenotypes identified major virulence factors in *T. gondii* [56, 57]. Crosses in the cat can be used to map a specific phenotype to a single or multiple loci with the genome. This approach was instrumental in identifying *T. gondii* genes that determined well-known virulence and host response differences between different clonal *T. gondii* lineages. For example, the ROP (pseudo)kinases ROP18, ROP16, and ROP5 were found to be major genetic determinants of virulence in F1 progeny from a type II and type III cross. ROP18 functions by directly phosphorylating and inactivating host immunity-related GTPases (IRGs)14, while ROP16 directly phosphorylates host STAT3 [58] and STAT6 [7] leading to profound transcriptional changes including downregulating interleukin 12 (IL-12) [58–62]. Production of this proinflammatory cytokine, IL-12, plays a prominent role in anti-parasite immunity, as the IL-12-IFN-γ axis is a major immune response to inhibit *T. gondii* proliferation in mice [63–65]. Importantly, these progeny are readily available in multiple laboratories and at ATCC and have been used in studies investigating genetic loci driving strain-specific differences (as in [66, 67]). While this approach has been successful at identifying many strain-specific effectors and dramatically altered our understanding of virulence mechanisms in *T. gondii*, it is less sensitive to identifying core virulence factors that have similar impacts on virulence across all *T. gondii* strains.

3.1 CRISPR-Cas9: A Quantum Leap in Genetic Research

CRISPR-Cas9-based functional genomics screening has emerged as a powerful and transformative approach for high-throughput determination of gene function in a wide variety of models, and most importantly, non-model, species, and *T. gondii* is no exception. Prior to the emergence of CRISPR as a tool for genetic manipulation in *T. gondii*, mutating, knocking out, or replacing endogenous genes or inserting sequences at a specific locus was efficient enough to be of great use in the field but typically required long (1–2 kb) homologous flanking sequences to improve targeting to specific loci [68, 69]. The efficiency of gene targeting was improved in Ku80 knockout strains by eliminating much of the random insertion that occurs when introducing DNA into *T. gondii*, allowing for very efficient gene replacement and gene tagging [70, 71]. Although traditional approaches yielded tremendous insights into *T. gondii*'s biology, these strategies often required multiple steps for plasmid construction. In its role as a genetic tool, CRISPR utilizes an RNA-guided nuclease known as Cas9. This endonuclease can be directed by a 20-base-pair guide RNA (gRNA) complementary sequence, termed a "protospacer" to target a specific genomic region with precision. The gRNA and the Cas9 protein combine to form a complex, which together induces double-stranded DNA breaks at the targeted site within the genome [72–76]. These breaks can then be repaired by *T. gondii* through two primary mechanisms: nonhomologous end joining (NHEJ) or homology-directed repair (HDR). *T. gondii* appears to preferentially utilize the NHEJ pathway based on its high nonhomologous integration frequencies and low-end targeting frequencies observed experimentally [69, 77]. NHEJ relies on a complex of proteins including the heterodimer subunits Ku70/Ku80 that form a complex that is conserved functional and structurally and can be found in organisms ranging from bacteria to humans [78–80]. Disruption in the Ku70/Ku80 complex impacts the NHEJ pathway but does not affect the process of HDR. Therefore, in ΔKu80 *T. gondii* lines, HDR can reach >90% of target genes for tagging and gene knock-ins and knockouts [70, 71, 81].

CRISPR Cas9 enhanced gene targeting in *T. gondii* independent of the presence of the *Ku80* gene. Cas9 provides an effective means to introduce site-specific double-stranded DNA breaks in the genome at specific loci. The specificity of Cas9 made mutagenesis, insertion and deletion of genes, and epitope tagging using NHEJ and HDR feasible in any *T. gondii* strain. Because of these specific features and the easy-to-generate sgRNAs and CAS9-encoding constructs, many labs sought to generate knockout of individual genes in genome-scale functional screens which are discussed below.

3.2 Genome-Wide and Multi-Gene CRISPR-CAS9 Knockout Screens

CRISPR screens can employ various strategies depending on the phenotypes of interest. For instance, survival or proliferation-based screens can identify essential or dispensable genes and genes that modulate the host cell to survive under specific conditions. This type of screening remains the simplest and most scalable readout for pooled CRISPR screens, directly reflecting the essentiality of genes while playing a fundamental role in deciphering gene function.

The utilization of CRISPR-based functional genomics has heralded a new era in unraveling the intricacies of *T. gondii* biology, as demonstrated by several groundbreaking studies. One such pioneering endeavor devised the first CRISPR/Cas9-mediated genome-wide knockout strategy, significantly advancing our understanding of *T. gondii* proliferation mechanisms [82]. This approach involved the systematic introduction of a library of single-guide RNAs (sgRNAs), with 10 sgRNAs targeting each of the 8156 protein-coding genes within *T. gondii*, into Cas9-expressing parasites (TgRH) via transfection. The library was designed to target exon regions. The gRNAs were synthesized by CustomeArray with homology on either side of the pU6-DHFR integration site (Addgene, # 80329). The assembled constructs were transfected into *T. gondii*, and the resultant population of mutant parasites was cultured in human foreskin fibroblasts (HFFs) under pyrimethamine drug selection [77]. Subsequently, the genomic DNA from the generated parasite mutants was extracted and sequenced to assess the relative abundance of sgRNAs targeting genes essential for *T. gondii* growth/virulence. The phenotype of each gene was calculated by comparing the gRNA sequence reads from DNA extracted from the input DNA to the gRNA sequence reads extracted from transfected parasites grown in HFFs. The genome-wide library contains built-in barcodes that can be directly used to amplify libraries to target specific gene families, such as the dense granule GRA or the inner membrane complex sub-libraries (Addgene #80636).

The insights gleaned from this study have not only provided invaluable genetic resources but have also facilitated the refinement of our understanding of gene essentiality for overall parasite fitness *in vitro*, as evidenced by the incorporation of "phenotype" scores derived from changes in the relative abundance of sgRNA from the CRISPR screen onto ToxoDB (https://toxodb.org/). The results include ~200 previously uncharacterized fitness-conferring genes in tachyzoites. Moreover, this seminal work has laid the foundation for comprehensive investigations into the biology of *T. gondii*, thereby invigorating research efforts within the field. The application of CRISPR/Cas9 screens in numerous subsequent studies has proven instrumental in identifying genes involved in various aspects of *T. gondii* biology, including life cycle regulation, growth, virulence, morphology, and dissemination, further solidifying the pivotal role of CRISPR-based functional genomics in advancing our understanding of *T. gondii* biology.

There are limitations to using Cas9 in pooled CRISPR screens that are more prominent compared with generating individual mutants. Constitutive expression of Cas9 can lead to unintended consequences, including cellular toxicity, and poses challenges in deciphering key regulators that necessitate precise spatiotemporal control [83]. This can be circumvented by expressing a "decoy" gRNA to limit the toxicity associated with Cas9 activity directed by endogenous RNAs [82].

In contrast to genome-wide screens that require significant parasite numbers to avoid population bottlenecks, multi-gene screens targeting a subset of the genome are proving to also be useful *in vitro* and especially *in vivo*. In one study, Cas9 and sgRNA were expressed on the same construct [84]. In prior screens, CRISPR screens were done in parasite lines stably expressing Cas9 and a decoy sgRNA to limit the Cas9 toxicity. This limitation was circumvented by expressing both Cas9 and sgRNA from a single construct allowing the use of this vector in a wider range of

T. gondii strains. An additional advantage of this construct stems from the use of single-step cloning of single-stranded gRNA oligonucleotides to generate the libraries. This study identified multiple novel virulence factors that contributed to parasite growth in mice. The fitness phenotype effects of genes were evaluated by comparing the relative read gRNA counts between input and output DNA extracted from parasites. In addition, this study highlighted that mutants may play different roles in heterologous versus homogenous KO populations. For example, MYR1 is part of a translocon responsible for translocating multiple effector proteins important in host–pathogen interaction but appears only to be important in a homogenous KO population [85]. There is likely a compensatory gene knockout combination that could have circumvented the effect of this phenotype, which is yet to be identified.

Another *in vivo* study targeted 217 genes with 2170 sgRNA (10 sgRNA per gene) [86]. The gRNA library in this study was constructed using the same libraries as in Sidik et al [82] but using sub-libraries amplified using specific sub-library targeting primers for putative dense granule proteins GRAs. The Sidik et al [82] whole-genome sgRNA library contains built-in barcodes that can be used to amplify eight sub-libraries targeting specific genes, such as genes encoding the micronemes, rhoptries, and dense granule proteins. The dense granule GRA sgRNA library in this study was transfected into parasites constitutively expressing Cas9 and passaged in the drug-selective medium in HFFs. Post-selection, parasite mutants were injected intraperitoneally in mice to evaluate the contribution of the tested genes on parasite survival and virulence *in vivo* (CD-1 mice) during the acute phase of infection. At ~4 days post-infection, parasites were harvested from the peritoneum (site of infection) and non-peritoneal organs (lungs, heart, brain, and eyes) allowing the identification of genes crucial for parasite survival and dissemination to distant organs. TgWIP, a secreted rhoptry protein, was identified as necessary for parasite dissemination because it was fitness-conferring in non-peritoneal organs such as the lung, heart, brain, and eye compared with the site peritoneal [86]. This study described the mechanism that governs the TgWIP function via interaction with the host WAVE regulatory complex. This effector is responsible for modulating the host actin dynamics to induce dendritic cell hypermotility. In addition, the study identified multiple genes that were required for fitness in the peritoneum and non-peritoneal organs. This study expanded our understanding of *T. gondii*'s genetic repertoire and its intricate interplay with the *in vivo* host environment.

The utilization of CRISPR screens has also emerged as a powerful technique for identifying essential genes under specific conditions including genes important for parasite survival within interferon-gamma (IFNγ)-activated murine cells [87]. IFNγ is a key cytokine that orchestrates host defense mechanisms to restrict *T. gondii* proliferation within host cells by activating IRGs and guanylate-binding proteins [88–93]. These proteins bind to and disrupt the parasitophorous vacuole membrane [94], thereby leading to parasite death. Furthermore, IFNγ induces the production of nitric oxide (NO) and reactive oxygen species [95], which impair parasite viability and restrict growth within murine macrophages. This CRISPR screen identified 160 genes required for parasite fitness within IFNγ-activated murine macrophages and 193 genes that were essential in naïve macrophages. Of the top hits, several genes (six) were validated to be required in IFNγ-activated murine macrophages using a single-gene knockout approach. The function of these genes reflects their importance in sustaining parasite growth within the harsh environment orchestrated by IFNγ. The observation suggests the critical role of these genes in facilitating parasite survival within IFNγ-activated peritoneal macrophages in mice.

Among the genes identified, GRA45 emerged as a notable virulence factor implicated in the translocation of multiple *T. gondii* effector proteins into the host cell [96]. This gene exhibited a significant impact on fitness in IFNγ-activated bone marrow-derived macrophages (BMDMs), coupled with aberrant localization of exported proteins. Such findings underscore the multifaceted role of GRA45 in *T. gondii* pathogenesis, particularly in evading host immune responses orchestrated by IFNγ. Moreover, this approach not only identified essential genes but also shed light on the intricate mechanisms underlying *T. gondii* survival within IFNγ-activated macrophages.

Fluorescence-activated cell sorting-based screens using reporter cell lines can be useful in studying specific signaling pathways to identify the regulators involved in these pathways. A bradyzoite reporter strain and a focused genetic screen were used to identify regulators of the tachyzoite to bradyzoite transition [32]. In this study, a CRISPR screen was performed using a differentiation reporter strain that conditionally expressed a fluorescent marker under the regulation of the canonical bradyzoite-specific *BAG1* promoter. Using the combination of cell sorting and gRNA cassette sequencing, BFD1 was identified as a master regulator of *T. gondii* bradyzoite differentiation. This was exciting as the loss of BFD1 blocked *T. gondii* differentiation *in vitro* and *in vivo*. This was a breakthrough as differentiation from tachyzoites to bradyzoites allows *T. gondii* to establish a chronic infection, further highlighting the advancements in molecular approaches and their impact on unraveling complex functional genomics.

The pooled screens described above allowed the identification of fitness-conferring genes based on constitutive expression of Cas9. To perform phenotypic screens in *T. gondii* using a conditional regulation system, splitCas9 has been adapted for use in *T. gondii* [97]. This innovative system capitalizes on the modular nature of the Cas9 protein, which can be divided into two distinct domains, Cas9(N) and Cas9(C) [98]. These Cas9(N) and Cas9(C) domains are fused to FRB (FK506-binding protein-rapamycin-binding) and FKBP (FK506-binding protein), respectively. With the addition of rapamycin, the two domains reconstitute the full and functional Cas9 protein via the fusion of FRB and FKBP, thereby restoring the protein's activity. This study identified genes encoding a novel conoid-associated protein and a putative neurotransmitter symporter. These two genes are important for *T. gondii*'s egress from the host cell. Precisely, this approach has shown that these two genes are critical for different periods of egress based on time-lapse video microscopy. This approach has a great application in allowing us to study the function of essential genes using precise spatiotemporal regulated CRISPR-Cas9 activity.

3.3 dCas9

Another advancement of CRISPR Cas9 in *T. gondii* is the generation of dead Cas9 which enables tuning the expression of genes without genome editing [99]. This uses a catalytically inactive form of Cas9, named dead Cas9 (dCas9), which retains its ability to bind to target DNA without exerting its nuclease activity. This strategic modification allows dCas9 to function as a potent tool for gene regulation by interfering with transcriptional processes [99]. For example, dCas9 was tested in a reporter strain expressing mNeonGreen driven by the surface antigen1promoter (SAG1), where gRNA targeted the SAG1 upstream regions leading to repression of the fluorescence. The gene repression is reproducible and stable over time, while the efficiency appears to be dependent on gRNA position.

dCas9 binding to a target gene disrupts its transcriptional machinery and effectively prevents the expression of the targeted gene. This approach is known as CRISPR-interference or CRISPRi, which enables precise and reversible gene knockdowns without permanently altering the genetic sequence. This innovative approach provides the ability to modulate gene expression levels dynamically, facilitating the study of gene function with fine-tuned control. This system can be applied to target multiple genes simultaneously to explore the intricate interplay between different genes within the *T. gondii* genome [99]. This capability opens new avenues for investigating complex gene interaction networks and elucidating the mechanisms underlying specific functions, such as dynamic networks of signaling pathways, regulators of cell cycle, and virulence.

3.4 Perturb-Seq

The pioneering of cutting-edge CRISPR-mediated single-cell screening, Perturb-seq, has tremendous potential to revolutionize our understanding of functional genomics in *T. gondii* [100]. This technology has been used in a targeted screen aiming to study the mechanisms of *T. gondii* secreted effector-mediated host manipulation [100]. To establish an infection, *T. gondii* secretes several proteins from specialized organelles known as Rhoptry and dense granules, to reprogram the host gene expression profile, modulate immune responses, and evade restriction/clearance. This study targeted 256 genes encoding rhoptry and dense granule proteins with a pooled CRISPR sgRNA [100]. The mutant populations generated from this library were evaluated for their role in inducing transcriptional change in infected host cells. Noteworthy is that this screen not only affirmed the roles of previously identified parasite proteins in shaping host transcriptional responses but also identified novel effectors pivotal to *T. gondii's* manipulation of host cells. Among those effectors, GRA59, a dense granule protein, emerged as a key player in orchestrating the trafficking of other parasite-secreted proteins into the host cells and the identification of TgSOS1, which is a novel effector protein essential for sustaining STAT6 phosphorylation through the infection. TgSOS1 also modulates the expression of pro-inflammatory responses in macrophages via M2 polarization which is implicated with dampening anti-parasitic immune responses. Therefore, Perturb-seq not only sheds new light on the intricate interplay between *T. gondii* and its host cell but also represents a significant leap forward in the technology that can give an in-depth exploration of *T. gondii* genome and potentially identify genes/pathways implicated in the course of infection. This will be particularly interesting when applied to the parasite transcriptome, especially in the context of mutating candidate regulators of stage conversion, including tachyzoite to bradyzoite transitions and sexual development.

4 Conclusion

Understanding the intricate functions of specific genes in *T. gondii* is paramount for identifying potential drug targets and advancing vaccine development against toxoplasmosis. Various genetic tools have been devised for studying *T. gondii*. While these traditional approaches have yielded tremendous insights into the parasite's biology, they possess inherent limitations, particularly in their capacity to generate mutants on a large scale. The development of CRISPR-Cas9 in *T. gondii* started a new era in genome engineering, offering high versatility, efficiency, and simplicity in generating custom libraries. High-throughput functional analysis of *T. gondii*'s genome can be easily generated by making parasite mutant libraries. Although this system is still in its infancy, CRISPR-Cas9 has already proven to be an effective tool for precisely disrupting target genes within the parasite at various scales. The systematic-based analysis approaches will help us understand the unique biology of *T. gondii* and broaden the scope.

References

1. Ander SE et al (2018) Human placental syncytiotrophoblasts restrict *Toxoplasma gondii* attachment and replication and respond to infection by producing immunomodulatory chemokines. mBio 9:1–14
2. Montoya J, Liesenfeld O (2004) Toxoplasmosis. Lancet 363:1965–1976
3. Weiss LM, Dubey J, P. (2009) Toxoplasmosis: a history of clinical observations. Int J Parasitol 39:895–901
4. Alexander DL, Mital J, Ward GE, Bradley P, Boothroyd JC (2005) Identification of the moving junction complex of *Toxoplasma gondii*: a collaboration between distinct secretory organelles. PLoS Pathog 1:0137–0149
5. Besteiro S, Dubremetz JF, Lebrun M (2011) The moving junction of apicomplexan parasites: a key structure for invasion. Cell Microbiol 13:797–805
6. Koshy AA et al (2010) Toxoplasma secreting Cre recombinase for analysis of host-parasite interactions. Nat Methods 7:307
7. Koshy AA et al (2012) Toxoplasma co-opts host cells it does not invade. PLoS Pathog 8:e1002825
8. Boyle JP, Saeij JPJ, Harada SY, Ajioka JW, Boothroyd JC (2008) Expression quantitative trait locus mapping of Toxoplasma genes reveals mul-

tiple mechanisms for strain-specific differences in gene expression. Eukaryot Cell 7:1403
9. Singh U, Brewer JL, Boothroyd JC (2002) Genetic analysis of tachyzoite to bradyzoite differentiation mutants in *Toxoplasma gondii* reveals a hierarchy of gene induction. Mol Microbiol 44:721–733
10. Cleary MD, Meiering CD, Jan E, Guymon R, Boothroyd JC (2005) Biosynthetic labeling of RNA with uracil phosphoribosyltransferase allows cell-specific microarray analysis of mRNA synthesis and decay. Nat Biotechnol 23:232
11. Cleary MD, Singh U, Blader IJ, Brewer JL, Boothroyd JC (2002) *Toxoplasma gondii* asexual development: identification of developmentally regulated genes and distinct patterns of gene expression. Eukaryot Cell 1:329
12. Boyle JP, Saeij JPJ, Cleary MD, Boothroyd JC (2006) Analysis of gene expression during development: lessons from the Apicomplexa. Microbes Infect 8(6):1623–1630. https://doi.org/10.1016/j.micinf.2005.11.017
13. Buchholz KR et al (2011) Identification of tissue cyst wall components by transcriptome analysis of *in vivo* and *in vitro Toxoplasma gondii* bradyzoites. Eukaryot Cell 10:1637
14. Behnke MS et al (2010) Coordinated progression through two subtranscriptomes underlies the tachyzoite cycle of *Toxoplasma gondii*. PLoS One 5:e12354
15. Behnke MS, Zhang TP, Dubey JP, Sibley LD (2014) *Toxoplasma gondii* merozoite gene expression analysis with comparison to the life cycle discloses a unique expression state during enteric development. BMC Genomics 15:350
16. Bahl A et al (2010) A novel multifunctional oligonucleotide microarray for *Toxoplasma gondii*. BMC Genomics 11
17. Gajria B et al (2008) ToxoDB: an integrated *Toxoplasma gondii* database resource. Nucleic Acids Res 36:D553
18. White MW, Radke JR, Radke JB (2014) Toxoplasma development - turn the switch on or off? Cell Microbiol 16:466–472
19. Farhat DC, Hakimi MA (2022) The developmental trajectories of Toxoplasma stem from an elaborate epigenetic rewiring. Trends Parasitol 38:37–53
20. Sokol-Borrelli SL, Coombs RS, Boyle JP (2020) A comparison of stage conversion in the Coccidian Apicomplexans *Toxoplasma gondii*, *Hammondia hammondi*, and *Neospora caninum*. Front Cell Infect Microbiol 10
21. Angel SO, Vanagas L, Alonso AM (2024) Mechanisms of adaptation and evolution in *Toxoplasma gondii*. Mol Biochem Parasitol 258:111615
22. Adomako-Ankomah Y, Wier GM, Boyle JP (2012) Beyond the genome: recent advances in *Toxoplasma gondii* functional genomics. Parasite Immunol 34:80–89
23. Sharma J, Rodriguez P, Roy P, Guiton PS (2020) Transcriptional ups and downs: patterns of gene expression in the life cycle of *Toxoplasma gondii*. Microbes Infect 22. https://doi.org/10.1016/j.micinf.2020.09.001
24. Jones A et al (2021) High-molecular weight DNA extraction, clean-up and size selection for long-read sequencing. PLoS One 16:e0253830
25. Prall TM et al (2021) Consistent ultra-long DNA sequencing with automated slow pipetting. BMC Genomics 22:182
26. Bunnik EM et al (2019) Comparative 3D genome organization in apicomplexan parasites. Proc Natl Acad Sci USA 116:3183
27. Xia J et al (2021) Third-generation sequencing revises the molecular karyotype for *Toxoplasma gondii* and identifies emerging copy number variants in sexual recombinants. Genome Res 31:834
28. Lee VV et al (2021) Direct nanopore sequencing of mRNA reveals landscape of transcript isoforms in Apicomplexan parasites. mSystems 6
29. Macosko EZ et al (2015) Highly parallel genome-wide expression profiling of individual cells using nanoliter droplets. Cell 161:1202–1214
30. Rosenberg AB et al (2018) SPLiT-seq reveals cell types and lineages in the developing brain and spinal cord. Science 360:176. https://doi.org/10.1126/science.aam8999
31. Xue Y et al (2020) A single-parasite transcriptional atlas of *Toxoplasma gondii* reveals novel control of antigen expression. Elife 9
32. Waldman BS et al (2020) Identification of a master regulator of differentiation in Toxoplasma. Cell 180:359–372.e16
33. Naumov AV et al (2022) Restriction checkpoint controls Bradyzoite development in *Toxoplasma gondii*. Microbiol Spectr 10
34. Vizcarra EA et al (2023) An ex vivo model of Toxoplasma recrudescence reveals developmental plasticity of the bradyzoite stage. mBio 14:e0183623
35. Farhat DC et al (2020) A MORC-driven transcriptional switch controls Toxoplasma developmental trajectories and sexual commitment. Nat Microbiol 5:570
36. Di Genova BM, Wilson SK, Dubey JP, Knoll LJ (2019) Intestinal delta-6-desaturase activity determines host range for Toxoplasma sexual reproduction. PLoS Biol 17
37. Antunes AV et al (2024) *In vitro* production of cat-restricted Toxoplasma pre-sexual stages. Nature 625:366
38. Speer CA, Dubey JP (2005) Ultrastructural differentiation of *Toxoplasma gondii* schizonts (types B to E) and gamonts in the intestines of cats fed bradyzoites. Int J Parasitol 35:193
39. Sokol-Borrelli SL et al (2023) A transcriptional network required for bradyzoite development in *Toxoplasma gondii* is dispensable for recrudescent disease. Nat Commun 14:6078

40. Wheeler AM et al (2023) Achieving a deeper understanding of drug metabolism and responses using single-cell technologies. Drug Metab Dispos 51:350–359
41. Jenkins C, Orsburn BC (2023) Simple tool for rapidly assessing the quality of multiplexed single cell proteomics data. J Am Soc Mass Spectrom 34:2615–2619
42. Orsburn BC (2023) Metabolomic, proteomic, and single-cell proteomic analysis of cancer cells treated with the KRASG12D inhibitor MRTX1133. J Proteome Res 22:3703–3713
43. Straubhaar J, D'Souza A, Niziolek Z, Budnik B (2024) Single cell proteomics analysis of drug response shows its potential as a drug discovery platform. Mol Omics 20:6–18
44. Pfefferkorn ER, Kasper LH (1983) *Toxoplasma gondii*: genetic crosses reveal phenotypic suppression of hydroxyurea resistance by fluorodeoxyuridine resistance. Exp Parasitol 55:207–218
45. Pfefferkorn ER, Borotz SE (1994) *Toxoplasma gondii*: characterization of a mutant resistant to 6-thioxanthine. Exp Parasitol 79:374–382
46. Gubbels M-J et al (2008) Forward genetic analysis of the apicomplexan cell division cycle in *Toxoplasma gondii*. PLoS Pathog 4:e36
47. Su C, Howe DK, Dubey JP, Ajioka JW, Sibley LD (2002) Identification of quantitative trait loci controlling acute virulence in *Toxoplasma gondii*. Proc Natl Acad Sci USA 99:10753–10758
48. Garrison E et al (2012) A forward genetic screen reveals that calcium-dependent protein kinase 3 regulates egress in Toxoplasma. PLoS Pathog 8:e1003049
49. Pfefferkorn ER, Pfefferkorn LC (1976) Arabinosyl nucleosides inhibit *Toxoplasma gondii* and allow the selection of resistant mutants. J Parasitol 62:993–999
50. Pfefferkorn ER, Pfefferkorn LC (1977) *Toxoplasma gondii*: characterization of a mutant resistant to 5-fluorodeoxyuridine. Exp Parasitol 42:44–55
51. Pfefferkorn ER, Borotz SE, Nothnagel RF (1992) *Toxoplasma gondii*: characterization of a mutant resistant to sulfonamides. Exp Parasitol 74:261–270
52. Brown KM et al (2014) Forward genetic screening identifies a small molecule that blocks *Toxoplasma gondii* growth by inhibiting both host- and parasite-encoded kinases. PLoS Pathog 10:1–14
53. Black MW, Arrizabalaga G, Boothroyd JC (2000) Ionophore-resistant mutants of *Toxoplasma gondii* reveal host cell permeabilization as an early event in egress. Mol Cell Biol 20:9399–9408
54. Knoll LJ, Boothroyd JC (1998) Isolation of developmentally regulated genes from *Toxoplasma gondii* by a gene trap with the positive and negative selectable marker hypoxanthine-xanthine-guanine phosphoribosyltransferase. Mol Cell Biol 18:807–814
55. Bradley PJ, Li N, Boothroyd JC (2004) A GFP-based motif-trap reveals a novel mechanism of targeting for the Toxoplasma ROP4 protein. Mol Biochem Parasitol 137:111–120
56. Saeij JPJ et al (2006) Polymorphic secreted kinases are key virulence factors in Toxoplasmosis. Science 1979(314):1780–1783
57. Behnke MS et al (2012) The polymorphic pseudokinase ROP5 controls virulence in *Toxoplasma gondii* by regulating the active kinase ROP18. PLoS Pathog 8:e1002992
58. Yamamoto M et al (2009) A single polymorphic amino acid on *Toxoplasma gondii* kinase ROP16 determines the direct and strain-specific activation of Stat3. J Exp Med 206:2747–2760
59. Saeij JPJ et al (2007) Toxoplasma co-opts host gene expression by injection of a polymorphic kinase homologue. Nature 445:324–327
60. Ong YC, Reese ML, Boothroyd JC (2010) Toxoplasma Rhoptry protein 16 (ROP16) subverts host function by direct tyrosine phosphorylation of STAT6. J Biol Chem 285:28731–28740
61. Jensen KDC et al (2011) Toxoplasma polymorphic effectors determine macrophage polarization and intestinal inflammation. Cell Host Microbe 9:472–483
62. Butcher BA et al (2011) *Toxoplasma gondii* rhoptry kinase ROP16 activates STAT3 and STAT6 resulting in cytokine inhibition and arginase-1-dependent growth control. PLoS Pathog 7:e1002236
63. Yap G, Pesin M, Sher A (2000) Cutting edge: IL-12 is required for the maintenance of IFN-γ production in T cells mediating chronic resistance to the intracellular pathogen, *Toxoplasma gondii*. J Immunol 165:628–631
64. Scharton-Kersten TM et al (1996) In the absence of endogenous IFN-gamma, mice develop unimpaired IL-12 responses to *Toxoplasma gondii* while failing to control acute infection. J Immunol 157:4045–4054
65. Yap GS, Sher A (1999) Effector cells of both nonhemopoietic and hemopoietic origin are required for interferon (IFN)-γ– and tumor necrosis factor (TNF)-α–dependent host resistance to the intracellular pathogen, *Toxoplasma gondii*. J Exp Med 189:1083–1092
66. Glausen TG et al (2021) The Toxoplasma polymorphic effector GRA15 mediates seizure induction by modulating Interleukin-1 signaling in the brain. mBio 12
67. Gov L, Karimzadeh A, Ueno N, Lodoen MB (2013) Human innate immunity to *Toxoplasma gondii* is mediated by host caspase-1 and ASC and parasite GRA15. mBio 4
68. Donald RG, Roos DS (1994) Homologous recombination and gene replacement at the dihydrofolate reductase-thymidylate synthase locus in *Toxoplasma gondii*. Mol Biochem Parasitol 63:243–253
69. Kim K, Soldati D, Boothroyd JC (1993) Gene replacement in *Toxoplasma gondii* with chloramphenicol acetyltransferase as selectable marker. Science 1979(262):911–914
70. Fox BA, Ristuccia JG, Gigley JP, Bzik DJ (2009) Efficient gene replacements in *Toxoplasma gondii* strains deficient for nonhomologous end joining. Eukaryot Cell 8:520–529
71. Huynh MH, Carruthers VB (2009) Tagging of endogenous genes in a *Toxoplasma gondii* strain lacking Ku80. Eukaryot Cell 8:530–539

72. Jinek M et al (2012) A programmable dual-RNA–guided DNA endonuclease in adaptive bacterial immunity. Science 1979(337):816–821
73. Shaojun L, Qiuling W, David SL (2016) Analysis of noncanonical calcium-dependent protein kinases in *Toxoplasma gondii* by targeted gene deletion using CRISPR/Cas9. Infect Immun 84:1262–1273
74. Behnke MS, Khan A, Sibley LD (2015) Genetic mapping reveals that sinefungin resistance in *Toxoplasma gondii* is controlled by a putative amino acid transporter locus that can be used as a negative selectable marker. Eukaryot Cell 14:140–148
75. Shen B, Brown KM, Lee TD, David Sibley L (2014) Efficient gene disruption in diverse strains of *Toxoplasma gondii* using CRISPR/CAS9. mBio 5:e01114
76. Sidik SM, Hackett CG, Tran F, Westwood NJ, Lourido S (2014) Efficient genome engineering of *Toxoplasma gondii* using CRISPR/Cas9. PLoS One 9:1–8
77. Donald RGK, Roos DS (1993) Stable molecular transformation of *Toxoplasma gondii*: a selectable dihydrofolate reductase-thymidylate synthase marker based on drug-resistance mutations in malaria. Proc Natl Acad Sci USA 90:11703–11707
78. Walker JR, Corpina RA, Goldberg J (2001) Structure of the Ku heterodimer bound to DNA and its implications for double-strand break repair. Nature 412:607–614
79. Downs JA, Jackson SP (2004) A means to a DNA end: the many roles of Ku. Nat Rev Mol Cell Biol 5:367–378
80. Aravind L, Koonin EV (2001) Prokaryotic homologs of the eukaryotic DNA-end-binding protein Ku, novel domains in the Ku protein and prediction of a prokaryotic double-strand break repair system. Genome Res 11:1365–1374
81. Fox BA et al (2011) Type II *Toxoplasma gondii* KU80 knockout strains enable functional analysis of genes required for cyst development and latent infection. Eukaryot Cell 10:1193–1206
82. Sidik SM et al (2016) A genome-wide CRISPR screen in Toxoplasma identifies essential apicomplexan genes. Cell 166:1423–1435.e12
83. Kuscu C, Arslan S, Singh R, Thorpe J, Adli M (2014) Genome-wide analysis reveals characteristics of off-target sites bound by the Cas9 endonuclease. Nat Biotechnol 32:677–683
84. Young J et al (2019) A CRISPR platform for targeted *in vivo* screens identifies *Toxoplasma gondii* virulence factors in mice. Nat Commun 10:3963
85. Franco M et al (2016) A novel secreted protein, MYR1, is central to toxoplasma's manipulation of host cells. mBio 7:1–16
86. Sangaré LO et al (2019) *In vivo* CRISPR screen identifies TgWIP as a Toxoplasma modulator of dendritic cell migration. Cell Host Microbe 26:478–492.e8
87. Wang Y et al (2020) Genome-wide screens identify *Toxoplasma gondii* determinants of parasite fitness in IFNγ-activated murine macrophages. Nat Commun 11
88. Murillo-León M et al (2019) Molecular mechanism for the control of virulent *Toxoplasma gondii* infections in wild-derived mice. Nat Commun 10:1233
89. Taylor GA et al (2000) Pathogen-specific loss of host resistance in mice lacking the IFN-gamma-inducible gene IGTP. Proc Natl Acad Sci USA 97:751–755
90. Degrandi D et al (2013) Murine guanylate binding protein 2 (mGBP2) controls *Toxoplasma gondii* replication. Proc Natl Acad Sci USA 110:294–299
91. Yamamoto M et al (2012) A cluster of interferon-γ-inducible p65 GTPases plays a critical role in host defense against *Toxoplasma gondii*. Immunity 37:302–313
92. Lykens JE et al (2010) Mice with a selective impairment of IFN-gamma signaling in macrophage lineage cells demonstrate the critical role of IFN-gamma-activated macrophages for the control of protozoan parasitic infections *in vivo*. J Immunol 184:877–885
93. Suzuki Y, Orellana MA, Schreiber RD, Remington JS (1988) Interferon-gamma: the major mediator of resistance against *Toxoplasma gondii*. Science 240:516–518
94. Kravets E et al (2016) Guanylate binding proteins directly attack *Toxoplasma gondii* via supramolecular complexes. eLife 5
95. de Abreu Cabral GR, Wang ZT, Sibley LD, DaMatta RA (2018) Inhibition of nitric oxide production in activated macrophages caused by *Toxoplasma gondii* infection occurs by distinct mechanisms in different mouse macrophage cell lines. Front Microbiol 9:1–11
96. Cygan AM et al (2020) Coimmunoprecipitation with MYR1 identifies three additional proteins within the *Toxoplasma gondii* parasitophorous vacuole required for translocation of dense granule effectors into host cells. mSphere 5:1–17
97. Li W et al (2022) A splitCas9 phenotypic screen in *Toxoplasma gondii* identifies proteins involved in host cell egress and invasion. Nat Microbiol 7:882. https://doi.org/10.1038/s41564-022-01114-y
98. Zetsche B, Volz SE, Zhang F (2015) A split-Cas9 architecture for inducible genome editing and transcription modulation. Nat Biotechnol 33:139–142
99. Markus BM, Boydston EA, Lourido S (2021) CRISPR-mediated transcriptional repression in *Toxoplasma gondii*. mSphere 6:e0047421. https://doi.org/10.1128/msphere.00474-21
100. Butterworth S et al (2023) High-throughput identification of *Toxoplasma gondii* effector proteins that target host cell transcription. Cell Host Microbe 31:1748–1762.e8

Toxoplasma gondii's Endosymbionts: An Insight into Their Genomic Secrets

Luisa Berná, Natalia Rego, Paula Faral-Tello, and Maria E. Francia

Abstract

The Apicomplexa phylum is composed entirely of obligate intracellular parasites. Among them, *Toxoplasma gondii* is both a prominent pathogen, causative agent of toxoplasmosis, and a model organism. A peculiar feature shared by most apicomplexans is the presence of two distinct DNA-containing organelles of endosymbiotic origin: the mitochondrion and the apicoplast. Particular attributes of these organelles, including their morphology, ultrastructure, and segregation processes, underpinned by their bacterial origins, represent attractive targets for antiparasitic drug development due to their divergence from mammalian host cells. They showcase several peculiarities. One of them is their significant genomic reductions, characteristics likely associated with adaptations to their ecological niches. Despite minimized genomes, they retain essential bacterial-like translation mechanisms and have evolved sophisticated protein import systems to maintain critical functions.

Recent breakthroughs in long-read sequencing technologies have provided invaluable insights into the complexity of coccidian mitogenomes, especially the *T. gondii* mitochondrial genome. This genome is characterized by a specific arrangement of sequence blocks that encode essential components for the electron transport chain. However, the complete architecture of the mitochondrial genome remains elusive, including the presence of potentially multiple mitochondrial genomes within a single individual and whether post-transcriptional processing occurs or not.

The apicoplast—a relic of a red algal endosymbiont—underscores the evolutionary trajectory of *T. gondii*, with its anabolic ability to

L. Berná (✉)
Laboratory of Apicomplexan Biology, Institut Pasteur de Montevideo, Montevideo, Uruguay

Laboratorio de Genómica Evolutiva, Facultad de Ciencias, Universidad de la República, Montevideo, Uruguay
e-mail: lberna@pasteur.edu.uy

N. Rego
Laboratorio de Genómica Evolutiva, Facultad de Ciencias, Universidad de la República, Montevideo, Uruguay

Bioinformatics Unit, Institut Pasteur de Montevideo, Montevideo, Uruguay

P. Faral-Tello
Laboratory of Apicomplexan Biology, Institut Pasteur de Montevideo, Montevideo, Uruguay

M. E. Francia
Laboratory of Apicomplexan Biology, Institut Pasteur de Montevideo, Montevideo, Uruguay

Departamento de Parasitologia y Micologia, Facultad de Medicina, Universidad de la República, Montevideo, Uruguay

J. D. Ramírez González (ed.), *Recent Advances in Parasitomics*,
https://doi.org/10.1007/978-3-031-70591-5_9

sustain vital metabolic pathways. Despite its compact size and low GC content, the apicoplast genome presents a conserved structure across species.

Furthermore, the coordination of organelle dynamics with nuclear DNA replication reveals the deep interplay between endosymbionts and the life cycle of *T. gondii*, demonstrating a delicate balance between evolutionary conservation and innovation. Further understanding the role of these divergent organelles and the essential genes they encode in the process of adaptation to parasitism will open new opportunities for therapeutic interventions directed against the core biological processes of apicomplexan parasites.

Keywords

Apicomplexa · Endosymbiotic origin · Mitochondrion · Apicoplast · Mitochondrial genome architecture · Antiparasitic drug development · Genomic reduction

1 Introduction

The phylum Apicomplexa is known for harboring parasites that cause critical public health concerns such as malaria (*Plasmodium* spp.) cryptosporidiosis (*Cryptosporidium* spp.) and toxoplasmosis (*Toxoplasma gondii*), as well as veterinary pathogens of renown, such as *Neospora caninum*. This diverse eukaryotic phylum is an integral member of the superphylum Alveolata which also includes Chrompodellids, Dinoflagellates, and Ciliates. With over 6000 identified species, the true diversity of apicomplexans is thought to be underestimated.

Apicomplexans have complex life cycles that involve different parasite stages and multiple modes of transmission. In particular, *T. gondii* is a widespread parasite capable of infecting virtually any nucleated cell within any warm-blooded animal. Intermediate hosts (humans, sheep, pigs, etc.) can become infected by interacting with infectious oocysts or consuming tissue cysts through carnivorism. As a result of a primary infection or reactivation of a chronic infection, toxoplasmosis can cause acute encephalitis, cerebral calcifications, chorioretinitis, and blindness. Additionally, fetuses can acquire the infection through vertical transmission from an immune-naïve mother infected during pregnancy, causing congenital disease or fetal loss.

The scientific journey to understand apicomplexans has significantly advanced through studies on model organisms such as *Plasmodium* and *T. gondii*. A major feature shared by most apicomplexans is the presence of two organelles of endosymbiotic origin: a single mitochondrion, originating from an alpha-proteobacterium, and the apicoplast, a derivative of red algal endosymbiosis [1, 2]. Despite long-standing interest in the biology and drug-targeting potential of these organelles, recent technological advancements have revitalized their study, particularly from a genomics standpoint, shedding light onto the reduced mitochondrial genome and the evolutionary transition from the photosynthetic ancestors to the current obligate parasitic lifestyle [3].

The differentiation between mitochondrion and apicoplast as separate DNA-containing organelles in apicomplexans marks a major scientific milestone in the field. Initial misconception attributed the apicoplast's 35-kb genome to the mitochondrion's. However, sequencing efforts in *Plasmodium* rectified this misconception but also uncovered the "6 kb element," containing the crucial mitochondrial genes, such as cytochrome oxidase subunit I (*coxI*), cytochrome oxidase subunit III (*coxIII*), and cytochrome b (*cob*), and ribosomal RNA (rRNA) fragments. This discovery provided a fresh perspective on mitochondrial genetics [4, 5]. Simultaneously, the sequencing of a 35-kb DNA fragment, which may have originated from chloroplast DNA, contributed to distinguishing apicoplast DNA from the mitochondrial genome [6–8]. This was confirmed with the complete sequencing of the apicoplast genome [9, 10]. The latter represented a hallmark in *Plasmodium* genomics that was rapidly expanded across apicomplexans, shedding light onto the intricate evolutionary narrative of these parasites and their organelles, fueling ongoing scientific interest in this field [10].

T. gondii and other close coccidians have undergone distinct evolutionary paths characterized by genomic reductions and reorganizations, recently elucidated by new sequencing technologies [11–14]. At a functional level, these genomic changes have endowed these parasites with unique and specialized biological functions that account for their complex and divergent roles in the cellular machinery. Several reviews have delved into the biology, biochemistry, and functionality of *T. gondii*'s mitochondrion and apicoplast [15–18]. This chapter will focus on our current understanding of the genomes of these endosymbionts, with a particular emphasis on those found in *T. gondii*.

2 From Endosymbionts to Organelles: The Mitochondrion and Apicoplast of Apicomplexa

Apicomplexans harbor two DNA-containing and membrane-bound organelles, acquired through two independent endosymbiotic events. The mitochondrion, originating from a primary endosymbiosis between an alpha-proteobacterium and an ancestral cell associated with the Asgard archaea, marks a foundational event for all eukaryotic life. This event has been extensively studied. Ongoing research, powered by technological advancements and a multidisciplinary approach, continues to refine our understanding of eukaryotic evolution [19]. Contrary to earlier beliefs that mitochondrial acquisition was linked to Earth's oxygenation, recent insights suggest that eukaryogenesis likely occurred under anoxic conditions, when low oxygen levels were found only at the oceans' surfaces [20]. Mitochondria retention across eukaryotes is attributed to its efficiency in ATP generation through aerobic respiration, oxidizing organic compounds, carbohydrates, amino acids, and lipids. This evolutionary milestone was accompanied by several processes such as the incorporation of transporters, gene transfer to the nucleus, the emergence of new mechanisms for replication and segregation, and genomic reduction [21].

Apicomplexans and dinoflagellates, sharing a common ancestor within the Alveolata supergroup, exhibit a marked reduction in mitochondrial genome complexity. This simplification, characterized by the transfer of most genetic material to the nucleus and the retention of only a few rRNA fragments alongside three essential genes for the respiratory chain, predates the speciation of both groups. This massive genomic reduction, coupled with specific metabolic modifications, has been associated with a strategic adaptation to distinct ecological niches [22].

Secondary endosymbiosis, where an eukaryotic cell engulfs another eukaryote that has already undergone primary endosymbiosis, leads to the acquisition of the apicoplast. As more than two membranes is a hallmark of secondary endosymbiosis, this event was elucidated by the identification of circular DNA inside additional membranes enveloping the endosymbiont and was finally confirmed by phylogenetic studies. Discoveries around the apicoplast radically shifted our understanding of apicomplexan evolution, cell biology, and metabolism. The consensus, supported by the study of organisms like *Chromera velia*, is that the apicoplast originated from red algae before the divergence of dinoflagellates and Apicomplexa [2, 15].

Unlike other groups that have retained plastids by secondary endosymbiosis—and retained parts of the nuclear genome of the algal symbiont—apicomplexans have lost the nucleus of their engulfed algal ancestor. During their evolution, the remaining plastid relocated many of its genes into the host's nuclear genome and lost many genes, particularly those linked to photosynthesis. Nevertheless, the apicoplast has retained essential metabolic pathways crucial to the parasite, underscoring the fundamental role of the organelle in its biology [3, 16]. As we understand it today, Apicomplexan apicoplast contains a 35-kb genome featuring an inverted repeat of around 10 kb. The apicoplast genome retained only genes that are essential for its replication and transcription, and two additional genes: *SufB*, critical for sulfur activation during iron-sulfur cluster assembly, and *ClpC*, a subunit of an ATP-dependent Clp protease, with potential chaperone activity [23, 24].

In summary, both the mitochondria and the apicoplast, acquired through separate endosymbiotic events, exhibit remarkable genomic reduction in size and content. This process involved the translocation of a significant portion of their genetic material to the nucleus and the loss of redundant and non-essential genes. Originating from bacterial ancestors, these organelles retain bacterial-like translation machinery and have evolved complex protein import mechanisms to maintain essential functionality with minimized genomes. This interdependence between nuclear and organelle genomes underlies the complexity of intracellular communication and coordination. The genomic characteristics of these organelles illuminate their evolutionary history and emphasize their pivotal roles within the cell's metabolic and regulatory networks. Finally, the unique features of apicomplexan endosymbionts highlight these organisms as fascinating subjects for continued research, particularly in understanding cellular evolution and organellar function in their parasitic lifestyle and in identifying potential drug targets.

3 Beyond Powerhouses: Expanding Roles of Mitochondria

Mitochondria are universally recognized as the cellular powerhouses, orchestrating the synthesis of ATP along with the production of essential nutrients and metabolites critical for cellular survival. However, our historical understanding of mitochondrial biology has been largely shaped by studies in a limited range of eukaryotic organisms in which mitochondria typically exist in multiple copies. Nonetheless, we currently understand that, in some cases, evolutionary processes have led to a reduction in mitochondrial numbers, highlighting the complexity of mitochondrial evolution and adaptation [25, 26]. How these reductions impact mitochondrial function is currently under investigation.

Alongside other apicomplexans, *T. gondii* harbors one mitochondrion per cell, with a unique genomic structure that has a significantly reduced sequence and, consequently, gene content. Despite most of their components being encoded in the nuclear genome, *T. gondii*'s mitochondria retain their essential role in classic energy-providing pathways such as the tricarboxylic acid cycle, the electron transport chain, and oxidative phosphorylation. However, the mitochondrion also plays a pivotal role in the biosynthesis of iron-sulfur clusters, heme, and pyrimidines, which are vital for the parasite's survival and virulence [22, 27, 28].

The critical role of mitochondria in the biology of *T. gondii* has earned it a prominent place as a target for therapeutic interventions. A number of drugs, known to impact the electron transport chain, have been validated across the Apicomplexa, with specific compounds demonstrating marked efficacy against *T. gondii* [18, 29, 30]. Such is the case of atovaquone, a drug that is highlighted also for its usefulness in treating malaria and *T. gondii* infections in immunocompromised patients [29]. However, the development of *Plasmodium* resistance to atovaquone, driven by mutations in the mitochondrial gene encoding for *coxIII*, underscores the urgent need for new therapeutic targets and the study and understanding of resistance-acquiring mutations [31]. Recent proteomic and complexome analyses of *T. gondii* have elucidated the protein composition of several mitochondrial complexes, showing that many of these identified proteins have no homologs in animals [32, 33]. These findings highlight the unique evolutionary path in Apicomplexa and confirm the mitochondrion as a promising reservoir for future drug target discovery.

4 The Divergent and Dynamic Mitochondria of *T. gondii*: Organization Through the Parasite's Life Cycle

In addition to a reduced genomic content, the single mitochondrion of *T. gondii* exhibits multiple ultrastructural peculiarities such as the shape of its cristae. Unlike the lamellar cristae found in mammalian cells, *T. gondii* mitochondria display bulbous cristae, given by an unusual pyramidal arrangement of cristae-embedded ATP synthase hexamers [34]. This divergent structure is further underscored by a remarkable divergence among the components of the ATP synthase complex itself, featuring multiple units that lack homology beyond the apicomplexan [35].

Beyond this, the mitochondrion displays various morphologies, including lasso, sperm-like, and collapsed forms [36, 37]. In intracellular parasites, mitochondrial morphology typically remains stable in a peripheral lasso configuration. However, extracellular parasites can display any of the three morphologies. Among extracellular parasites not exhibiting the lasso shape, the predominant morphology is sperm-like, featuring a rounded bundle directed toward the proximal end of the cell. A minority of extracellular parasites exhibit collapsed mitochondria, a morphology that transitions to a sperm-like shape upon reaching the periphery. These morphologies are observed with similar frequency during processes such as gliding, host cell invasion, and entry, suggesting that they are not specifically associated with any particular stage of the lytic cycle. Instead, they likely represent distinct morphological states between intra- and extracellular parasites [36].

The clinically relevant form of *T. gondii*, the fast-dividing tachyzoite, proliferates by a process known as endodyogeny. Unlike mammalian cell division, which is preceded by duplication of all organelles, endodyogeny entails *de novo* formation of many cellular components. Accompanying this process, the mitochondrion's localization and morphology vary dynamically [36, 38]. Daughter cell formation initiates early after nuclear genome duplication. At this stage, mitochondria duplication begins by branching at multiple sites [36, 39, 40]. Time-lapse microscopy of fluorescent markers targeted to various organelles evidenced that the mitochondrion is the last organelle to be segregated [40].

The mechanisms of mitochondrial segregation have recently begun to emerge. In other systems, mitochondria severing and segregation rely on dynamin-related GTPases, which assemble fission complexes leading to its constriction and division. Three dynamin-related proteins, namely, DrpA, DrpB, and DrpC, can be identified in the nuclear genome of *T. gondii*. DrpA plays a role in the fission of the apicoplast (see sections on the apicoplast below for further details), while DrpB is involved in the biogenesis of secretory organelles, specifically the rhoptries. TgDrpC is important for mitochondrial biogenesis, as its depletion or dominant negative expression results in interconnected mitochondria [39]. However, it was recently shown that DrpC appears to be involved in other canonical dynamin functions, namely, vesicle trafficking and endocytosis, and might not mediate fission directly [41].

Fis1 homologs recruit fission mediating dynamins. A homolog of the Fis1p protein of yeast, TgFis1, was shown to interact with an outer mitochondrial membrane protein named lasso-maintenance factor 1 (TgLMF1). Disruption of lmf1 in *T. gondii* severely impacts mitochondrial morphology and impairs proper organellar segregation [42]. LMF1 has additionally been shown to interact with the alveolin-containing pellicle protein IMC10, providing a mechanistic link, namely, the components of the membrane contact site, between a mitochondrial-resident protein and the parasite periphery [43]. IMC10 knockdown phenocopies disruption of lmf1, whereby parasites display altered mitochondrial morphology in addition to segregation defects into daughter cells. The latter revealed that membrane contact sites likely mediate the segregation of severed mitochondria into individual daughter cells during cell division [43, 44].

5 Mitochondrial Genome of *T. gondii*: Structure, Function, and Evolution

In recent years, various strategies were successfully implemented to sequence the nuclear genomes of different *T. gondii* strains, yet they did not succeed in determining the complete sequence and structure of the mitochondrial genome [45–48]. Until recently, our understanding of *T. gondii*'s mitogenome was primarily based on research conducted on *Plasmodium*.

The early identification of a tandem duplicate known as the "6 kb element" [49], followed by sequencing in various species [50–54], unveiled what was at that time, one of the smallest mitochondrial genomes ever discovered. This pioneering effort revealed a small genome amounting to only 5967 base pairs in length encoding solely three genes. Such discoveries laid the vital groundwork for further exploring the mitochondrial DNA of *Plasmodium* and, more broadly, apicomplexans. These findings also shed light on the compact nature and critical functions of mitochondrial genomes in these parasites.

We now appreciate that all apicomplexan mitochondrial genomes underwent the same genomic reduction and that their content and size are very similar. The mitochondrial genomes of all Apicomplexa encode only three genes, namely: cytochrome oxidase subunit I (*coxI*), cytochrome oxidase subunit III (*coxIII*), and cytochrome b (*cob*). In addition, they display fragmented rRNA genes; several of the fragments that would be required to assemble complete ribosomes have not been identified or were lost [49, 55–57] (see below). However, despite sharing these features, apicomplexan mitogenomes differ wildly in structure and genomic organization.

As mentioned above, Dinoflagellates, a sister group to apicomplexans, possess a similarly reduced mitochondrial genome, harboring only three genes alongside fragmented rRNA gene segments. However, their mitogenome is distinguished by independent and numerous amplifications and recombination events, resulting in a fragmented structure and intricate transcriptional processes underlying the expression of mitochondrial genes [58].

The challenging task of assembling coccidian mitogenomes has been linked to both their repetitive nature and the presence of nuclearly-encoded sequence fragments of mitochondrial origin (NUMTs) within the nuclear genome [13]. However, as will be discussed below, NUMTs likely played a secondary role in the complexities associated with clarifying the definitive mitogenome of *T. gondii* as bioinformatic tools have been developed allowing their distinction from bona fide mitochondrial resident genes [59]. Most likely, it has been its repetitive and fragmented nature that has precluded its definitive assembly to date.

5.1 Using Long-Read Sequencing to Dissect the Unusual Mitogenome of *T. gondii*

Recently, the group led by Dr. Jessica Kissinger and our team have implemented long-read sequencing technologies to elucidate the mitochondrial genome of *T. gondii* [11, 13]. These state-of-the-art sequencing methods significantly increase the length of reads, from 150 bp to an average of 15 kb and longer. Such extended reads are crucial for navigating through complex genomic regions, including repetitive or low-complexity sequences, facilitating the resolution or enhancement of various complex genomes [11, 60, 61]. Despite these technological advances, fully resolving the structure of the *T. gondii* mitochondrial genome has remained elusive.

Using both PacBio and Oxford Nanopore Technologies (ONT) methods, our team identified several reads belonging to the mitochondrion, distinguished by their significantly lower GC content (~36%) compared with the nuclear genome (~55%) [11]. Surprisingly, neither our efforts nor those of other groups have succeeded in assembling these reads into a single mitochondrial chromosome. However, Namasivayam et al. further determined that the mitochondrial genome of *T. gondii* comprises various rearrangements of the same 5909-bp sequence [13]. Moreover, they described that the mitogenome is structured into 21 different sequence blocks, ranging from 40 to 1050 bp each, arranged in diverse configurations, thereby creating complex structures reminiscent of those seen in dinoflagellates (Fig. 1).

These configurations are evident in the assembly of multiple distinct contigs, with no homolo-

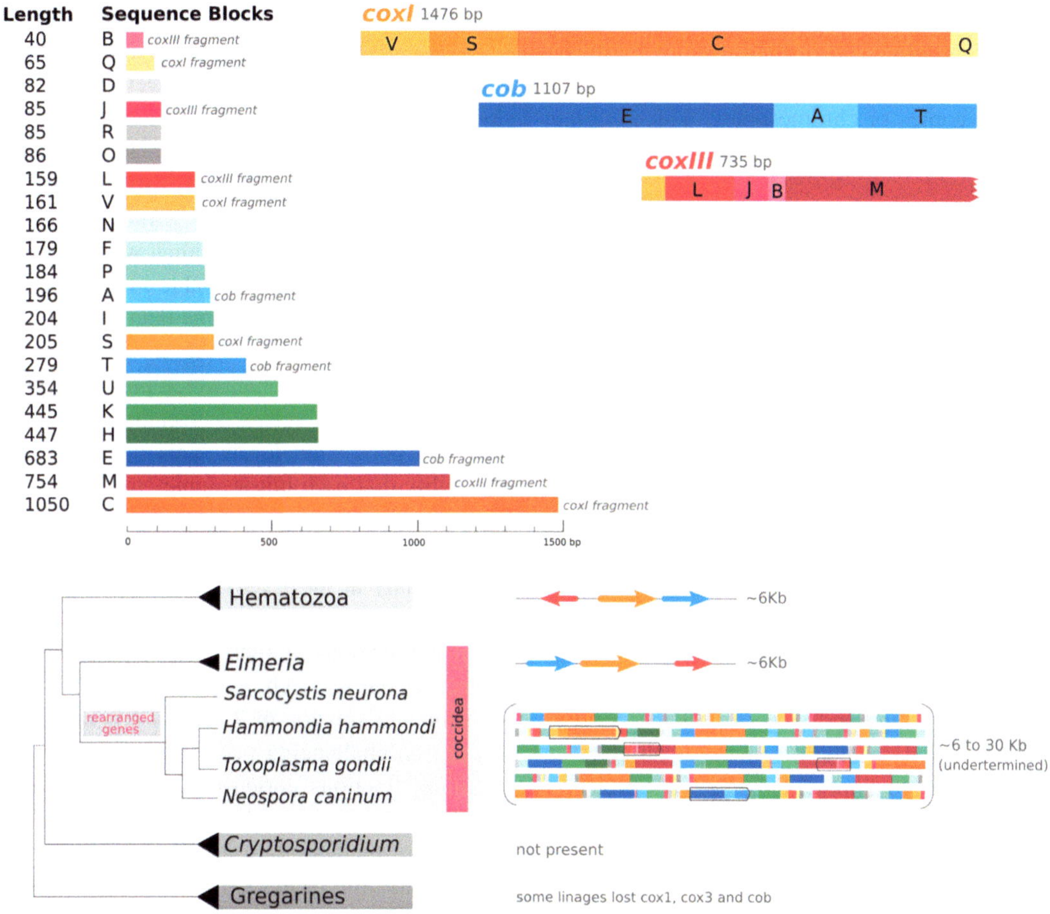

Fig. 1 Composition of the *T. gondii* mitogenome. The mitochondrial genome (mitogenome) comprises 21 distinct sequence blocks ranging from 40 to 1050 bp, labeled with letters from A to V. Blocks integral to specific genes are highlighted in orange for the *coxI* gene, in blue for the *cob* gene, and in pink for the *coxIII* gene. Notably, block V forms part of the *coxI* gene, yet a segment of this block is also present at the start of *coxIII*, reflecting the shared initial sequences (the first bases) of these genes. The diagram accurately represents the scale of sequence blocks and genes. Below, the mitochondrial genome architecture of *T. gondii* is illustrated through the arrangements of blocks into longer sequences, extending to approximately 30 kb. In this structure, complete genes are rarely observed and are indicated as gray boxes. Unlike the conserved 6 kb element characteristic of Hematozoa (piroplasmids and hemosporidians) and Eimeria, which maintains the three coding genes in a fixed arrangement occasionally with terminal inverted repeats (refer to [62], the final genome structure of *T. gondii* remains undetermined. Other coccidians, distinct from Eimeria, such as *Hammondia hammondi* and *Neospora caninum*, exhibit a similar configuration of rearranged blocks in their mitogenomes. This figure is adapted from Figs. 1 and 7 in [13, 62], respectively, and Fig. 1 in Berná et al. [11, 62]

gous segment among them exceeding 2–3 kb in length. Yet, the arrangement of these sequence blocks follows a non-random pattern; although their configuration demonstrates considerable diversity, it is subject to constraints [13]. Designated alphabetically from A to V, these sequence blocks encapsulate the mitochondrion's entire genetic repository, including genes and ribosomal fragments, and are subject to transcriptional activity (depicted in the top left of Fig. 1). Crucially, the presence of specific contiguous blocks is imperative for gene encoding (in the top right of Fig. 1). For instance, blocks E, A, and T facilitate the encoding of *cob*, while a specific sequence arrangement of blocks V, L, J, B, and M is essential for encoding *coxIII* [13].

In summary, the genome of *T. gondii* mitochondria is composed of finite recombination of identified sequence blocks that combine to form distinct copies or chromosomes, which contain the information to encode the characteristic genes.

The distinctive organization observed in the *T. gondii* mitochondrial genome is not unique to this species; it is also found in other related coccidians. Sequence blocks akin to those identified in *T. gondii* have been detected in *Neospora caninum*, *Hammondia hammondi*, and even the more distantly related *Sarcocystis neurona* (in the bottom of Fig. 1) [13, 62]. These findings underscore a shared genomic architecture among these species. However, this configuration is not universally present across all coccidians; for example, *Eimeria tenella* exhibits a more "canonical" mitochondrial genome structure [63] resembling that of *Plasmodium* species. These variations highlight the evolutionary diversity within the coccidian lineage and suggest that the peculiar mitochondrial genome organization observed in some species, but not all, might be a derived trait among certain members of this group.

5.2 Is There a Single Mitogenome in *T. gondii?*

The concept of heteroplasmy, which describes the coexistence of different mitochondrial genomes within a single individual, has been extensively examined in human biology for its potential functional implications and association with various diseases. This phenomenon, resulting from mechanisms such as *de novo* mutations, intracellular transfer, recombination, or the inheritance of diverse mitochondrial DNAs, emphasizes the complexity of mitochondrial genetics [64]. In *T. gondii*, the sequence blocks constituting the mitochondrial genome do not exhibit heteroplasmy in terms of point mutations among the analyzed copies. However, there is significant variability in how these blocks are arranged, leading to a vast heterogeneity in their length and overall organization. In fact, we and others have rarely observed the exact same read twice during sequencing. This can be ascribed to the low coverage of the mitochondrial genome in whole-cell DNA sequencing. However, as these studies have not been carried out at a single cell level, it is also possible that each read represents a distinct configuration of block ordering present at the individual cellular level within a population, or that every single individual within a population displays multiple variable mitochondrial genomes. Notably, all available data are derived from the genomic sequencing of populations, either polyclonal or clonal, which precludes precise differentiation between diversity at the cellular and population levels. This opens up the possibility that each mitogenome could be composed of unique configurations of sequence block organization, thereby indicating the presence of heteroplasmy at the level of structural arrangement, rather than at point mutations. The functional implications of such diversity are current subjects of our research.

Although isolation of *T. gondii* mitochondria has proven challenging [40], methodologies for obtaining fractions enriched with these organelles have been developed and recently optimized [65]. These methodological advances facilitated the identification of the mitochondrial ribosome and enabled the characterization of several essential proteins involved in mitochondrial translation. Furthermore, with the advent of new single-cell technologies applied to organelle research, we anticipate that detailed insights into the structure and organization of the mitochondrial genome of *T. gondii* will be imminently achieved.

5.3 From Genes to Proteins: Decoding Mitochondrial Function in *T. gondii*

While the nucleus encodes the vast majority of mitochondrial components, the three cytochrome genes essential for the electron transport chain are encoded within the mitochondrial genome and are uniquely synthesized by and within the mitochondria. These pivotal genes generally undergo transcription and translation within the mitochondrion in other species, and *T. gondii* as well as other apicomplexan parasites mirror this process. In general, mitochondrial ribosomes (known as mitoribosomes) consist of two primary macromolecular complexes: the large subunit (LSU) and the small subunit (SSU), alongside various rRNA molecules. Notably, in *T. gondii* and apicomplexans at large, the nuclear genome is responsible for encoding all mitoribosomal proteins [66–68]. The composition of these complexes, including the specific proteins and the quantity of rRNA, differs among species.

The laboratory led by Dr. Lilach Sheiner has recently provided central insights into mitochondrial ribosome assembly in *T. gondii*, identifying several key proteins that constitute the LSU and SSU, marking a significant advance in our understanding of *T. gondii* mitochondrial biology [65, 69]. By reciprocally querying microarray data of cell cycle-synchronized parasites, Lacombe and colleagues identified and validated the localization of ten novel components of the mitochondrial proteome; four of which are predicted mitoribosome components and were previously predicted by bioinformatic approaches by Gupta and colleagues in *Plasmodium* [70]. Namely, TGME49_201830/226280/312680 are predicted to encode the mitochondrial LSU components 54, 28, and 41, respectively. The gene products were named TgmL54, TgbL28m, and TgmL41. In addition, the hypothetical protein coding gene TGME49_203620 is a putative homolog of the mitochondrial SSU 35 (mS35).

The mitoribosome is characterized by its rRNA components, which are fragmented and distributed across the mitochondrial genome, encoded on both strands. In *Plasmodium*, 34 small RNAs have been identified; 12 corresponding to SSU and 15 to LSU RNAs, representing one of the most extreme cases of rRNA fragmentation among Apicomplexa [55, 56]. In *T. gondii*, initial findings by Namasivayam et al. [13] reported 11 rRNA fragments, and more recently, an additional 23 fragments were identified through small RNA sequencing of mitochondria-enriched fractions [57]. This brings the total to 34 rRNA fragments, mirroring the number found in *Plasmodium* and, more importantly, confirming their expression [57] demonstrating the complex organization of mitochondrial rRNA transcription also in this organism.

Apart from ribosomal proteins and rRNAs, the translation of proteins encoded by the mitochondrial genome necessitates specific tRNAs. However, the mitochondrial genome does not encode tRNAs; instead, these molecules are nuclearly encoded and subsequently imported into the mitochondria. Elegant experiments by Esseiva and colleagues [71] demonstrated that *T. gondii* imports all of its tRNAs, with these molecules being present in both the cytosol and the mitochondria [71]. A notable exception is the initiator methionine tRNA ($tRNA^{Met-i}$), which is encoded by the nucleus and is specific to the cytosol, not being imported into the mitochondria. This indicates that the only $tRNA^{Met}$ present in the mitochondria is the one responsible for inserting methionine at internal positions ($tRNA^{Met-e}$ for elongation). Additionally, findings by Pino et al. [72]—which revealed the absence of aminoacyl-tRNA synthetases within the mitochondria—support the notion that tRNAs are likely imported in their charged (aminoacylated) form [72].

As mentioned, the *T. gondii* mitogenome features complete copies of three genes, interspersed with segments of these genes (sequence blocks) distributed throughout other regions of the genome in a complex pattern. Transcription of these genes has been confirmed using RT-PCR [13] and direct ONT mRNA sequencing [12]. While ONT mRNA sequencing has identified sequences covering the entire length of the genes [13], it also revealed transcriptional activity for parts of the genes confined to specific genomic blocks, all truncated at consistent positions [62]. The low abundance of full gene copies, together with the sparse detection of full-length gene transcripts in ONT mRNA data, points to a limited representation of these complete sequences in each mitogenome. Combined with the transcription of gene segments from specific blocks, this observation raises the intriguing possibility that the mitogenome might not contain enough complete copies of these genes to fulfill the transcriptional needs of the mitochondria. Consequently, this scenario hints at a speculative, yet innovative, mechanism within *T. gondii* mitochondria for the reassembly of primarily transcribed RNA segments into functional, full-length coding RNAs, a process that remains to be fully elucidated.

Similar to the intriguing post-transcriptional processing mechanisms speculated in *T. gondii*, other alveolates exhibit unique mitochondrial gene configurations. For example, Dinoflagellates showcase a mitogenome where genes are fragmented and dispersed. Specifically, the detection of polyadenylated *coxIII* gene fragments implies the existence of a novel mechanism of splicing [73]. Moreover, the complexity of dinoflagellate mitochondria extends to phenomena such as gene editing and fusion, presenting a rich field of study with abundant unanswered questions [58, 74].

6 The Apicoplast Genome of *T. gondii*

The apicomplexan plastid (apicoplast), derived from an engulfed red alga, represents a critical evolutionary acquisition that occurred before the divergence of Dinoflagellates and Apicomplexa ([75, 2]). However, the journey to arrive at this conclusion was anything but straightforward. This process has been meticulously chronicled in reviews by various authors, illustrating the complex and intricate path of discovery that underpins our current understanding of apicoplast origins [16, 24].

6.1 The Anabolic Capacity of the Apicoplast as a Source of Potential Drug Targets

The apicoplast has the anabolic capacity to carry out essential metabolic pathways of great importance for the survival of both the acute and chronic forms of *T. gondii* [76]. Specifically, four metabolic pathways have been described: heme biosynthesis, type II fatty acid synthesis (FASII), isoprenoid biosynthesis, and the sulfur utilization factor (SUF) pathway. These pathways have been shown to be important for the life of *T. gondii* and therefore represent an interesting source of potential drug targets currently under investigation; we will briefly describe their main functions.

Heme is essential for numerous critical biological functions, facilitated by iron's ability to fluctuate between reduced and oxidized states. This property allows heme and heme-containing proteins to participate in a variety of electron transport and redox reactions. Heme biosynthesis is a sophisticated process involving the cytosol, mitochondria, and apicoplast. *De novo* heme production is crucial for the intracellular proliferation and virulence of *T. gondii*, highlighting its importance in the pathogenic mechanisms of the parasite [77].

The FASII pathway encompasses a series of enzymes that convert acetyl-CoA into fatty acids. This process takes place within the apicoplast, playing a critical role in the survival of *T. gondii*. The absence of FASII activity results in a marked growth defect, primarily due to the loss of the apicoplast. Nonetheless, the presence of this pathway is not uniform across all apicomplexans; it has been specifically identified in coccidians and haemosporidia, highlighting a divergence in metabolic capabilities within the phylum [78, 79].

Isoprenoid synthesis in apicomplexans involves two distinct pathways that converge to produce isoprenoids, utilizing different precursors and unrelated enzymes. The cytosolic mevalonate (MVA) pathway is widespread among eukaryotes, while the 2-C-methyl-D-erythritol 4-phosphate (MEP) pathway, localized within the apicoplast of apicomplexans, represents a unique biosynthetic route. In the MEP pathway, pyruvate and glyceraldehyde-3-phosphate undergo a seven-step process to generate the precursors for isoprenoids [80]. While the overall essentiality of this pathway remains to be fully elucidated, specific genes within the MEP pathway have been identified as critically important for the survival of the parasite, underscoring their potential as targets for therapeutic intervention [3].

The SUF pathway is involved in the biosynthesis of iron-sulfur cofactors. These cofactors are crucial for the viability of proteins involved in important metabolic reactions, including plastidial pathways for isoprenoid and fatty acid synthesis [81]. Importantly, the gene *SufB*, which is integral to this pathway, is still located within the apicoplast genome, highlighting its significance in the apicoplast's metabolic functions.

Many of the genes and proteins involved in these biosynthetic processes are not found in the mammalian host and are therefore either validated drug targets or potential candidates for drug development [82]. This discrepancy between host and parasite metabolic pathways offers a unique opportunity to design selective and effective therapeutic agents aimed at tackling *T. gondii* without harming the host [15, 81, 83].

6.2 The Conserved Genome Structure of a Divergent Organelle

Through the course of evolution, the apicoplast has followed distinct paths characterized by independent gene losses and convergences of metabolic pathways [3]. However, unlike the mitogenome, whose complexity could not be fully defined for *T. gondii*, the apicoplast genome exhibits a conserved general structure. This includes the presence of an inverted repeat containing SSU and LSU rRNAs, arranged tail to tail, which is characteristic of plastids and is also present in other apicomplexans [6, 16]. Additionally, there is minimal variation in gene content among apicomplexans. Despite this structural conservation, variations exist among this phylum at the levels of base content and genomic arrangements [84]. For instance, the apicoplast genes of Eimeria show little homology with those of other species. *Babesia* species lack the inverted repeat found in many plastidial genomes, whereas *Theileria* has a larger genome, over 40 kb. These differences highlight the diverse evolutionary pathways and structural adaptations of apicoplast genomes within the apicomplexan [85, 86].

The apicoplast genome is remarkably compact, encompassing merely 35 kb with an unusually low GC content of 21.5%. This is in stark contrast to the 59.2% GC content of the *T. gondii* nuclear genome. This minute DNA segment harbors an inverted repeat of approximately 10 kb in length comprising a total of 60 genes. These genes predominantly pertain to replication and transcription processes, with only two exceptions outside these functional categories: *SufB* (a subunit of the Fe-S cluster assembly) and *ClpC* (a subunit of the ATP-dependent Clp protease). The retention of the latter two is strongly associated with apicoplast genome retention [87]. On the contrary, the assortment of maintenance genes within the apicoplast genome includes ribosomal genes, RNA polymerase, the elongation factor (tufA), several tRNAs, and five open reading frames of undetermined function. Apart from these, all other genetic instructions are relegated to the nuclear genome, underscoring a highly specialized and streamlined genetic architecture within the apicoplast. This genomic configuration highlights the intricate balance of gene retention and loss that shapes the functionality and evolutionary history of the apicoplast in *T. gondii*.

The *T. gondii* apicoplast genome displays the notable feature of TGA codons being present within the coding sequences of 17 genes, a trait also observed in Eimeria, *Theileria*, and *Babesia* [16]. These STOP-signaling codons are located at sites of conserved tryptophans [88]. The utilization of TGA for tryptophan implies that TGA does not code for the canonical stop codon; hence, these apicoplast genes of *T. gondii* invariably terminate either with TAA or TAG [89]. Interestingly, *Plasmodium* has lost the TGA codons in apicoplast genes, both internally and terminally. The tryptophan tRNA encoded by the *T. gondii* plastid genome carries a CCA anticodon that cannot pair with UGA, and the mechanism employed for the incorporation/modification of tryptophan by *T. gondii* remains unknown. Finally, certain genes, such as *rpoc2* and *rps8*, exhibit bona fide stop codons within their coding sequences, raising questions about their functionality as pseudogenes or the existence of additional—yet to be clarified—mechanisms for their utilization.

7 Apicoplast Segregation Through the Parasite's Life Cycle

Apicoplast nucleoid division and segregation initiate at the beginning of the S-phase and are coupled to the kinetics of nuclear DNA replication. Both processes conclude during mitosis, with critical participation of the apicoplast-encoded polymerase and gyrase underpinning the process within the organelle [90]. During endodyogeny, elongation of the Golgi apparatus, together with centrosome duplication, and elongation of the apicoplast are among the first observable morphological changes. The Golgi apparatus is the first organelle to be encapsulated as the daughter cell's cytoskeleton scaffold grows toward the mother cell's basal end. Further extension of the cytoskeletal scaffold results in a partitioning of the apicoplast, the nucleus, and the endoplasmic reticulum. The segregation of the apicoplast into daughter cells occurs by its physical association with the centrosome [91]. In particular, it has been shown that myosin F, an essential motor participating also in centrosomes positioning, is involved in apicoplast inheritance; depletion of MyoF results in parasites losing the apicoplast [92]. The dynamin-related protein named DrpA (mentioned above) assembles multimeric spirals that upon GTP hydrolysis mediate the fission of the apicoplast before cytokinesis [93].

8 Endosymbiont Genomes as Modulators of the Nuclear Genome

Beyond the metabolic benefits and biochemical roles offered by the endosymbionts, their streamlined genomes play a crucial role in fostering nuclear genome variability and supplying genetic material that may enable diverse neofunctionalizations within the *T. gondii* genome [59]. These recent discoveries have highlighted the Nuclear DNA of Organellar Origin (NUOTs), encompassing mitochondrial DNA (NUMTs) and apicoplast DNA (NUPTs), as perpetually integrating into the nuclear genome, constituting 1.61% of it. These small sequences correspond to fragments shorter than 200 bp, with most of them being 50–100 bp. Some of these fragments present in the nuclear genome are 100% identical to the endosymbiont genomic source, indicating an actively occurring process [59]. These incorporations serve as a source of genetic variation, acting as modulators of genomic architecture and aiding in the parasite's potential adaptation and phenotypic evolution in an ongoing process.

9 Closing Remarks

In summary, the distinct and specialized biological functions of *T. gondii* endosymbionts, together with their crucial role in parasite survival and their importance in the ongoing evolution of the *T. gondii* tripartite genome, make them a topic of interest to address with several yet unexplored or unknown areas. This is a rich source of inspiration for scientific research, underscoring the potential for discoveries in understanding the intricate interplay between these organelles and the biology of the parasite.

References

1. Burki F, Flegontov P, Oborník M et al (2012) Re-evaluating the green versus red signal in eukaryotes with secondary plastid of red algal origin. Genome Biol Evol 4:626–635
2. Janouškovec J, Horák A, Oborník M et al (2010) A common red algal origin of the apicomplexan, dinoflagellate, and heterokont plastids. Proc Natl Acad Sci 107:10949–10954
3. Mathur V, Salomaki ED, Wakeman KC et al (2023) Reconstruction of plastid proteomes of apicomplexans and close relatives reveals the major evolutionary outcomes of cryptic plastids. Mol Biol Evol 40. https://doi.org/10.1093/molbev/msad002
4. Feagin JE (1992) The 6-kb element of *Plasmodium* falciparum encodes mitochondrial cytochrome genes. Mol Biochem Parasitol 52:145–148
5. Suplick K, Morrisey J, Vaidya AB (1990) Complex transcription from the extrachromosomal DNA encoding mitochondrial functions of *Plasmodium* yoelii. Mol Cell Biol 10:6381–6388
6. Feagin JE (1994) The extrachromosomal DNAs of apicomplexan parasites. Ann Rev Microbiol 48:81–104

7. Gardner MJ, Feagin JE, Moore DJ et al (1993) Sequence and organization of large subunit rRNA genes from the extrachromosomal 35 kb circular DNA of the malaria parasite *Plasmodium* falciparum. Nucleic Acids Res 21:1067–1071
8. Gardner MJ, Goldman N, Barnett P et al (1994) Phylogenetic analysis of the rpoB gene from the plastid-like DNA of *Plasmodium* falciparum. Mol Biochem Parasitol 66:221–231
9. Köhler S, Delwiche CF, Denny PW et al (1997) A plastid of probable green algal origin in apicomplexan parasites. Science 275:1485–1489
10. McFadden GI, Reith ME, Munholland J, Lang-Unnasch N (1996) Plastid in human parasites. Nature 381:482
11. Berná L, Marquez P, Cabrera A et al (2021) Reevaluation of the *Toxoplasma gondii* and *Neospora caninum* genomes reveals misassembly, karyotype differences, and chromosomal rearrangements. Genome Res 31:823–833
12. Lee VV, Judd LM, Jex AR et al (2021) Direct nanopore sequencing of mRNA reveals landscape of transcript isoforms in apicomplexan parasites. mSystems 6. https://doi.org/10.1128/mSystems.01081-20
13. Namasivayam S, Baptista RP, Xiao W et al (2021) A novel fragmented mitochondrial genome in the protist pathogen *Toxoplasma gondii* and related tissue coccidia. Genome Res 31:852–865
14. Xia J, Venkat A, Bainbridge RE et al (2021) Third-generation sequencing revises the molecular karyotype for *Toxoplasma gondii* and identifies emerging copy number variants in sexual recombinants. Genome Res 31:834–851
15. McFadden GI (2011) The apicoplast. Protoplasma 248:641–650
16. Seeber F, Feagin JE, Parsons M, van Dooren GG (2020) Chapter 11 - The apicoplast and mitochondrion of *Toxoplasma gondii*. In: Weiss LM, Kim K (eds) *Toxoplasma gondii*, 3rd edn. Academic Press, pp 499–545
17. Sheiner L, Striepen B (2014) The apicoplast: a parasite's symbiont. In: Plastid biology. Springer, New York, pp 209–238
18. Usey MM, Huet D (2022) Parasite powerhouse: a review of the *Toxoplasma gondii* mitochondrion. J Eukaryot Microbiol 69:e12906
19. Gabaldón T (2021) Origin and early evolution of the eukaryotic cell. Ann Rev Microbiol 75:631–647
20. Mills DB, Boyle RA, Daines SJ et al (2022) Eukaryogenesis and oxygen in earth history. Nat Ecol Evol 6:520–532
21. Roger AJ, Muñoz-Gómez SA, Kamikawa R (2017) The origin and diversification of mitochondria. Curr Biol 27:R1177–R1192
22. Seeber F, Limenitakis J, Soldati-Favre D (2008) Apicomplexan mitochondrial metabolism: a story of gains, losses and retentions. Trends Parasitol 24:468–478
23. Florentin A, Cobb DW, Fishburn JD et al (2017) PfClpC is an essential Clp chaperone required for plastid integrity and Clp protease stability in *Plasmodium* falciparum. Cell Rep 21:1746–1756
24. Salomaki ED, Kolisko M (2019) There is treasure everywhere: reductive plastid evolution in Apicomplexa in light of their close relatives. Biomol Ther 9. https://doi.org/10.3390/biom9080378
25. Karnkowska A, Vacek V, Zubáčová Z et al (2016) A eukaryote without a mitochondrial organelle. Curr Biol 26:1274–1284
26. Maciszewski K, Karnkowska A (2019) Should I stay or should I go? Retention and loss of components in vestigial endosymbiotic organelles. Curr Opin Genet Dev 58-59:33–39
27. Hortua Triana MA, Cajiao Herrera D, Zimmermann BH et al (2016) Pyrimidine pathway-dependent and -independent functions of the *Toxoplasma gondii* mitochondrial dihydroorotate dehydrogenase. Infect Immun 84:2974–2981
28. van Dooren GG, Kennedy AT, McFadden GI (2012) The use and abuse of Heme in apicomplexan parasites. Antioxid Redox Signal 17:634–656
29. Dunay IR, Gajurel K, Dhakal R et al (2018) Treatment of Toxoplasmosis: historical perspective, animal models, and current clinical practice. Clin Microbiol Rev 31. https://doi.org/10.1128/CMR.00057-17
30. Hegewald J, Gross U, Bohne W (2013) Identification of dihydroorotate dehydrogenase as a relevant drug target for 1-hydroxyquinolones in *Toxoplasma gondii*. Mol Biochem Parasitol 190:6–15
31. Siregar JE, Syafruddin D, Matsuoka H et al (2008) Mutation underlying resistance of *Plasmodium* berghei to atovaquone in the quinone binding domain 2 (Qo(2)) of the cytochrome b gene. Parasitol Int 57:229–232
32. Maclean AE, Bridges HR, Silva MF et al (2021) Complexome profile of *Toxoplasma gondii* mitochondria identifies divergent subunits of respiratory chain complexes including new subunits of cytochrome bc1 complex. PLoS Pathog 17:e1009301
33. Seidi A, Muellner-Wong LS, Rajendran E et al (2018) Elucidating the mitochondrial proteome of *Toxoplasma gondii* reveals the presence of a divergent cytochrome c oxidase. Elife 7. https://doi.org/10.7554/eLife.38131
34. Mühleip A, Kock Flygaard R, Ovciarikova J et al (2021) ATP synthase hexamer assemblies shape cristae of Toxoplasma mitochondria. Nat Commun 12:120
35. Huet D, Rajendran E, van Dooren GG, Lourido S (2018) Identification of cryptic subunits from an apicomplexan ATP synthase. Elife 7. https://doi.org/10.7554/eLife.38097
36. Ovciarikova J, Lemgruber L, Stilger KL et al (2017) Mitochondrial behaviour throughout the lytic cycle of *Toxoplasma gondii*. Sci Rep 7:42746
37. Voleman L, Doležal P (2019) Mitochondrial dynamics in parasitic protists. PLoS Pathog 15:e1008008
38. Melo EJ, Attias M, De Souza W (2000) The single mitochondrion of tachyzoites of *Toxoplasma gondii*. J Struct Biol 130:27–33

39. Melatti C, Pieperhoff M, Lemgruber L et al (2019) A unique dynamin-related protein is essential for mitochondrial fission in *Toxoplasma gondii*. PLoS Pathog 15:e1007512
40. Nishi M, Hu K, Murray JM, Roos DS (2008) Organellar dynamics during the cell cycle of *Toxoplasma gondii*. J Cell Sci 121:1559–1568
41. Heredero-Bermejo I, Varberg JM (2019) TgDrpC, an atypical dynamin-related protein in *Toxoplasma gondii*, is associated with vesicular transport factors and parasite division. Mol Microbiol 111(1):46–64
42. Jacobs K, Charvat R, Arrizabalaga G (2020) Identification of Fis1 interactors in *Toxoplasma gondii* reveals a novel protein required for peripheral distribution of the mitochondrion. mBio 11. https://doi.org/10.1128/mBio.02732-19
43. Oliveira Souza RO, Jacobs KN, Back PS et al (2022) IMC10 and LMF1 mediate mitochondrial morphology through mitochondrion-pellicle contact sites in *Toxoplasma gondii*. J Cell Sci 135. https://doi.org/10.1242/jcs.260083
44. Huet D, Moreno SNJ (2023) Interorganellar communication through membrane contact sites in *Toxoplasma gondii*. Contact 6:25152564231189064
45. Lau Y-L, Lee W-C, Gudimella R et al (2016) Deciphering the draft genome of *Toxoplasma gondii* RH strain. PLoS One 11:e0157901
46. Lorenzi H, Khan A, Behnke MS et al (2016) Local admixture of amplified and diversified secreted pathogenesis determinants shapes mosaic *Toxoplasma gondii* genomes. Nat Commun 7:10147
47. Minot S, Melo MB, Li F et al (2012) Admixture and recombination among *Toxoplasma gondii* lineages explain global genome diversity. Proc Natl Acad Sci U S A 109:13458–13463
48. Yucesan B, Guldemir D, Babur C et al (2021) Whole-genome sequencing of a *Toxoplasma gondii* strain from a Turkish isolate using next-generation sequencing technology. Acta Trop 218:105907
49. Vaidya AB, Arasu P (1987) Tandemly arranged gene clusters of malarial parasites that are highly conserved and transcribed. Mol Biochem Parasitol 22:249–257
50. Aldritt SM, Joseph JT, Wirth DF (1989) Sequence identification of cytochrome b in *Plasmodium* gallinaceum. Mol Cell Biol 9:3614–3620
51. Feagin JE, Werner E, Gardner MJ et al (1992) Homologies between the contiguous and fragmented rRNAs of the two *Plasmodium* falciparum extrachromosomal DNAs are limited to core sequences. Nucleic Acids Res 20:879–887
52. Joseph JT (1991) The mitochondrial microgenome of malaria parasites. Harvard University
53. Vaidya AB, Akella R, Suplick K (1989) Sequences similar to genes for two mitochondrial proteins and portions of ribosomal RNA in tandemly arrayed 6-kilobase-pair DNA of a malarial parasite. Mol Biochem Parasitol 35:97–107
54. Vaidya AB, Lashgari MS, Pologe LG, Morrisey J (1993) Structural features of *Plasmodium* cytochrome b that may underlie susceptibility to 8-aminoquinolines and hydroxynaphthoquinones. Mol Biochem Parasitol 58:33–42
55. Feagin JE, Harrell MI, Lee JC et al (2012) The fragmented mitochondrial ribosomal RNAs of *Plasmodium* falciparum. PLoS One 7:e38320
56. Feagin JE, Mericle BL, Werner E, Morris M (1997) Identification of additional rRNA fragments encoded by the *Plasmodium* falciparum 6 kb element. Nucleic Acids Res 25:438–446
57. Tetzlaff S, Hillebrand A, Drakoulis N et al (2023) Small RNAs from mitochondrial genome recombination sites are incorporated into *T. gondii* mitoribosomes. bioRxiv. https://doi.org/10.1101/2023.09.02.556019
58. Gagat P, Mackiewicz D, Mackiewicz P (2017) Peculiarities within peculiarities - dinoflagellates and their mitochondrial genomes. Mitochondrial DNA B Resour 2:191–195
59. Namasivayam S, Sun C, Bah AB et al (2023) Massive invasion of organellar DNA drives nuclear genome evolution in toxoplasma. Proc Natl Acad Sci U S A 120:e2308569120
60. Marx V (2023) Method of the year: long-read sequencing. Nat Methods 20:6–11
61. Zhang T, Zhou J, Gao W et al (2022) Complex genome assembly based on long-read sequencing. Brief Bioinform 23. https://doi.org/10.1093/bib/bbac305
62. Berná L, Rego N, Francia ME (2021) The elusive mitochondrial genomes of apicomplexa: where are we now? Front Microbiol 12:751775
63. Hikosaka K, Nakai Y, Watanabe Y-I et al (2011) Concatenated mitochondrial DNA of the coccidian parasite Eimeria tenella. Mitochondrion 11:273–278
64. Johnston IG, Burgstaller JP (2019) Evolving mtDNA populations within cells. Biochem Soc Trans 47:1367–1382
65. Lacombe A, Maclean AE, Ovciarikova J et al (2019) Identification of the *Toxoplasma gondii* mitochondrial ribosome, and characterisation of a protein essential for mitochondrial translation. Mol Microbiol 112:1235–1252
66. Gupta K, Gupta A, Habib S (2018) Characterization of a *Plasmodium* falciparum rRNA methyltransferase. Mol Biochem Parasitol 223:13–18
67. Gupta K, Gupta A, Haider A, Habib S (2018) Characterization of mitochondrion-targeted GTPases in *Plasmodium* falciparum. Parasitology 145:1600–1612
68. Ke H, Dass S, Morrisey JM et al (2018) The mitochondrial ribosomal protein L13 is critical for the structural and functional integrity of the mitochondrion in *Plasmodium* falciparum. J Biol Chem 293:8128–8137
69. Shikha S, Silva MF, Sheiner L (2022) Identification and validation of *Toxoplasma gondii* mitoribosomal large subunit components. Microorganisms 10. https://doi.org/10.3390/microorganisms10050863
70. Gupta A, Shah P, Haider A et al (2014) Reduced ribosomes of the apicoplast and mitochondrion of *Plasmodium* spp. and predicted interactions with antibiotics. Open Biol 4:140045

71. Esseiva AC, Naguleswaran A, Hemphill A, Schneider A (2004) Mitochondrial tRNA import in *Toxoplasma gondii*. J Biol Chem 279:42363–42368
72. Pino P, Aeby E, Foth BJ et al (2010) Mitochondrial translation in absence of local tRNA aminoacylation and methionyl tRNA met formylation in Apicomplexa. Mol Microbiol 76:706–718
73. Jackson CJ, Norman JE, Schnare MN et al (2007) Broad genomic and transcriptional analysis reveals a highly derived genome in dinoflagellate mitochondria. BMC Biol 5:41
74. Nash EA, Nisbet RER, Barbrook AC, Howe CJ (2008) Dinoflagellates: a mitochondrial genome all at sea. Trends Genet 24:328–335
75. More RB, Oborník M, Janouskovec J et al (2008) A photosynthetic alveolate closely related to apicomplexan parasites. Nature 451(7181):959-63. https://doi.org/10.1038/nature06635
76. Sanchez SG, Bassot E, Cerutti A et al (2023) The apicoplast is important for the viability and persistence of *Toxoplasma gondii* bradyzoites. Proc Natl Acad Sci U S A 120:e2309043120
77. Bergmann A, Floyd K, Key M et al (2020) *Toxoplasma gondii* requires its plant-like heme biosynthesis pathway for infection. PLoS Pathog 16:e1008499
78. Krishnan A, Kloehn J, Lunghi M et al (2020) Functional and computational genomics reveal unprecedented flexibility in stage-specific Toxoplasma metabolism. Cell Host Microbe 27:290–306.e11
79. Xu X-P, Elsheikha HM, Liu W-G et al (2021) The role of type II fatty acid synthesis enzymes FabZ, ODSCI, and ODSCII in the pathogenesis of *Toxoplasma gondii* infection. Front Microbiol 12:703059
80. Henkel S, Frohnecke N, Maus D et al (2022) *Toxoplasma gondii* apicoplast-resident ferredoxin is an essential electron transfer protein for the MEP isoprenoid-biosynthetic pathway. J Biol Chem 298:101468
81. Renaud EA, Pamukcu S, Cerutti A et al (2022) Disrupting the plastidic iron-sulfur cluster biogenesis pathway in *Toxoplasma gondii* has pleiotropic effects irreversibly impacting parasite viability. J Biol Chem 298:102243
82. Patrick GL (2020) Chapter 9 - Drug targets in the apicoplast. In: Patrick GL (ed) Antimalarial agents. Elsevier, pp 319–374
83. Ralph SA, van Dooren GG, Waller RF et al (2004) Tropical infectious diseases: metabolic maps and functions of the *Plasmodium* falciparum apicoplast. Nat Rev Microbiol 2:203–216
84. Cinar HN, Qvarnstrom Y, Wei-Pridgeon Y et al (2016) Comparative sequence analysis of Cyclospora cayetanensis apicoplast genomes originating from diverse geographical regions. Parasit Vectors 9:611
85. Kappmeyer LS, Thiagarajan M, Herndon DR et al (2012) Comparative genomic analysis and phylogenetic position of Theileria equi. BMC Genomics 13:603
86. Liu Q, Yu L, Jiang F et al (2020) Annotation and characterization of Babesia gibsoni apicoplast genome. Parasit Vectors 13:209
87. Janouškovec J, Tikhonenkov DV, Burki F et al (2015) Factors mediating plastid dependency and the origins of parasitism in apicomplexans and their close relatives. Proc Natl Acad Sci U S A 112:10200–10207
88. Denny P, Preiser P, Williamson D, Wilson I (1998) Evidence for a single origin of the 35 kb plastid DNA in apicomplexans. Protist 149:51–59
89. Duarte I, Nabuurs SB, Magno R, Huynen M (2012) Evolution and diversification of the organellar release factor family. Mol Biol Evol 29:3497–3512
90. Martins-Duarte ÉS, Sheiner L, Reiff SB et al (2021) Replication and partitioning of the apicoplast genome of *Toxoplasma gondii* is linked to the cell cycle and requires DNA polymerase and gyrase. Int J Parasitol 51:493–504
91. Striepen B, Crawford MJ, Shaw MK et al (2000) The plastid of *Toxoplasma gondii* is divided by association with the centrosomes. J Cell Biol 151:1423–1434
92. Jacot D, Daher W, Soldati-Favre D (2013) *Toxoplasma gondii* myosin F, an essential motor for centrosomes positioning and apicoplast inheritance. EMBO J 32:1702–1716
93. Breinich MS, Ferguson DJP, Foth BJ et al (2009) A dynamin is required for the biogenesis of secretory organelles in *Toxoplasma gondii*. Curr Biol 19:277–286

Cyclospora Comparative Genomics

Joel Barratt, Anna Peterson, and Yvonne Qvarnstrom

Abstract

Human-infecting *Cyclospora* species have been a recognized cause of seasonal food-borne outbreaks of diarrheal illness in the United States and Canada since the 1990s. Prior to 2013, outbreak investigations were almost exclusively based on epidemiologic data because there were no laboratory methods to distinguish between outbreaks caused by distinct *Cyclospora* strains. Due to the challenges in obtaining sufficient parasite-derived material for research, the first draft genome was not published until 2015. To date, 36 samples have been genome sequenced; however, many of the published genomes suffer from low quality as they are fragmented (i.e., are not chromosome-level assemblies), were generated from pooled fecal specimens from multiple individuals, and contain contaminating sequences from other gastrointestinal microbes. The genome is estimated to be around 44 MBp large with a GC content of 52%.

Initial efforts to develop a genotyping method to aid in outbreak investigations focused on organelle genomes, as they are small, present in high copy numbers, and are not subject to heterozygosity. However, both the mitochondrial and the apicoplast genomes are highly conserved and offer only limited genetic variability among isolates of human-infecting *Cyclospora*. Thus, nuclear markers should be included, and unique data analysis considerations are needed to account for the heterogeneity of nuclear loci.

Multi-locus-sequence typing (MLST) is considered the most accessible strategy for routine *Cyclospora* genotyping. Since 2018, a targeted amplicon deep sequencing method based on eight markers has been in routine use at the Centers for Disease Control and Prevention. The resultant genotypes are used to compute a genetic distance via a heuristic algorithm that scores genetic relationships based on the number of shared haplotypes between genotypes and the frequency of these haplotypes in the study population. An analysis of 2841 genotypes obtained by this method revealed a population structure of two groups that seemed to be reproductively isolated and thus constitute distinct species (*Cyclospora cayetanensis* and *Cyclospora ashfordi*). The analysis also indicated that some genotypes recurred across multiple years. This pattern could be due to the environmental persistence of these genotypes within produce-growing regions. Alternatively, this could be an artifact

J. Barratt · A. Peterson · Y. Qvarnstrom (✉)
Division of Parasitic Diseases and Malaria, National Center for Emerging and Zoonotic Infectious Diseases, Centers for Disease Control and Prevention, Atlanta, GA, USA
e-mail: bvp2@cdc.gov

J. D. Ramírez González (ed.), *Recent Advances in Parasitomics*,
https://doi.org/10.1007/978-3-031-70591-5_10

from the limited discriminatory power of the method; parasites with high similarity in the eight markers may in fact be genetically distinct when expanding the analysis to additional markers. To this end, the Food and Drug Administration has recently developed an expanded panel consisting of 52 target loci. This method also includes a hybridization capture method to enrich for target DNA to increase sensitivity, which is advantageous for the genotyping of *Cyclospora* from environmental samples.

Our understanding of the *Cyclospora* population structure and dynamics has dramatically increased after the introduction of routine genotyping since 2018. However, improved tools to investigate *Cyclospora* genomics are needed to refine our understanding of the temporal and geographical trends of *Cyclospora* in human transmission and in the environment.

Keywords

Cyclospora cayetanensis · *Cyclospora ashfordi* · *Cyclospora henanensis* · Food-borne pathogen · Multi-locus sequence typing · Targeted deep amplicon sequencing

1 Introduction

The genus *Cyclospora* sits within the kingdom Protista, phylum Apicomplexa, subclass Coccidiasina, order Eucoccidiorida, and family Eimeriidae. A total of 23 *Cyclospora* species have been described to date (Table 1), the first being *Cyclospora glomericola* identified in millipedes by Schneider [1]. Members of the *Cyclospora* genus were once considered parasites exclusively of animals. However, in 1979, R. W. Ashford observed a novel coccidian in human fecal specimens while working in a diagnostic laboratory in Papua New Guinea [2], and

Table 1 *Cyclospora* species identified to date[a]

Species	Main host	Year of first description	References
C. glomericola	Millipede	1881	Schneider[1]
C. caryolytica	Mole	1902	Levine [6]
C. viperae	Snake	1923	Levine [6]
C. babaulti	Snake	1924	Levine [6]
C. scinci	Snake	1924	Levine [6]
C. tropidonoti	Snake	1924	Levine [6]
C. zamenis	Snake	1924	Levine [6]
C. niniae	Snake	1965	Levine [6]
C. talpae	Mole	1968	Levine [6]
C. megacephali	Mole	1988	Ford and Duszynski [7]
C. ashtabulensis	Mole	1989	Ford and Duszynski [7]
C. parascalopi	Mole	1989	Ford and Duszynski [7]
C. angimurinensis	Rodent	1990	Ford and Duszynski [7]
C. cayetanensis	Human	1994	Ortega et al. [5]
C. colobi	Primate	1999	Eberhard et al. [8]
C. cercopitheci	Primate	1999	Eberhard et al. [8]
C. papionis	Primate	1999	Eberhard et al. [8]
C. schneideri	Snake	2005	Lainson [9]
C. macacae	Primate	2015	Li et al. [10]
C. yatesi	Mole	2018	McAllister and Hnida [11]
C. duszynskii	Mole	2018	McAllister and Hnida [11]
C. ashfordi	Human	2022	Barratt et al. [12]
C. henanensis	Human	2022	Barratt et al. [12]

[a]Adapted from Shields and Olson [13] and Lainson [9]

reports of similar organisms infecting people from Haiti, Nepal, and Peru appeared shortly after [3]. In 1993, Ortega et al. [4] assigned this "new" human-infecting coccidian to the genus *Cyclospora* on the basis that its spherical oocysts possessed two sporocysts each containing two twin sporozoites, which are morphological features placing this organism within the *Cyclospora* genus. This human-infecting *Cyclospora* species was named *Cyclospora cayetanensis* after the Universidad Peruana Cayetano Heredia in Lima, Peru, where much of the initial work to describe this organism took place [5].

Since the 1990s, *C. cayetanensis* became recognized as a cause of seasonal food-borne outbreaks of diarrheal illness in the United States of America (USA) and Canada, and by 1999, cyclosporiasis was declared a nationally notifiable disease in the USA. There has been a marked increase in the number of cyclosporiasis cases reported to the Centers for Disease Control and Prevention (CDC) since then, from below 100 per year in the early 2000s to 2408 cases in 2019. After a drop in annual cases in 2020–2022 due to the global COVID-19 pandemic, USA case counts returned to pre-pandemic trends in 2023, with 2272 cyclosporiasis cases reported to the CDC by October 24, 2023.

Human cyclosporiasis is endemic to many tropical and subtropical parts of the world, including Central and South America, Asia, and Africa, and the disease is always highly seasonal in endemic countries. In non-endemic countries, cases of cyclosporiasis are primarily associated with international travel to areas where cyclosporiasis is endemic or to outbreaks caused by consumption of contaminated food imported from endemic areas. Food items associated with cyclosporiasis outbreaks mostly include fresh produce, such as berries, herbs, and leafy greens. Earlier cyclosporiasis outbreaks in the USA and Canada (circa 1996–1998) were linked to raspberries imported from Guatemala [3, 14]. More recently, USA outbreaks have been linked to fresh produce imported from Mexico, including basil (2019), cilantro (2013 and 2014), and romaine lettuce (2013) [15–17]. In Germany, an outbreak of cyclosporiasis was epidemiologically linked to mixed salads with many ingredients, most of which were imported from southern Europe [18]. Sweden experienced its first detected cyclosporiasis outbreak in 2010, linked to sugar snap peas imported from Guatemala [19]. Cases of cyclosporiasis were reported in the United Kingdom and Canada in 2015, associated with travel to the Riviera Maya region in Mexico [20].

Increasing interest in human-infecting *Cyclospora* stems partly from its propensity to cause relatively large outbreaks of diarrheal disease in non-endemic countries. Despite this, the precise routes by which *Cyclospora* contaminates large batches of produce to cause these outbreaks remain poorly understood. The difficulty in detecting *Cyclospora* in contaminated produce implies that it has a low infectious dose, possibly within the range of 10–100 oocysts [21]. Cyclosporiasis outbreak dynamics also suggest that sporulated *Cyclospora* oocysts attach strongly to surfaces of plant material and remain attached after several washes of fresh produce.

2 Biology of Human-Infecting *Cyclospora*

Oocysts of human-infecting *Cyclospora* are distinguished from other human-infecting coccidia such as *Cryptosporidium*, *Cystoisospora*, and *Sarcocystis* by their size, shape, and number of sporozoites and sporocysts within it (Fig. 1). *Cyclospora* oocysts are spherical, approximately 8–10 μm in diameter, and when sporulated possess two sporocysts with two sporozoites each. Sporulated and unsporulated oocysts of *Cyclospora* (and oocysts of other coccidia) autofluoresce with a characteristic blue-green outer ring when examined under ultraviolet (UV) light.

Cyclospora is an obligate intracellular parasite, completing its lifecycle within the epithelial cells in the gastrointestinal tract, typically in the jejunum, of the host [22]. Transmission of human-infecting *Cyclospora* species occurs when sporulated oocysts are ingested through contaminated food or water. In the intestines, the oocysts excyst with the assistance of bile, trypsin, and other proteolytical factors, releasing crescent-shaped sporozoites (1.2 × 9.0 μm in size) that invade jejunal epithelial cells. *Cyclospora* then

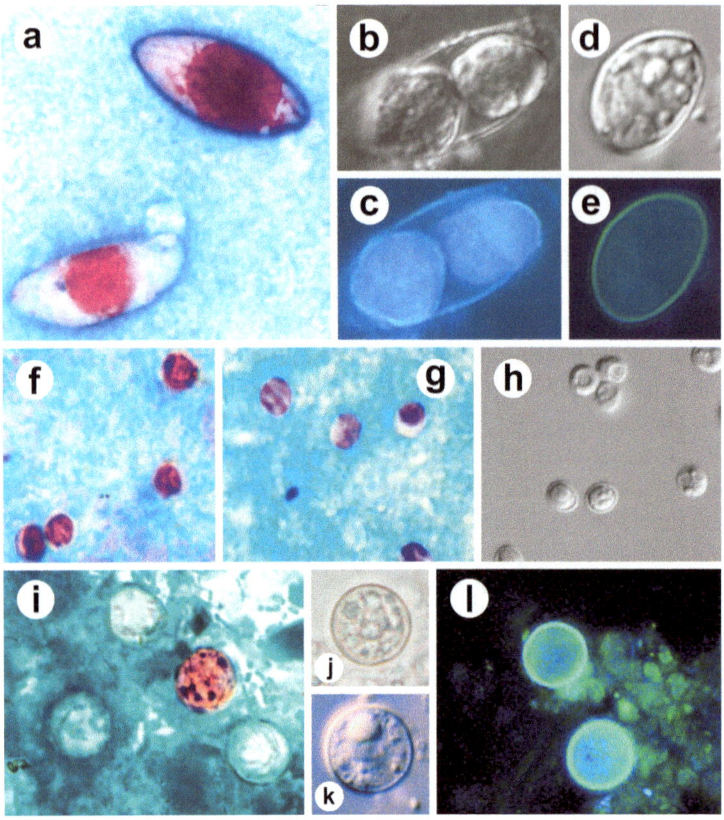

Fig. 1 Microscopic images of coccidian oocysts. Panels (**a–c**): *Cystoisospora belli* oocysts (25–30 μm in length) displaying their typical ellipsoidal shape stained with an acid-fast stain (**a**), viewed under differential interference contrast (DIC) microscopy (**b**), and displaying autofluorescence when microscopically viewed under UV light (**c**). Panels (**d, e**): A free or liberated oocyst of *Sarcocystis* species (15–25 μm in length) viewed under DIC microscopy (**d**) and displaying autofluorescence when viewed under UV light (**e**). Panels (**f–h**): *Cryptosporidium* species oocysts (4.2–5.4 μm in diameter) stained with an acid-fast stain (**f, g**), and viewed under DIC microscopy (**h**). Panels (**i–l**): *Cyclospora* oocysts (unsporulated) from human stool (7.5–10 μm in diameter) stained with an acid-fast stain (**i**), viewed under bright field microscopy (**j**), viewed under DIC microscopy (**k**) and displaying autofluorescence when viewed under UV light (**l**)

undergoes both asexual and sexual reproduction within parasitophorous vacuoles inside host cells [23]. Asexual reproduction (merogony) generates large numbers of merozoites, that can infect new host cells to initiate more cycles of merogony or proceed to sexual reproduction via gametogony, where merozoites turn into male microgamonts or female macrogamonts. The fusion of male microgamonts and female macrogamonts produces zygotes that mature into oocysts that are excreted in the feces of infected individuals.

To become infectious, *Cyclospora* oocysts must undergo sporulation, where the sporont divides into two sporocysts, each containing two sporozoites [24]. Sporulation requires humidity, oxygenation, and the correct temperature, but the precise factors influencing sporulation in the natural environment remain obscure. Optimal sporulation temperatures are between 22 and 30 °C; oocysts sporulate poorly at 37 °C and not at all when refrigerated. Sporulation can be enhanced by placing unsporulated oocysts in 2.5% potassium dichromate, which is a technique commonly used to induce sporulation in other coccidians. Under favorable laboratory conditions, sporulation takes 7–13 days. However, all attempts so far to experimentally infect laboratory animals and human volunteers with laboratory-produced sporulated oocysts have failed to establish infection [25, 26]. This raises the possibility that additional processes are needed to induce infectious *Cyclospora* oocyst production in the laboratory. Similarly, it is currently unknown how long sporulated oocysts remain viable and infectious in the environment. Notably, because oocysts of human-infecting *Cyclospora* are unsporulated and thus non-infectious when first excreted, they are incapable of direct fecal–oral transmission.

Human-infecting *Cyclospora* species are thought to be strict human pathogens without zoonotic potential [27]. Oocysts resembling *C. cayetanensis* have been detected in the feces of poultry and dogs, but these studies failed to distinguish passive carriage (i.e., ingested oocysts passing through the intestinal tract unaltered) from true infection (where intracellular life stages occur) [28]. It is possible that certain animals serve as transport hosts that contribute to the spread of cyclosporiasis [29]. While three human-infecting species of *Cyclospora* have been proposed, *Cyclospora* oocysts almost identical morphologically to those from humans have been detected in non-human primates from Tanzania [30], Ethiopia [8], Equatorial Guinea [31], Kenya [32], and China [33]. The host specificity of these non-human primate *Cyclospora* species is currently not well-explored, and it is not known if non-human primates can be infected with human-infecting *Cyclospora* [31]. In regions where human and non-human primates overlap in their distribution, detection of *Cyclospora* in either host would benefit from *Cyclospora* species confirmation by DNA sequencing.

3 The *Cyclospora* Genome

As mentioned above, human-infecting *Cyclospora* cannot be propagated in the laboratory. Thus, the only source of material for genomic research is fecal material from naturally infected humans. In addition, infected humans may shed oocysts intermittently and in low numbers. Early genomic sequencing techniques (e.g., Sanger shotgun sequencing) were therefore unavailable for *Cyclospora* research due to the requirement for large amounts of high-quality genomic DNA. Even when methods to sequence minute amounts of DNA became available, most genomic datasets generated for *Cyclospora* were incomplete and highly fragmented because of substantial levels of fecal contaminants. However, the relatively small and high copy number of organelle genomes could be extracted from these datasets and studied.

3.1 Mitochondrial Genomes

The mitochondrial genome from human-infecting *Cyclospora* species is similar in structure, gene content, and base composition to genomes from related species in Eimeriidae [34–36]. Unique among eukaryotes, certain coccidians—including *Cyclospora*—possess a linear (not a circular) mitochondrial genome, which exists as multiple linear concatemers arranged in a head-to-tail configuration [34, 36]. Between the concatemers is a hypervariable region that has been called the terminal region, the polymorphic link region, the distal region, or the junction, which consists of a complex pattern of 15 mer repeats varying in number and terminating in one of at least four possible end motifs. The mitochondrial genome is 6.3 kb in length and codes for three complete genes, *cox1*, *cox3*, and *cytb*. Although the copy number is unknown in *Cyclospora*, estimates range from 67 to 513 mitochondria genome copies per haploid genome, based on whole-genome sequencing (WGS) read coverage [36, 37].

Several studies have suggested that whole mitochondrial sequencing could provide sufficient information for geographic tracing, analogous to what has been performed for *Plasmodium* parasites [36–39]. The majority of sequence variation is concentrated in the highly polymorphic repeat junction, which has been utilized as a genotyping marker [40, 41]; see later section for more details about this. In contrast, the core mitochondrial genome (the terminal region excluded) has very little sequence variation. A comparative study of 33 mitochondrial genomes resulted in 13 haplotypes with only 9 single-nucleotide polymorphism (SNP) sites (0.1% diversity) and no clear geographic pattern [42]. Another study of 47 mitochondrial genomes (including the 33 in the previous study) identified 15 haplotypes with only 0–0.3% differences between isolates. The predicted *cytb* coding region was 100% identical in all genomes, while the predicted *cox1* and *cox3* genes had two and three SNPs, respectively. There was no apparent geographic pattern among the compared genomes, with the possible exception that all samples from Indonesia were in the same clade. However, two of the Indonesian samples displayed a mitochondrial sequence type identical to several outbreak-related samples from the USA. The most divergent sequence came from a sample collected in China. A total of 25 SNP sites were identified in alignment with the Chinese sample included, while only 12 SNP sites remained when it was excluded. This is in agreement with other studies that consistently place Chinese samples separate from samples from other countries, and supports the designation of a distinct species, *C. henanensis* [12, 43–45].

Mitochondrial genomes from two closely related *Cyclospora* species that infect non-human primates (*C. papionis* and *C. cercopitheci*) have been sequenced. The non-human primate *Cyclospora* species and human-infecting *Cyclospora* share complete gene synteny and about 96% sequence identity of their mitochondrial genomes. Among the protein-coding genes, *cox3* has the most sequence divergence with 8.5% (64 SNPs in 753 base pairs), followed by *cox1* with 4.7% (68 SNPs in 1449 bp) and *cytb* with 3.4% (37 SNPs in 1080 base pairs). The genomes are AT-rich, with a GC content of around 34%.

A phylogenetic analysis of the mitochondrial genomes from *Cyclospora* and *Eimeria* isolates demonstrates that whole mitochondrial genomes are an excellent phylogenetic and taxonomic marker for distinguishing species within the family Eimeriidae (Fig. 2). The tree supports that primate-infecting *Cyclospora* species are monophyletic, that *Cyclospora* and *Eimeria* are closely related sister genera, and that among the *Eimeria* species, *E. dispersa* is the closest known relative to the *Cyclospora* clade [34, 36]. The human-infecting *Cyclospora* genomes cluster tightly together due to the limited intra-species sequence variability, while the two non-human primate *Cyclospora* species form their own, distinct clade.

3.2 Apicoplast Genomes

Apicoplast genomes are highly conserved among apicomplexan parasites. The *Cyclospora* apicoplast genome is 34 kb large with 22% GC content, with complete gene synteny and 85% sequence identity to *Eimeria* species [36]. The genome codes for 29 proteins, 33 tRNAs, and 4 rRNAs. Cinar et al. [46] extracted apicoplast reads from WGS datasets from 12 biosamples and identified 21 SNPs and some indels throughout the genome, resulting in a sequence variation of only 0.1%. A phylogenetic study did not reveal any clear clustering based on geographic or epidemiologic metadata.

3.3 Whole Genomes

The first full *Cyclospora* genome was released in 2015 [47]. It originated from a human stool sample collected in 2011 from a patient in Henan, China. Extensive purification of the oocysts, whole-genome amplification of the genomic DNA, and multiple rounds of sequencing using both Roche 454 and Illumina techniques were performed, resulting in a draft genome with 2297 contigs. The methods for generating full-coverage whole genomes from *Cyclospora* oocysts have since been refined, facilitating the generation of genomic datasets with as few as 313 contigs [48–50]. However, chromosomal level assemblies have not yet been achieved due to the difficulties of getting sufficient high-molecular-weight DNA for longer-read sequencing.

A total of 36 biosamples (originated from China, Nepal, Indonesia, Mexico, Guatemala, and outbreaks in the USA and Canada) of human-infecting *Cyclospora* have been sequenced to date. The genome of *Cyclospora* is around 44 MBp large with a GC content of 52%. Compared with the closely related *Eimeria* species, its genome is slightly smaller, the GC content is the

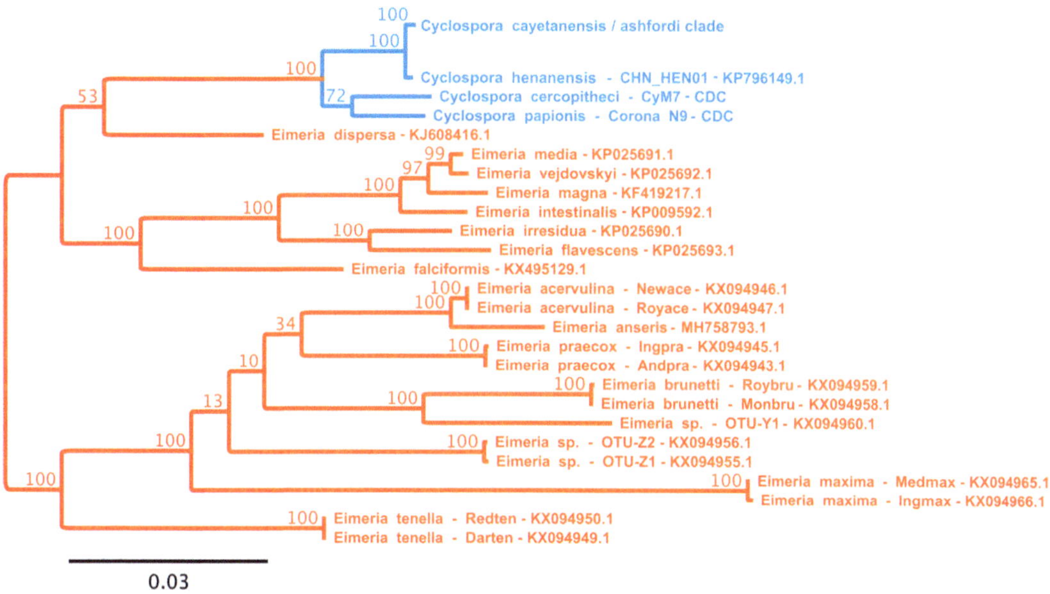

Fig. 2 Phylogenetic analysis of *Cyclospora* and *Eimeria* mitochondrial genomes. A total of 40 mitochondrial genome sequences from various *Eimeria* and *Cyclospora* species were aligned using MAFFT. Following alignment, the repetitive mitochondrial junction region was manually trimmed from each *Cyclospora* sequence, and the homologous region of the *Eimeria* mitochondrial genomes, resulting in sequences of approximately 6.1 kb. The phylogeny shown was generated using the Phangorn R package. Genetic distances were calculated using the dist.ml function and a neighbor-joining (NJ) tree was first constructed from these distances. Next, a maximum-likelihood tree was optimized from the NJ tree using the pml and optim.pml functions, applying the NNI rearrangement model (log-likelihood: −29569.62). The bootstrap.pml function was used to calculate non-parametric bootstrap values across 1000 replicates. Bootstrap values are indicated on nodes, and the scale bar represents the number of substitutions per site. Specimen names begin with the species, followed by a strain name (if available) and then a GenBank Accession number. The tree was rooted at the node between the *Eimeria tenella* and *Eimeria falciformis* clades. The branches shown in blue indicate clades occupied by *Cyclospora* species, while those shown in orange indicate clades occupied by *Eimeria* species. The topmost branch labeled as the "*Cyclospora cayetanensis/ashfordi* clade" is a collapsed node comprised of 15 *Cyclospora* mitochondrial genomes from this group

same, and the genome organization and gene content are broadly similar. A comparison of the predicted protein-coding genes in *Cyclospora henanensis* with *Eimeria tenella* revealed notable differences in carbon and amino acid metabolism, post-translational modifications, and surface antigens [47].

Gonzalez-Gomez et al. [42] performed a genome-wide comparison study of the 36 available *Cyclospora* genomes. They identified a core genome of 2966 genes and, among them, selected 47 markers for phylogenetic analysis. The resulting tree did not reveal any clear pattern regarding geographic origin or epidemiologic outbreak clustering. The authors suspected that available genomes are of poor quality due to long storage of samples prior to sequencing and/or fecal contaminants and suggested that additional genomes from fresh samples are required for more informative comparative analysis.

4 Genotyping and Population Structure

Following the spike in USA cyclosporiasis cases in the 2000s, the CDC sought to develop a *Cyclospora* genotyping tool to complement epidemiological outbreak investigations. As discussed above, WGS of *Cyclospora* is complicated by multiple factors. Thus, MLST was the most accessible strategy for routine *Cyclospora* genotyping, where multiple genetic markers would be PCR amplified from stool DNA extracts and sequenced to construct genotypes for clustering.

An early iteration of a *Cyclospora* MLST approach involved Sanger sequencing of three markers identified by comparing 11 draft-quality *Cyclospora* genomes [51]. One marker, referred to as "MSR," is encoded in the mitochondrial genome and captures four SNP sites within the 16S ribosomal RNA gene. The two additional markers—"360i2" and "378"—are nuclear encoded; the 360i2 locus includes a genomic region of undefined function capturing at least 20 SNPs, while 378 captures roughly 15 SNPs and includes part of a Sec14 family protein gene [51]. Sanger sequencing of these three loci from 88 specimens revealed that genetic heterogeneity is a common feature of *Cyclospora* infections, which are characteristically heterozygous or multi-clonal. Sanger chromatograms of 360i2 and 378 often required haplotype phasing due to evidence of heterozygous (double) peaks. Notably, the heterogeneity of *Cyclospora* infections precluded alignment-based phylogenetic analyses for defining genetic relationships between genotypes, as many *Cyclospora* genotypes possessed two or even three underlying haplotypes at their 360i2 and 378 loci. To address this, two algorithms were developed to facilitate the clustering of these heterogeneous *Cyclospora* genotypes. These algorithms, one based on Bayesian probability theory and the other a heuristic based on set theory and propositional logic, compare all genotype pairs within a dataset to identify the haplotypes shared between them (i.e., their intersect). A genetic distance is then assigned to each genotype pair based on their degree of intersect, and a pairwise genetic distance matrix is computed. This matrix is finally clustered to produce a tree to visualize the various genetic relationships [51].

Nascimento et al. [41] explored the genotyping utility of the mitochondrial junction repeat region by Sanger sequencing this locus from 132 *Cyclospora*-positive fecal specimens and discovered 14 unique variants. Using a workflow developed to identify segregating SNPs from whole-genome sequence data, Houghton et al. [52] identified four additional *Cyclospora* genotyping markers. These markers, referred to as CDS-1 through CDS-4, captured 13 segregating SNPs total in three protein-encoding loci plus one locus of uncharacterized function. When CDS-1 through CDS-4 were Sanger sequenced from 57 *Cyclospora*-positive fecal specimens, 19 unique genotypes were identified [52].

Nascimento et al. [53] next evaluated the genetic distance computation approach from Barratt et al. [51] on an eight-marker genotyping panel combining the "MSR," "378," and "360i2" markers [51], the mitochondrial junction [41], and CDS-1 through CDS-4 [52]. Nascimento et al. [53] also described methods for next-generation sequencing (NGS) of amplicons generated from these eight markers on the Illumina MiSeq platform, which offered higher throughput than Sanger sequencing and facilitated detection of all alleles comprising a given PCR amplicon. Using this approach, 648 *Cyclospora*-positive fecal specimens collected during the US cyclosporiasis peak season of 2018 were genotyped, and numerous haplotype combinations were identified. Given the characteristic genetic heterogeneity of *Cyclospora* infections—attributed to its sexual mode of reproduction—genotypes assigned to the same epidemiologic source were typically found to be very similar but not always identical. Thus, guided by epidemiologic data, the cluster dendrogram generated by Nascimento et al. [53] was dissected into ten populations, each representing a set of genotypes sharing a comparatively large number of alleles. The resolution observed using this procedure was a substantial improvement on previous methods. In a later study combining *Cyclospora* genotypes compiled from the cyclosporiasis peak periods of 2018–2022, hundreds of unique haplotype combinations were observed, comprising up to 52 major genetic groupings [54]. Given its utility, this approach was deployed routinely at the CDC during subsequent cyclosporiasis peak periods starting in 2018, to support epidemiologic outbreak investigations in near real time [55–57].

Several years of routine *Cyclospora* genotyping utilizing the Illumina sequencing approach of the eight-marker panel ("MSR," "360i2," "378," the mitochondrial junction, and CDS-1 through CDS-4) led to the accumulation of more than 5000 *Cyclospora* genotypes, thus improving our understanding of the *Cyclospora* population structure and revealing several patterns. Notably, a two-type population structure emerged when these *Cyclospora* genotypes were clustered, prompting an investigation into the genetic factors driving this structure [12]. This investigation revealed that the two-type structure (*Cyclospora* populations A and B) was driven predominantly by certain allele combinations observed at the 360i2 locus. Specifically, five 360i2 alleles were observed among *Cyclospora* isolates causing infections in the USA and Canada, and these alleles were separated into two groups: group A and group B. Alleles within each group were phylogenetically related while the two groups themselves were distinct. Group A 360i2 alleles included alleles A1 and A2, while group B alleles included B1 through B3, and close examination of *Cyclospora* genotypes clustering within population A revealed that they possessed one of the 360i2 genotypes A1, A2, or A1 + A2, while genotypes in population B possessed either B1, B2, B3, or a combination of these B-type alleles. Remarkably, genotypes comprising mixtures of A-type and B-type alleles were extremely rare and were typically only observed in complex genotypes attributed to multi-strain infections [12]. The marked rarity of A-type and B-type allele combinations within *Cyclospora* infections suggested that these two populations were reproductively isolated and therefore constitute distinct species. This culminated in the retention of *C. cayetanensis* as the name for population A, while population B was assigned the name *Cyclospora ashfordi* [12].

Barratt et al. [12] also noted that a *Cyclospora* isolate from Henan province in China was genetically distinct from those causing illness in the US and Canada. This isolate from China, referred to as strain CHN_HEN01, shared some affinity with *Cyclospora* population B (*C. ashfordi*) as it possessed B-type 360i2 alleles, including a fourth B-type allele not observed in any other genotype. Strain CHN_HEN01 also possessed haplotypes at the MSR and mitochondrial junction loci not observed among the thousands of genotypes sequenced in the USA. The genome sequence of strain CHN_HEN01 was publicly available [47], facilitating a comparison of its mitochondrial and apicoplast genome, and various nuclear genes, to those from strains originated in other geographical locations. This comparison supported that strain CHN_HEN01 is distinct from both *C. cayetanensis* and *C. ashfordi*, where the former two species represent recently diverged sister taxa and strain CHN_HEN01 represents an outgroup. On this basis, strain CHN_HEN01 was assigned the name *Cyclospora henanensis* [12].

Recently, Jacobson et al. [58] explored changes in *C. cayetnanensis* and *C. ashfordi* population dynamics over time by examining genotypes sequenced from 2018 to 2022. A combined *C. ashfordi* and *C. cayetanensis* population dataset comprising 2841 genotypes was divided into 30 distinct genetic groups (referred to as strains in that study for convenience—Fig. 3), and several features of these "strains" were highlighted, including the observation that some were highly abundant and were recurrent across multiple years. A notable outcome of the study by Jacobson et al. [58] was the observation of seemingly rapid selection for specific mitochondrial haplotypes in both *C. cayetanensis* and *C. ashfordi,* favoring shorter junction repeat lengths and the near disappearance of one of three major MSR haplotypes [58].

Other genotyping approaches have been described for *Cyclospora.* Methods to sequence whole mitochondrial and apicoplast genomes of *Cyclospora* have been described, and these were considered as possible approaches for *Cyclospora* genotyping. A benefit of this is that these genomes are not subject to heterozygosity, which simplifies clustering and other downstream analyses. They also have a high copy number within each cell, which is particularly useful when testing plant material for the presence of *Cyclospora*, as oocysts tend to contaminate fresh produce at very low levels. A major downside of utilizing apicoplast and mitochondrial genomes for genotyping is that they are highly conserved in *Cyclospora*, greatly limiting their genotyping utility. Another

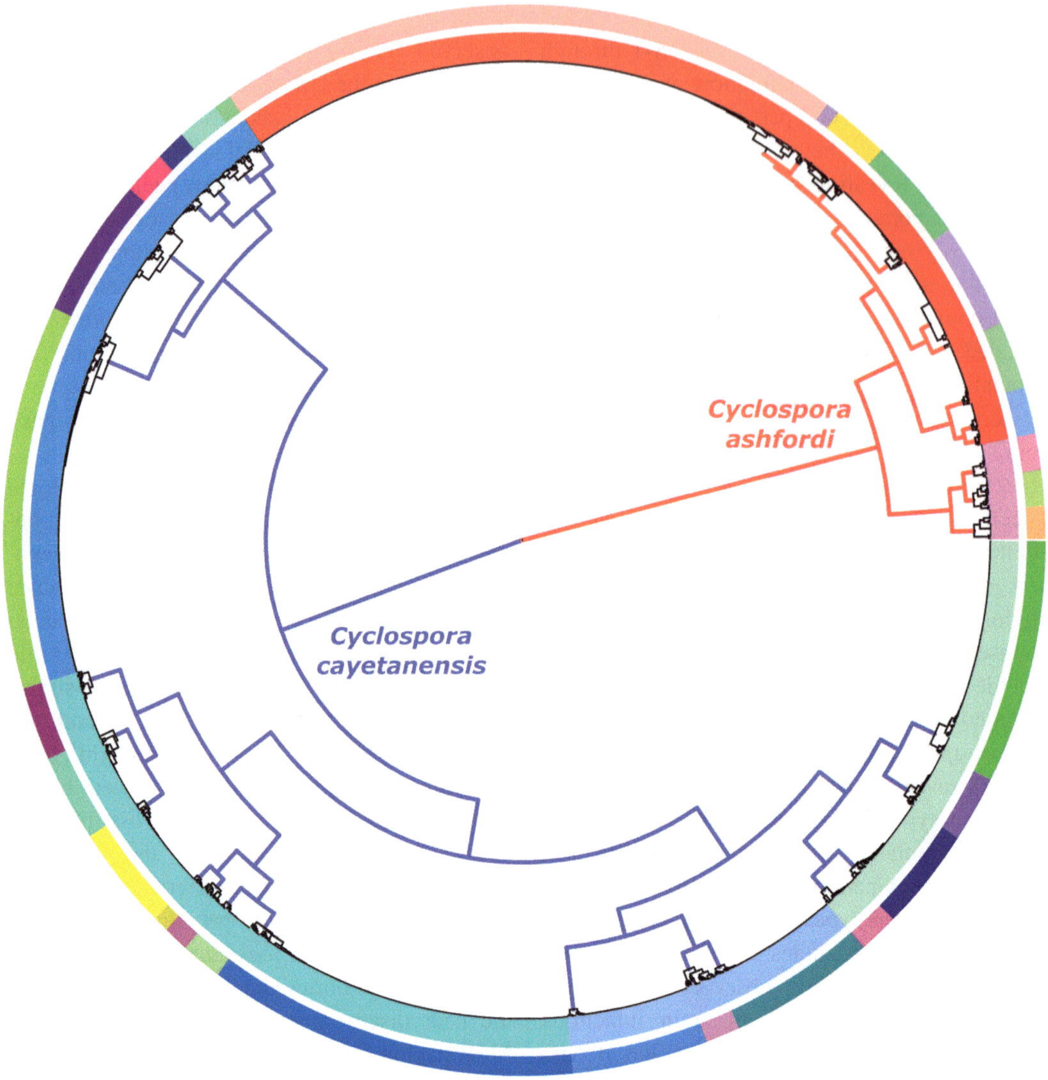

Fig. 3 Circular cluster dendrogram presenting an overview of the population structure of human-infecting *Cyclospora* causing illness in North America. This figure presents a clustering analysis of 2841 *Cyclospora* genotypes sequenced from 2018 to 2022, providing an overview of the genetic population structure of *Cyclospora* causing illness in North America over that time period. This tree is an adaptation of the tree presented by Jacobson et al. [54, 58], and was generated using the eight-marker NGS-based genotyping system, as described in that study [54, 58]. The two major populations within this structure reflect the relationship between *C. cayetanensis* and *C. ashfordi*, as shown in the blue and red branches and text, respectively. Two rings of colored bars encircle this tree. The innermost bars reflect major "subpopulations" within each species, and the outermost circle of colored bars reflects various "strains," as defined by Jacobson et al. [54, 58] using the eight-marker genotyping system

genotyping method based on a set of *Cyclospora* microsatellite repeat markers was also proposed [44, 45]. This microsatellite method displayed promise as it divided *Cyclospora* isolates from China and the Americas into several discrete groups, which prompted a subsequent evaluation of this approach [59]. Hofstetter et al. [59] Sanger sequenced four of the five microsatellite loci (excluding one as it showed no variability in a prior study [44]) and found that these repetitive markers sequenced poorly, owing to high heterozygosity that greatly limited the utility of tools.

Markers comprised of repetitive homopolymer runs remain a challenge for many high-throughput genome sequencing chemistries. Fragment length analysis by capillary electrophoresis may represent a solution to this, though that approach has limited throughput compared with NGS approaches.

When comparing the genetic clustering from the eight-marker Illumina-based NGS approach with epidemiologic data, it was noted that *Cyclospora* infections associated with distinct contamination sources sometimes conflated into a single genetic grouping [50, 56]. Ultimately, a *Cyclospora* genotyping approach involving NGS of numerous additional SNP-based markers encoded predominantly in the nuclear genome is required. Future developments will therefore focus on sequencing additional SNP-based nuclear markers that expand on the original eight-marker panel. Such future developments are discussed in the following section.

5 New and Future Developments for Genotyping

The current set of eight markers used to differentiate clusters of *Cyclospora* from clinical stool samples only represents a relatively small portion of the full *Cyclospora* genome. Approaches that capture a broader suite of genetic variability within and among human-infecting *Cyclospora* species are needed to better refine cluster identification and can provide enhanced insight into patterns of *Cyclospor*a infection over time and space. As mentioned in the previous section, recent work has revealed that some of the most prevalent strains of *Cyclospora* are observed in outbreaks recurrently year after year [58]. This observation led to the introduction of "temporal genetic clusters" as a concept, where a system of further dividing genetic clusters into sub-clusters using temporal information was implemented [55]. The tendency for a few strains to recur and cause outbreaks over multiple years has a few different potential explanations. The first is that if a few *Cyclospora* strains are geographically widespread within growing regions from which produce is imported to USA, produce items from multiple growing regions could be contaminated with the same few strains year after year. Alternatively, if a large portion of produce imported into the USA originates from a relatively small geographic area, a single strain could be endemic to this region and therefore frequently observed in case patients on a recurrent basis. Further work to understand the geographic distribution of *Cyclospora* strains across the endemic region (particularly in growing regions for produce bound for USA distribution) is needed to better clarify these patterns.

However, a third explanation for the pattern of recurrent observed strains could simply be a function of the limited discriminatory power of the eight genotyping markers used in this analysis [58]. It is possible these recurrent strains are in fact genetically distinct, but the current methods may be insufficient to identify these strains as belonging to separate genetic clusters. Clearly, a key step in better understanding the temporal and geographic patterns of *Cyclospora* parasites is to ensure that these large genetic clusters do, in fact, represent a single strain of *Cyclospora.*

Researchers at the Food and Drug Administration (FDA) in collaboration with the company Chapter Diagnostics recently developed an expanded amplicon panel of genotyping markers, with methods that are suitable for routine typing from both clinical and non-clinical *Cyclospor*a specimens [60]. That panel encompasses 396 known SNP sites and includes 52 target loci (49 of which are nuclear and 2 mitochondrial). The authors compared the phylogenetic trees developed from the SNP panel targeting their 52 loci to phylogenetic trees generated from SNPs from publicly available *Cyclospora* genomes and found that the trees closely resemble one another—a promising indication of the utility of this panel. Further work is currently underway to continue to test and refine the markers used in this broader amplicon panel. As this is done, it will be key to identify a suite of markers that can provide better resolution among *Cyclospora* strains than the current set of eight markers but is not so refined that related specimens no longer cluster together. Careful efforts to incorporate epidemiological data will help to identify the "base" or "background" level of genetic diversity that is expected even among closely related *Cyclospora* strains. Using epidemiologic data in the evaluation of genotyping markers can ensure that specimens from case patients known to be linked to the same epidemiological cluster (e.g., those from patients with known shared food exposures) will cluster together genetically as well.

Advancements that allow for high-quality nuclear genotyping data to be generated from both clinical and non-clinical (environmental) specimens can allow for more direct linkages between *Cyclospora* causing illness and potential food sources of *Cyclospora* exposure. To increase sequencing success from environmental samples, the approach developed by the FDA utilizes a hybridization capture method (also known as capture enrichment sequencing or bait-capture) to enrich target DNA segments allowing for improved sequencing libraries. These bait-capture methods work through the use of sequence-specific single-stranded oligonucleotides (baits) that are combined and hybridized with DNA from target loci, "capturing" the target fragments while remaining non-target DNA fragments are washed away and ultimately depleted [60]. This enrichment is particularly needed when attempting to genotyping *Cyclospora* from non-clinical specimens, such as produce or environmental samples, as target parasite DNA (especially nuclear DNA) is often too low to enable use of the same metagenomic amplicon sequencing methods used for genotyping *Cyclospora* from clinical specimens. The use of surrogate parasite species that are closely related to *Cyclospora* but are able to be more easily obtained through animal models (such as *Eimeria*) has been suggested as a tool to help evaluate such capture enrichment sequencing methods for *Cyclospora*—as has been done for parasite species such as *Cryptosporidium* [61].

6 Concluding Remarks

Before 2013, very little was known about *Cyclospora* genomics. Attempts to find potential genotyping markers had failed to identify loci suitable for strain differentiation. Our understanding of the *Cyclospora* population structure and dynamics in the USA has dramatically increased after the introduction of routine genotyping since 2018.

However, much is left to be discovered. One area that warrants attention is the genetic variation of *Cyclospora* circulating among humans in countries where cyclosporiasis is endemic. Another area of high interest is the genetic characterization of environmental samples, such as contaminated produce, irrigation water, soil, and other potential sources of contamination. As better, more sensitive tools to investigate *Cyclospora* genomics are being developed, they will undoubtedly reveal new insights about this relatively new human pathogen and lead to a better understanding of the temporal and geographical trends of *Cyclospora* strains in human transmission and in the environment.

References

1. Schneider A (1881) Sur les psorospermies oviformes ou coccidies. Espécies nouvelles ou peu connues. Arch Zool Exp Gen 9:387–404
2. Ashford RW (1979) Occurrence of an undescribed coccidian in man in Papua New Guinea. Ann Trop Med Parasitol 73:497–500
3. Herwaldt BL (2000) Cyclospora cayetanensis: a review, focusing on the outbreaks of cyclosporiasis in the 1990s. Clin Infect Dis 31:1040–1057
4. Ortega YR, Sterling CR, Gilman RH, Cama VA, Diaz F (1993) Cyclospora species--a new protozoan pathogen of humans. N Engl J Med 328:1308–1312
5. Ortega YR, Gilman RH, Sterling CR (1994) A new coccidian parasite (Apicomplexa: Eimeriidae) from humans. J Parasitol 80:625–629
6. Levine ND (2018) The protozoan phylum Apicomplexa. Routledge
7. Ford PL, Duszynski DW (1989) Coccidian parasites (Apicomplexa: Eimeriidae) from insectivores. VII. Six new species from the hairy-tailed mole, Parascalops breweri. J Parasitol 75:508
8. Eberhard ML, Da Silva AJ, Lilley BG, Pieniazek NJ (1999) Morphologic and molecular characterization of new Cyclospora species from Ethiopian monkeys:

C. cercopitheci sp.n., C. colobi sp.n., and C. papionis sp.n. Emerg Infect Dis 5:651–658
9. Lainson R (2005) The genus Cyclospora (Apicomplexa: Eimeriidae), with a description of Cyclospora schneideri n.sp. in the snake Anilius scytale scytale (Aniliidae) from Amazonian Brazil--a review. Mem Inst Oswaldo Cruz 100:103–110
10. Li N, Ye J, Arrowood MJ, Ma J, Wang L, Xu H, Feng Y, Xiao L (2015) Identification and morphologic and molecular characterization of Cyclospora macacae n. sp. from rhesus monkeys in China. Parasitol Res 114:1811–1816
11. Mcallister CT, Hnida JA (2023) A new Eimerian (Apicomplexa: Eimeriidae) from eastern mole, Scalopus Aquaticus (Mammalia: Eulipotyphla: Talpidae), in Central Arkansas, with additional information on Cyclospora Yatesi Mcallister, Motriuk-Smith, and Kerr. J Parasitol 109:21–26
12. Barratt JLN, Shen J, Houghton K, Richins T, Sapp SGH, Cama V, Arrowood MJ, Straily A, Qvarnstrom Y (2023) Cyclospora cayetanensis comprises at least 3 species that cause human cyclosporiasis. Parasitology 150:269–285
13. Shields JM, Olson BH (2003) Cyclospora cayetanensis: a review of an emerging parasitic coccidian. Int J Parasitol 33:371–391
14. Chalmers RM, Nichols G, Rooney R (2000) Foodborne outbreaks of cyclosporiasis have arisen in North America. Is the United Kingdom at risk? Commun Dis Public Health 3:50–55
15. Abanyie F, Harvey RR, Harris JR, Wiegand RE, Gaul L, Desvignes-Kendrick M, Irvin K, Williams I, Hall RL, Herwaldt B, Gray EB, Qvarnstrom Y, Wise ME, Cantu V, Cantey PT, Bosch S, Aj DAS, Fields A, Bishop H, Wellman A, Beal J, Wilson N, Fiore AE, Tauxe R, Lance S, Slutsker L, Parise M, Multistate Cyclosporiasis Outbreak Investigation Team (2015) 2013 multistate outbreaks of Cyclospora cayetanensis infections associated with fresh produce: focus on the Texas investigations. Epidemiol Infect 143:3451–3458
16. Anonymous (2024) Outbreak investigations and updates [Online]. US Centers for Disease Control and Prevention. https://www.cdc.gov/parasites/cyclosporiasis/outbreaks/index.html
17. Buss BF, Joshi MV, O'Keefe AL, Allensworth CD, Garvey A, Obbink K, Mandernach S, Safranek TJ (2016) Regional investigation of a cyclosporiasis outbreak linked to imported romaine lettuce - Nebraska and Iowa, June-August 2013. Epidemiol Infect 144:1807–1817
18. Doller PC, Dietrich K, Filipp N, Brockmann S, Dreweck C, Vonthein R, Wagner-Wiening C, Wiedenmann A (2002) Cyclosporiasis outbreak in Germany associated with the consumption of salad. Emerg Infect Dis 8:992–994
19. Insulander M, Svenungsson B, Lebbad M, Karlsson L, De Jong B (2010) A foodborne outbreak of Cyclospora infection in Stockholm, Sweden. Foodborne Pathog Dis 7:1585–1587
20. Nichols GL, Freedman J, Pollock KG, Rumble C, Chalmers RM, Chiodini P, Hawkins G, Alexander CL, Godbole G, Williams C, Kirkbride HA, Hamel M, Hawker JI (2015) Cyclospora infection linked to travel to Mexico, June to September 2015. Euro Surveill 20
21. Prescott LM, Harley JP, Klein DA (2005) Human disease caused by fungi and protozoa. In: Microbiology, 6th edn. McGraw Hill, New York
22. Almeria S, Cinar HN, Dubey JP (2019) Cyclospora cayetanensis and Cyclosporiasis: an update. Microorganisms 7
23. Dubey JP, Almeria S, Mowery J, Fortes J (2020) Endogenous developmental cycle of the human coccidian Cyclospora cayetanensis. J Parasitol 106:295–307
24. Dubey JP, Khan A, Rosenthal BM (2022) Life cycle and transmission of Cyclospora cayetanensis: knowns and unknowns. Microorganisms 10
25. Alfano-Sobsey EM, Eberhard ML, Seed JR, Weber DJ, Won KY, Nace EK, Moe CL (2004) Human challenge pilot study with Cyclospora cayetanensis. Emerg Infect Dis 10:726–728
26. Eberhard ML, Ortega YR, Hanes DE, Nace EK, Quy Do R, Robl MG, Won KY, Gavidia C, Sass NL, Mansfield K, Gozalo A, Griffiths J, Gilman R, Sterling CR, Arrowood MJ (2000) Attempts to establish experimental cyclospora Cayetanensis infection in laboratory animals. J Parasitol 86:577–582
27. Eberhard ML, Nace EK, Freeman AR, Streit TG, Da Silva AJ, Lammie PJ (1999) Cyclospora cayetanensis infections in Haiti: a common occurrence in the absence of watery diarrhea. Am J Trop Med Hyg 60:584–586
28. Sterling CR, Ortega YR (1999) Cyclospora: an enigma worth unraveling. Emerg Infect Dis 5:48–53
29. Totton SC, O'Connor AM, Naganathan T, Martinez BAF, Sargeant JM (2021) A review of Cyclospora cayetanensis in animals. Zoonoses Public Health 68:861–867
30. Smith HV, Paton CA, Girdwood RW, Mtambo MM (1996) Cyclospora in non-human primates in Gombe, Tanzania. Vet Rec 138:528
31. Eberhard ML, Owens JR, Bishop HS, De Almeida ME, Da Silva AJ (2014) Cyclospora spp. in drills, Bioko Island, Equatorial Guinea. Emerg Infect Dis 20:510–511
32. Eberhard ML, Njenga MN, Dasilva AJ, Owino D, Nace EK, Won KY, Mwenda JM (2001) A survey for Cyclospora Spp. in Kenyan primates, with some notes on its biology. J Parasitol 87:1394–1397
33. Zhao GH, Cong MM, Bian QQ, Cheng WY, Wang RJ, Qi M, Zhang LX, Lin Q, Zhu XQ (2013) Molecular characterization of Cyclospora-like organisms from golden snub-nosed monkeys in Qinling Mountain in Shaanxi province, northwestern China. PLoS One 8:e58216
34. Cinar HN, Gopinath G, Jarvis K, Murphy HR (2015) The complete mitochondrial genome of the foodborne

parasitic pathogen *Cyclospora cayetanensis*. PLoS One 10:e0128645
35. Ogedengbe ME, Qvarnstrom Y, Da Silva AJ, Arrowood MJ, Barta JR (2015) A linear mitochondrial genome of *Cyclospora cayetanensis* (Eimeriidae, Eucoccidiorida, Coccidiasina, Apicomplexa) suggests the ancestral start position within mitochondrial genomes of eimeriid coccidia. Int J Parasitol 45:361–365
36. Tang K, Guo Y, Zhang L, Rowe LA, Roellig DM, Frace MA, Li N, Liu S, Feng Y, Xiao L (2015) Genetic similarities between *Cyclospora cayetanensis* and cecum-infecting avian *Eimeria* spp. in apicoplast and mitochondrial genomes. Parasit Vectors 8:358
37. Cinar HN, Gopinath G, Almeria S, Njoroge JM, Murphy HR, Da Silva A (2022) Targeted next generation sequencing of Cyclospora cayetanensis mitochondrial genomes from seeded fresh produce and other seeded food samples. Heliyon 8:e11575
38. Cinar HN, Gopinath G, Murphy HR, Almeria S, Durigan M, Choi D, Jang A, Kim E, Kim R, Choi S, Lee J, Shin Y, Lee J, Qvarnstrom Y, Benedict TK, Bishop HS, Da Silva A (2020) Molecular typing of Cyclospora cayetanensis in produce and clinical samples using targeted enrichment of complete mitochondrial genomes and next-generation sequencing. Parasit Vectors 13:122
39. Gopinath GR, Cinar HN, Murphy HR, Durigan M, Almeria M, Tall BD, Dasilva AJ (2018) A hybrid reference-guided de novo assembly approach for generating Cyclospora mitochondrion genomes. Gut Pathog 10:15
40. Guo Y, Wang Y, Wang X, Zhang L, Ortega Y, Feng Y (2019) Mitochondrial genome sequence variation as a useful marker for assessing genetic heterogeneity among Cyclospora cayetanensis isolates and source-tracking. Parasit Vectors 12:47
41. Nascimento FS, Barta JR, Whale J, Hofstetter JN, Casillas S, Barratt J, Talundzic E, Arrowood MJ, Qvarnstrom Y (2019) Mitochondrial junction region as genotyping marker for Cyclospora cayetanensis. Emerg Infect Dis 25:1314–1319
42. Gonzalez-Gomez JP, Lozano-Aguirre LF, Medrano-Felix JA, Chaidez C, Gerba CP, Betancourt WQ, Castro-Del Campo N (2023) Evaluation of nuclear and mitochondrial phylogenetics for the subtyping of Cyclospora cayetanensis. Parasitol Res 122:2641–2650
43. Guo Y, Li N, Ortega YR, Zhang L, Roellig DM, Feng Y, Xiao L (2018) Population genetic characterization of Cyclospora cayetanensis from discrete geographical regions. Exp Parasitol 184:121–127
44. Guo Y, Roellig DM, Li N, Tang K, Frace M, Ortega Y, Arrowood MJ, Feng Y, Qvarnstrom Y, Wang L, Moss DM, Zhang L, Xiao L (2016) Multilocus sequence typing tool for *Cyclospora cayetanensis*. Emerg Infect Dis 22:1464–1467
45. Li J, Chang Y, Shi KE, Wang R, Fu K, Li S, Xu J, Jia L, Guo Z, Zhang L (2017) Multilocus sequence typing and clonal population genetic structure of Cyclospora cayetanensis in humans. Parasitology 144:1890–1897
46. Cinar HN, Qvarnstrom Y, Wei-Pridgeon Y, Li W, Nascimento FS, Arrowood MJ, Murphy HR, Jang A, Kim E, Kim R, Da Silva A, Gopinath GR (2016) Comparative sequence analysis of Cyclospora cayetanensis apicoplast genomes originating from diverse geographical regions. Parasit Vectors 9:611
47. Liu S, Wang L, Zheng H, Xu Z, Roellig DM, Li N, Frace MA, Tang K, Arrowood MJ, Moss DM, Zhang L, Feng Y, Xiao L (2016) Comparative genomics reveals Cyclospora cayetanensis possesses coccidia-like metabolism and invasion components but unique surface antigens. BMC Genomics 17:316
48. Qvarnstrom Y, Wei-Pridgeon Y, Li W, Nascimento FS, Bishop HS, Herwaldt BL, Moss DM, Nayak V, Srinivasamoorthy G, Sheth M, Arrowood MJ (2015) Draft genome sequences from Cyclospora cayetanensis oocysts purified from a human stool sample. Genome Announc 3
49. Qvarnstrom Y, Wei-Pridgeon Y, Van Roey E, Park S, Srinivasamoorthy G, Nascimento FS, Moss DM, Talundzic E, Arrowood MJ (2018) Purification of Cyclospora cayetanensis oocysts obtained from human stool specimens for whole genome sequencing. Gut Pathog 10:45
50. Yanta CA, Pollo SMJ, Barta JR, Reiling SJ, Wasmuth JD, Dixon BR, Guy RA (2022) Draft hybrid genome assembly of a Canadian Cyclospora cayetanensis isolate. Microbiol Resour Announc 11:e0107221
51. Barratt JLN, Park S, Nascimento FS, Hofstetter J, Plucinski M, Casillas S, Bradbury RS, Arrowood MJ, Qvarnstrom Y, Talundzic E (2019) Genotyping genetically heterogeneous Cyclospora cayetanensis infections to complement epidemiological case linkage. Parasitology 146:1275–1283
52. Houghton KA, Lomsadze A, Park S, Nascimento FS, Barratt J, Arrowood MJ, Vanroey E, Talundzic E, Borodovsky M, Qvarnstrom Y (2020) Development of a workflow for identification of nuclear genotyping markers for Cyclospora cayetanensis. Parasite 27:24
53. Nascimento FS, Barratt J, Houghton K, Plucinski M, Kelley J, Casillas S, Bennett CC, Snider C, Tuladhar R, Zhang J, Clemons B, Madison-Antenucci S, Russell A, Cebelinski E, Haan J, Robinson T, Arrowood MJ, Talundzic E, Bradbury RS, Qvarnstrom Y (2020) Evaluation of an ensemble-based distance statistic for clustering MLST datasets using epidemiologically defined clusters of cyclosporiasis. Epidemiol Infect 148:e172
54. Jacobson DK, Low R, Plucinski MM, Barratt JLN (2023) An improved framework for detecting discrete epidemiologically meaningful partitions in hierarchically clustered genetic data. Bioinform Adv 3:vbad118
55. Ahart L, Jacobson D, Rice M, Richins T, Peterson A, Zheng Y, Barratt J, Cama V, Qvarnstrom Y, Montgomery S, Straily A (2023) Retrospective evalu-

ation of an integrated molecular-epidemiological approach to cyclosporiasis outbreak investigations - United States, 2021. Epidemiol Infect 151:e131
56. Barratt J, Ahart L, Rice M, Houghton K, Richins T, Cama V, Arrowood M, Qvarnstrom Y, Straily A (2022) Genotyping Cyclospora cayetanensis from multiple outbreak clusters with an emphasis on a cluster linked to bagged salad mix-United States, 2020. J Infect Dis 225:2176–2180
57. Barratt J, Houghton K, Richins T, Straily A, Threlkel R, Bera B, Kenneally J, Clemons B, Madison-Antenucci S, Cebelinski E, Whitney BM, Kreil KR, Cama V, Arrowood MJ, Qvarnstrom Y (2021) Investigation of US Cyclospora cayetanensis outbreaks in 2019 and evaluation of an improved Cyclospora genotyping system against 2019 cyclosporiasis outbreak clusters. Epidemiol Infect 149:e214
58. Jacobson DK, Peterson AC, Qvarnstrom Y, Barratt JLN (2023) Novel insights on the genetic population structure of human-infecting Cyclospora spp. and evidence for rapid subtype selection among isolates from the USA. Curr Res Parasitol Vector Borne Dis 4:100145
59. Hofstetter JN, Nascimento FS, Park S, Casillas S, Herwaldt BL, Arrowood MJ, Qvarnstrom Y (2019) Evaluation of multilocus sequence typing of Cyclospora cayetanensis based on microsatellite markers. Parasite 26:3
60. Leonard SR, Mammel MK, Gharizadeh B, Almeria S, Ma Z, Lipman DJ, Torrence ME, Wang C, Musser SM (2023) Development of a targeted amplicon sequencing method for genotyping Cyclospora cayetanensis from fresh produce and clinical samples with enhanced genomic resolution and sensitivity. Front Microbiol 14:1212863
61. Tucker MS, Khan A, Jenkins MC, Dubey JP, Rosenthal BM (2022) Hastening progress in Cyclospora requires studying Eimeria surrogates. Microorganisms 10

An Overview of *Babesia* Species Biology Based on Omics Studies

Laura Cuy-Chaparro, Jeimmy Pilar Martínez-González, and Darwin Andrés Moreno-Pérez

Abstract

Babesiosis, caused by parasites of the *Babesia* genus transmitted by ticks, is a disease that significantly impacts animal and human health worldwide. Understanding the biological aspects related to the pathogenesis of *Babesia* is critical, given that no vaccine is available to date, and multiple strains of the species are developing resistance to drugs. Advances in sequencing have greatly improved the understanding of the pathogenic dynamics of different species and have provided invaluable information for the search for alternative therapies, management, and perhaps the elimination of the disease. Consequently, this chapter summarizes the main omics findings (genomics, transcriptomics, proteomics, immunomics, and metabolomics) that have been carried out in recent years to understand the biology of *Babesia* species, emphasizing their usefulness for the search for key molecules in diagnosis, pathogenesis, pathophysiology, and vaccine development. A comprehensive view of the biology of *Babesia* is provided, and new perspectives for the development of control and treatment strategies are discussed, addressing the challenges and future directions in the fight against babesiosis.

Keywords

Babesiosis · *Babesia* species · Omics · Diagnosis · Pathogenesis · Control strategies

1 Introduction

Babesiosis is a disease that represents a significant threat to health worldwide and continues to be a major concern and challenge for veterinary and human medicine as it affects many ruminants (both domestic and wild) and humans [1]. The disease is caused by protozoan parasites taxonomically classified in the phylum Apicomplexa, class Piroplasmea, order Piroplasmida, family Babasidae, and genus *Babesia* which are transmitted by ixodid ticks of the genus Rhipicephalus whose presence is limited to tropical and subtropical regions [1, 2]. With no totally effective vaccines available and a significant global epidemiological impact, babesiosis is one of the most important arthropod-transmitted diseases, causing high morbidity and mortality. Additionally, some species are responsible for causing significant economic losses and health challenges worldwide [3, 4].

L. Cuy-Chaparro · J. P. Martínez-González · D. A. Moreno-Pérez (✉)
Veterinary Medicine Program, Universidad de Ciencias Aplicadas y Ambientales [U.D.C.A], Bogotá, Colombia
e-mail: lcuyc@udca.edu.co; jeimmartinez@udca.edu.co; darmoreno@udca.edu.co

J. D. Ramírez González (ed.), *Recent Advances in Parasitomics*,
https://doi.org/10.1007/978-3-031-70591-5_11

The control and prevention of bovine babesiosis represent important biological challenges that drive continuous research efforts worldwide. To date, various control and prevention strategies have been employed, such as the elimination of ticks with acaricides or parasites through the induction of an active immune response in the host using live attenuated vaccines, recombinant protein-based vaccines, or synthetic peptide vaccines [5]. Despite these efforts, these strategies have not been efficient, and furthermore, the parasites are developing resistance. To improve the identification of vaccine components, reverse vaccinology has been developed, which, in combination with omics technologies, has provided unprecedented knowledge about the biology of *Babesia* parasites and their interactions with the host [5, 6]. Exploration of these parasites at the levels of genes (genomics), transcripts (transcriptomics), proteins (proteomics), antigenic response (immunomics), and metabolites (metabolomics) has shed light on various aspects related to genetic diversity and the molecular mechanisms underlying the pathogenesis of various *Babesia* species, which remain crucial for the future development of effective control strategies [7].

This chapter provides an overview of the current state of omics studies advanced to study different *Babesia* species (*B. bovis, B. bigemina, B. divergens, B. microti, B. ovata, B. duncani, B. xinjiang, B. ovis, B. gibsoni, B. caballi, B. orientalis*, and *B. ovata*), discussing their usefulness in understanding biology, strategies to identify possible pharmacological targets or vaccines, and the development of new therapeutic interventions. Future directions and challenges in the application of omics technologies to develop control strategies against babesiosis will be highlighted.

2 *Babesia* Lyfe Cycle

Members of the *Babesia* species have a complex life cycle that requires a vector, where the parasite reproduces sexually, as well as a vertebrate host, where asexual multiplication occurs (Fig. 1) [8]. When the vector bites, it injects *Babesia* sporozoites into the host's bloodstream, which directly invade red blood cells through molecular interactions. Once inside a cell, sporozoites move freely within the cytoplasm and transform into trophozoites within a few hours. This marks the beginning of the merogony phase, where the parasite reproduces asexually by replicating its nucleus within the host cell. The parasite then induces cell division through binary fission to produce two to four infectious merozoites (according to the *Babesia* species) that lyse red blood cells to enter the host's circulation and invade other erythrocytes. This process of invasion and lysis of the erythrocytes occurs continuously, leading to the successive production of asynchronous merozoites (Consequently, different stages of the parasite can be found in the host's bloodstream), and the presentation of clinical symptoms between 7 and 35 days after the initial inoculation of the parasite [9].

There has been an ongoing debate about the development of pregametocytes in the bloodstream of a vertebrate host, with no experimental evidence supporting this phenomenon in some *Babesia* species. However, when ticks feed on blood, the development of male and female gametes begins in the intestinal lumen, where they fuse and form zygotes (known as motile ookinetes) (Fig. 1). These invade the intestinal cells of the tick and transform into kinetes by meiotic division, which migrate to the hemolymph and invade the peripheral tissues of the vector, including the salivary glands. In some species, the kinetes can migrate to the ovaries, resulting in transovarian and transstadial transmission to the tick's offspring [8, 10, 11]. After dispersal to the salivary glands, the kinetes transform into sporoblasts through asexual multiplication via sporogony, producing 5000–10,000 sporozoites asynchronously. These sporozoites are continuously released into the bloodstream of a vertebrate host during tick feeding. Successful invasion of the parasite depends on the interaction of its extracellular transmembrane proteins with receptors on the surface of the target host cell, a mechanism observed in other Apicomplexa.

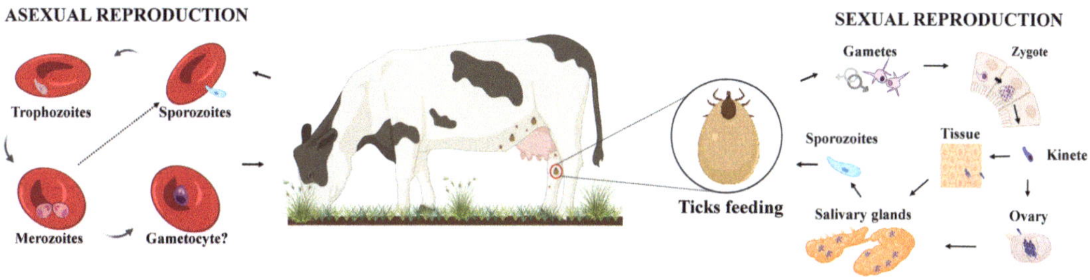

Fig. 1 Babesia life cycle. In this process, two hosts play essential roles: one acts as the vector (tick), where the sexual phase of the parasite's life cycle occurs (right), while the other serves as the vertebrate host (animal or human), where asexual multiplication takes place (left). Figure created with BioRender.com

3 Genomic Insights Into *Babesia* Species

Knowledge of the life cycle of species is essential to understanding their stages of development, interaction with the host, and survival mechanisms. While this basic information is fundamental, it is crucial to delve deeper into the biology of the parasite to establish appropriate control strategies. The genome of a species is fundamental to understanding its biology, evolution, and adaptation as it constitutes the complete set of genetic instructions that determine the characteristics and functions of an organism [12]. Therefore, studying the genome enables us to predict possible mechanisms of pathogenicity, adaptation to the host, and inter- and intra-species evolutionary relationships in different species.

In *Babesia*, the genome is characterized by having a maximum of four chromosomes, which were identified mainly in *B. bovis* and *B. bigemina* in the 1990s [13, 14]. However, with the development of high-throughput sequencing technologies and advanced bioinformatics analyses, useful genomic information is available for 12 *Babesia* species (Fig. 2), allowing for a closer understanding of their biology (Table 1) [15–25].

Analysis of the reported information shows that the genome sizes of different *Babesia* species vary significantly, ranging from 6.4 Mb in *B. microti* to 14.5 Mb in *B. ovata* (Table 1). This variability reflects differences in genomic complexity and quantity, suggesting that species have evolved to survive and adapt in their respective vertebrate and arthropod niches. This is further supported by the co-adaptations driven by the coevolution of parasites with their hosts, as seen in intracellular Apicomplexans [26]. This variability also aligns with differences in the genomes of the apicoplast and mitochondria (Table 1), indicating the specialization of parasitic functions and adaptations, given these organelles' roles in synthesizing key metabolic products, and forming functional parasitophorous vacuoles, which are crucial for the survival of different Apicomplexans [27, 28].

Interestingly, *in silico* comparative and predictive analyses of genomes have also contributed to identifying multigenic families that play a role in the evolutionary dynamics of different parasite species, suggesting their critical role in establishing host–pathogen interactions. Notably, all species contain multigenic families related to pathogenicity (Table 1). The most abundant of these is the family of genes encoding VESA (variable erythrocyte surface antigen), present in all members of the genus *Babesia* that infect animals but not in those that affect humans (*B. microti* and *B. duncani*). These proteins are highly variable and are grouped into alpha and beta types based on sequence composition, length, and architecture [23]. It has been proposed that VESAs are synthesized in the merozoite and then exported to the surface of erythrocytes to enable cytoadhesion and capillary sequestration, evading spleen-mediated elimination and promoting parasite persistence and pathogenesis within the host [29]. This cytoadhesion function is mediated by the CKRD (cysteine-rich domain) and VDCS (Von Willebrand factor D and

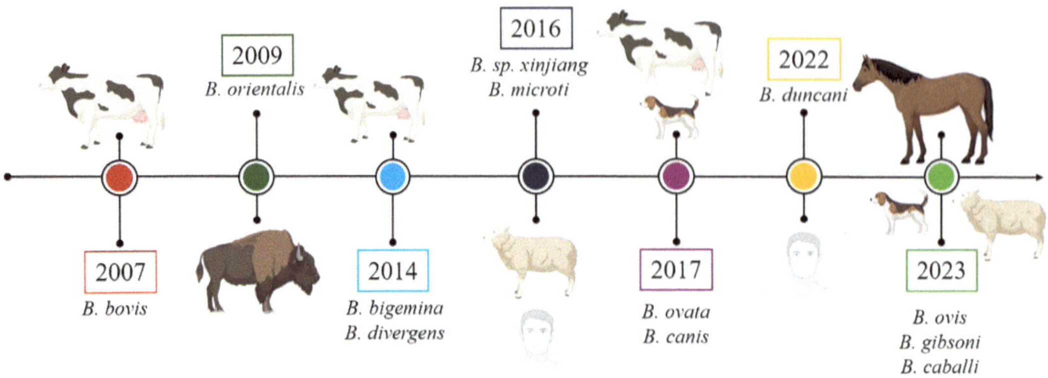

Fig. 2 Timeline of *Babesia* species genome sequencing. The species and the vertebrate host (animal or human) are shown. Figure created with BioRender.com

cysteine-rich domain) domains, which are only present in the VESA proteins of *B. bovis*, the only species with virulent strains capable of causing serious brain damage in animals [25]. *Ves* genes have also been found to be structurally variable between species and are thus given different nomenclature: *ves2* in *B. bigemina* and *B. divergens*, *ves1δ* in *B. bovis*, and *ves1c* in *B. caballi* (Table 1). This suggests that *ves* genes have evolved from an ancestral gene and have been modified according to host adaptation needs, as suggested in a previous study [25].

Other gene families have been identified. For example, the gene families that encode SmORFs (small open reading frames), SBP (spherical body protein), VMSA (variable merozoite surface antigen), MFS (major facilitator superfamily), BMN (Babesia microti immunodominant antigen), GPI-AP (glycosylphosphatidylinositol-anchored protein), SEA (serine esterase), SA (secreted antigens), PK (protein kinase), and Bc28 (a surface protein member). The *smorf* genes are found in subtelomeric regions of the genome, and although they were considered important for modulating *ves* gene expression, they have been shown to have differential expression and undergo variation within strains in a lineage-specific pattern independent of strain specificity [30]. The *sbp* genes are expressed in the spherical bodies of *B. bovis* merozoites, with some coding products involved in reducing cytoadhesion, suggesting their role as key modulators of parasite virulence [31, 32]. The *vmsa* genes encode proteins targeted by antibodies capable of reducing the parasite's invasion of target cells, underscoring their importance as modulators of the host immune response [33]. The *mfs* genes encode multi-transmembrane proteins, which, while known to be involved in the transport of a wide range of substrates across the plasma membrane [34], have an unknown function in *Babesia*. Other antigens such as GPI-AP, SEA, and Bc28 are expressed on the surface of merozoites and play important roles in immune adhesion and evasion given their high variability [35, 36]. These findings related to genetic diversity among species and different mutagenic families within the same species are undoubtedly an evolutionary footprint, likely dependent on the adaptation and survival needs in different hosts.

Genome sequencing has revolutionized the understanding of the genomic architecture of species, particularly in elucidating their genetic composition and identifying key genes associated with virulence and pathogenicity. Continued research leveraging genomic data from different *Babesia* species will enable us to unravel the molecular basis of their pathogenesis, drug resistance mechanisms, and host adaptation strategies. This knowledge will ultimately be useful for developing novel strategies for controlling Babesiosis.

Table 1 Comparative analysis of *Babesia* spp. genomes reported to date

Host	Etiologic agent	Strain	Country	Genome size (kbp)	Apicoplast size (kbp)	Mitochondria genome (kbp)	Protein coding gene	Multigene family	References
Cattle	*B. bovis*	T2Bo	Texas	8.2	33	6	3706	*Ves1 (ves1α, ves1β), smorf, sbp2, vmsa*	[15]
	B. bigemina	BOND clone C7-1	Argentina	13.8	NR	8	4457	*Ves 1(ves1a, ves1b), ves2, msf*	[25]
	B. divergens	1802A	France	8.9	NR	6	4134	*Ves1, ves2 (ves2a, ves2b), mfs*	
	B. ovata	Miyake	Japan	14.5	30	6	5031	*Ves 1(ves1a, ves1b)*	[22]
Sheep	*B. sp. Xinjiang*	Bxj	China	8.4	31	6	3066	*Ves1* (VESA1_N domain I-IV)	[17]
Human	*B. microti*	R1	Unknown	6.4	29	11	3494	*Bmn*	[20]
	B.duncani	PRA-302	USA	7.9	34	5.9	3759	*Gpi-ap, sea*	[21]
Dog	*B. canis*	BcH-CHIPZ	Switzerland	7	NR	NR	3467	*Ves1, DnaJ, SA, PK, Bc28*	[16]
	B. gibsoni	WH58	China	7.9	52	5.9	3552	*Ves1* (*ves1*-like genes)	[24]
Buffalo	*B. orientalis*	Wuhan	NR	7.8	33	6	4199	*Ves 1*	[18]
Horse	*B. caballi*	USDA-D6B2	USA	12.9	32	6	5910	*Ves1c* (*ves1*-like genes)	[19]
Sheep and goats	*B. ovis*	Wildly infected sheep	Turkey	7.8	36	6	3419	*Ves1δ* (*ves1*-like genes)	[23]

NR, not reported; *ves*, variant erythrocyte surface antigen; *Smorf*, small open reading frame; *sbp2*, spherical bodysuit protein 2; *vmsa*, merozoite variable surface antigen; *msf*, major facilitator superfamily; *bmn*, *B. microti* immunodominant antigen; *gpi-ap*, glycosylphosphatidylinositol-anchored protein; *sea*, serine esterase; *DnaJ*, protein-containing DnaJ domain; *SA*, secreted antigen; *pk*, protein kinase; *Bc28*, a surface protein

4 Gene Expression Dynamics in *Babesia*

Strict gene regulation is essential for all forms of life, particularly for microorganisms, as it enables them to invade, replicate, and survive in different microenvironments. This correlates with the upregulated or downregulated expression of a subset of genes encoding proteins necessary for each function [37].

Despite the progress in analyzing the molecular composition of the genomes of different *Babesia* species, especially in identifying genes related to the pathogenesis these microorganisms induce in their hosts, this information is not completely useful as it is not known which genes are active at different stages of the parasite's life cycle. Transcriptomic analyses thus offer a comprehensive view of the dynamics of gene expression in *Babesia* parasites during different stages of infection in their mammalian hosts [15–17, 19–21, 23–25, 30, 38]. By profiling the transcriptomes of the *Babesia* parasite, key molecular pathways involved in immune evasion, parasite adhesion and/or invasion, and the host response to infection have been identified. These results are thus very important for identifying new therapeutic targets, vaccine candidates, and diagnostic biomarkers for babesiosis.

Compared with the advances in knowledge of the genome of the 12 *Babesia* species, to date, 11 transcriptomes of blood-stage merozoites from mammalian hosts evaluated by RNA sequencing technologies have been reported (Table 2). The results show a wide diversity in gene expression among the different *Babesia* species, reflecting each species' adaptation to its specific environment and its strategies for evading the host immune system as was previously mentioned. For example, the interspecies variation of *ves* genes supports the idea that this family undergoes expansion, probably associated with the complexity of the specific adaptations of each lineage in the host and interactions with the environment [30] (Fig. 3). In contrast, the genes that code for MSPs (merozoite surface proteins) and SA (secreted antigens) are present in multiple species, emphasizing their importance in pathogenesis and immune evasion. Furthermore, the presence of specific genes such as ApiAP2 (apicomplexan apetala 2 integrase DNA binding domain) and components of the mobile junction (AMA-1 (apical merozoite antigen 1) and RON2 (rhoptry neck protein 2)) suggests the conservation of certain invasion mechanisms among apicomplexans that are important for precisely regulating the stages of parasite development [39] as well as for the cell adhesion/invasion process [40, 41].

The results also provide invaluable broad-scale data for selecting and studying useful target genes to be studied as diagnostic targets (due to their exclusivity in each species), pharmacological targets (given that they are part of metabolic pathways exclusive to the parasite), indicators of pathogenicity (belonging to multigenic families related to immune evasion), and genes encoding antigens to be used in methods of prevention and control of the disease (as they are secreted and facilitate host–pathogen interactions such as adhesion and/or invasion) (Table 2). Consequently, identifying specific genes and their differential expression in blood stages is indispensable for designing control strategies aimed at inhibiting critical functions of the parasite without affecting the host.

Another representative finding is the intraspecies comparison, which provides information on molecular mechanisms associated with the phenotypic adaptation of various strains depending on the environmental context. This offers an invaluable source of information for the search and identification of specific markers associated with diagnosis, virulence, or possible candidate vaccine molecules (Table 3) [20, 42–44]. For example, in the case of *B. bigemina*, 25 genes were found to be upregulated in the virulent strain, including membrane proteins and the exonuclease of the exosome complex, which have been described as involved in the adhesion and invasion of the parasite to their target cells or in modulating gene expression through RNA degradation to adapt to intracellular conditions, respectively [44]. Compared with the attenuated strain, only three genes were downregulated, including two endonuclease reverse transcriptases, which

Table 2 Pathogenesis-related genes determined by the transcriptome of various *Babesia* species

Babesia species	Protein coding gene number	Functionally annotated by gene ontology terms	Hypothetical proteins	Species-specific genes number[a]	Genes related to pathogenesis	References
B. bovis	3671	NR	NR	174	*Ves1a, ves1b, vesγ, ves2, smorf A and B genes, mtm A* and *B* genes	[15, 25, 30]
B. bigemina	4457	NR	NR	356	*Ves1a, ves1b, ves1ba, ves2* Six-cysteine protein-coding genes	[25, 38]
B. microti	3494	2485	1009	59 (47 up-regulated in hamster isolates, and 12 in SCID mice)	*Gpi-ap, sec, associated with the membrane, rap, ama-1, ron2, ron4, ron5*	[20]
B. duncani	3759	1636	2123	223 (215 with blood-stage expression)[b]	*Blfl, gpi-ap, sea, trap*	[21]
B. sp. *Xinjiang*	3066	984	NR	172	*Ves1* (derived from N-domain I-IV groups)	[17]
B. ovis	3419	2772	647	281	*Ves1δ* genes, genes encoding proteins with a TR	[23]
B. canis	3467	1030[a]	469[a]	480[b]	*Ves1a, ves1b, msp, sa, rap,* six-cysteine protein-coding genes	[16]
B. gibsoni	3552	2641	137[b]	22	*Ves1*-like genes, *sa, ApiAP2, ama-1, ron2, rap1, sbp*	[24]
B. caballi	5910	NR	NR	*B. caballi* repetitive sequence 1 (*Bcrs1*)	*Ves1c, genes encoding proteins with TR*	[19]
B. ovata	5031	3660	1371	1788	*Ves1* and *ves2, msp, ap2-g,* six-cysteine protein-coding genes	[22]

NR, not reported; *ves*, variant erythrocyte surface antigen; *Smorf*, small open reading frame; *mtm*, multitrasmembrane; *blf1*, *B. duncani* largest family 1; *gpi-ap*, glycosylphosphatidylinositol-anchored protein; *sec*, soluble secreted proteins; *sea*, serine esterease; *trap*, thrombospondin-related anonymous protein; *TR*, transmembrane region; *rap*, rhoptry-associated proteins; *ama-1*, apical merozoite antigen 1; *ron*, rhoptry neck proteins; *msp*, merozoite surface proteins; *sa*, secreted antigens; *ApiAP2*, apicomplexan apetala 2 gene family; *sbp*, spherical body proteins. Results derived from *B. divergens* transcriptome were not described given that the paper is not available online

[a]Predicted from the paper's supplementary dataset

[b]Derived from the prediction of secreted proteins

may suggest a lower capacity of the parasite to manipulate its own genome and that of the host, potentially contributing to its attenuation. These findings highlight the importance of membrane proteins and exonucleases in the virulence of *B. bigemina*, suggesting that they could be potential drug targets for designing new treatments.

In *B. bovis*, there is a marked difference in gene expression between virulent and attenuated strains of different isolates (T2Bo and LV17) (Table 3). In the virulent T2Bo strain, 18 *ves* genes were upregulated, while a reduced expression of these genes (only five) was observed in the attenuated T2Bo strain. This may indicate a

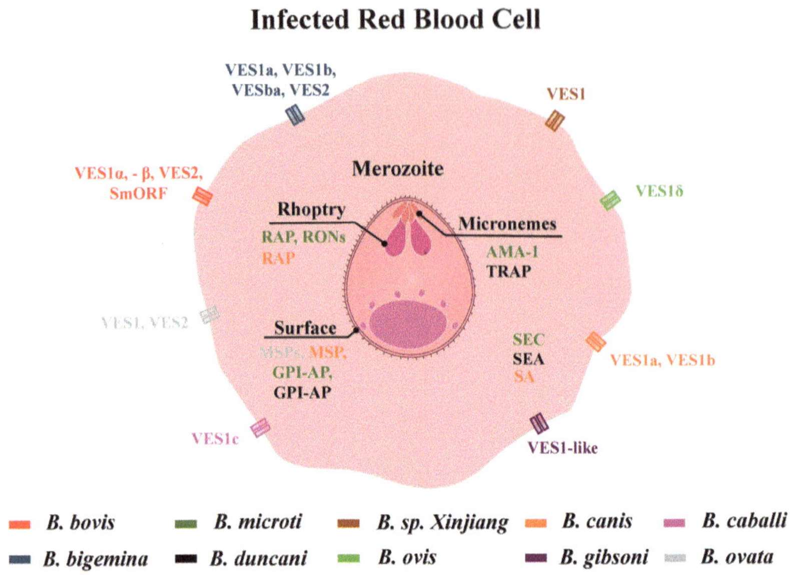

Fig. 3 Upregulated gene in merozoites of different *Babesia* species infecting erythrocytes. The figure shows some genes encoding products related to pathogenicity by species

Table 3 Differential gene expression in *Babesia* strains subjected to different environments

Babesia species	Strain characteristic	Number of differentially expressed genes		Most important terms derived from genes up/downregulated	References
B. bovis	T2B0 Virulent	761[a]	25 upregulated	*Ves* genes (18), *smorf* (2), membrane proteins (3)	[43]
	T2Bo Attenuated		36 upregulated	*Ves* genes (5), *smorf* (13), SBP2 (4)	
	LV17 Virulent	484	116 upregulated	*Ves* genes (30), *smorf* (18), hypothetical (38)	
	LV17 Attenuated		16 upregulated	Hypothetical (20), putative membrane (12), *smorf* (9), *ves* (3), *sbp* (4)	
B. bigemina	Virulent	519	25 upregulated	Membrane proteins (2), CS domain family protein (1), exome complex exonuclease (1)[b]	[44]
	Attenuated		3 downregulated	Endonuclease-reverse transcriptase (2), retrovirus-related Pol poly LINE-1, putative	
B. gibsoni[+]	*In vitro* cultured	711	377 upregulated	*Merozoite surface protein 2, microneme protein 14, SBP 3*	[42]
			344 downregulated	Hap2, *ap2-g*	
B. microti	Isolate from Hamster	59	47 upregulated	Proteins with unknown function (27), secretome (12; 3 *BmS48/45* genes), protein expression (6)	[20]
	Isolate from SCID mice		12 upregulated		

+, Results are derived from the comparison with *in vivo* cultured strain; *ves*, variant erythrocyte surface antigen; *smorf*, small open reading frame; *sbp*, spherical bodysuit protein; *hap2*, hook-associated protein 2; *ap2-g*, AP2 transcription factor

[a]Average among two assays involving T2Bo virulent versus attenuated strains

[b]Derived from supplementary information of the paper

direct role in reducing the parasite's ability to evade the host immune system, thus contributing to its attenuation [43]. In the virulent LV17 strain, 30 *ves* genes were upregulated, along with 18 *smorf* genes and 38 hypothetical genes, suggesting that many genes with uncharacterized functions could be participating in the pathogenesis. In the attenuated LV17 strain, there were fewer downregulated *ves* (three) and *smorf* (nine) genes, supporting the notion that virulence directly depends on the expression of several genes derived from these families. The differential expression of *ves, smorf,* and *sbp2* genes between virulent and attenuated strains in both *B. bovis* isolates highlights their importance in virulence and differential genetic regulation specific to individual strains in a similar environmental setting. These findings suggest that identifying and characterizing these molecular targets could be vital for developing new treatments. Furthermore, it means that pathogen attenuation could be a viable mechanism as a disease control strategy, mainly focused on developing a vaccine.

Other studies verify the influence of the environment on alterations in the blood transcriptional profile of some *Babesia* species, which is thought to be related to the survival and spread of the parasites [20, 42]. Transcriptomic analysis of *B. microti* reveals significant differences in gene expression regarding the host. In parasites isolated from hamsters, there is notable upregulation of secretome genes, including those encoding BmS48/45, suggesting that these could be important for evading the hamster's immune system. On the contrary, the lower number of downregulated genes in parasites isolated from SCID mice highlights the influence of the host's immune status on parasite gene expression. SCID mice lack an adaptive immune response, likely reducing selective pressure on the parasite to express immune evasion genes. In the case of *B. gibsoni*, the overexpression of genes such as those encoding MSP2, MIC14, and SBP3 indicates that the parasite specializes its cellular machinery for adhesion, invasion, and intracellular maintenance, while the downregulation of HAP2 and AP2-G suggests strict, phase-specific regulation for these genes, which are not necessary in the merozoite stage cultured *in vitro*. These findings highlight the importance of differentiation and specialization of the stages of the parasite's life cycle, providing crucial information to understand its biology and pathogenesis.

The *Babesia* life cycle is thus of fundamental importance for understanding its biology and pathogenesis. As found in different studies, gene expression allows the validation of the activation of specific genes at critical moments in the parasite's life cycle, providing information about its specific functions. This validation is especially relevant for genes related to processes such as adherence, invasion, pathogenicity, and metabolism, as well as other key aspects of the parasite-host interaction. Understanding the role of each gene in specific stages of the life cycle of different *Babesia* species will open new opportunities for developing specific and effective control and treatment strategies against these microorganisms.

5 Proteomic and Immunological Exploration in *Babesia*

The ability of pathogens to invade, survive, and replicate within different cell types is critical to maintaining their life cycle. However, knowledge of the transcriptome is not sufficient to ensure that a gene is being expressed and fulfilling essential functions in the parasite. Proteomic studies have sought to fill this gap by providing information on the repertoire of proteins expressed by *Babesia* parasites throughout their life cycle and within the host environment [16, 20, 25, 45–47]. However, despite advances in genome and transcriptome studies, research on the proteome has been more limited, and even more scarce in the case of immunome analysis (Fig. 4, Tables 2 and 4). However, the data have allowed us to understand how several proteins can be used in the development of different control methods.

The protein expression in different stages of the parasite is essential to understanding the

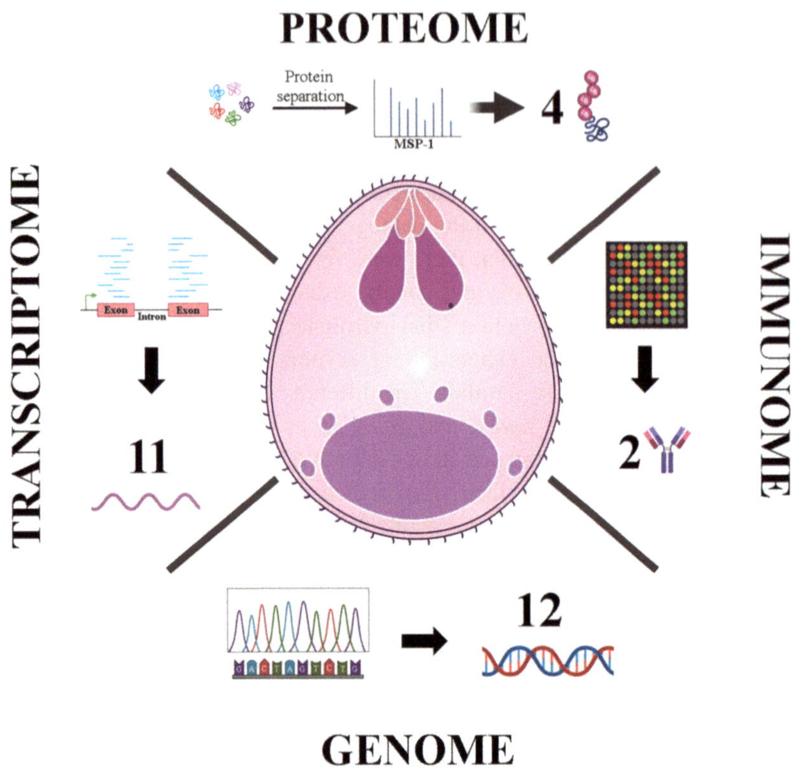

Fig. 4 State of the art of the number of omics studies in *Babesia*. Figure created with BioRender.com

interactions that occur with the host. For example, in *B. bovis*, 38 proteins were identified in the membrane of iRBCs (infected red blood cells), including MTM (multitransmembrane) family proteins and VESA export-associated proteins. The expression reduction of the latter one resulted in a decrease in the growth rate, reduction in the number of ridges on the surface of iRBCs, wrong localization of VESA1, and cytoadhesion abrogation to endothelial cells, highlighting the role of VEAP as a virulence factor in *B. bovis* [16]. In the case of *B. bigemina* and *B. canis*, the VESA2, VESA, and MSP proteins are considered potentially essential as they are expressed in the blood stages; however, more studies are required to comprehensively determine their function [25, 45].

Recently, an integrated approach has been attempted to jointly study proteomics and immunomics to precisely identify proteins that are not only expressed but also play important roles in the parasite due to their localization and functional activity. Characterization of the immunoproteome thus provides compelling evidence for the identification of virulence factors, surface molecules, and proteins involved in the host–parasite interaction. In accordance with the above, the search for vaccines has focused on an integrated approach that consists of the identification of immunodominant antigens that induce an immune response to inhibit the development of the parasite in its vector agent or within the mammal [4]. For example, immunodetected proteins, such as those found in *B. bovis* (SEC-independent translocase, GCC2, GCC3) and *B. microti* (SA1, GPI-10, RONs, among others), are promising candidates to evaluate their usefulness as components of a blood-stage or transmission-blocking vaccine [45, 46]. The results presented also underline the complexity and specialization of *Babesia* species during their life cycle. Therefore, to advance the understanding of the pathogenic mechanisms of the different *Babesia* species, it is critical to expand proteomic and immunomic studies to determine the specific functions of the identified proteins and investigate how some proteins (e.g., GCC2/GCC3 and SNAP proteins)

Table 4 *Babesia* species proteins identified by proteomics and immunoproteomics

Babesia species	Parasite stage	Identified proteins	Predicted proteins as secreted	Immunodetected proteins	References
B. bovis	iRBC membrane	38	SmORF, SBP1, SBP2, SBP3, MTM (Bb60-MTM and Bb11920-MTM)[a], VEAP[a]	NE	[47]
	Blood stage	10	Hypothetical (7), SNAP (1)	SEC-independent translocase, GCC2, GCC3	[46]
	Kinetes		SEC-independent translocase[b], GCC2/GCC3[b]		
B. bigemina	Blood stage	366	VESA1*b* (2), VESA2 (15)	NE	[25]
B. microti	Blood stage	106	15 Surface/SA	SA1, S1/P1 nuclease, conserved protein BMR1, GPI-10, N1–15, MDN1, BMN2 family, R1	[45]
		420	GPI-AP (19), secreted proteins (SEC = 196), proteins with a TR (205)	5 GPI, 6 SEC, 3 TM, GPI12/SA1, S1/P1 nuclease, RON2, RON-5, hypothetical (2), GPI-9, GPI-10, BMN2	[20]
B. canis	iRBC membrane	11	VESA (2), MSP (2), hypothetical (5), H2a (1), SA1 (1)	NE	[16]

iRBC, infected red blood cell; *NE*, not evaluated; *SmORF*, small open reading frame; *SBP*, spherical body proteins; *MTM*, multitrasmembrane; *VEAP*, VESA export-associated protein (VEAP)*; *SNAP*, soluble NSF attachment protein; *GCC*, gripped coiled-coil domain; *SA*, secreted antigens; *GPI-AP*, glycosylphosphatidylinositol-anchored protein; *TR*, transmembrane region; *MSP*, merozoite surface protein; *H*, histone

[a]These were located on the iRBC membrane

[b]Proteins detected only in kinete stages

modulate host–parasite interactions and facilitates the invasion and immune evasion process.

The integration of proteomic and immunomic data is a more precise and comprehensive approach to identifying key proteins that may be therapeutic targets or vaccine candidates. However, it is suggested to increase immunoproteomic studies on these parasites to complete and analyze the information more exhaustively. This additional effort will help fill gaps in our knowledge and significantly improve strategies to search for important molecules that may be useful in the study of *Babesia*.

6 Metabolomics in *Babesia*: Keys to Diagnosis and Treatment

Metabolomic profiling provides a comprehensive picture of *Babesia* metabolic pathways and their modulation in response to host factors and environmental cues. This technology has clarified essential metabolic pathways required for the survival and replication of parasites and has identified new biomarkers to facilitate timely diagnosis and treatment, thereby controlling disease progression.

In *Babesia*, the identification of the metabolic profile in some species has shed light on the metabolic adaptations that both the parasite and the host develop during the dynamics of infection, thus providing additional information on the pathophysiology of the disease (Table 5) [48–51]. In the case of the parasite, metabolites in *B. divergens* participate in different metabolic pathways important for the functionality of the merozoites during their invasion process. These could be used as pharmacological targets given their high production and specificity in the parasite [48]. On the contrary, some metabolites have been identified in the host that can serve as diagnostic markers since they are exclusively produced during infection. For example, in the case

Table 5 *Babesia* metabolic pathways modulated in response to host factors and environmental signals

Babesia species	Clinical manifestation	Source	Number of dysregulated metabolites	Significantly affected metabolic pathways	Affected metabolites	Metabolism modulation response	References
B. canis	Kidney dysfunction	Urine	58	Amino acid	Phenylalanine, tyrosine, tryptophan, valine, leucine, isoleucine, alanine, aspartate, glutamate, glutamine, proline, histidine, arginine	Protein biosynthesis alteration, oxidative stress	[49]
				Energy	Acylcarnitine	Proinflammatory signaling, oxidative stress, increased levels of acylcarnitines	
				Fatty acid			
				Urea cycle, ammonia detoxication	Aspartate, arginine, alanine, glutamate, glutamine	Oxidative stress	
	Fever, lethargy, pale mucus membranes, anemia, jaundice, hemoglobinuria or hematuria, splenomegaly, tachycardia, anorexia, vomiting	Serum	64	Glutathione	5-Oxoproline—glycine—putrescine	Oxidative stress—low immunomodulatory response—parasite proliferation	[50]
				Alanine, aspartate, glutamate	Citrate—aspartate—pyruvate	Disrupt cell signaling, interrupt cellular differentiation, apoptosis—altered metabolic pathways—oxidative stress	
				Glyoxylate and dicarboxylate	Pyruvate—glycine—citrate	Oxidative stress—low immunomodulatory response—disrupt cell signaling, interrupt cellular differentiation, apoptosis	
				Cysteine and methionine	Cysteine—methionine—pyruvate	Low immunomodulatory response—antioxidant—oxidative stress	
				Arginine and proline	Proline—putrescine—pyruvate	Antioxidant—parasite proliferation—oxidative stress	
				Arginine biosynthesis	Citrulline—aspartate	Antioxidant—altered metabolic pathways	
				Citrate cycle	Pyruvate—citrate	Oxidative stress—disrupt cell signaling, interrupt cellular differentiation, apoptosis	
				Phenylalanine, tyrosine, and tryptophan biosynthesis	Phenylalanine	Protein synthesis	
B. divergens	NE	Free merozoites	NR	Lipids	Monounsaturated fatty acids, polyunsaturated fatty acids, phospholipids, phosphatidylcholine, among others	Maintaining both membrane deformability and fluidity	[48]
				Amino acids and peptides	Citrulline	Protein synthesis, redox regulatory	
				Carbohydrates, energy	Malate, fumarate, pyruvate	Mitochondrial aerobic respiration	
B. microti	NR	Serum	110	Taurine and hypotaurine	Taurocholic acid, L-cysteic acid	Fat breakdown	[51]
				Histidine	Anserine	Antioxidant	
				Arachidonic acid	Arachidonic acid, leukotriene B4, 20-HETE	Antiparasitic	

NR, not reported; *NE*, not evaluated

of *B. canis*, a wide range of metabolic alterations were identified in both urine and serum, highlighting oxidative stress and alterations in protein synthesis and cellular signaling as underlying mechanisms in the pathology caused in the animals [49, 50]. Other metabolites, such as taurine, histidine, and arachidonic acid, stand out in *B. microti* infection [51]. The findings highlight the importance of oxidative stress and alterations in metabolic and cell signaling pathways as critical factors in the pathogenesis of *Babesia* infections. These data also underscore the complexity of *Babesia* infections and the need to use therapeutic approaches that address oxidative stress, altered cell signaling, and specific metabolic pathways affected by infection.

The analysis of metabolites secreted during *Babesia* infections has provided invaluable information on specific biomarkers that can be used for the diagnosis and monitoring of the disease in different animal species. Given that some *Babesia* species possess essential metabolic pathways that define their survival in the host, additional metabolomics studies are required to provide integrated functional profiles and thus select better therapeutic targets. The combination of monitoring and therapeutic strategies based on these findings could significantly improve the management and control of *Babesia* infections, which is of clinical relevance in the veterinary field.

7 Perspectives and Future Directions

Invaluable advances related to the biology of *Babesia* have been made over the last decade through omics techniques and bioinformatics integration, which have essentially enabled the analysis of a large volume of data. Consequently, a few important and complex patterns of molecules involved in the pathogenicity of some members of the genus have been deciphered.

Omics data sets allow for a holistic understanding of *Babesia* species biology, host–parasite interactions, and disease pathogenesis. Therefore, based on the information described in this book chapter, the study of omics is required for most species in terms of proteome, immunome, and interactome. With this information available, bioinformatics analysis strategies could be designed to reduce the major bottleneck in the identification of key molecular players and elucidate the molecular basis of the susceptibility and resistance of *Babesia* species during infection. Likewise, accelerated continuity can be given to the development of innovative mechanisms for the control and prevention of the disease, such as the search for molecular biomarkers for the diagnosis of infection in different stages of the disease, the monitoring and identification of parasite-exclusive targets to evaluate therapeutic alternatives, and the search for immunogenic proteins involved in pathogenicity, adhesion, and/or invasion to continue advancing the development of completely efficient control strategies.

The integration of omics data through deep *in silico* analysis is undoubtedly one of the most revolutionary and effective strategies that provide valuable information that could be exploited to improve timely and successful treatment rates, favor the creation of new effective pharmacological medications against resistant strains of parasites, and guide the design of effective multi-antigenic vaccines. Multidisciplinary collaboration and future technological advancements will be essential to overcome the challenges and maximize the opportunities that omics offer in the veterinary field.

References

1. Homer MJ, Aguilar-Delfin I, Telford SR, Krause PJ, Persing DH (2000) Babesiosis. Clin Microbiol Rev [Internet] 13(3):451–469. https://journals.asm.org/doi/10.1128/CMR.13.3.451
2. Sonenshine DE (2018) Range expansion of tick disease vectors in North America: implications for spread of tick-borne disease. Int J Environ Res Public Health [Internet] 15(3):478. https://pubmed.ncbi.nlm.nih.gov/29522469/
3. Rodríguez-Vivas RI, Grisi L, De León AAP, Villela HS, De Jesús Torres-Acostaa JF, Sánchez HF et al (2017) Potential economic impact assessment for cattle parasites in Mexico. Rev Mex Cienc Pecu 8(1):61–74

4. Suarez CE, Noh S (2011) Emerging perspectives in the research of bovine babesiosis and anaplasmosis. Vet Parasitol [Internet] 180(1–2):109–125. https://pubmed.ncbi.nlm.nih.gov/21684084/
5. Florin-Christensen M, Suarez CE, Rodriguez AE, Flores DA, Schnittger L (2014) Vaccines against bovine babesiosis: where we are now and possible roads ahead. Parasitol Int 141(12):1563–1592. https://pubmed.ncbi.nlm.nih.gov/25068315/
6. Cuy-Chaparro L, Reyes C, Díaz-Guiot EV, Moreno-Pérez DA, Patarroyo MA (2023) Developing Anti-Babesia bovis blood stage vaccines: a new perspective regarding synthetic vaccines. Int J Mol Sci [Internet] 24(6) https://pubmed.ncbi.nlm.nih.gov/36982294/
7. Santos JHM, Siddle HV, Raza A, Stanisic DI, Good MF, Tabor AE (2023) Exploring the landscape of Babesia bovis vaccines: progress, challenges, and opportunities. Parasit Vectors [Internet] 16(1):1–12. https://parasitesandvectors.biomedcentral.com/articles/10.1186/s13071-023-05885-z
8. Jalovecka M, Sojka D, Ascencio M, Schnittger L (2019) Babesia life cycle - when phylogeny meets biology. Trends Parasitol [Internet] 35(5):356–368. https://pubmed.ncbi.nlm.nih.gov/30733093/
9. Chauvin A, Moreau E, Bonnet S, Plantard O, Malandrin L (2009) Babesia and its hosts: adaptation to long-lasting interactions as a way to achieve efficient transmission. Vet Res [Internet] 40(2) https://pubmed.ncbi.nlm.nih.gov/19379662/
10. Ravindran R, Hembram PK, Kumar GS, Kumar KGA, Deepa CK, Varghese A (2023) Transovarial transmission of pathogenic protozoa and rickettsial organisms in ticks. Parasitol Res 122(3):691–704. https://link.springer.com/article/10.1007/s00436-023-07792-9
11. Jalovecka M, Hajdusek O, Sojka D, Kopacek P, Malandrin L (2018) The complexity of piroplasms life cycles. Front Cell Infect Microbiol [Internet] 8:248. https://pubmed.ncbi.nlm.nih.gov/30083518/
12. Del Giacco L, Cattaneo C (2012) Introduction to genomics. Methods Mol Biol [Internet] 823:79–88. https://pubmed.ncbi.nlm.nih.gov/22081340/
13. Ray BK, Bailey CW, Jensen JB, Andrew CC (1992) Chromosomes of Babesia bovis and Babesia bigemina. Mol Biochem Parasitol [Internet] 52(1):123–126. https://pubmed.ncbi.nlm.nih.gov/1625699/
14. Jones SH, Lew AE, Jorgensen WK, Barker SC (1997) Babesia bovis: genome size, number of chromosomes and telomeric probe hybridisation. Int J Parasitol [Internet] 27(12):1569–1573. https://pubmed.ncbi.nlm.nih.gov/9467743/
15. Brayton KA, Lau AOT, Herndon DR, Hannick L, Kappmeyer LS, Berens SJ et al (2007) Genome sequence of Babesia bovis and comparative analysis of apicomplexan hemoprotozoa. PLoS Pathog 3(10):1401–1413
16. Eichenberger RM, Ramakrishnan C, Russo G, Deplazes P, Hehl AB (2017) Genome-wide analysis of gene expression and protein secretion of Babesia canis during virulent infection identifies potential pathogenicity factors. Sci Rep 7(1):3357
17. Guan G, Korhonen PK, Young ND, Koehler AV, Wang T, Li Y et al (2016) Genomic resources for a unique, low-virulence Babesia taxon from China. Parasit Vectors 9(1):1–8
18. He L, Liu Q, Yao B, Zhou Y, Hu M, Fang R et al (2017) A historical overview of research on Babesia orientalis, a protozoan parasite infecting water buffalo. Front Microbiol 8:1323
19. Ochi A, Kidaka T, Hakimi H, Asada M, Yamagishi J (2023) Chromosome-level genome assembly of Babesia caballi reveals diversity of multigene families among Babesia species. BMC Genomics 24(1):483
20. Silva JC, Cornillot E, McCracken C, Usmani-Brown S, Dwivedi A, Ifeonu OO et al (2016) Genome-wide diversity and gene expression profiling of Babesia microti isolates identify polymorphic genes that mediate host-pathogen interactions. Sci Rep 6:35284
21. Wang J, Chen K, Yang J, Zhang S, Li Y, Liu G et al (2022) Comparative genomic analysis of Babesia duncani responsible for human babesiosis. BMC Biol 20(1):153
22. Yamagishi J, Asada M, Hakimi H, Tanaka TQ, Sugimoto C, Kawazu S, ichiro. (2017) Whole-genome assembly of Babesia ovata and comparative genomics between closely related pathogens. BMC Genomics 18(1)
23. Yamagishi J, Ceylan O, Xuan X, Sevinc F (2023) Whole genome sequence and diversity in multigene families of Babesia ovis. Front Cell Infect Microbiol 13:13
24. Liu Q, Guan XA, Li DF, Zheng YX, Wang S, Xuan XN et al (2023) Babesia gibsoni whole-genome sequencing, assembling, annotation, and comparative analysis. Microbiol Spectr 11(4):e0072123
25. Jackson AP, Otto TD, Darby A, Ramaprasad A, Xia D, Echaide IE et al (2014) The evolutionary dynamics of variant antigen genes in Babesia reveal a history of genomic innovation underlying host-parasite interaction. Nucleic Acids Res [Internet] 42(11):7113–7131. https://pubmed.ncbi.nlm.nih.gov/24799432/
26. Morrison DA (2009) Evolution of the Apicomplexa: where are we now? Trends Parasitol [Internet] 25(8):375–382. https://pubmed.ncbi.nlm.nih.gov/19635681/
27. Lim L, McFadden GI (2010) The evolution, metabolism and functions of the apicoplast. Philos Trans R Soc Lond B Biol Sci 365(1541):749–763. https://royalsocietypublishing.org/doi/10.1098/rstb.2009.0273
28. Berná L, Rego N, Francia ME (2021) The elusive mitochondrial genomes of apicomplexa: where are we now? Front Microbiol 12:751775. https://pubmed.ncbi.nlm.nih.gov/34721355/
29. O'Connor RM, Allred DR (2000) Selection of Babesia bovis-infected erythrocytes for adhesion to endothelial cells coselects for altered variant erythrocyte surface antigen isoforms. J Immunol 164(4):2037–2045. https://pubmed.ncbi.nlm.nih.gov/10657656/
30. Ferreri LM, Brayton KA, Sondgeroth KS, Lau AOT, Suarez CE, McElwain TF (2012) Expression and strain variation of the novel "small open reading

frame" (smorf) multigene family in Babesia bovis. Int J Parasitol 42(2):131–138
31. Gallego-Lopez GM, Cooke BM, Suarez CE (2019) Interplay between attenuation- and virulence-factors of Babesia bovis and their contribution to the establishment of persistent infections in cattle. Pathogens [Internet] 8(3) https://pubmed.ncbi.nlm.nih.gov/31277392/
32. Hakimi H, Yamagishi J, Kawazu SI, Asada M (2022) Advances in understanding red blood cell modifications by Babesia. PLoS Pathog [Internet] 18(9) https://pubmed.ncbi.nlm.nih.gov/36107982/
33. Suarez CE, Florin-Christensen M, Hines SA, Palmer GH, Brown WC, McElwain TF (2000) Characterization of allelic variation in the Babesia bovis merozoite surface antigen 1 (MSA-1) locus and identification of a cross-reactive inhibition-sensitive MSA-1 epitope. Infect Immun [Internet] 68(12):6865–6870. https://pubmed.ncbi.nlm.nih.gov/11083806/
34. Quistgaard EM, Löw C, Guettou F, Nordlund P (2016) Understanding transport by the major facilitator superfamily (MFS): structures pave the way. Nat Rev Mol Cell Biol [Internet] 17(2):123–132. https://pubmed.ncbi.nlm.nih.gov/26758938/
35. Yang YS, Murciano B, Moubri K, Cibrelus P, Schetters T, Gorenflot A et al (2012) Structural and functional characterization of Bc28.1, major erythrocyte-binding protein from Babesia canis merozoite surface. J Biol Chem [Internet] 287(12):9495–9508. https://pubmed.ncbi.nlm.nih.gov/22294693/
36. Delbecq S (2022) Major surface antigens in zoonotic Babesia. Pathogens [Internet]. 11(1) https://pubmed.ncbi.nlm.nih.gov/35056047/
37. García-Nieto PE, Wang B, Fraser HB (2022) Transcriptome diversity is a systematic source of variation in RNA-sequencing data. PLoS Comput Biol [Internet] 18(3):e1009939. https://journals.plos.org/ploscompbiol/article?id=10.1371/journal.pcbi.1009939
38. Yamagishi J, Asada M, Hakimi H, Tanaka TQ, Sugimoto C, Kawazu S-I (2017) Whole-genome assembly of Babesia ovata and comparative genomics between closely related pathogens. BMC Genomics [Internet] 18(1):1–9. https://bmcgenomics.biomedcentral.com/articles/10.1186/s12864-017-4230-4
39. Balaji S, Madan Babu M, Iyer LM, Aravind L (2005) Discovery of the principal specific transcription factors of Apicomplexa and their implication for the evolution of the AP2-integrase DNA binding domains. Nucleic Acids Res [Internet] 33(13):3994–4006. https://pubmed.ncbi.nlm.nih.gov/16040597/
40. Lamarque MH, Roques M, Kong-Hap M, Tonkin ML, Rugarabamu G, Marq JB et al (2014) Plasticity and redundancy among AMA-RON pairs ensure host cell entry of Toxoplasma parasites. Nat Commun [Internet] 5. https://pubmed.ncbi.nlm.nih.gov/24934579/
41. Ben Chaabene R, Lentini G, Soldati-Favre D (2021) Biogenesis and discharge of the rhoptries: key organelles for entry and hijack of host cells by the Apicomplexa. Mol Microbiol [Internet] 115(3):453–465. https://pubmed.ncbi.nlm.nih.gov/33368727/
42. Han Z, Zheng Y, Shi Y, Chen F, Wu C, Wang L et al (2023) Transcriptional variation in Babesia gibsoni (Wuhan isolate) between in vivo and in vitro cultures in blood stage. Parasit Vectors 16(1):268
43. Pedroni MJ, Sondgeroth KS, Gallego-Lopez GM, Echaide I, Lau AOT (2013) Comparative transcriptome analysis of geographically distinct virulent and attenuated Babesia bovis strains reveals similar gene expression changes through attenuation. BMC Genomics [Internet] 14(1) https://pubmed.ncbi.nlm.nih.gov/24195453/
44. Santamaria RM, Estrada K, López ME, Rojas E, Martínez G, Alcalá Y et al (2024) Comparative transcriptome analysis of Babesia bigemina attenuated vaccine and virulent strains of Mexican origin. Vaccines (Basel) 12(3)
45. Magni R, Luchini A, Liotta L, Molestina RE (2019) Analysis of the Babesia microti proteome in infected red blood cells by a combination of nanotechnology and mass spectrometry. Int J Parasitol 49(2):139–144
46. Johnson WC, Taus NS, Reif KE, Bohaliga GAR, Kappmeyer LS, Ueti MW (2017) Analysis of stage-specific protein expression during Babesia Bovis development within female Rhipicephalus Microplus. J Proteome Res 16(3):1327–1338
47. Hakimi H, Templeton TJ, Sakaguchi M, Yamagishi J, Miyazaki S, Yahata K et al (2020) Novel Babesia bovis exported proteins that modify properties of infected red blood cells. PLoS Pathog [Internet] 16(10) https://pubmed.ncbi.nlm.nih.gov/33017449/
48. Beri D, Singh M, Rodriguez M, Goyal N, Rasquinha G, Liu Y et al (2023) Global metabolomic profiling of host red blood cells infected with Babesia divergens reveals novel antiparasitic target pathways. Microbiol Spectr 11(2):e0468822
49. Kuleš J, Rubić I, Beer Ljubić B, Bilić P, Barić Rafaj R, Brkljačić M et al (2021) Combined untargeted and targeted metabolomics approaches reveal urinary changes of amino acids and energy metabolism in canine Babesiosis with different levels of kidney function. Front Microbiol 12:715701
50. Rubić I, Burchmore R, Weidt S, Regnault C, Kuleš J, Rafaj RB et al (2022) Multi platforms strategies and metabolomics approaches for the investigation of comprehensive metabolite profile in dogs with Babesia canis infection. Int J Mol Sci 23(3)
51. Shen L, Wang C, Wang R, Hu X, Liao S, Liu W et al (2023) Serum metabolomic profiles in BALB/c mice induced by Babesia microti infection. Front Cell Infect Microbiol 13:1179967

Part IV

Helminths

From Genes to Worms: A Deep Dive into Helminth Omics

Sergio Castañeda and Juan David Ramírez

Abstract

Helminths, including nematodes and platyhelminths, pose a significant global health threat, infecting over 1 billion people and causing diseases such as lymphatic filariasis, schistosomiasis, and soil-transmitted helminthiasis. These infections are especially prevalent in low- and middle-income countries, where they contribute to malnutrition, cognitive impairment, and stunted growth. Despite historical underrepresentation in research due to their complex life cycles, recent advances in genomics, transcriptomics, and proteomics have revolutionized our understanding of these parasites. Genomics has enabled the identification of essential genes and metabolic pathways, facilitating targeted drug discovery. Transcriptomics has revealed how parasites respond to various environments, aiding in the development of precise control strategies and vaccine candidates. Proteomics has illuminated protein interactions and potential diagnostic markers, enhancing early detection and treatment monitoring. Additionally, the interplay between helminths, the host, and the gut microbiome introduces new complexities and therapeutic opportunities. This chapter reviews cutting-edge research in helminth genomics, transcriptomics, and proteomics, detailing advancements in genome sequencing, gene expression analysis, and protein functionality. Understanding these interactions is crucial for developing innovative diagnostics, drugs, and vaccines, ultimately aiming to alleviate the global burden of helminth infections.

Keywords

Helminths · Genomics · Transcriptomics · Proteomics · Gut microbiome · Parasite–host interactions

S. Castañeda (✉)
Centro de Investigaciones en Microbiología y Biotecnología-UR (CIMBIUR), Faculty of Natural Sciences, Universidad del Rosario, Bogotá, Colombia
e-mail: sergio.castaneda@urosario.edu.co

J. D. Ramírez
Centro de Investigaciones en Microbiología y Biotecnología-UR (CIMBIUR), Faculty of Natural Sciences, Universidad del Rosario, Bogotá, Colombia

Department of Pathology, Molecular and Cell-based Medicine, Icahn School of Medicine at Mount Sinai, New York, USA

1 Introduction

Helminths, which are parasitic worms belonging to the phyla Nematoda (roundworms) and Platyhelminthes (flatworms) (Trematoda and

J. D. Ramírez González (ed.), *Recent Advances in Parasitomics*,
https://doi.org/10.1007/978-3-031-70591-5_12

Cestoda), represent a significant global public health threat, infecting over 1 billion individuals worldwide [1–3]. These diverse organisms cause a spectrum of debilitating diseases, including lymphatic filariasis, schistosomiasis, soil-transmitted helminthiasis, and taeniasis/cysticercosis, among others [1, 4–7]. Notably, the burden of these diseases disproportionately impacts low- and middle-income countries, contributing to malnutrition, cognitive impairment, and stunted growth. According to the latest prevalence estimates from the Global Burden of Disease (GBD) Project in 2016 (GBD, 2016 Disease and Injury Incidence and Prevalence Collaborators, 2017), there are approximately 800 million cases of ascariasis, 450 million cases of hookworm, 435 million cases of trichuriasis, 190 million cases of schistosomiasis, and 75 million cases of food-borne trematodiasis worldwide. Due to intensive control efforts over the past three to four decades, lymphatic filariasis (29 million cases) and onchocerciasis (15 million cases) have become less common. Newly completed GBD estimates for the prevalence of neurocysticercosis (NCC, 2.7 million cases) and cystic echinococcosis (CE, 1 million cases) now underscore the significant burden of these debilitating diseases. Consequently, helminth infections act as potent impediments to economic development and poverty reduction efforts [1, 8–13].

Historically, helminths have been underrepresented in research and development endeavors, partially due to their complex life cycles and intricate host–parasite interactions [14]. However, recent breakthroughs in genomics, transcriptomics, and proteomics have revolutionized our understanding of helminth parasites, leading to significant progress in the fight against helminth-caused diseases. These tools provide unparalleled insights into parasite biology by revealing vulnerabilities in their genomes, elucidating the dynamics of gene expression, and unraveling the intricate machinery of protein interactions. This knowledge unlocks a treasure trove of potential for public health advancements [15]. Genomics enables targeted drug discovery by identifying essential metabolic pathways and enzymes unique to helminths. By pinpointing these vulnerabilities, researchers can develop new drugs with higher efficacy and reduced side effects, contributing to more effective treatment and reduced disease burden. Similarly, transcriptomics sheds light on how parasites respond to different environments and drugs, informing the development of more precise control strategies and vaccine candidates. By understanding the genes crucial for parasite development and host interaction, researchers can design vaccines that block these vital processes, ultimately preventing infection and reducing disease transmission. Finally, proteomics unveils the functional networks of proteins within parasites, identifying potential drug targets and diagnostic markers. This allows for the development of better diagnostic tools for early detection and improved monitoring of treatment response, leading to a more efficient and targeted approach to disease control [15]. These "-omics" technologies are not merely scientific marvels; they represent powerful tools in the public health arsenal against helminth-caused diseases. By illuminating the inner workings of these parasitic worms, they pave the way for innovative diagnostics, drugs, and vaccines, ultimately contributing to a future where millions are free from the burden of these debilitating infections.

This chapter comprehensively reviews the state-of-the-art in helminth genomics, transcriptomics, and proteomics research. We will delve into the following:

I. Helminth Genomics: This section will explore the latest advances in genome sequencing, assembly, annotation, and comparative analysis, with a focus on the insights these advancements have provided into helminth evolution, drug resistance mechanisms, and virulence factors.
II. Helminth Transcriptomics: We will delve into the application of RNA-seq technologies, differential expression analysis, and the role of non-coding RNAs (ncRNAs) in helminth biology, providing valuable information regarding gene regulation and the intricate dynamics of host–parasite interactions.
III. Helminth Proteomics: This section will explore the utilization of proteomic techniques

for identifying and characterizing helminth proteins and secreted proteins (including exovesicles), elucidating protein–protein interactions, and analyzing protein expression changes under various conditions. These valuable data will be instrumental in facilitating drug and vaccine development efforts.

IV. Helminths–Host–Microbiome Interactions: The human gut encompasses a dynamic interplay involving helminths, their hosts, and the microbial component of the gut microbiome. Unraveling this intricate play reveals how helminth infections introduce an additional layer of complexity to the interplay between host health and microbial composition, unveiling novel challenges and potential therapeutic strategies.

2 Helminth Genomics: Unraveling the Secrets of Parasitic Worms

The past decade has witnessed a remarkable surge in helminth genome sequencing projects, offering unparalleled insights into the genetic underpinnings of these parasitic worms [15–18]. However, the journey from DNA sequence to biological understanding remains fraught with challenges. This section delves into the intricate world of helminth genomics, dissecting the accomplishments and limitations associated with genome sequencing, assembly, annotation, and comparative analysis.

2.1 Genome Sequencing and Assembly: Charting the Genomic Landscape

Helminth genomes display an intriguing diversity in size and complexity. Following the sequencing of the genome of the first parasitic nematode, *Brugia malayi*, in 2004, numerous endeavors have been undertaken to expand the repertoire of available genomes [17–20]. Indeed, contemporary genomic research on helminths benefits significantly from repositories that comprise extensive genomic information (Fig. 1). Foremost among these repositories is WormBase Parasite (https://parasite.wormbase.org/), which currently encompasses a wealth of data, including more than 240 genomes and annotations from over 181 species of nematodes and flatworms (Fig. 1) [18]. This repository serves as a comprehensive resource, incorporating genomic information from some of the most pathogenic helminth species affecting humans, animals, and plants (Tables 1 and 2). Additionally, it includes data from closely related species, facilitating comparative analyses [18].

Through these efforts, it has become evident that the intricate nature of these organisms necessitates a specialized approach. Owing to variations in the sizes of helminth genomes (the median genome assembly size is 122.18 Mb and ranges from 38.18 Mb for *Meloidogyne graminicola* to 1.25 Gb for *Spirometra erinaceieuropaei*), such as *Ascaris lumbricoides* with a genome size of approximately 290 Mbp and *Fasciola hepatica* with a genome size of around 1 Gbp, each species poses distinct challenges in terms of sequencing and assembly (Fig. 2) [17, 21]. Short-read technologies, while cost-effective, often yield fragmented assemblies riddled with gaps and misassemblies, especially in repetitive regions. Conversely, long-read sequencing platforms, like PacBio and Oxford Nanopore, offer the promise of contiguous assemblies but remain hampered by higher costs and error rates [22].

Hybrid approaches that integrate both short-read and long-read data have emerged as a potent solution. This is exemplified by the near-complete assembly of the *Schistosoma mansoni* genome, a prominent causative agent of schistosomiasis [23]. In this instance, a hybrid approach utilizing Illumina and PacBio technologies facilitated the assembly of a 391.4 Mb genome of *S. mansoni* into individual chromosomes within a single scaffold, encompassing Z and W. Furthermore, the incorporation of bioinformatics processes, including efficient manual curation, yielded significantly improved gene annotation. This comprehensive approach successfully addressed challenges related to gene sets, repeats, trans-splicing, and nearly all untranslated regions (UTRs) [23].

Despite these advances, technical and bioinformatics challenges remain. For instance,

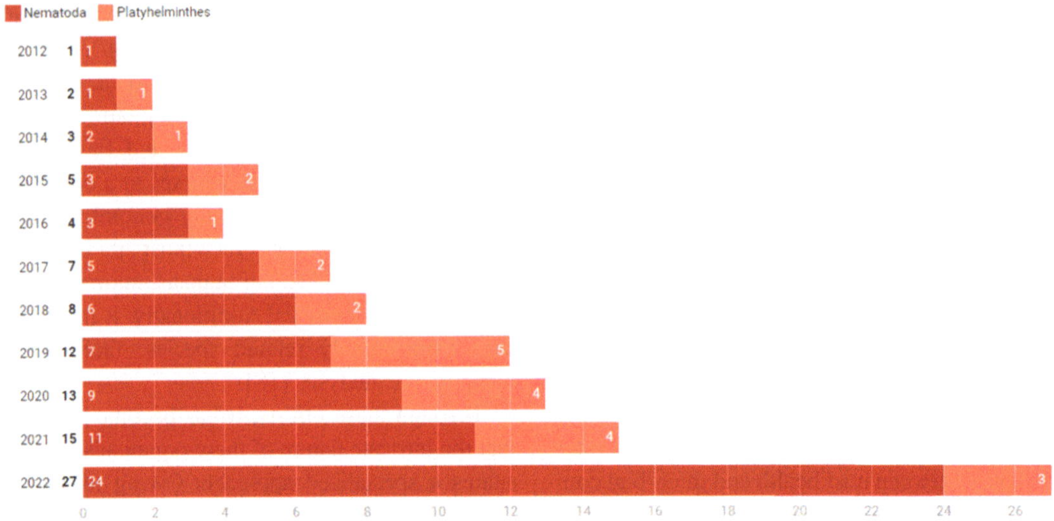

Fig. 1 Number of helminth genomes published and stored in WormBase Parasite per year. Source: WormBase Parasite Release 18

Table 1 Genome list of Nematoda (roundworms) species available in WormBase Parasite (177 genomes available). The table includes information related to the species name, provider, BioProject ID, N50 value, genome size, the number of scaffolds, and the number of coding genes estimated after the annotation process. Source: WormBase Parasite Release 18

Species name	Provider	BioProject ID	N50	Genome size	Number of Scaffolds	Number of coding genes
Acanthocheilonema viteae	University of Edinburgh	PRJEB1697	25752	77350906	6796	10397
Acrobeloides nanus	University College London	PRJEB26554	19572	248054631	30759	32984
Ancylostoma caninum	McDonnell Genome Institute	PRJNA72585	255299	465750606	25339	30198
Ancylostoma ceylanicum	McDonnell Genome Institute	PRJNA72583	372147	348994891	8098	17509
Ancylostoma ceylanicum	Cornell University	PRJNA231479	668412	313092710	1736	36687
Ancylostoma duodenale	McDonnell Genome Institute	PRJNA72581	10112	332878295	69984	27485
Angiostrongylus cantonensis	Sun Yat-sen University	PRJNA350391	860760	293306699	26577	10314
Angiostrongylus cantonensis	Wellcome Sanger Institute	PRJEB493	43712	253242791	18635	14520
Angiostrongylus costaricensis	Wellcome Sanger Institute	PRJEB494	111998	262782696	6384	13417
Angiostrongylus vasorum	University of Zurich	PRJNA663250	1673247	279948744	468	13764
Anisakis simplex	Wellcome Sanger Institute	PRJEB496	9229	126869778	42005	20971
Aphelenchoides besseyi	Academia Sinica	PRJNA834627	17779264	47428222	39	11794
Aphelenchoides besseyi	Academia Sinica	PRJNA834627	16068654	46759715	32	11239

(continued)

Table 1 (continued)

Species name	Provider	BioProject ID	N50	Genome size	Number of Scaffolds	Number of coding genes
Aphelenchoides besseyi	Academia Sinica	PRJNA834627	16939329	44715554	28	11081
Aphelenchoides besseyi	Academia Sinica	PRJNA834627	2203724	44700143	80	12200
Aphelenchoides bicaudatus	Academia Sinica	PRJNA834627	11905581	46428382	106	10675
Aphelenchoides fujianensis	Academia Sinica	PRJNA834627	553027	143834322	780	17477
Ascaris lumbricoides	Wellcome Sanger Institute	PRJEB4950	55909	316975410	31710	23604
Ascaris suum	University of Colorado Denver	PRJNA62057	4646302	298028455	415	16778
Ascaris suum	University of Melbourne	PRJNA80881	407675	272782664	29831	18542
Brugia malayi	WormBase	PRJNA10729	14214749	88235797	197	10878
Brugia pahangi	Wellcome Sanger Institute	PRJEB497	65530	90545113	14029	14674
Brugia timori	Wellcome Sanger Institute	PRJEB4663	4903	64930714	23963	16203
Bursaphelenchus okinawaensis	University of Miyazaki	PRJEB40023	11641800	69956301	7	14593
Bursaphelenchus xylophilus	University of Miyazaki	PRJEB40022	12794883	78274012	11	15884
Bursaphelenchus xylophilus	Wellcome Sanger Institute	PRJEA64437	949830	74561461	5527	17704
Caenorhabditis angaria	California Institute of Technology	PRJNA51225	79665	105997628	34621	27970
Caenorhabditis auriculariae	University of Miyazaki	PRJEB40642	948694	109583113	433	16279
Caenorhabditis becei	BANG	PRJEB28243	767505	87867412	1565	23926
Caenorhabditis bovis	BANG	PRJEB34497	7557458	62726283	35	13128
Caenorhabditis brenneri	WormBase	PRJNA20035	377250	190369721	3305	30705
Caenorhabditis briggsae	WormBase	PRJNA10731	17485439	108384165	367	20821
Caenorhabditis briggsae	NORTHWESTERN UNIVERSITY	PRJNA784955	17304409	106953292	7	19749
Caenorhabditis briggsae	NORTHWESTERN UNIVERSITY	PRJNA784955	17080301	106196058	7	19957
Caenorhabditis elegans	WormBase	PRJNA13758	17493829	100286401	7	19981
Caenorhabditis inopinata	University of Miyazaki	PRJDB5687	20594552	122994328	6	21443
Caenorhabditis japonica	WormBase	PRJNA12591	94114	166256191	18817	29935
Caenorhabditis latens	University of Oregon	PRJNA248912	366678	117124750	1858	24855
Caenorhabditis nigoni	Cornell University	PRJNA384657	20390332	129488540	155	29167
Caenorhabditis panamensis	BANG	PRJEB28259	487150	79045575	985	22310

(continued)

Table 1 (continued)

Species name	Provider	BioProject ID	N50	Genome size	Number of Scaffolds	Number of coding genes
Caenorhabditis parvicauda	BANG	PRJEB12595	44393	93733926	5716	16412
Caenorhabditis quiockensis	BANG	PRJEB11354	139266	100371645	4890	22238
Caenorhabditis remanei	WormBase	PRJNA53967	435512	145442736	3670	31438
Caenorhabditis remanei	University of Oregon	PRJNA577507	21501900	130480874	187	26308
Caenorhabditis remanei	University of Oregon	PRJNA248911	1765890	124541912	912	24867
Caenorhabditis remanei	University of Oregon	PRJNA248909	1522088	118549266	1591	24977
Caenorhabditis sinica	University of Edinburgh	PRJNA194557	25151	131797386	15261	34696
Caenorhabditis sulstoni	BANG	PRJEB12601	136669	65098870	2044	18192
Caenorhabditis tribulationis	BANG	PRJEB12608	224519	101228815	3276	24760
Caenorhabditis tropicalis	McDonnell Genome Institute	PRJNA53597	20921866	79321433	665	22326
Caenorhabditis uteleia	BANG	PRJEB12600	175547	104046868	3222	27614
Caenorhabditis waitukubuli	BANG	PRJEB12602	15045	91416478	21203	30089
Caenorhabditis zanzibari	BANG	PRJEB12596	91045	101086038	3127	22197
Cercopithifilaria johnstoni	Wellcome Sanger Institute	PRJEB47283	98884	76938708	2092	10438
Cylicostephanus goldi	Wellcome Sanger Institute	PRJEB498	1235	173362437	154509	13906
Dictyocaulus viviparus	University of Edinburgh	PRJEB5116	22491	169388535	17715	14306
Dictyocaulus viviparus	McDonnell Genome Institute	PRJNA72587	225234	161038511	7175	13514
Diploscapter coronatus	National Institute of Genetics, Japan	PRJDB3143	1007652	170470382	511	34063
Diploscapter pachys	New York University	PRJNA280107	124169	157719873	10775	27341
Dirofilaria immitis	University of Edinburgh	PRJEB1797	71050	88309529	16061	12872
Dirofilaria immitis	CIBIO â€“ Research Centre in Biodiversity and Genetic Resources—InBIO Associate Laboratory	PRJNA723804	4253153	86813779	110	9741
Ditylenchus destructor	Huazhong Agricultural University	PRJNA312427	555026	111138200	1761	13931
Ditylenchus dipsaci	Agriculture and Agri-Food Canada	PRJNA498219	287009	227234012	1394	26428
Dracunculus medinensis	Wellcome Sanger Institute	PRJEB500	643251	103750625	1347	10969

(continued)

Table 1 (continued)

Species name	Provider	BioProject ID	N50	Genome size	Number of Scaffolds	Number of coding genes
Elaeophora elaphi	Wellcome Sanger Institute	PRJEB502	25422	82568297	8078	10410
Enterobius vermicularis	Wellcome Sanger Institute	PRJEB503	20546	150050672	19832	12895
Globodera pallida	Wellcome Sanger Institute	PRJEB123	120163	123625196	6873	16403
Globodera pallida	University of St Andrews	PRJNA702104	2225126	119582986	173	7762
Globodera pallida	Wageningen University and Research	PRJNA764088	2862751	113220295	163	18071
Globodera rostochiensis	Wageningen University and Research	PRJNA695196	2102824	101587200	135	19459
Globodera rostochiensis	University of Edinburgh	PRJEB13504	88495	95876286	4281	14308
Globodera rostochiensis	Wageningen University and Research	PRJNA695196	3274370	92682755	88	17856
Gongylonema pulchrum	Wellcome Sanger Institute	PRJEB505	4421	322334493	128401	27158
Haemonchus contortus	University of Melbourne	PRJNA205202	56143	319640208	14419	23610
Haemonchus contortus	Wellcome Sanger Institute	PRJEB506	47382676	283439308	7	19623
Haemonchus placei	Wellcome Sanger Institute	PRJEB509	37562	259125910	24923	21928
Halicephalobus mephisto	American University, Washington, D.C.	PRJNA528747	313311	61428783	880	12546
Halicephalobus sp. NKZ332	Indiana University	PRJNA555616	60483	47438423	5084	10997
Heligmosomoides polygyrus	University of Edinburgh	PRJEB15396	179069	696954138	23647	23471
Heligmosomoides polygyrus	Wellcome Sanger Institute	PRJEB1203	35732	560662278	44726	27459
Heterodera glycines	Iowa State University	PRJNA381081	17907690	157978452	9	22465
Heterodera schachtii	Wageningen University and Research	PRJNA767548	500876	190153211	705	29851
Heterodera schachtii	The H. schachtii genome sequencing consortium	PRJNA722882	1273070	179246932	395	26739
Heterorhabditis bacteriophora	McDonnell Genome Institute	PRJNA13977	312328	77007652	1263	20964
Litomosoides sigmodontis	University of Edinburgh	PRJEB3075	45733	64813410	3165	10246
Loa loa	Institute for Genome Sciences	PRJNA246086	180288	96405388	2250	12473
Loa loa	Broad Institute	PRJNA37757	172043	91373458	5773	14908
Meloidogyne arenaria	RIKEN	PRJNA438575	204341	284032373	2224	76700
Meloidogyne arenaria	INRA	PRJEB8714	16417	258067405	26196	101269
Meloidogyne arenaria	University of Hull	PRJNA340324	10460	163746389	46436	27548
Meloidogyne chitwoodi	Northwest Potato Research Consortium	PRJNA666745	2451023	47477280	30	12295
Meloidogyne enterolobii	University of Hull	PRJNA340324	10529	162967259	42008	29473

(continued)

Table 1 (continued)

Species name	Provider	BioProject ID	N50	Genome size	Number of Scaffolds	Number of coding genes
Meloidogyne floridensis	University of Edinburgh	PRJEB6016	3688	96673063	58696	49938
Meloidogyne floridensis	University of Hull	PRJNA340324	13230	74846012	8887	12762
Meloidogyne graminicola	IARI	PRJNA411966	20427	38184958	4304	10895
Meloidogyne hapla	North Carolina State University	PRJNA29083	37501	53017507	3452	14419
Meloidogyne incognita	INRA	PRJEB8714	38543	183531997	12091	43718
Meloidogyne incognita	University of Hull	PRJNA340324	16480	121964221	33351	21830
Meloidogyne javanica	INRA	PRJEB8714	10373	235798407	31341	97208
Meloidogyne javanica	University of Hull	PRJNA340324	14103	150345270	34316	24679
Mesorhabditis belari	ENS Lyon	PRJEB30104	462517	163453184	1721	27588
Micoletzkya japonica	Max Planck Institute for Developmental Biology	PRJEB27334	188443	201900956	137965	19855
Necator americanus	McDonnell Genome Institute	PRJNA72135	211860	244075060	11864	15728
Nippostrongylus brasiliensis	Wellcome Sanger Institute	PRJEB511	33446	294400206	29375	22796
Oesophagostomum dentatum	McDonnell Genome Institute	PRJNA72579	19257	489973581	99770	25291
Onchocerca flexuosa	Wellcome Sanger Institute	PRJEB512	2932	86175476	45472	16119
Onchocerca flexuosa	McDonnell Genome Institute	PRJNA230512	540294	67740367	1604	8140
Onchocerca ochengi	University of Edinburgh	PRJEB1465	12275	95513350	24057	13990
Onchocerca ochengi	Wellcome Sanger Institute	PRJEB1204	16140	91660559	20243	12816
Onchocerca volvulus	WormBase	PRJEB513	25485961	96427137	708	12109
Oscheius tipulae	Wellcome Sanger Institute	PRJNA644888	10817478	60923166	7	16356
Oscheius tipulae	University of Edinburgh	PRJEB15512	1203411	59468623	191	14626
Panagrellus redivivus	California Institute of Technology	PRJNA186477	262414	65093147	940	24249
Panagrolaimus davidi	University College London	PRJEB32708	4441	118052509	39598	25806
Panagrolaimus sp. 1159	University College London	PRJEB32708	9884	84957092	17628	24802
Panagrolaimus sp. ES5	University College London	PRJEB32708	6292	90240976	25284	24678
Panagrolaimus sp. JU765	University College London	PRJEB32708	10830	64262255	13268	20756
Panagrolaimus superbus	University College London	PRJEB32708	1889	76664811	53192	19887

(continued)

Table 1 (continued)

Species name	Provider	BioProject ID	N50	Genome size	Number of Scaffolds	Number of coding genes
Parapristionchus giblindavisi	Max Planck Institute for Developmental Biology	PRJEB53331	790724	251641265	735	22488
Parascaris equorum	Wellcome Sanger Institute	PRJEB514	5333	185376996	60399	14472
Parascaris univalens	University of Colorado Denver	PRJNA386823	1825986	253353821	1274	14323
Parastrongyloides trichosuri	Wellcome Sanger Institute	PRJEB515	836942	42486966	1810	15010
Plectus sambesii	MRC London Institute of Medical Sciences	PRJNA390260	23404	186666232	18974	39064
Pristionchus arcanus	Max Planck Institute for Developmental Biology	PRJEB27334	269953	202757162	4263	39331
Pristionchus entomophagus	Max Planck Institute for Developmental Biology	PRJEB27334	367513	264161588	72721	31150
Pristionchus exspectatus	MAX PLANCK INSTITUTE FOR DEVELOPMENTAL BIOLOGY	PRJEB46690	25511976	169583105	24	31021
Pristionchus fissidentatus	Max Planck Institute for Developmental Biology	PRJEB27334	441872	247442930	56870	28283
Pristionchus japonicus	Max Planck Institute for Developmental Biology	PRJEB27334	447930	223093336	33291	34293
Pristionchus maxplancki	Max Planck Institute for Developmental Biology	PRJEB27334	308537	265592641	69506	37393
Pristionchus mayeri	Max Planck Institute for Developmental Biology	PRJEB27334	234324	297448349	84599	32719
Pristionchus pacificus	WormBase	PRJNA12644	23915096	158500396	47	26342
Rhabditophanes sp. KR3021	Wellcome Sanger Institute	PRJEB1297	537195	47267908	471	13496
Romanomermis culicivorax	University of Cologne	PRJEB1358	17583	322765761	62537	48376
Setaria digitata	University of Colombo	PRJNA479729	120860	78770088	1879	10531
Soboliphyme baturini	Wellcome Sanger Institute	PRJEB516	19742	218332512	22336	13193
Steinernema carpocapsae	University of California, Irvine	PRJNA202318	299566	86127942	1578	28307
Steinernema carpocapsae	University of California, Irvine	PRJNA202318	7362381	84512609	16	30931
Steinernema feltiae	Dept of Entomology, College of Agricultural, Human, and Natural Resource	PRJNA353610	60158	121603157	4678	32303
Steinernema feltiae	California Institute of Technology	PRJNA204661	47472	82627153	5839	33457
Steinernema glaseri	California Institute of Technology	PRJNA204943	37382	92932499	7514	34137
Steinernema monticolum	California Institute of Technology	PRJNA205067	11533	89356173	14332	36005
Steinernema scapterisci	California Institute of Technology	PRJNA204942	90783	79776474	2877	31376

(continued)

Table 1 (continued)

Species name	Provider	BioProject ID	N50	Genome size	Number of Scaffolds	Number of coding genes
Strongyloides papillosus	Wellcome Sanger Institute	PRJEB525	86067	60447969	4703	18456
Strongyloides ratti	Wellcome Sanger Institute	PRJEB125	11693564	43166851	136	12464
Strongyloides stercoralis	Wellcome Sanger Institute	PRJEB528	431128	42674647	800	13123
Strongyloides venezuelensis	Wellcome Sanger Institute	PRJEB530	715404	52178999	587	16904
Strongylus vulgaris	Wellcome Sanger Institute	PRJEB531	2377	291096078	167310	21071
Syphacia muris	Wellcome Sanger Institute	PRJEB524	60614	99093317	3879	11117
Teladorsagia circumcincta	McDonnell Genome Institute	PRJNA72569	46814	700614188	81734	25572
Thelazia callipaeda	Wellcome Sanger Institute	PRJEB1205	50894	75392092	5778	10912
Toxocara canis	University of Melbourne	PRJNA248777	374962	317115901	22857	18596
Toxocara canis	Wellcome Sanger Institute	PRJEB533	31084	299788016	29185	20264
Trichinella britovi	University of Melbourne	PRJNA257433	147150	51504795	7991	16058
Trichinella murrelli	University of Melbourne	PRJNA257433	106269	49016728	5242	14852
Trichinella nativa	McDonnell Genome Institute	PRJNA179527	12611	49070491	21024	10175
Trichinella nativa	University of Melbourne	PRJNA257433	141195	47933238	4443	13545
Trichinella nelsoni	University of Melbourne	PRJNA257433	293867	47436005	2969	13229
Trichinella papuae	University of Melbourne	PRJNA257433	222396	46844510	2542	11837
Trichinella patagoniensis	University of Melbourne	PRJNA257433	154103	49773209	6623	15319
Trichinella pseudospiralis	University of Melbourne	PRJNA257433	235426	49162916	7221	12659
Trichinella pseudospiralis	University of Melbourne	PRJNA257433	287133	49152658	6594	12455
Trichinella pseudospiralis	University of Melbourne	PRJNA257433	234172	48398615	5956	14621
Trichinella pseudospiralis	University of Melbourne	PRJNA257433	112255	48106863	7462	13717
Trichinella pseudospiralis	University of Melbourne	PRJNA257433	167180	46055013	1379	11005
Trichinella sp. T6	University of Melbourne	PRJNA257433	158103	50853738	5951	15241
Trichinella sp. T8	University of Melbourne	PRJNA257433	239129	49331761	4124	14919
Trichinella sp. T9	University of Melbourne	PRJNA257433	212690	49075249	5752	13111
Trichinella spiralis	McDonnell Genome Institute	PRJNA12603	6373445	63525422	6863	16380
Trichinella spiralis	University of Melbourne	PRJNA257433	212546	50025957	7652	14737
Trichinella zimbabwensis	University of Melbourne	PRJNA257433	205645	50795114	10839	14764
Trichuris muris	WormBase	PRJEB126	28941788	111837642	803	14995
Trichuris suis	University of Melbourne	PRJNA208415	503034	74234559	4292	14436
Trichuris suis	University of Melbourne	PRJNA208416	443734	71056402	3284	14261
Trichuris suis	McDonnell Genome Institute	PRJNA179528	1322386	63839715	306	9831
Trichuris trichiura	Wellcome Sanger Institute	PRJEB535	11299416	80573711	113	12671
Wuchereria bancrofti	Case Western Reserve University	PRJNA275548	56656	90325107	5105	11068
Wuchereria bancrofti	Wellcome Sanger Institute	PRJEB536	9872	76991470	13520	13058

Table 2 Genome list of Platyhelminthes (flatworms) species available in WormBase Parasite (63 genomes available). The table includes information related to the species name, provider, BioProject ID, N50 value, genome size, the number of scaffolds, and the number of coding genes estimated after the annotation process. Source: WormBase Parasite Release 18

Species name	Provider	BioProject ID	Clade	N50	Genome size	Number of Scaffolds	Number of coding genes
Atriophallophorus winterbourni	Swiss Federal Institute of Technology in Zurich	PRJNA636673	Trematoda (Flukes)	39978	601728533	26114	11499
Clonorchis sinensis	Sun Yat-sen University	PRJDA72781	Trematoda (Flukes)	415842	547288241	4348	13634
Clonorchis sinensis	The University of Melbourne	PRJNA386618	Trematoda (Flukes)	168711085	558124894	78	13489
Dibothriocephalus latus	Wellcome Sanger Institute	PRJEB1206	Cestoda (Tapeworms)	6726	531434409	140294	19966
Dicrocoelium dendriticum	Wellcome Sanger Institute	PRJEB44434	Trematoda (Flukes)	117106881	1889995958	19456	13685
Echinococcus canadensis	BMHid-Biologia Molecular de Hidatidosis	PRJEB8992	Cestoda (Tapeworms)	74230	115024622	9330	11432
Echinococcus granulosus	Wellcome Sanger Institute	PRJEB121	Cestoda (Tapeworms)	5228736	114538160	1288	10245
Echinococcus granulosus	Chinese National Human Genome Center	PRJNA182977	Cestoda (Tapeworms)	682596	110837706	957	11319
Echinococcus multilocularis	Wellcome Sanger Institute	PRJEB122	Cestoda (Tapeworms)	13762452	114963242	1217	10663
Echinococcus oligarthrus	IMPAM-UBA-CONICET	PRJEB31222	Cestoda (Tapeworms)	10175222	86214087	3764	8243
Echinostoma caproni	Wellcome Sanger Institute	PRJEB1207	Trematoda (Flukes)	26853	834556378	86083	18607
Fasciola gigantica	The Genome Institute	PRJNA230515	Trematoda (Flukes)	178720	1127753889	16040	12669
Fasciola hepatica	CGR, University of Liverpool	PRJEB25283	Trematoda (Flukes)	1901411	1203652780	2816	9731
Fasciola hepatica	The Genome Institute	PRJNA179522	Trematoda (Flukes)	160607	1138329001	23604	11217
Fasciolopsis buski	McDonnell Genome Institute	PRJNA284521	Trematoda (Flukes)	180282	748157630	11829	11837
Gyrodactylus bullatarudis	Adam Mickiewicz University	PRJNA532341	Monogenea	308184	84346884	4331	10738

(continued)

Table 2 (continued)

Species name	Provider	BioProject ID	Clade	N50	Genome size	Number of Scaffolds	Number of coding genes
Gyrodactylus salaris	Natural History Museum, University of Oslo	PRJNA244375	Monogenea	18344	67380660	6075	15436
Heterobilharzia americana	Wellcome Sanger Institute	PRJEB44434	Trematoda (Flukes)	85344987	685355840	736	12183
Heterobilharzia americana	Wellcome Sanger Institute	PRJEB44434	Trematoda (Flukes)	86847596	665588900	393	10892
Hydatigera taeniaeformis	Wellcome Sanger Institute	PRJEB534	Cestoda (Tapeworms)	12412	103664560	30091	11649
Hymenolepis diminuta	University of Warmia and Mazury	PRJEB30942	Cestoda (Tapeworms)	2331359	177074070	719	15165
Hymenolepis diminuta	Wellcome Sanger Institute	PRJEB507	Cestoda (Tapeworms)	49752	165878914	13910	11271
Hymenolepis microstoma	Wellcome Sanger Institute	PRJEB124	Cestoda (Tapeworms)	25815276	168943216	7	10139
Hymenolepis nana	Wellcome Sanger Institute	PRJEB508	Cestoda (Tapeworms)	19213	162943536	16212	13777
Macrostomum lignano	Schatz Laboratory	PRJNA284736	Rhabditophora	36597	1040124789	49174	60534
Macrostomum lignano	University Medical Center Groningen	PRJNA371498	Rhabditophora	244885	764410778	5269	49013
Mesocestoides corti	Wellcome Sanger Institute	PRJEB510	Cestoda (Tapeworms)	65816	117181740	7068	14704
Opisthorchis felineus	Institute of Cytology and Genetics SB RAS	PRJNA413383	Trematoda (Flukes)	619338	679245959	13306	11427
Opisthorchis viverrini	University of Melbourne	PRJNA222628	Trematoda (Flukes)	1347703	620452578	42216	16379
Paragonimus westermani	University of Queensland	PRJNA454344	Trematoda (Flukes)	134838	922849782	30456	12783
Protopolystoma xenopodis	Wellcome Sanger Institute	PRJEB1201	Monogenea	2893	617344661	316398	37906
Schistocephalus solidus	Wellcome Sanger Institute	PRJEB527	Cestoda (Tapeworms)	31505	539424504	56778	20228
Schistosoma bovis	University of Queensland	PRJNA451066	Trematoda (Flukes)	201915	373478075	4774	11576

(continued)

Table 2 (continued)

Species name	Provider	BioProject ID	Clade	N50	Genome size	Number of Scaffolds	Number of coding genes
Schistosoma bovis	Wellcome Sanger Institute	PRJEB44434	Trematoda (Flukes)	47558671	391812492	78	9589
Schistosoma bovis	Wellcome Sanger Institute	PRJEB44434	Trematoda (Flukes)	50257572	406292391	132	9673
Schistosoma curassoni	Wellcome Sanger Institute	PRJEB44434	Trematoda (Flukes)	46058712	396952245	370	10084
Schistosoma curassoni	Wellcome Sanger Institute	PRJEB519	Trematoda (Flukes)	13843	344202215	60140	23546
Schistosoma guineensis	Wellcome Sanger Institute	PRJEB44434	Trematoda (Flukes)	48012402	379339432	75	9715
Schistosoma haematobium	The University of Melbourne	PRJNA78265	Trematoda (Flukes)	48328128	400271889	163	9431
Schistosoma haematobium	Wellcome Sanger Institute	PRJEB44434	Trematoda (Flukes)	48644961	390696044	78	9911
Schistosoma haematobium	Wellcome Sanger Institute	PRJEB44434	Trematoda (Flukes)	53926266	431589308	188	10339
Schistosoma intercalatum	Wellcome Sanger Institute	PRJEB44434	Trematoda (Flukes)	46762690	386083432	299	9676
Schistosoma intercalatum	Wellcome Sanger Institute	PRJEB44434	Trematoda (Flukes)	45688419	382498898	652	10022
Schistosoma japonicum	Chinese National Human Genome Center	PRJEA34885	Trematoda (Flukes)	174764	402743189	25048	12738
Schistosoma japonicum	Fudan University	PRJNA520774	Trematoda (Flukes)	1093989	369900518	1789	10089
Schistosoma mansoni	Wellcome Sanger Institute	PRJEA36577	Trematoda (Flukes)	49793710	391422299	10	9896
Schistosoma margrebowiei	Wellcome Sanger Institute	PRJEB44434	Trematoda (Flukes)	48858358	394936775	36	9780
Schistosoma mattheei	Wellcome Sanger Institute	PRJEB44434	Trematoda (Flukes)	48524730	389351336	98	10771
Schistosoma rodhaini	Wellcome Sanger Institute	PRJEB44434	Trematoda (Flukes)	48179494	388488086	91	9927
Schistosoma rodhaini	Wellcome Sanger Institute	PRJEB44434	Trematoda (Flukes)	48789089	413541355	203	10242

(continued)

Table 2 (continued)

Species name	Provider	BioProject ID	Clade	N50	Genome size	Number of Scaffolds	Number of coding genes
Schistosoma spindale	Wellcome Sanger Institute	PRJEB44434	Trematoda (Flukes)	47457101	390101833	113	9330
Schistosoma turkestanicum	Wellcome Sanger Institute	PRJEB44434	Trematoda (Flukes)	47759033	425029217	3497	11134
Schmidtea mediterranea	McDonnell Genome Institute	PRJNA12585	Rhabditophora	40740	901626601	43294	29850
Schmidtea mediterranea	Max Planck Institute of Molecular Cell Biology and Genetics	PRJNA885486	Rhabditophora	270168396	839952239	656	21392
Schmidtea mediterranea	Max Planck Institute of Molecular Cell Biology and Genetics	PRJNA885486	Rhabditophora	268961546	819865861	432	21310
Spirometra erinaceieuropaei	Wellcome Sanger Institute	PRJEB1202	Cestoda (Tapeworms)	4624	1258717181	482396	39557
Taenia asiatica	Wellcome Sanger Institute	PRJEB532	Cestoda (Tapeworms)	51866	135977618	24137	10331
Taenia asiatica	Chinese Academy of Agricultural Sciences	PRJNA299871	Cestoda (Tapeworms)	342420	168679183	6900	13322
Taenia multiceps	Chinese Academy of Agricultural Sciences	PRJNA307624	Cestoda (Tapeworms)	44815576	240610560	738	12890
Taenia saginata	Chinese Academy of Agricultural Sciences	PRJNA71493	Cestoda (Tapeworms)	585232	169104283	3626	13161
Taenia solium	National University of Mexico	PRJNA170813	Cestoda (Tapeworms)	67829	122393951	11237	12467
Trichobilharzia regenti	Wellcome Sanger Institute	PRJEB44434	Trematoda (Flukes)	125840967	954258533	573	14478
Trichobilharzia szidati	Wellcome Sanger Institute	PRJEB44434	Trematoda (Flukes)	142264673	1116403795	1278	14843

repetitive elements abound in helminth genomes, requiring specialized tools and strategies for accurate assembly. Additionally, heterozygosity in certain species further complicates assembly and subsequent analyses. Yet, ongoing technological advancements in sequencing and assembly algorithms offer hope for overcoming these hurdles, paving the way

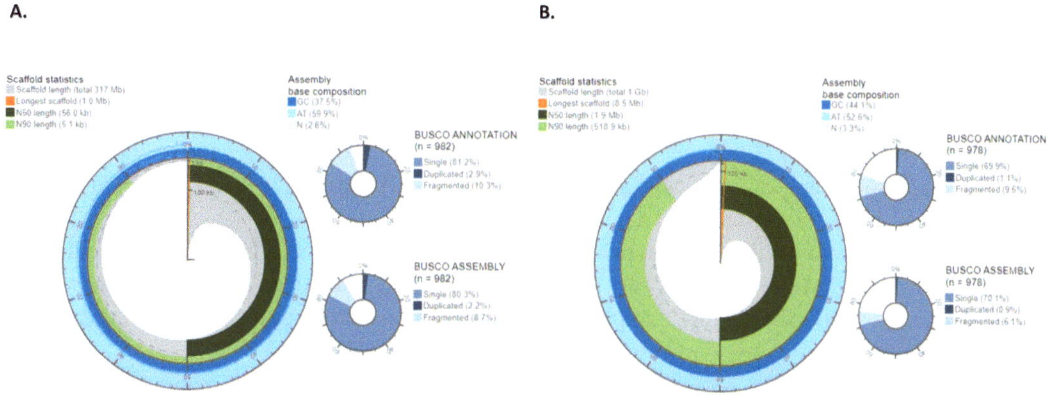

Fig. 2 Genome assembly characteristics of (**a**) *Ascaris lumbricoides* and (**b**) *Fasciola hepatica*. The figure displays statistics and features, including scaffold size, N50, completeness (BUSCO), and %GC, among others. Source: WormBase Parasite—a comprehensive resource for helminth genomics. Source: WormBase Parasite Release 18

for more accurate and comprehensive helminth genome reconstructions.

In a 2022 review conducted by Doyle, notable limitations to the generation of appropriate assemblies and, consequently, the acquisition of high-quality genomes were highlighted [21]. Given these observed limitations and issues across several studies, it is apparent that various factors contribute to these challenges:

- Morphological Differences and Tissue Composition: The diverse morphological disparities and tissue compositions among different helminths and their various life stages may necessitate distinct approaches to extraction processes. This variability could potentially impact the quality of genetic material obtained [24].
- Availability and Recovery of Parasites: The accessibility of parasites in their different stages and the successful retrieval from living organisms or biological models prove challenging. For certain helminths, limitations associated with geographic distribution can impede the recovery of specimens and, consequently, hinder the obtainment of a sufficient amount of DNA [13, 25].
- Genetic Heterogeneity: A crucial factor with the potential to impede bioinformatic analyses is the substantial genetic heterogeneity observed among individuals of many helminth species. If sample pooling is necessary to generate adequate DNA for sequencing, this diversity may lead to assembly artifacts, complicating the accurate representation of the genome [17].
- Heterogeneity in Bioinformatic Analysis Protocols: In addition to the challenges proposed by Doyle [21], it is crucial to highlight that the attainment of high-quality assemblies and genomes is influenced by the diversity of bioinformatics approaches [26, 27]. Various tools constitute the bioinformatics pipeline used to obtain these assemblies, employing distinct strategies and algorithms grounded in statistical models, graph theory, and machine learning, among others (e.g., SPAdes, Megahit, Unicycler, IDBA, SOAP, AutoHIC, etc.) [17, 21, 22, 26–29]. These diverse strategies may yield results that are not consistently homogeneous, potentially leading to assemblies of varying quality. It is imperative to establish a consensus encouraging the simultaneous use of different tools for the same purpose, thereby facilitating the acquisition of superior assemblies, and ensuring the reproducibility of bioinformatic analyses.
- Limited Helminth Genomic Data From Low-Income Countries: It is crucial to consider the restricted representation of genomic data from low-income countries, a prevalent situation in many South American nations [18, 21, 30]. Challenges such as the unavailability of

molecular detection strategies and inadequate infrastructure and resources for extraction and sequencing, among other factors, contribute to underestimating the prevalence of these infections in these regions. Consequently, there exists an insufficient amount of genomic information, emphasizing the need for attention to enhance data available in repositories like WormBase Parasite [18, 21, 30]. This improvement is vital for a more comprehensive understanding of omics aspects associated with specific sociodemographic components [18, 21, 30].

This underscores the ongoing challenges in achieving comprehensive functional annotations for helminth genomes, highlighting the need for continued advancements in bioinformatics methodologies to bridge the existing gaps in protein identification and characterization.

2.2 Genome Annotation and Functional Analysis: Decoding the Blueprint of Life

The assembly process, while initially challenging, is only the first step in the broader bioinformatics workflow. Various tools such as SPADES, Megahit, and Unicycler, among others, have been implemented to optimize this phase and yield optimal results [17]. However, a significant hurdle arises in the subsequent annotation process, where the identification of coding regions, non-coding regions, and repetitive regions requires the evaluation and implementation of bioinformatics algorithms [15, 21].

Accurate annotation of the helminth genome is crucial for unraveling the functions encoded within its genes. To achieve this, a variety of annotation tools have been applied, each with its unique strengths and methodologies. These tools suggested by Helminths Genome Consortium pipeline are instrumental in the systematic identification and prediction of coding regions, non-coding regions, and repetitive elements within the genome [17]. Notable tools in this domain include the following [17, 21]:

- *Augustus:* Augustus is a widely used tool that employs probabilistic models to predict genes in eukaryotic genomes. It incorporates training data to enhance accuracy and is particularly useful for refining gene predictions [31].
- *Braker:* Braker is a hybrid approach that combines GeneMark-ET, which predicts genes, and AUGUSTUS, which refines predictions using RNA-Seq data. This dual strategy aims to improve accuracy, especially in non-model organisms [32].
- *Maker:* Maker is an annotation pipeline that integrates evidence from multiple sources, including *ab initio* predictions, protein homology, and transcript evidence. It aims to produce comprehensive and accurate annotations [33].
- *PASA (Program to Assemble Spliced Alignments):* PASA integrates evidence from transcript alignments to improve gene structure predictions. It is valuable for refining gene models based on experimental data [34].

These tools collectively contribute to the precision and reliability of gene annotations, aiding researchers in deciphering the functional elements embedded in the helminth genome. The utilization of diverse tools allows for a comprehensive and thorough annotation process, enabling a deeper understanding of the genomic landscape and the biological processes encoded within these parasitic organisms (Tables 1 and 2) [17, 21, 35].

Despite considerable efforts to enhance these predictions, a substantial gap persists in the identification of proteins within the helminths. Consequently, a considerable portion of these proteins remains categorized as "hypothetical." Recent estimations indicate that over 47% of gene families lack any functional annotation, underscoring the prevalence of hypothetical proteins [21]. Even those proteins with functional information often rely heavily on computational inference based on homology with closely related species [21]. Computational tools and resources such as Expressed Sequence Tags libraries and RNA-seq data, paired with experimental approaches such as gene knockouts and protein–

protein interaction studies, are increasingly employed to unravel gene function [17, 21].

Indeed, the dedicated efforts in helminth genomics have led to exciting discoveries, for instance, in models like *Brugia malayi* and *Schistosoma mansoni*, shedding light on potential avenues for novel treatments and interventions [23, 36–39]. In the case of *Brugia malayi*, a filarial worm responsible for lymphatic filariasis, researchers have identified genes encoding novel drug targets and immunomodulatory molecules [36–38]. It has significantly enhanced gene models through the extension of UTRs, identification of novel splice junctions, and correction of mispredicted junctions. Furthermore, it has validated and identified operons, prioritized, and assessed potential drug targets, such as *Bma-slo-1*, which exhibits sex-specific splice patterns likely contributing to the differential response to emodepside. These findings carry transformative potential for the development of new treatment modalities. Additionally, the discovery of immunomodulatory molecules opens up possibilities for modulating the host immune response, offering novel avenues for therapeutic interventions. This emphasizes the critical need for accurate gene models and isoform resolution, which are essential for cloning, expression, and target-based screening efforts in the development of new antifilarial drugs and therapies [36–38].

Similarly, to assess the genomic impact of repeated mass drug administration on *Schistosoma mansoni* and gain insights into the genetic diversity and potential gene flow among natural *S. mansoni* populations in Uganda, a study conducted whole-genome sequencing on 198 *S. mansoni* larvae collected from 34 Ugandan children residing in regions with varying levels of praziquantel exposure. The study revealed that a single round of treatment did not diminish the population diversity of parasites in children from a transmission hotspot. The findings indicated evidence of positive selection acting on gene families associated with praziquantel action. However, no high-frequency, functionally impactful variants were detected in genes such as the Ca2+-permeable transient receptor potential channel (Smp_246790), which was recently identified as a likely target of praziquantel activity. This study establishes a foundation for genomic surveillance of *S. mansoni*, contributes to efforts aimed at eliminating schistosomiasis, and presents opportunities for developing interventions that specifically target these crucial processes [23, 39].

The exploration of helminth genomics not only deepens our understanding of these parasites at a molecular level but also unlocks opportunities for developing targeted interventions. Such interventions hold the potential to disrupt key pathways in the parasite's lifecycle, thereby paving the way for more effective and specific treatment strategies. Understanding the intricacies of helminth genomics has practical applications in various fields. In medicine, it can aid in the identification of novel drug targets and the development of more precise therapeutic approaches. Furthermore, insights gained from helminth genomics contribute to the design of effective control measures in public health initiatives, allowing for the strategic management and prevention of helminth infections. Additionally, this knowledge plays a crucial role in agriculture by informing strategies to control parasitic infections in livestock, enhancing animal health and productivity.

2.3 Comparative Genomics: Weaving Tales of Evolution and Divergence

Comparative analyses across helminth species reveal captivating narratives of adaptation. For instance, for the genus *Schistosoma*, notorious for its blood-fluke exploits, it witnessed an astounding expansion of gene families dedicated to tegumental protein synthesis. This elaborate coat of armor serves as their frontline defense against the host's immune system, highlighting its critical role in survival and success [40]. Furthermore, studies involving micro-exon gene (MEG) protein products and venom allergen-like (VAL) proteins, which have been identified in schistosome secretions or associated with glands, have led to the hypothesis that these proteins play crucial roles in the molecular interaction between the parasite and the host [41]. Phylogenetic and structural analyses of MEG and VAL genes in three different schistosome species revealed a recent species-specific expansion of gene numbers in each species. Notably, the enrichment of transposable elements in MEG and VAL genes in *Schistosoma mansoni* suggests a potential mechanism for the preferential expansion of gene numbers in these families. Moreover, the analysis of the ratio of synonymous and nonsynonymous substitution rates (dN/dS) in schistosome orthologs of these genes demonstrated significantly higher dN/dS values compared with a set of control genes. This implies that these genes are under strong selective pressure. Researchers observed that exposure of the proteins to the definitive host immune system is a determining factor in these elevated dN/dS values, leading to the hypothesis that the parasite is responding to evolutionary pressure imposed by the host immune system. The expansion of these genes and the high dN/dS values indicate their critical role in the parasite's ability to evade the host immune system. These findings suggest that MEG and VAL genes could be potential targets for therapeutic interventions against schistosomiasis, offering new avenues for the development of effective treatments [41].

The ongoing research on filarial nematodes, like *Brugia malayi*, explores the complex interplay between the parasite and its endosymbiosis *Wolbachia* bacteria, potentially influencing both virulence and drug susceptibility [42]. This study developed a compartmentalized metabolic model (iDC625) of *B. malayi* and utilized it to analyze metabolic pathway usage, identifying essential reactions crucial for the worm's survival. iDC625 comprises 1266 total reactions involving 1252 metabolites. Among these reactions, 1011 represent enzymes, with 849 associated with 625 genes. Specifically, 575 enzymatic reactions are linked to the cytosolic compartment, 166 to the mitochondria, and 270 to *Wolbachia*. The remaining 226 reactions are associated with transport across compartments, including 37 representing metabolite exchange between the mitochondria and cytosol, 80 between *Wolbachia* and the cytosol, and the remaining 109 facilitating metabolite exchange between the cytosol and the extracellular milieu. This research significantly contributes to the field by offering a new understanding of the metabolic interplay between the parasitic worm and its bacterial endosymbiont. Additionally, it identifies potential drug targets crucial for the treatment of lymphatic filariasis. The compartmentalized metabolic model, iDC625, provides a comprehensive framework for investigating the intricate metabolic pathways and essential reactions that underpin the survival of *B. malayi*, offering valuable insights for the development of targeted therapeutic strategies [42].

Comparative studies within the broader eukaryotic tapestry allow us to identify gene families unique to parasitic worms, providing crucial clues into adaptations intricately linked to their parasitic lifestyle [17, 30]. Conversely, shared conserved genes with free-living nematodes, like *Caenorhabditis elegans*, offer glimpses into their shared ancestry and evolutionary trajectories. This comparative framework serves as a cornerstone for comprehending the distinctive biology and adaptations that define the helminth world [17, 30].

Examining helminth genomes from diverse geographical locations unlocks yet another layer of understanding—local adaptation and population structure. Drawing on the example of the *Schistosoma mansoni* population in Africa, comparative analyses have unveiled distinct geographic clusters attributed to host-specific adaptations [43]. Several studies underscore the importance of population genetic research in identifying reservoir hosts, assessing the potential for the development of drug resistance, understanding recruitment patterns in hosts, and unraveling the genetic structure and gene flow within the context of disease control. These analyses emphasize the adaptive and evolving nature of *Schistosoma* species, illustrating their ability to swiftly adapt to new environments and challenges. This comprehension is pivotal for global initiatives aimed at controlling, eliminating, and ultimately eradicating schistosomiasis.

Furthermore, it underlines the significance of intra-species hybridization concerning the natural epidemiology of co-endemic species. This raises crucial questions about species integrities and their impact on control and elimination efforts. Such knowledge not only offers valuable insights into the emergence of drug resistance but also lays the foundation for tailored control strategies specific to different regions. The ability to recognize and address the nuances of genetic diversity and adaptation within *Schistosoma* populations is essential for the success of comprehensive and region-specific approaches to combat schistosomiasis [43].

The field of comparative helminth genomics is in its dynamic nascent stage, continually enhancing our comprehension of these remarkable organisms. From intricate adaptations to fascinating evolutionary leaps, each newly sequenced genome contributes another brushstroke to the ever-evolving panorama of helminth biology. As we delve further into these genetic archives, we unveil not only the secrets of their past but also the keys to their future, laying the groundwork for more effective control strategies and potentially novel therapeutic avenues.

3 Helminth Transcriptomics: Elucidating the Transcriptomic Dynamics of Gene Expression in Helminths

The intricacies of gene expression in helminths, formerly enigmatic, are now elucidated through the robust tools of transcriptomics. RNA-seq technologies have revolutionized our understanding of these parasitic worms, offering a dynamic window into their responses to diverse stimuli, developmental stages, and host environments [21, 30, 44]. This section delves into the realm of helminth gene expression, tracing the evolution from microarrays to cutting-edge RNA-seq, elucidating the potency of differential expression analysis, and highlighting the emerging role of ncRNAs in this intricate interface.

3.1 Microarrays: The First Steps

The initial stages of helminth transcriptomics involved microarrays, providing insight into gene expression patterns. Through the hybridization of fluorescently labeled cDNA to known gene sequences on a chip, researchers could concurrently measure the relative abundance of thousands of genes [45].

One of the pioneering transcriptomic studies leveraging microarray technology concentrated on *Schistosoma mansoni*, aiming to delineate differentially expressed genes gender-associated [46]. For example, the study revealed 16 genes with novel sex-specific associations in adult schistosomes, comprising 12 female-associated and 4 male-associated transcripts [46]. The research demonstrated that adult female parasites exhibit a significantly higher transcription rate of the mRNA for a tyrosinase ortholog compared with adult males. This finding suggests that this enzyme may be responsible for the phenol oxidase/tyrosinase activities previously described in *S. mansoni* [46]. On the contrary, three genes displaying male-associated expression are tropomyosin, actin, and a dynein light-chain ortholog. The expression of tropomyosin and actin in adult male schistosomes likely stems from their intimate interaction. Actin polymers provide mechanical stability for the cytoskeleton of eukaryotes and act as tracks for motor proteins like tropomyosin. The upregulation of these genes may be related to the larger size of male schistosomes, their increased tegument and muscle volume, and higher physical activity. Moreover, the preferential transcription of tropomyosin, actin, and dynein light chains in males is suggested to play a significant developmental role in copulating worm pairs [46].

Despite its groundbreaking contributions, the microarray approach encountered limitations due to its probe-based nature, restricting analyses to known genes and inadvertently overlooking the extensive potential of the transcriptome. This highlights the need for more comprehensive approaches to capture the full complexity of gene expression patterns and regulatory mechanisms in schistosomes.

3.2 Advancements in Helminth Transcriptomics: A Comprehensive Exploration Through RNA-Seq

The introduction of RNA-seq has revolutionized helminth transcriptomics, exposing the dynamics of gene expression [47, 48]. Whether through poly(A)-selected RNA-seq, which captures messenger RNA (mRNA), or total RNA-seq, encompassing the entire transcriptome, each approach provides a distinctive perspective on the helminth's genetic repertoire (Table 3).

- *Poly(A)-Selected RNA-seq:* The implementation of mRNA capture strategies has enabled a targeted approach to transcriptome sequencing, specifically focusing on the differential expression patterns of coding genes. This methodology has proven particularly valuable in helminth models, yielding insightful information that enhances our understanding of the

Table 3 Comparison between RNA-Seq approaches: Poly(A)-selected RNA-seq, Total RNA-seq, and Single-cell (sc) RNA-seq, focusing on several analytic, technical, and methodological characteristics

Feature	Poly(A)-selected RNA-seq	Total RNA-seq	Single-cell (sc)RNA-seq
RNA capture	Only polyadenylated (poly(A)) RNA, which mostly represents protein-coding mRNAs	All RNA molecules, including non-coding RNAs (ncRNAs), ribosomal RNAs (rRNAs), and pre-mRNAs	Individual cells captured and lysed, with RNA extracted from each
Transcriptome coverage	Biased toward protein-coding mRNAs, missing non-coding RNAs and low-abundance transcripts	More comprehensive, capturing both coding and non-coding RNAs, but with varying sensitivity	Highly specific, capturing the transcriptome of individual cells, revealing cellular heterogeneity
Cell resolution	Bulk population, averaging the transcriptomes of many cells	Bulk population, averaging the transcriptomes of many cells	Single cell, resolving transcriptomic differences between individual cells
Quantification	Both absolute and relative quantification possible	Both absolute and relative quantification possible	Relative quantification due to technical limitations and variations in capture efficiency
Technical replicates	Required for accurate gene expression measurements	Required for accurate gene expression measurements	Optional, but highly recommended to account for technical variability
Biological replicates	Optional, but recommended for robust analysis	Optional, but recommended for robust analysis	Required to capture biological variability between individuals or samples
Cost	Moderate, lower than total RNA-seq and scRNA-seq	High due to library preparation methods	Very high due to single-cell isolation and library preparation complexity
Data analysis complexity	Moderate, standard RNA-seq analysis pipelines	Moderate, similar to Poly(A)-seq but may require additional tools for ncRNA analysis	High, requires specialized tools and expertise for dimensionality reduction and cell clustering
Applications	Gene expression profiling, differential expression analysis, identification of differentially expressed genes (DEGs)	Gene expression profiling, differential expression analysis, non-coding RNA analysis, alternative splicing analysis	Cell type identification, cell state analysis, trajectory inference, studying cellular heterogeneity, understanding cell-cell interactions
Limitations	Misses non-coding RNAs and low-abundance transcripts, biased toward highly expressed genes	Higher technical variability, rRNA depletion needed, may not capture rare cell types	Lower sensitivity, technical challenges like batch effects, limited capture efficiency for some cell types

biology of these parasites [29]. A noteworthy example is the application of this approach to *Ascaris lumbricoides*, where differential expression patterns of genes associated with protein phosphorylation, the TGF signaling system, and the oxytocin signaling pathway were elucidated [49]. These gene expressions were linked to defects in germline development, embryogenesis, and reproductive behavior in both male and female somatic and reproductive tissue samples. Beyond offering comprehensive transcriptome profiles, these findings highlight specific genes that likely play crucial roles in the functions of vital tissues. The identification of such genes holds substantial promise in pinpointing genetic targets for therapeutic exploitation, with potential implications for the future sustained control of ascariasis [49]. This nuanced understanding of the molecular underpinnings of helminth biology lays the groundwork for targeted interventions, contributing to the ongoing efforts to develop effective strategies for combating parasitic infections.

- *Total RNA-seq:* In addition to exploring polyadenylated mRNA transcripts, RNA-Seq offers the versatility to investigate various RNA populations, including total RNA, pre-mRNA, and ncRNA such as microRNA (miRNA) and long non-coding RNA (lncRNA). This broadened scope is achieved through comprehensive or whole RNA sequencing, commonly referred to as total or whole RNA-Seq [47]. This approach enables a holistic examination of the entire RNA landscape, providing a nuanced understanding of not only coding genes but also regulatory elements and ncRNA species [47]. Such inclusivity in RNA-Seq methodologies enhances the depth and breadth of molecular insights gained from these analyses. It has been established that numerous ncRNAs play pivotal roles in post-transcriptional regulatory mechanisms, thereby serving as key contributors to the intricate life cycles of parasitic organisms [44, 50]. For instance, a study conducted in *Trichinella spiralis* utilized total RNA-Seq sequencing to unveil the presence of stage-specific miRNAs and endogenous small interfering RNAs (endo-siRNAs) at various developmental stages. The identification of stage-specific miRNAs and endo-siRNAs suggests a significant involvement of miRNAs, in particular, in the developmental processes of the parasite, providing a foundational basis for a more profound understanding of the molecular regulation and functional evolution of miRNAs in parasitic nematodes, shedding light on their roles in shaping the complex life cycles of these parasites [51].

  The choice of RNA-seq platform plays a crucial role. Short-read sequencing, while cost-effective, provides snapshots of gene expression but faces challenges with low-abundance transcripts. Long-read platforms, such as PacBio and Oxford Nanopore, capture full-length transcripts and offer valuable insights into alternative splicing and complex gene structures. Additionally, it is essential to consider both sequence depth and the quantity of reads to ensure the recovery of sufficient small RNAs, not solely those associated with mRNA. In this context, a recent study on *Ascaris lumbricoides*, employing Nanopore sequencing, identified novel transcripts and alternative splicing events, underscoring the capability of this technology to unveil hidden layers of gene regulation [17, 21, 36, 52, 53].
- *Single-Cell RNA-seq:* This powerful technology allows us to listen to the individual voices within the helminth chorus. Single-cell RNA sequencing (scRNA-seq) technologies have revolutionized transcriptomic studies by enabling the dissection of gene expression at a single-cell resolution [15, 54]. This breakthrough approach has significantly advanced our understanding of complex mechanisms associated with parasite biology and interactions with hosts. Several scRNA-seq protocols have been developed, each possessing unique characteristics [54, 55]. However, due to both technical and biological heterogeneity, scRNA-seq data tend to be noisier and more complex compared with bulk RNA-seq data, presenting intriguing computational challenges [54, 55]. In response to this complexity,

a growing number of bioinformatics methods are being proposed to analyze and interpret scRNA-seq data. These methods are designed to enhance the accuracy and reproducibility of results, considering the nuances inherent in single-cell analyses [15, 54–56]. As scRNA-seq continues to be a transformative tool in advancing our understanding of parasitic biology, the development and refinement of bioinformatics methodologies play a pivotal role in extracting meaningful insights from the intricate and dynamic nature of single-cell transcriptomic data.

In the context of transcriptomic advancements, a comprehensive study has been undertaken using the *Schistosoma mansoni* model. Leveraging scRNA-Seq, this groundbreaking research has enabled the characterization of 43642 cells from the adult schistosome [53]. Notably, the study identified 68 distinct cell populations, encompassing specialized stem cells responsible for maintaining the blood-digesting intestine of the parasite. This effort has resulted in the construction of one of the most exhaustive transcriptomic atlases available, providing unprecedented insights into the intricate cellular landscape of *Schistosoma mansoni* [53]. The utilization of scRNA-Seq has proven instrumental in unraveling the intricacies of cellular diversity and functions within this parasitic organism, contributing significantly to our understanding of its biology.

3.3 Differential Expression Analysis in Transcriptomics: Unveiling Adaptive Strategies and Vulnerabilities of Helminths

Differential expression analysis, a pivotal facet of transcriptomics, serves as a cornerstone for identifying genes that undergo changes in expression in response to diverse conditions or across different life stages, offering valuable insights into the adaptive strategies and vulnerabilities of helminths [55]. For instance, to unearth additional hookworm vaccine antigens, a comprehensive differential expression analysis was conducted on the hookworm *Ancylostoma ceylanicum* [57]. This analysis revealed upregulated genes encoding proteolytic enzymes, including *A. ceylanicum* APR-1, with cysteine-, serine-, and metalloproteases being the most common. This information proves instrumental in identifying proteins involved in the blood-feeding process, marking an initial step toward a reverse vaccinology approach for developing a human hookworm vaccine [57].

Similarly, a comparative analysis of transcriptome profiles across four *Schistosoma japonicum* larval stages, encompassing eggs, miracidia, sporocysts, and cercariae, elucidated distinct biological and physical features associated with each stage [58]. This analysis unveiled that miracidia and cercariae may employ different groups of proteases for host invasion and suggested the existence of diverse populations of germinal cells in larval stages. The observed high expression of tandem-duplicated cathepsin B-like proteases during the cercariae stage may contribute to the broad definitive host range of *S. japonicum* [58]. Additionally, the study noted the pivotal role of alternative splicing in regulating gene expression in *S. japonicum*, with skip exon events being the most predominant mechanism. These findings collectively contribute to a deeper understanding of the molecular intricacies underpinning the developmental stages and adaptation strategies of helminths [57, 58].

The bioinformatic analysis, encompassing differential expression evaluation, has proven successful in identifying genes crucial to various biological processes in helminth parasites. Nevertheless, the analysis of such data poses considerable challenges, especially in helminth genomes characterized by a plethora of repetitive elements [21, 44, 48, 55, 59]. This emphasizes the critical necessity for precise transcript assembly and annotation. Effectively addressing biological variability and minimizing technical noise demands the implementation of robust statistical approaches. These challenges underscore the complexity of analyzing helminth transcriptomic data, highlighting the need for advanced computational methods and bioinformatics tools to ensure accurate and meaningful insights into the molecular intricacies of these parasites [55].

To confront these challenges, various tools based on diverse algorithms and mathematical models have been developed and implemented for performing differential analysis [21, 59]. It is common practice to use multiple tools simultaneously with the aim of properly identifying differentially expressed genes. For example, tools like edgeR and baySeq implement the negative binomial model as their primary approach, while others like NOIseq and SAMseq employ non-parametric methods. Some methods, such as EBSeq and Cuffdiff2, are based on transcript detection, designed not only to identify known transcripts or isoforms but also to unveil unknown ones [59].

Presently, there is no consensus on which methodology is most appropriate or which approach ensures the validity of results in terms of robustness, accuracy, and reproducibility. This area of research in bioinformatics is still evolving. Despite these challenges, ongoing progress in bioinformatics tools and pipelines is continually improving the accuracy and reliability of differential expression analyses. This reflects a dynamic and evolving landscape, with the collective goal of refining methodologies to extract meaningful insights from complex omic data.

3.4 Exploring the Regulatory Landscape: ncRNAs as Conductors of Gene Expression in Helminths

Beyond the protein-coding genes, a fascinating world of ncRNAs has emerged as conductors of gene expression in helminths. ncRNAs are a diverse class of RNA molecules that do not encode proteins. They play a wide range of roles in gene regulation, including transcription, translation, RNA processing, and post-transcriptional modification [60]. These small regulatory molecules fine-tune the expression of protein-coding genes, adding an extra layer of complexity to the transcriptional landscape [35, 51, 53, 55, 61].

ncRNAs can be classified into several different types based on their length, function, and location in the genome [60].

- *Small ncRNAs:* Small ncRNAs are typically 20–50 nucleotides in length. They include miRNAs, which are involved in post-transcriptional gene silencing, and piRNAs, which are involved in transposon silencing [60].
- *Medium-Sized ncRNAs:* Medium-sized ncRNAs are typically 50–200 nucleotides in length. They include small nucleolar RNAs, which are involved in RNA processing, and transfer RNAs, which are involved in protein translation [60].
- *Long ncRNAs:* Long ncRNAs are typically more than 200 nucleotides in length. They include lncRNAs, which can regulate gene expression at multiple levels, and antisense transcripts (asRNAs), which are complementary to coding RNAs and can regulate gene expression by blocking translation [60].

In helminths, ncRNAs have been implicated in a variety of biological processes, including:

- *Development:* ncRNAs have been shown to regulate the expression of genes involved in helminth development, including cell differentiation, organogenesis, and growth [60]. As an exemplar, a recent study conducted on the liver fluke *Fasciola hepatica* revealed dynamic changes in the expression of lncRNAs during different stages of parasite development, mirroring the patterns observed in protein-coding mRNAs [62]. This observation suggests pivotal roles for lncRNAs in the growth and development of the parasite. Additionally, the study identified certain lncRNAs with expression patterns that correlate developmentally with mRNAs encoding crucial parasite proteins. This correlation implies a potential interaction between these lncRNAs and mRNAs, indicating an essential regulatory function in the intricate processes governing parasite development [62].
- *Host–parasite interaction:* ncRNAs also play a role in host–parasite interaction. Indeed,

ncRNAs have been demonstrated to play crucial roles in regulating the expression of genes involved in various aspects of parasite biology, including immune evasion, parasite nutrition, and host–parasite signaling [60, 62, 63]. In the context of parasite–host interactions, efforts have been made in diverse models to uncover the potential key functions of ncRNAs. For instance, in the study of the parasite *Echinococcus granulosus*, the focus was placed on ncRNAs stored and transported within exosome-like vesicles (ELVs), which have emerged as pivotal mediators in both parasite–parasite intercommunication and parasite–host interactions [63]. The examination of ncRNAs within ELVs yielded insights into the mechanisms through which these non-coding molecules contribute to the dynamic interplay between parasites and their hosts. The study identified specific ncRNA profiles in ELVs that may be involved in host immunity and pathogenesis. These findings provide valuable resources for further analysis of the regulatory potential of ncRNAs, particularly miRNAs, within ELVs at the parasite–host interface [63]. This approach significantly advances our understanding of the intricate molecular communication systems employed by parasites, unveiling their strategies for interacting with other parasites and their host environments.

- *Resistance to Drugs:* ncRNAs can also be essential in drug resistance. In the case of *Schistosoma mansoni*, the ribonucleoside analog 5-azacitidine (5-AzaC) is recognized as an epigenetic drug that exerts inhibitory effects on oviposition and ovarian development in this parasite [64, 65]. The inhibition is achieved by interfering with parasite transcription, translation, and stem cell activities. Consequently, the investigation of downstream pathways affected by 5-AzaC in *S. mansoni* holds the potential to unveil new drug targets [64, 65].

  Notably, lncRNAs have been implicated in drug resistance, and a study conducted in the *S. mansoni* model demonstrated that numerous lncRNAs exhibit differential expression following *in vitro* treatment with 5-AzaC in females [64]. This set of lncRNAs includes intergenic, antisense, and sense lncRNAs. A significant proportion of these lncRNAs is associated with co-expression network modules linked to male metabolism and displays differential expression in unpaired versus paired females and ovaries [64]. Intriguingly, half of these identified lncRNAs bear histone marks at their genomic loci, signifying regulation by histone modification. This observation underscores that many lncRNAs potentially involved in the reproductive biology of *S. mansoni* are modulated by 5-AzaC, emphasizing their relevance to the parasite's response to drug treatments [64]. This insight contributes to the ongoing exploration of epigenetic interventions and their implications for the development of novel therapeutic strategies against schistosomiasis [64, 65].

3.5 Organoids to Transcriptomics Research

Transcriptomic studies aimed at unraveling helminth biology and the interactions between parasites and hosts have traditionally relied on naturally infected hosts and animal models [21, 66]. However, the considerable variability introduced by host-associated factors, including genetic and immunological background, susceptibility, co-morbidities, and co-infections, among others, can lead to heterogeneity in findings and challenges in result reproducibility. Consequently, there is a growing need to explore alternatives that can effectively control both technical and biological variations in experimental setups [21, 29].

Indeed, the revolutionary and currently implemented approach to address challenges in helminth research involves the use of organoids [20, 67–69]. Organoids are *in vitro* multicellular clusters derived from either primary tissue stem cells or embryonic or induced pluripotent stem cells. These cells are cultured in conditions that simulate the stem cell niche (SCN) of a specific organ, allowing for the generation of various differentiated cell types. Organoids exhibit characteristics such as self-renewal, organization, and some level of functionality and architecture similar to the tissue of origin. Essentially, they represent 2D–3D miniature versions of organs or tissues that can be grown in the laboratory [67].

The use of organoids provides a more controlled and reproducible system for studying host–parasite interactions. Researchers can create three-dimensional structures that mimic the complexity of tissues or organs *in vitro* [67]. This innovative technique offers a valuable tool to dissect the intricacies of helminth biology and the host response while minimizing the confounding effects of host-associated variables. Organoid-based transcriptomic studies contribute to more accurate and reproducible insights into the molecular mechanisms underlying host–helminth interactions. The adoption of organoids in helminth research represents a significant advancement in experimental design, providing a platform that allows for more controlled and reliable investigations into the complex interactions between parasites and their hosts [68–70]. Organoids offer several advantages over traditional *in vitro* culture systems. They are more complex and realistic than cell culture, and they can recapitulate the *in vivo* environment of the parasite. This makes them ideal for studying gene expression in the context of development, host–parasite interaction, and drug resistance.

An exemplary study involved the infection by *Heligmosomoides polygyrus* (Hp), a mouse co-evolved pathosymbiont, to assess the epithelial response to mucosal barrier rupture. This parasite alters tissue integrity when it penetrates the duodenal mucosa, developing surrounded by a multicellular granulomatous infiltrate [71]. By utilizing organoids to study the interaction between this helminth and the host, it was possible to identify that larval invasion of the submucosal tissue results in the breakage of the epithelial barrier. This breach leads to a change in stem cell phenotype from Lgr5+ to Sca1+, resembling a fetal intestinal epithelial profile that is proliferative or indicative of wound healing [71].

The observed helminth-induced changes in crypt epithelia suggest a utilization of some of the functional capacities of the developing fetal intestine. This highlights a novel repair mechanism in the intestinal crypt involving infection-induced developmental plasticity. The use of organoids in this context provides a controlled and reproducible system for studying these intricate host–parasite interactions, revealing insights into the adaptive responses of the host's intestinal epithelium to helminth infection and the associated repair mechanisms [71]. Organoids are a promising new tool for helminth transcriptomics research. They offer the potential to provide new insights into the biology of these parasites and to develop novel strategies for their control [67–70].

Helminth transcriptomics has provided an intriguing glimpse into the dynamic realm of gene expression in these parasitic worms. From the diverse tools of RNA-seq to the complex interplay of differential expression and ncRNA regulation, each revelation contributes to the comprehensive understanding of helminth biology. This expanding knowledge presents significant potential for the innovation of novel diagnostics, drugs, and vaccines, ultimately contributing to the elimination of helminth-associated diseases and enhancing global health outcomes.

As research progresses, we can expect even more sophisticated techniques and tools to emerge, allowing us to listen to the faintest whispers of the helminth transcriptome and unravel the secrets hidden within its melodies. This journey promises not only a deeper understanding of these fascinating parasites but also a brighter future free from the burden of helminth-associated diseases.

4 Proteomic Insights into Helminth Functionality: Deciphering the Interplay of Proteins for Illuminating Parasite Biology and Enabling Novel Interventions

While gene expression in helminths provides invaluable insights, the actual functional performance relies on proteins. Helminth proteomics, employing diverse techniques, deciphers this interplay, unveiling the intricate relationships among proteins, their abundance, modifications, and interactions. This process ultimately illuminates parasite biology and lays the groundwork for novel interventions [15, 72–74].

4.1 Protein Identification and Characterization: Revealing the Key Players

The identification and characterization of helminth proteins represent a pivotal initial phase in understanding their biology. Mass spectrometry (MS) stands out as a dominant technique, offering diverse approaches with unique strengths. Shotgun proteomics, involving the fragmentation of proteins and analysis of resulting peptides, excels in identifying a broad spectrum of proteins within complex mixtures. In contrast, gel-based methods, such as 2D-PAGE, contribute valuable insights into protein isoforms and post-translational modifications [15, 29, 30, 72–74].

The advent of high-resolution MS instruments and sophisticated bioinformatics tools further elevates the capabilities of protein identification and characterization. These advancements empower researchers to explore the helminth proteome more comprehensively, providing a nuanced understanding of the diverse proteins involved in helminth biology. Whether unraveling intricate protein networks or deciphering post-translational modifications, the integration of various MS-based approaches facilitates a deeper exploration of the molecular landscape within helminths [15, 72–74].

For instance, in a study, high-resolution two-dimensional polyacrylamide gel electrophoresis (2-DE) coupled with matrix-assisted laser desorption/ionization time-of-flight/time-of-flight MS (MALDI-TOF/TOF MS) was employed to characterize and compare the proteomic profiles of *Ascaris lumbricoides* and *Ascaris suum*. This study successfully identified and compared proteins between the two nematode species, offering a detailed analysis of their proteomes [75].

The results of the study revealed a significant increase in the number of protein spots obtained (approximately threefold higher) compared with traditional methods. More than 630 and 750 protein spots were reproducibly detected among the six gels examined. Peptide mass fingerprints were acquired successfully for 17 out of 22 differential 2-DE spots, corresponding to 12 unique proteins involved in several metabolic, molecular, and structural functions. Functional analysis indicated a low level of difference among the differentially expressed proteins (DEPs) of the two nematode species, suggesting their close taxonomic relationship [75].

The study concludes that the combined 2-DE and MALDI-TOF/TOF MS approaches enable the identification of proteins with different expression levels between *Ascaris species*. The findings suggest a close taxonomic relationship between *A. lumbricoides* and *A. suum*, providing proteomic evidence for cross-host transmission and hybrid populations. This research underscores the value of the combined 2-DE and MS approach in identifying DEPs in *Ascaris* species, offering valuable insights into their proteomic diversity and potential implications for host interactions (Fig. 3) [75].

4.2 Protein Expression Profiling: The Dynamics of the Performance

Protein expression in helminths is a dynamic process, subject to fluctuations under diverse conditions and life stages. The study of protein expression profiling, employing techniques such as 2D-PAGE or quantitative MS, allows for the exploration of these dynamic changes. For instance, a quantitative proteomic analysis of *Fasciola hepatica* somatic proteins in different growth phases was conducted using high-performance liquid chromatography-tandem MS [76]. In this study, a total of 6228 proteins were identified across various growth phases of *F. hepatica*. Gene ontology analysis unveiled that DEPs participated in diverse cellular processes, including transport, metabolism, enzyme regula-

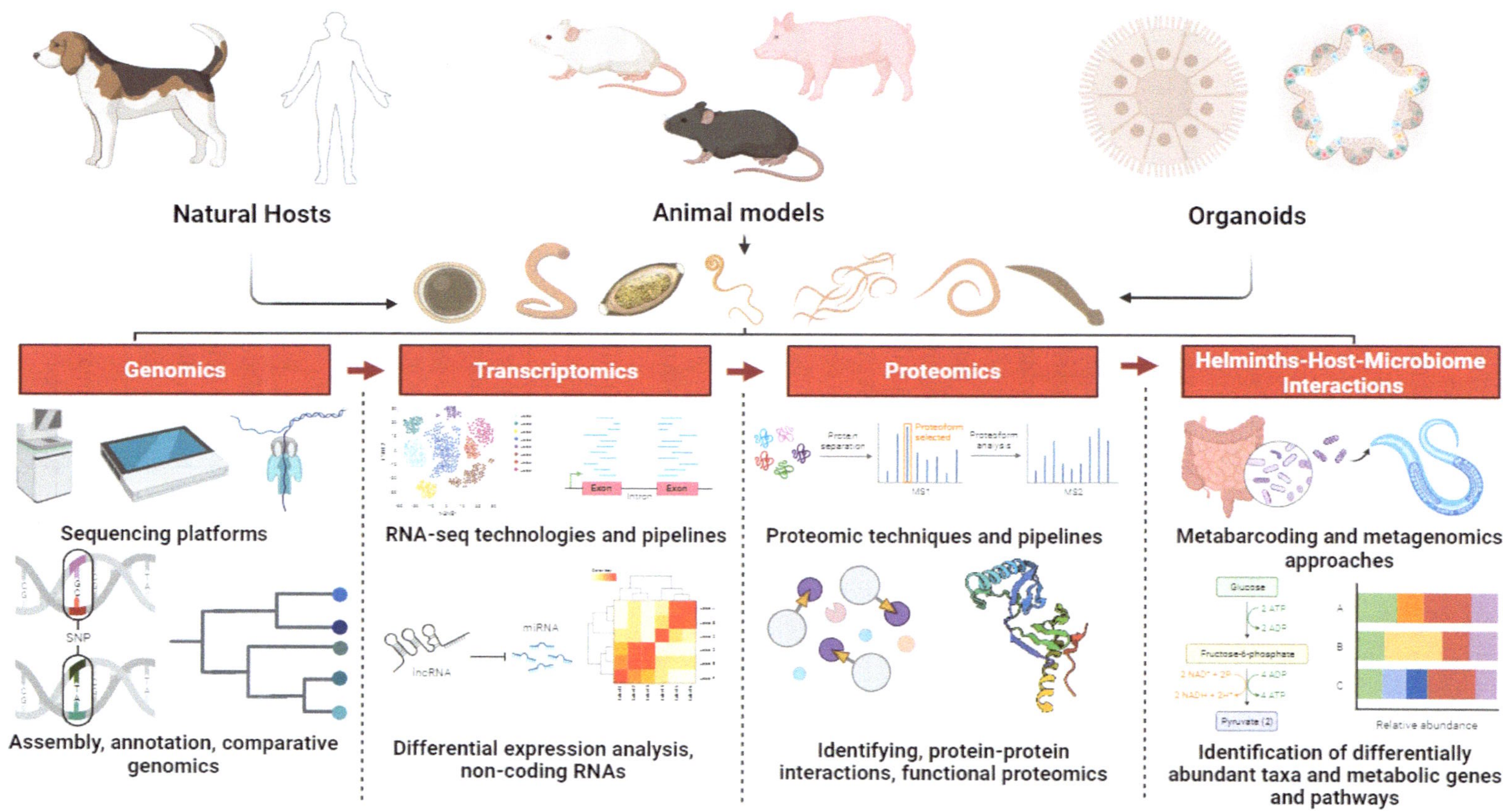

Fig. 3 Typical workflow for mass spectrometry-based proteomic analyses of helminths. This process commences with obtaining a sample from specimens of parasites, excretory/secretory (ES) products, or exovesicles (EVs). Subsequently, the sample is subjected to extraction and homogenization, followed by the separation of proteins through techniques such as liquid chromatography (LC), and then mass spectrometry (MS) analysis. Finally, the obtained data undergo several bioinformatics protocols, encompassing protein identification and quantification, statistical and enrichment analysis, and the identification of metabolic pathways and networks. It is important to note that there are various strategies that may vary between protocols, including the avoidance or inclusion of specific steps

tion, and protein folding and binding. The DEPs were found to be enriched in cells, intracellular components, organelles, cytoplasm, vesicles, and membranes. Additionally, the Kyoto Encyclopedia of Genes and Genomes pathway analysis demonstrated the involvement of DEPs in metabolism, genetic information processing, environmental information processing, cellular processes, organismal systems, and other processes. The outcomes of this study provide a theoretical foundation for vaccine development and the establishment of early diagnostic methods for *F. hepatica*. By comprehensively characterizing the proteomic changes in different growth phases, the research contributes valuable insights that can inform strategies for controlling and combating infections caused by *F. hepatica* [76].

In a recent study, the first comparative analysis of sex-specific tegument proteins in paired or virgin *Schistosoma mansoni* was conducted [40]. The researchers employed a novel and highly sensitive workflow for protein analysis, enabling the detection of low-abundance proteins. In total, 1519 tegument proteins for male and female worms were identified and categorized by sex. Bioinformatic analysis of the identified proteins revealed specific functions for female-specific and male-specific tegument proteins. These functions encompass involvement in signaling pathways, cellular processes, antioxidant mechanisms, phosphorylation, and signal transduction. The study provides a foundational understanding of the sex-specific tegument protein profiles in *S. mansoni*, offering insights into the strategies employed by the two worm sexes to evade the immune system. This research lays the groundwork for further investigations aimed at comprehending the intricacies of sex-specific biology in schistosomes. Such understanding is crucial for the development of new drugs and interventions for infection control, contributing to the ongoing efforts to combat schistosomiasis [40].

4.3 Protein–Protein Interactions: The Network of Collaboration

Proteins seldom act in isolation; rather, they form intricate networks through interactions. Helminth proteomics is increasingly employed to unravel these networks, offering insights into cellular processes and potential drug targets [15, 21, 30, 74]. Techniques such as co-immunoprecipitation and protein interaction assays identify protein partners, while yeast two-hybrid systems and affinity chromatography further elucidate the nature of these interactions [73, 74]. Moreover, bioinformatic and statistical models have been designed to explore these interactions, enabling the evaluation of large datasets to understand complex protein–protein interaction networks [77].

For instance, a study conducted in a tapeworm model describes the construction of a probabilistic, integrated network of protein–protein interactions for the tapeworm *Hymenolepis microstoma* [78]. A third of the gene models in the tapeworm's genome could be assigned interaction data, and the resulting network indicates that topologically important proteins are related to essential cellular pathways. The network clusters into biologically meaningful components and exhibits parameters similar to interaction networks of other model organisms. This study serves as a valuable resource for systems-level analysis in parasitic flatworms and informs future research on helminths.

In this field, it is crucial to understand these protein–protein networks and their implications in parasite–host interactions. For instance, a study characterizing the immunomodulatory protein of the *Brugia malayi* parasite, known as BmCRT, and its interaction with the human host immune component C1q, reveals that BmCRT inhibits the classical complement pathway activation by interacting with C1q [79]. This interaction is confirmed through the inhibition of C1q-dependent lysis of red blood cells. The study also identifies the amino acids involved in the interaction between BmCRT and C1q and explores the role of calcium in the formation of the BmCRT–C1q complex. These findings provide valuable information on the architecture and chemistry of the BmCRT–C1q interaction, offering insights that may aid in developing strategies to block the interaction and prevent parasite infection [79].

4.4 Functional Proteomics: Decoding the Protein's Role

Functional proteomics in the context of helminths aims to decode the roles of proteins. Various techniques, such as enzyme activity assays, ligand binding assays, and protein localization studies, are employed to elucidate how these molecules contribute to the biology of the parasite. These methodologies provide valuable insights into the functional aspects of helminth proteins, offering a deeper understanding of their roles and paving the way for potential therapeutic interventions and control strategies [29, 30, 73, 74].

An exemplary study found that serine proteases secreted by the nematode *Trichuris muris* degrade the mucus barrier in the intestine [80]. These proteases act on the major intestinal mucin called Muc2, causing depolymerization of Muc2 polymers and making the mucus barrier more porous. However, the respiratory/gastric mucin Muc5ac, induced during a TH2-mediated worm expulsion response, is protected from degradation by the nematode-secreted proteases. The study also revealed that serine protease inhibitors (Serpins) are highly expressed in mice resistant to chronic infection, indicating their role in protecting mucins, especially Muc2, from depolymerization. Overall, the study demonstrates how nematodes modify the mucus barrier to prevent clearance from the host or facilitate efficient mating and egg laying, highlighting the protective mechanisms of serpins, Muc5ac, and increased levels of Muc2 during a TH2-mediated worm expulsion response [80].

Similarly, ligand binding assays in *Onchocerca volvulus* revealed proteins interacting with host immune molecules, providing insights into parasite immune evasion strategies [81]. This study identified Ov58GPCR as a G-protein coupled receptor (GPCR) antigen conserved in related nematodes. It demonstrated that synthetic peptides predicted to be in the extracellular domain of Ov58GPCR are indeed immunogenic epitopes in actively infected individuals. The study also showed that synthetic peptide cocktails can discriminate between actively-infected individuals, treated individuals, and healthy controls. Moreover, it found that Ov58GPCR is transcribed in both larvae and adult parasite stages and observed that IgG and IgE responses to the recombinant extracellular domain of Ov58GPCR decline with ivermectin treatment. These findings suggest that the extracellular domain and synthetic peptides of Ov58GPCR, as well as the specific immune response generated, could be used for disease diagnosis and surveillance. This has significant implications for the development of tools for mapping endemic regions and monitoring treatment success in onchocerciasis control programs [81].

In conclusion, helminth proteomics has emerged as a potent tool, shedding light on the intricate interplay of proteins in these parasitic worms. From the identification and characterization of players to understanding their dynamic expression, interactions, and functions, each discovery contributes a new dimension to our understanding of helminth biology. This knowledge holds significant promise for the development of innovative therapeutic interventions, ultimately leading to the eradication of helminth-associated diseases and improving global health outcomes.

5 Helminths–Host–Microbiome Interactions: Unveiling the Hidden Players: A Deep Dive into Helminth–Host–Microbiome Relationships

Interactions between hosts and helminth parasites are intricate yet crucial for comprehending the biology, dynamics, and pathophysiology of these infections. Among these interactions, extensive research has been dedicated to understanding the immunomodulatory capabilities of helminths within the host. Additionally, there has been a growing interest in recent years regarding how helminths engage with the host's intestinal microbiome. This interest underscores the essential role this relationship plays in the initial infection, survival, and persistence of the parasite, as well as in the development of chronic infections [82–86].

Examining the alterations induced by helminths in the composition and structure of the host's intestinal microbiome, through metabarcoding and metagenomics approaches, constitutes a field of significant scientific interest. This exploration holds promise in offering crucial and actionable insights for designing effective control and therapeutic strategies [82–86]. This section delves into the intricate interactions between helminths, the microbiome, and the host, unraveling the multifaceted interplay that shapes parasite biology, host susceptibility, and potential therapeutic approaches.

5.1 Helminth–Microbiome Interactions: Active Coordination and Structural Reshaping of the Host's Gut Microbiota

Far from being passive inhabitants, helminths actively coordinate alterations in the gut microbial landscape through a multifaceted strategy. Numerous studies, employing cross-sectional analyses in natural hosts and diverse animal models, have showcased the helminth's capacity to reshape the structure of the host's microbiome [82–86].

One illuminating study evaluated the impact of *Hymenolepis diminuta* colonization on the cecal microbiome of rats. The findings revealed significant changes, particularly within the Bacillota (synonym Firmicutes) phylum. Approximately 20% of the total bacteria exhibited alterations, predominantly marked by a transition from Bacilli to Clostridia species in the presence of the helminth. These results suggest intricate ecological relationships within different components of the biome and propose potential mechanisms through which the helminth may exert therapeutic effects [87].

Employing a similar methodological approach, a study aimed at assessing the infection of pigs with *Ascaris suum*, a roundworm, revealed a notable decrease in microbial diversity within the gut microbiome independently of the number of worms present in the colon [88]. The infection induced significant alterations in the abundance of certain genera in the colon contents. Additionally, it prompted a shift in the metabolic potential of the gut microbiome, manifesting changes in the abundance of metabolic pathways related to carbohydrate and amino acid metabolism. In infected pigs, the microbial co-occurrence network exhibited a high degree of modularity. Specific modules demonstrated negative correlations with fecal butyrate concentrations, while others displayed negative correlations with fecal acetate, propionate, and total short-chain fatty acids. Notably, a strong positive correlation was identified between certain operational taxonomic units assigned to the Proteobacteria family and fecal acetate and propionate levels. Concurrently, the family Porphyromonadaceae exhibited a positive correlation with fecal egg counts. In summary, the study concluded that *Ascaris* infection exerted a significant impact on the gut microbiome, particularly in the proximal colon [88]. These findings offer valuable insights into the potential pathological consequences of gastrointestinal nematode infections [86, 88].

5.2 Microbiome-Mediated Host Immunity

Conversely, the gut microbiome plays a pivotal role in determining an individual's susceptibility and immune response to helminth infection. To illustrate, a study focusing on the impact of chronic infection with the large-intestinal nematode *Trichuris muris* on the murine intestinal microbiota, using 16S ribosomal-RNA gene-based sequencing, revealed significant alterations induced by the infection. These changes included a pronounced decrease in diversity and an elevation in the abundance of *Lactobacillus*. Moreover, the infection led to a notable shift in the balance between regulatory and inflammatory T cells within the intestinal adaptive immune system, favoring inflammatory cells [89]. This research underscores the intricate and dynamic interactions among parasite infection, the microbiota, and the immune system. Such complexities emphasize the need for a holistic understanding of these relationships to comprehend the nuanced mechanisms governing host–helminth interactions.

In the context of the same helminth model, a recent and notably conclusive delved into the *Trichuris*-host microbiota regulation hypothesis [90]. Using a Swiss Webster mouse model, the authors substantiated the occurrence of changes in the intestinal microbiota and elucidated the pivotal immunomodulatory effects induced by *Trichuris muris*. These effects play a crucial role in the establishment, progression, and persistence of the parasite's life cycle, facilitating the development of chronic infections [90]. Moreover, the study monitored a subset of dexamethasone-induced tolerant mice, mimicking a stage of chronic infection, to document changes in the intestinal microenvironment. Notably, *T. muris*-infected mice exhibited weight loss, anemia, an imbalance in the intestinal microbiota marked by a significant increase in *Escherichia coli* and *Bacteroides* bacteria, heightened tissue damage at the intestinal level, and alterations in the Th1/Th2 immune profile [90]. These findings underscore the essential role of specific bacterial taxa in facilitating the helminth's development while concurrently creating a permissive environment for its thriving. They also emphasize the imperative need for a deeper understanding of these interactions to unveil the mechanisms linked to the clinical progression of the infection. Ultimately, such insights are critical for the evaluation and implementation of novel control and treatment strategies.

In the case of the roundworm *Ascaris*, it has been observed that excretory-secretory (E/S) products, including extracellular vesicles and other metabolites, may elicit an immunomodulatory response linked to indirect interactions with the host's intestinal microbiota [91]. It has been identified various E/S products that interact in diverse ways with the host microbiota, potentially playing a pivotal role in both parasite biology and the pathophysiology of infection [91]. Studies in this context have specifically confirmed that *Ascaris suum* possesses the capability to induce the production of *Ascaris suum* antibacterial factor type of antimicrobial peptides and microbial peptides of the cecropin family (P1 to P4) in the host [84, 86, 92, 93]. These substances have demonstrated bactericidal effects against a broad spectrum of microbes, including Gram-positive bacteria (*Staphylococcus aureus*, *Bacillus subtilis*, and *Micrococcus luteus*) and Gram-negative bacteria (*Pseudomonas aeruginosa*, *Salmonella typhimurium*, *Serratia marcescens*, and *Escherichia coli*) [84, 86, 92, 93]. This underscores the roundworm's capacity to produce a variety of nematode-derived metabolites capable of exerting significant modulating effects on the host's intestinal microbiota. Consequently, this aspect stands out as one of the most prominent and trending areas in the field of helminth biology.

The helminths–host–microbiome interaction is a complex and dynamic interplay with profound implications for human health and disease. By deciphering the configuration of this multifaceted interface, we can unlock novel therapeutic strategies that harmonize with the intricate ecosystem of the human gut. Ultimately, understanding and manipulating this tripartite correlation holds immense promise for enhancing human health and well-being in the face of helminth-associated challenges.

5.3 Helminth–Microbiome Interactions: Implications in Health and Disease

The intricate relationship between helminths and the microbiome carries significant implications for human health and disease. Despite the immunomodulatory properties of helminth infections, they paradoxically correlate with reduced rates of allergic diseases and autoimmune disorders [94, 95]. This phenomenon is partly ascribed to microbiome changes induced by helminths, which promote the presence of anti-inflammatory bacterial species. On the flip side, helminth infections can potentially exacerbate conditions such as inflammatory bowel disease, perhaps through dysbiosis triggered by the immune responses elicited by the parasites.

Exploring the intricate world of helminth–microbiome interactions unveils exciting avenues for future research and therapeutic interventions [20, 84, 86, 96–98]. Leveraging prebiotics and probiotics presents an opportunity to manipulate the gut microbiome, either disrupting parasite development or enhancing host immunity. Moreover, considering fecal microbiota transplantation from individuals naturally resistant to helminth infections emerges as a novel approach to parasite control [86].

The multidimensional interface between helminths and the gut microbiome presents an intriguing context of dynamic interactions, where parasites and their microbial cohabitants mutually influence each other. This intricate interplay holds immense potential for understanding and manipulating parasite biology to improve human health. Deciphering the nuances of this microbial interaction enables the development of innovative therapeutic strategies that align with the complex ecosystem of the human gut, promising a future free from the burden of helminth-associated diseases.

6 Applications of Helminth Genomics, Transcriptomics, and Proteomics: Toward a Brighter Future

The intricate tapestry of helminth biology, woven from the threads of genomics, transcriptomics, and proteomics, presents a treasure trove of knowledge with immense potential for practical applications. By harnessing the power of these advanced tools, we can illuminate novel pathways to combat helminth-associated diseases, ultimately leading to improved global health outcomes (Fig. 4).

6.1 Drug Discovery and Development: Targeting the Core

Helminth genomes offer a rich source of potential drug targets. By identifying essential genes, unique metabolic pathways, and virulence factors, researchers can pinpoint vulnerabilities in the parasite's defense mechanisms [15, 21, 30, 83]. For example, genomic, transcriptomic, and proteomic analyses of *Trichuris trichiura* and *Trichuris muris* have provided insights into the molecular host–parasite interactions of whipworm infections [99, 100]. These studies have uncovered genes expressed in a sex- or life stage-specific manner, characterized the transcriptional landscape of unique biological adaptations in whipworms, and described the regulated TH1-like immune response in whipworm-infected mice [99]. Potential drug targets against trichuriasis have been identified, and proteomic analysis of 231 proteins in the soluble non-embryonated egg extracts of *T. trichiura* has contributed valuable information about the parasite's proteome [100]. This knowledge enhances our understanding of the immunobiology and pathogenesis of human trichuriasis, providing potential targets for diagnosis and treatment, thus contributing to the management of this neglected tropical disease affecting millions of people worldwide [99, 100].

However, identifying targets is only the initial step. Functional validation and preclinical testing are crucial for ensuring efficacy and minimizing side effects. Helminth models, including *Caenorhabditis elegans* and *Schistosoma mansoni*, play a vital role in this process, allowing researchers to test candidate drugs and refine their strategies [19, 21, 39, 40, 53, 101–103]. Ultimately, the combined power of genomics, transcriptomics, and proteomics holds the promise of accelerating the discovery and development of novel, effective, and safe drugs against helminth infections.

6.2 Vaccine Development: A Shield Against Invasion

Vaccines present a long-term solution for controlling helminth infections by inducing protective immunity in the host. Insights from helminth genomics and transcriptomics play a crucial role in identifying parasite antigens and understand-

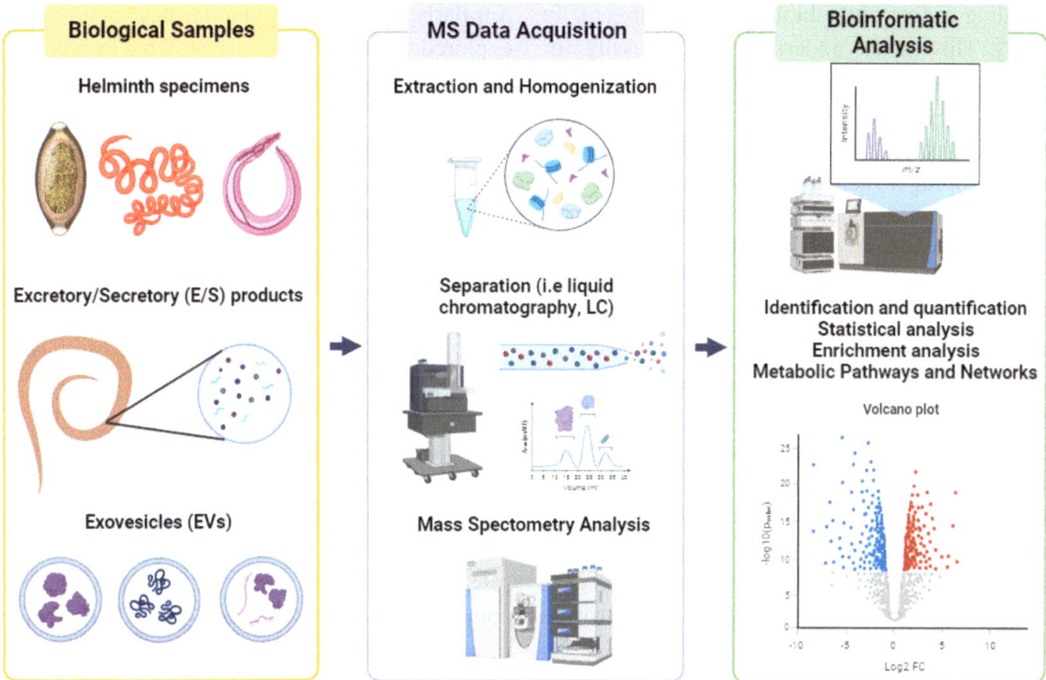

Fig. 4 Methodological approaches for studying helminths. Studies conducted in natural hosts, animal models (such as murine and porcine models), and organoids contribute significantly to advancing our understanding of helminths. The outlined approaches include the following: (i) genomic methods, employing second- and third-generation sequencing, enable the assembly, annotation, and comparative analysis of helminth genomes. These advancements shed light on helminth evolution, drug resistance mechanisms, and virulence factors. (ii) Transcriptomic approaches, utilizing RNA-seq technologies, allow for differential expression analysis and exploration of the role of non-coding RNAs in helminth biology. These methods provide valuable insights into gene regulation and the intricate dynamics of host-parasite interactions. (iii) Proteomics methodologies, utilizing techniques like mass spectrometry, facilitate the identification and characterization of helminth proteins. These methods also elucidate protein–protein interactions and analyze changes in protein expression under various conditions. (iv) The gut microbiome plays a vital role in parasite survival and host immunity. Understanding the intricate interplay between helminths and their microbial partners through metagenomic and metabarcoding strategies will offer new avenues for disease control and immunomodulatory strategies. The data generated through these approaches play a crucial role in advancing drug and vaccine development efforts

ing immunogenic pathways [104]. For instance, in the *Schistosoma mansoni* model, potential alternatives for the treatment and control of schistosomiasis are being explored [105, 106]. Current strategies face limitations due to the frequency of reinfection and the emergence of drug-resistant schistosomes. Recognizing the importance of developing effective vaccines aligns with the World Health Organization's goal of eliminating schistosomiasis as a public health problem by 2025 and interrupting transmission in low-prevalence regions [6, 12, 102]. Various studies, including genomic, transcriptomic, and proteomic analyses of *Schistosoma mansoni*, have identified surface proteins involved in host interaction and revealed antigens recognized by the human immune system. These findings pave the way for the development of subunit vaccines targeting specific parasite antigens [40, 105, 106].

Numerous advancements have been made in omics approaches aimed at developing new vaccines against various helminth infections [104]. The case of fasciolosis exemplifies the achievements and significance of omics tools in expediting vaccine discovery against these parasites, emphasizing the generation of extensive omics datasets for *Fasciola hepatica* and *Fasciola gigantica*, contributing to an improved understanding of liver fluke biology and pathogenicity [107]. This model underlines the necessity for rational, empirical, and collaborative approaches to identify vaccine candidates and conduct robust trials. It also underscores the challenges inherent in developing a vaccine against liver fluke infections, given the complexity of parasites and their extensive protein expression [107].

Furthermore, comprehending the molecular basis of host immunity is of equal importance. Helminth proteomics has illuminated host–parasite interactions, uncovering proteins involved in the immune evasion strategies employed by the parasite. This knowledge provides a foundation for designing vaccines that modulate the immune response and enhance host resistance against helminth infections [104]. Additionally, understanding the molecular basis of resistance to existing drugs is crucial for optimizing treatment strategies. Helminth genomics and transcriptomics can identify mutations and gene expression changes associated with drug resistance, allowing for personalized treatment regimens and the development of new drugs to combat resistance [5, 15, 21, 66].

While the road to developing effective helminth vaccines remains challenging, the combined power of genomics, transcriptomics, and proteomics is accelerating progress. By understanding the vulnerabilities of the parasite and the intricacies of host–parasite interactions, researchers can design innovative and targeted vaccines that offer lasting protection against these devastating diseases.

6.3 Diagnosis and Treatment: Precision at the Point of Care

Early and accurate diagnosis of helminth infections is crucial for effective treatment and preventing complications. Helminth genomics and proteomics offer promising avenues for developing rapid, sensitive, and specific diagnostic tools. For instance, the development of PCR-based assays targeting species-specific DNA sequences allows for the swift identification of helminth infections, even in asymptomatic individuals. Similarly, proteomic approaches are being explored to identify parasite proteins in blood or stool samples, offering the potential for non-invasive and rapid diagnosis [21, 29, 108–110].

An exemplary case involves the improvement of diagnostic tests for *Strongyloides stercoralis* through a comparative "omics" approach. By analyzing open-access "omics" datasets and utilizing protein analysis tools, researchers identified 22 potential candidate peptide/protein coproantigens for *S. stercoralis* diagnosis. These candidates hold the potential to be employed in the development of rapid diagnostic tests for *S. stercoralis*. This approach can also be extended to other parasitic infections, accelerating the development of specific diagnostic assays [111].

Likewise, in the *Schistosoma mansoni* model, researchers have applied computational approaches to choose candidate antigens for immunodiagnostic tests [112]. They selected seven proteins and identified one B-cell linear epitope from each protein. Synthetic peptides based on these epitopes were employed in ELISA assays to validate the in-silico selection strategy. The results indicated that the selected peptides could effectively distinguish between individuals with active infection and those without, showcasing their potential as diagnostic targets for schistosomiasis. The study also illustrated the feasibility of utilizing computational prediction to identify promising candidates for disease diagnosis [112].

By integrating these technologies into point-of-care settings, we can improve diagnosis, optimize treatment, and ultimately improve patient outcomes for individuals suffering from helminth infections.

The integration of helminth genomics, transcriptomics, and proteomics has woven a rich tapestry that illuminates the intricate world of parasitic worms. This comprehensive understanding opens up a plethora of applications, ranging from the development of targeted drugs and vaccines to precise diagnosis and optimized treatment strategies. The promise of these advancements is a future liberated from the debilitating burden of helminth-associated diseases, bringing hope for improved global health outcomes.

7 Prospects and Challenges in Helminth Biology: Navigating Unexplored Frontiers

The exploration of uncharted territories in helminth biology, guided by the illuminating tools of genomics, transcriptomics, and proteomics, has yielded a wealth of discoveries. Nevertheless, the horizon invites further investigation, offering promising avenues for future research and necessitating solutions to persisting challenges [15, 21, 113].

7.1 Future Directions: Advancing the Frontiers

- *Single-Cell Analysis*
  Delving deeper into the cellular heterogeneity within helminth tissues will provide crucial insights into development, differentiation, and host–parasite interactions. scRNA-seq and proteomics hold the key to unlocking the secrets of diverse cell populations within these complex organisms [21].
- *Functional Genomics and Proteomics*
  Beyond mapping genes and proteins, the future lies in understanding their functions. CRISPR-Cas9 and other gene editing tools offer opportunities to manipulate gene expression and elucidate the roles of specific genes in parasite biology. Similarly, functional proteomic assays will provide a deeper understanding of protein–protein interactions and enzyme activities.
- *Epigenetics and ncRNAs*
  The complex interplay between DNA methylation, histone modifications, and ncRNAs is emerging as a crucial regulatory layer in helminths. Exploring these mechanisms will shed light on parasite adaptation, drug resistance, and potential therapeutic targets [114].
- *Helminth–Microbiome Interactions*
  The gut microbiome plays a vital role in parasite survival and host immunity. Understanding the intricate interplay between helminths and their microbial partners will offer new avenues for disease control and immunomodulatory strategies [86, 97].
- *Comparative Helminthomics*
  By comparing the genomes, transcriptomes, and proteomes of diverse helminth species, we can identify conserved pathways and species-specific adaptations, leading to the development of broad-spectrum and targeted interventions [15, 21].

7.2 Challenges to Overcome: Charting the Course

- *Genome Assembly and Annotation*
  Many helminth genomes remain fragmented and incomplete, hindering downstream analyses. Continued advancements in sequencing technologies and assembly algorithms will be crucial for obtaining high-quality reference genomes. Additionally, accurate annotation of gene functions, particularly for novel genes, remains a significant challenge requiring innovative computational approaches and experimental validation [16, 17, 21].
- *Data Integration and Analysis*
  The sheer volume and complexity of data generated by these technologies necessitate robust bioinformatics tools and pipelines for data integration, analysis, and visualization. Developing user-friendly platforms for data sharing and collaborative analysis will be essential for accelerating research progress [22, 26, 28, 77, 115].
- *Translation to Applications*
  Bridging the gap between basic research and practical applications is crucial. Establishing strong collaborations between researchers, clinicians, and public health officials will be key to translating discoveries into effective diagnostics, drugs, and vaccines [15, 83, 104].
- *Funding and Infrastructure*
  Sustained funding and access to advanced research infrastructure, including high-performance computing facilities and specialized laboratories, are essential for

maintaining the momentum in helminth research and ensuring equitable access to these technologies for researchers in resource-limited settings [15, 21, 28, 54, 116–118].

As we navigate the uncharted seas of helminth biology, the combined power of genomics, transcriptomics, and proteomics provides a compass to guide our exploration. By embracing future directions and confronting challenges head-on, we can unlock the full potential of these technologies, ultimately charting a course toward a future free from the burden of helminth-associated diseases. By working together, researchers, clinicians, and public health officials can ensure that the knowledge gained from these voyages illuminates the path toward a healthier world for all.

8 Conclusion: Unveiling the Secrets, Unlocking the Solutions

The exploration across the chapters has revealed the remarkable capabilities of genomics, transcriptomics, and proteomics in elucidating the intricate world of helminths. From deciphering the genomic blueprint to observing the dynamic coordination of gene expression and the intricate interplay of protein interactions, each revelation sheds light on the strategies employed by these parasitic worms to thrive within their hosts.

Key findings throughout this exploration have painted a vivid picture of helminth biology. We have witnessed the remarkable diversity of their genomes. We have seen how RNA-seq unveils the dynamic landscape of gene expression, revealing the genes crucial for parasite development, survival, and host interaction. And we have marveled at the intricate networks of protein interactions, dictating everything from egg production to immune evasion.

The significance of these advancements extends far beyond mere knowledge. By unlocking the secrets of helminth biology, these technologies hold immense promise for the development of novel tools to combat the diseases these parasites inflict. From identifying and validating new drug targets to designing effective vaccines and improving diagnostic accuracy, the applications of this knowledge are vast and transformative.

However, the path forward is not without challenges. Overcoming the hurdles of genome assembly and annotation, developing robust data analysis tools, and translating basic research into practical applications remain critical steps in maximizing the impact of these technologies. Addressing these challenges requires continued investment in research infrastructure, fostering collaborations between scientists and public health officials, and ensuring equitable access to these powerful tools for researchers in resource-limited settings.

In conclusion, the journey through the world of helminth genomics, transcriptomics, and proteomics has been a testament to the transformative power of scientific exploration. By delving into the secrets of these parasitic worms, we have not only gained fundamental knowledge about their biology but also unlocked a treasure trove of potential solutions for the devastating diseases they cause. As we continue to navigate this exciting field, embracing future directions and overcoming challenges, we can confidently chart a course toward a future free from the burden of helminth-associated diseases, ultimately ensuring a healthier and more equitable world for all.

References

1. Ojha SC, Jaide C, Jinawath N, Rotjanapan P, Baral P (2014) Geohelminths: public health significance. J Infect Dev Ctries 8:5–16
2. Darwesh OM, El-Sayed HS, Darwesh OM, El-Sayed HS (2022) Perspective chapter: application of probiotics to inactivate helminth parasitic zoonosis. In: Parasitic Helminths and zoonoses - from basic to applied research. IntechOpen. https://doi.org/10.5772/INTECHOPEN.103744
3. WHO (2021) World Health Organization soil-transmitted helminth infections. https://www.who.int/news-room/fact-sheets/detail/soil-transmitted-helminth-infections. Accessed 15 Jul 2021
4. Jenkins TP, Rathnayaka Y, Perera PK, Peachey LE, Nolan MJ, Krause L, Rajakaruna RS, Cantacessi C (2017) Infections by human gastrointestinal hel-

minths are associated with changes in faecal microbiota diversity and composition. PLoS One 12:1–18
5. Yasuda K, Nakanishi K (2018) Host responses to intestinal nematodes. Int Immunol 30:93–102
6. Dejon-Agobé JC, Honkpehedji YJ, Zinsou JF et al (2020) Epidemiology of schistosomiasis and soil-transmitted helminth coinfections among schoolchildren living in Lambaréné, Gabon. Am J Trop Med Hyg 103:325–333
7. Houlden A, Hayes KS, Bancroft AJ, Worthington JJ, Wang P, Grencis RK, Roberts IS (2015) Chronic Trichuris muris infection in C57BL/6 mice causes significant changes in host microbiota and metabolome: effects reversed by pathogen clearance. PLoS One 10:e0125945. https://doi.org/10.1371/journal.pone.0125945
8. Chammartin F, Scholte RGC, Guimarães LH, Tanner M, Utzinger J, Vounatsou P (2013) Soil-transmitted helminth infection in South America: a systematic review and geostatistical meta-analysis. Lancet Infect Dis 13:507–518
9. King CH (2010) Parasites and poverty: the case of schistosomiasis. Acta Trop 113:95–104
10. Walker M, Hall A, Basáñez MG (2013) Ascaris lumbricoides: new epidemiological insights and mathematical approaches. In: Ascaris: the neglected parasite. Elsevier. https://doi.org/10.1016/B978-0-12-396978-1.00007-0
11. Jourdan PM, Lamberton PHL, Fenwick A, Addiss DG (2018) Soil-transmitted helminth infections. Lancet 391:252–265
12. McManus DP, Dunne DW, Sacko M, Utzinger J, Vennervald BJ, Zhou XN (2018) Schistosomiasis. Nat Rev Dis Prim 4:13
13. Specht S, Keiser J (2022) Helminth infections: enabling the World Health Organization road map. Int J Parasitol. https://doi.org/10.1016/J.IJPARA.2022.10.006
14. Hotez PJ, Brindley PJ, Bethony JM, King CH, Pearce EJ, Jacobson J (2008) Helminth infections: the great neglected tropical diseases. J Clin Invest 118:1311–1321
15. McVeigh P (2020) Post-genomic progress in helminth parasitology. Parasitology 147:835–840
16. Palevich N, Britton C, Kamenetzky L, Mitreva M, de Moraes MM, Bennuru S, Quack T, Scholte LLS, Tyagi R, Slatko BE (2018) Tackling hypotheticals in helminth genomes. Trends Parasitol 34:179–183
17. Coghlan A, Tyagi R, Cotton JA et al (2019) Comparative genomics of the major parasitic worms. Nat Genet 51:163–174
18. Howe KL, Bolt BJ, Shafie M, Kersey P, Berriman M (2017) WormBase ParaSite – a comprehensive resource for helminth genomics. Mol Biochem Parasitol 215:2–10
19. Ghedin E, Wang S, Foster JM, Slatko BE (2004) First sequenced genome of a parasitic nematode. Trends Parasitol 20:151–153
20. Gazzinelli-Guimaraes PH, Nutman TB, Ogretmen B et al (2020) Therapeutic helminth infection of macaques with idiopathic chronic diarrhea alters the inflammatory signature and mucosal microbiota of the colon. PLoS One 8:1–10
21. Doyle SR (2022) Improving helminth genome resources in the post-genomic era. Trends Parasitol 38:831–840
22. Amarasinghe SL, Su S, Dong X, Zappia L, Ritchie ME, Gouil Q (2020) Opportunities and challenges in long-read sequencing data analysis. Genome Biol 211(21):1–16
23. Buddenborg SK, Tracey A, Berger DJ et al (2021) Assembled chromosomes of the blood fluke Schistosoma mansoni provide insight into the evolution of its ZW sex-determination system. bioRxiv. https://doi.org/10.1101/2021.08.13.456314
24. Doyle SR, Sankaranarayanan G, Allan F et al (2019) Evaluation of DNA extraction methods on individual helminth egg and larval stages for whole-genome sequencing. Front Genet 10:1–12
25. Saboyá MI, Catalá L, Nicholls RS, Ault SK (2013) Update on the mapping of prevalence and intensity of infection for soil-transmitted helminth infections in Latin America and the Caribbean: a call for action. PLoS Negl Trop Dis 7:e2419. https://doi.org/10.1371/journal.pntd.0002419
26. Cao Y, Geddes TA, Yang JYH, Yang P (2020) Ensemble deep learning in bioinformatics. Nat Mach Intell 2:500–508
27. Korhonen PK, Young ND, Gasser RB (2016) Making sense of genomes of parasitic worms: tackling bioinformatic challenges. Biotechnol Adv 34:663–686
28. Allali I, Arnold JW, Roach J et al (2017) A comparison of sequencing platforms and bioinformatics pipelines for compositional analysis of the gut microbiome. BMC Microbiol 17:1–16
29. Jasmer DP, Rosa B, Tyagi R, Mitreva M (2019) Omics driven understanding of the intestines of parasitic nematodes. Front Genet 10:464556
30. Marzano V, Mancinelli L, Bracaglia G et al (2017) "Omic" investigations of protozoa and worms for a deeper understanding of the human gut "parasitome". PLoS Negl Trop Dis 11:1–19
31. Stanke M, Keller O, Gunduz I, Hayes A, Waack S, Morgenstern B (2006) AUGUSTUS: ab initio prediction of alternative transcripts. Nucleic Acids Res 34:W435–W439
32. Hoff KJ, Lomsadze A, Borodovsky M, Stanke M (2019) Whole-genome annotation with BRAKER. Methods Mol Biol 1962:65–95
33. Holt C, Yandell M (2011) MAKER2: an annotation pipeline and genome-database management tool for second-generation genome projects. BMC Bioinformatics 12:1–14
34. Haas BJ, Delcher AL, Mount SM et al (2003) Improving the Arabidopsis genome annotation using maximal transcript alignment assemblies. Nucleic Acids Res 31:5654–5666
35. Das C, Lucia MS HK and TJ (2017) Comparative genomics of the major parasitic worms. Physiol Behav 176:139–148

36. Wheeler N, Airs P, Zamanian M (2020) Long-read RNA sequencing of human and animal filarial parasites improves gene models and discovers operons. PLoS Negl Trop Dis 548902:1–22
37. Fauver JR, Martin J, Weil GJ, Mitreva M, Fischer PU (2019) De novo assembly of the Brugia malayi genome using long reads from a single MinION Flowcell. Sci Rep 91(9):1–10
38. Qing X, Kulkeaw K, Wongkamchai S, Tsui SKW (2021) Mitochondrial genome of Brugia malayi microfilariae isolated from a clinical sample. Front Ecol Evol 9:637805
39. Berger DJ, Crellen T, Lamberton PHL et al (2021) Whole-genome sequencing of Schistosoma mansoni reveals extensive diversity with limited selection despite mass drug administration. Nat Commun 12:1–14
40. Winkelmann F, Gesell Salazar M, Hentschker C, Michalik S, Macháček T, Scharf C, Reisinger EC, Völker U, Sombetzki M (2022) Comparative proteome analysis of the tegument of male and female adult Schistosoma mansoni. Sci Rep 121(12):1–10
41. Silva LL, Marcet-Houben M, Nahum LA, Zerlotini A, Gabaldón T, Oliveira G (2012) The Schistosoma mansoni phylome: using evolutionary genomics to gain insight into a parasite's biology. BMC Genomics 13:1–13
42. Curran DM, Grote A, Nursimulu N, Geber A, Voronin D, Jones DR, Ghedin E, Parkinson J (2020) Modeling the metabolic interplay between a parasitic worm and its bacterial endosymbiont allows the identification of novel drug targets. Elife 9:1–28
43. Rey O, Webster BL, Huyse T, Rollinson D, Van den Broeck F, Kincaid-Smith J, Onyekwere A, Boissier J (2021) Population genetics of African Schistosoma species. Infect Genet Evol 89:104727
44. Jex AR, Gasser RB, Schwarz EM (2019) Transcriptomic resources for parasitic nematodes of veterinary importance. Trends Parasitol 35:72–84
45. Blohm DH, Guiseppi-Elie A (2001) New developments in microarray technology. Curr Opin Biotechnol 12:41–47
46. Hoffmann KF, Johnston DA, Dunne DW (2002) Identification of Schistosoma mansonigender-associated gene transcripts by cDNA microarray profiling. Genome Biol 38(3):1–12
47. Ozsolak F, Milos PM (2010) RNA sequencing: advances, challenges and opportunities. Nat Rev Genet 122(12):87–98
48. Wang Z, Gerstein M, Snyder M (2009) RNA-Seq: a revolutionary tool for transcriptomics. Nat Rev Genet 101(10):57–63
49. Phuphisut O, Poodeepiyasawat A, Yoonuan T, Watthanakulpanich D, Chotsiri P, Reamtong O, Mousley A, Gobert GN, Adisakwattana P (2022) Transcriptome profiling of male and female Ascaris lumbricoides reproductive tissues. Parasit Vectors 15:1–18
50. Arora P, Moll JM, Andersen D, Workman CT, Williams AR, Kristiansen K, Brix S (2020) Body fluid from the parasitic worm Ascaris suum inhibits broad-acting pro-inflammatory programs in dendritic cells. Immunology 159:322–334
51. Liu X, Song Y, Lu H et al (2011) Transcriptome of small regulatory RNAs in the development of the zoonotic parasite Trichinella spiralis. PLoS One 6:e26448
52. Maloney JG, Molokin A, Santin M (2020) Use of Oxford Nanopore MinION to generate full-length sequences of the Blastocystis small subunit (SSU) rRNA gene. Parasit Vectors 13:1–8
53. Wendt G, Zhao L, Chen R, Liu C, O'Donoghue AJ, Caffrey CR, Reese ML, Collins JJ (2020) A single-cell RNA-seq atlas of Schistosoma mansoni identifies a key regulator of blood feeding. Science. https://doi.org/10.1126/SCIENCE.ABB7709/SUPPL_FILE/ABB7709_WENDT_SM.PDF
54. Saliba AE, Westermann AJ, Gorski SA, Vogel J (2014) Single-cell RNA-seq: advances and future challenges. Nucleic Acids Res 42:8845–8860
55. Chen G, Ning B, Shi T (2019) Single-cell RNA-seq technologies and related computational data analysis. Front Genet 10:441123
56. Ziegenhain C, Vieth B, Parekh S, Reinius B, Guillaumet-Adkins A, Smets M, Leonhardt H, Heyn H, Hellmann I, Enard W (2017) Comparative analysis of single-cell RNA sequencing methods. Mol Cell 65:631–643.e4
57. Wei J, Damania A, Gao X, Liu Z, Mejia R, Mitreva M, Strych U, Bottazzi ME, Hotez PJ, Zhan B (2016) The hookworm Ancylostoma ceylanicum intestinal transcriptome provides a platform for selecting drug and vaccine candidates. Parasit Vectors 9:1–14
58. Cheng S, Zhu B, Luo F et al (2022) Comparative transcriptome profiles of Schistosoma japonicum larval stages: implications for parasite biology and host invasion. PLoS Negl Trop Dis 16:e0009889
59. Costa-Silva J, Domingues D, Lopes FM (2017) RNA-Seq differential expression analysis: an extended review and a software tool. PLoS One 12:e0190152
60. Mattick JS, Makunin IV (2006) Non-coding RNA. Hum Mol Genet 15:R17–R29
61. Oberbauer V, Schaefer MR (2018) tRNA-derived small RNAs: biogenesis, modification, function and potential impact on human disease development. Genes (Basel). https://doi.org/10.3390/GENES9120607
62. McVeigh P, McCammick E, Robb E, Brophy P, Morphew RM, Marks NJ, Maule AG (2023) Discovery of long non-coding RNAs in the liver fluke, Fasciola hepatica. PLoS Negl Trop Dis 17:e0011663
63. Zhang X, Gong W, Cao S, Yin J, Zhang J, Cao J, Shen Y (2020) Comprehensive analysis of non-coding RNA profiles of exosome-like vesicles from the protoscoleces and hydatid cyst fluid of Echinococcus granulosus. Front Cell Infect Microbiol 10:529731
64. Amaral MS, Maciel LF, Silveira GO, Olberg GGO, Leite JVP, Imamura LK, Pereira ASA, Miyasato

PA, Nakano E, Verjovski-Almeida S (2020) Long non-coding RNA levels can be modulated by 5-azacytidine in Schistosoma mansoni. Sci Rep 101(10):1–17
65. Silveira GO, Coelho HS, Amaral MS, Verjovski-Almeida S (2021) Long non-coding RNAs as possible therapeutic targets in protozoa, and in Schistosoma and other helminths. Parasitol Res 1214(121):1091–1115
66. Klementowicz JE, Travis MA, Grencis RK (2012) Trichuris muris: a model of gastrointestinal parasite infection. Semin Immunopathol 34:815–828
67. Duque-Correa MA, Maizels RM, Grencis RK, Berriman M (2020) Organoids—new models for host–helminth interactions. Trends Parasitol 36:170–181
68. White R, Blow F, Buck AH, Duque-Correa MA (2022) Organoids as tools to investigate gastrointestinal nematode development and host interactions. Front Cell Infect Microbiol 12:976017
69. Smith D, Price DRG, Burrells A et al (2021) The development of ovine gastric and intestinal organoids for studying ruminant host-pathogen interactions. Front Cell Infect Microbiol 11:733811
70. Breyner NM, Hecht M, Nitz N, Rose E, Carvalho JL (2020) In vitro models for investigation of the host-parasite interface - possible applications in acute Chagas disease. Acta Trop 202:105262. https://doi.org/10.1016/j.actatropica.2019.105262
71. Nusse YM, Savage AK, Marangoni P, Rosendahl-Huber AKM, Landman TA, De Sauvage FJ, Locksley RM, Klein OD (2018) Parasitic helminthes induce fetal-like reversion in the intestinal stem cell niche. Nature 559:109
72. Aebersold R, Mann M (2003) Mass spectrometry-based proteomics. Nature 4226928(422):198–207
73. Patterson SD, Aebersold RH (2003) Proteomics: the first decade and beyond. Nat Genet 333(33):311–323
74. Cho WCS (2007) Proteomics technologies and challenges. Genomics Proteomics Bioinformatics 5:77–85
75. Xu MJ, Fu JH, Zhou DH, Elsheikha HM, Hu M, Lin RQ, Peng LF, Song HQ, Zhu XQ (2013) Ascaris lumbricoides and Ascaris suum: comparative proteomic studies using 2-DE coupled with mass spectrometry. Int J Mass Spectrom 339–340:1–6
76. Xu J, Wu L, Sun Y, Wei Y, Zheng L, Zhang J, Pang Z, Yang Y, Lu Y (2020) Proteomics and bioinformatics analysis of Fasciola hepatica somatic proteome in different growth phases. Parasitol Res 119:2837–2850
77. Sotillo J, Toledo R, Mulvenna J, Loukas A (2017) Exploiting Helminth–host interactomes through Big Data. Trends Parasitol 33:875–888
78. James K, Olson PD (2020) The tapeworm interactome: inferring confidence scored protein-protein interactions from the proteome of Hymenolepis microstoma. BMC Genomics 21:1DUMM
79. Yadav S, Gupta S, Selvaraj C, Doharey PK, Verma A, Singh SK, Saxena JK (2014) In silico and in vitro studies on the protein-protein interactions between Brugia malayi immunomodulatory protein Calreticulin and human C1q. PLoS One 9:e106413
80. Hasnain SZ, McGuckin MA, Grencis RK, Thornton DJ (2012) Serine protease(s) secreted by the nematode Trichuris muris degrade the mucus barrier. PLoS Negl Trop Dis 6:e1856
81. Shey RA, Ghogomu SM, Njume FN et al (2018) Prediction and validation of the structural features of Ov58GPCR, an immunogenic determinant of Onchocerca volvulus. PLoS One 13:e0202915
82. Glendinning L, Nausch N, Free A, Taylor DW, Mutapi F (2014) The microbiota and helminths: sharing the same niche in the human host. Parasitology 141:1255–1271
83. Whitman JD, Sakanari JA, Mitreva M (2021) Areas of metabolomic exploration for Helminth Infections. ACS Infect Dis 7:206–214
84. Midha A, Ebner F, Schlosser-Brandenburg J, Rausch S, Hartmann S (2021) Trilateral relationship: Ascaris, microbiota, and host cells. Trends Parasitol 37:251–262
85. Zaiss MM, Harris NL (2016) Interactions between the intestinal microbiome and helminth parasites. Parasite Immunol 38:5–11
86. Castañeda S, Paniz-Mondolfi A, Ramírez JD (2022) Detangling the crosstalk between Ascaris, Trichuris and gut microbiota: what's next? Front Cell Infect Microbiol 0:615
87. McKenney EA, Williamson L, Yoder AD, Rawls JF, Bilbo SD, Parker W (2015) Alteration of the rat cecal microbiome during colonization with the helminth Hymenolepis diminuta. Gut Microbes 6:182–193
88. Wang Y, Liu F, Urban JF, Paerewijck O, Geldhof P, Li RW (2019) Ascaris suum infection was associated with a worm-independent reduction in microbial diversity and altered metabolic potential in the porcine gut microbiome. Int J Parasitol 49:247–256
89. Holm JB, Sorobetea D, Kiilerich P, Ramayo-Caldas Y, Estellé J, Ma T, Madsen L, Kristiansen K, Svensson-Frej M (2015) Chronic Trichuris muris infection decreases diversity of the intestinal microbiota and concomitantly increases the abundance of lactobacilli. PLoS One 10:1–22
90. Schachter J, De OA, Miranda C, Alencar DB (2020) Whipworm infection promotes bacterial invasion, intestinal microbiota imbalance, and cellular immunomodulation. Infect Immun 88:1–24
91. Hansen EP, Fromm B, Andersen SD et al (2019) Exploration of extracellular vesicles from Ascaris suum provides evidence of parasite–host cross talk. J Extracell Vesicles 8. https://doi.org/10.1080/20013078.2019.1578116
92. Pillai A, Ueno S, Zhang H, Lee JM, Kato Y (2005) Cecropin P1 and novel nematode cecropins: a bacteria-inducible antimicrobial peptide family in the nematode Ascaris suum. Biochem J 390:207–214
93. Pillai A, Ueno S, Zhang H, Kato Y (2003) Induction of ASABF (Ascaris suum antibacterial factor)-type antimicrobial peptides by bacterial injection: novel

members of ASABF in the nematode Ascaris suum. Biochem J 371:663–668
94. Sobotková K, Parker W, Levá J, Růžková J, Lukeš J, Jirků Pomajbíková K (2019) Helminth therapy—from the parasite perspective. Trends Parasitol 35:501–515
95. Stiemsma L, Reynolds L, Turvey S, Finlay B (2015) The hygiene hypothesis: current perspectives and future therapies. Immunotargets Ther 4:143
96. Mutapi F (2015) The gut microbiome in the helminth infected host. Trends Parasitol 31:405–406
97. Rapin A, Harris NL (2018) Helminth–bacterial interactions: cause and consequence. Trends Immunol 39:724–733
98. Salloum PMM, Jorge F, Dheilly NMM, Poulin R (2023) Eco-evolutionary implications of helminth microbiomes. J Helminthol 97:e22
99. Foth BJ, Tsai IJ, Reid AJ et al (2014) Whipworm genome and dual-species transcriptome analyses provide molecular insights into an intimate host-parasite interaction. Nat Genet 46:693–700
100. Cruz K, Marcilla A, Kelly P, Vandenplas M, Osuna A, Trelis M (2021) Trichuris trichiura egg extract proteome reveals potential diagnostic targets and immunomodulators. PLoS Negl Trop Dis 15:e0009221
101. Wit J, Dilks CM, Andersen EC (2021) Complementary approaches with free-living and parasitic nematodes to understanding anthelmintic resistance. Trends Parasitol 37:240–250
102. da Paixão Siqueira L, Fontes DAF, Aguilera CSB, Timóteo TRR, Ângelos MA, Silva LCPBB, de Melo CG, Rolim LA, da Silva RMF, Neto PJR (2017) Schistosomiasis: drugs used and treatment strategies. Acta Trop 176:179–187
103. Danso-Appiah A, Olliaro PL, Donegan S, Sinclair D, Utzinger J (2013) Drugs for treating Schistosoma mansoni infection. Cochrane Database Syst Rev 2013:CD000528. https://doi.org/10.1002/14651858.CD000528.pub2
104. Daga V, Green E, Ravichandran P, Short M, May M, Daga V, Green E, Ravichandran P, Short M, May M (2022) Perspective chapter: multi-omic approaches to vaccine development against helminth diseases. In: Parasitic helminths and zoonoses - from basic to applied research. IntechOpen. https://doi.org/10.5772/INTECHOPEN.102621
105. Farias LP, Tararam CA, Miyasato PA, Nishiyama MY, Oliveira KC, Kawano T, Verjovski-Almeida S, Leite LCDC (2011) Screening the Schistosoma mansoni transcriptome for genes differentially expressed in the schistosomulum stage in search for vaccine candidates. Parasitol Res 108:123–135
106. Eyayu T, Zeleke AJ, Worku L (2020) Current status and future prospects of protein vaccine candidates against Schistosoma mansoni infection. Parasite Epidemiol Control 11:e00176
107. Cwiklinski K, Dalton JP (2022) Omics tools enabling vaccine discovery against fasciolosis. Trends Parasitol 38:1068–1079
108. Papaiakovou M, Pilotte N, Grant JR, Traub RJ, Llewellyn S, McCarthy JS, Krolewiecki AJ, Cimino R, Mejia R, Williams SA (2017) A novel, species-specific, real-time PCR assay for the detection of the emerging zoonotic parasite Ancylostoma ceylanicum in human stool. PLoS Negl Trop Dis 11:e0005734
109. Weerakoon KGADAD, Gobert GN, Cai P, McManus DP (2015) Advances in the diagnosis of human schistosomiasis. Clin Microbiol Rev 28:939–967
110. Sanprasert V, Kerdkaew R, Srirungruang S, Charuchaibovorn S, Phadungsaksawasdi K, Nuchprayoon S (2019) Development of conventional multiplex PCR: a rapid technique for simultaneous detection of soil-transmitted helminths. Pathogens 8. https://doi.org/10.3390/PATHOGENS8030152
111. Marlais T, Bickford-Smith J, Talavera-López C, Le H, Chowdhury F, Miles MA (2023) A comparative 'omics' approach for prediction of candidate Strongyloides stercoralis diagnostic coproantigens. PLoS Negl Trop Dis 17:e0010777
112. Carvalho GBF, Resende DM, Siqueira LMV, Lopes MD, Lopes DO, Coelho PMZ, Teixeira-Carvalho A, Ruiz JC, Fonseca CT (2017) Selecting targets for the diagnosis of Schistosoma mansoni infection: an integrative approach using multi-omic and immunoinformatics data. PLoS One 12:e0182299
113. Mitreva M (2022) Parasite OMICS, the grand challenges ahead. Front Parasitol 1:995302
114. Jaleta TG, Lok JB (2019) Advances in the molecular and cellular biology of Strongyloides spp. Curr Trop Med Rep 6:161–178
115. Midha A, Jarquín-Díaz VH, Ebner F, Löber U, Hayani R, Kundik A, Cardilli A, Heitlinger E, Forslund SK, Hartmann S (2022) Guts within guts: the microbiome of the intestinal helminth parasite Ascaris suum is derived but distinct from its host. Microbiome 10:1–21
116. Huttenhower C, Morgan XC (2014) Meta'omic analytic techniques for studying the intestinal microbiome. Gastroenterology 146:1437. https://doi.org/10.1053/j.gastro.2014.01.049
117. Papaiakovou M, Littlewood DTJ, Gasser RB, Anderson RM (2021) How qPCR complements the WHO roadmap (2021–2030) for soil-transmitted helminths. Trends Parasitol 37:698. https://doi.org/10.1016/J.PT.2021.04.005
118. Cools P, Vlaminck J, Verweij JJ, Levecke B (2021) Quantitative PCR in soil-transmitted helminth epidemiology and control programs: toward a universal standard. PLoS Negl Trop Dis 15:e0009134

Understanding the Extracellular Vesicles in Helminths

Antonio Marcilla, Christian M. Sánchez-López, Aránzazu González-Arce, Alba Cortés, and Dolores Bernal

Abstract

Since the re-description of extracellular vesicles (EVs) secreted by the trematodes *Fasciola hepatica* and *Echinostoma caproni* in 2012, these biological nanoparticles have been identified in the excretory/secretory products of numerous species of trematodes, cestodes, and nematodes. Recent studies have focused on their distinct origin in nematodes and flatworms, primarily due to their unique surface characteristics. Despite facing various challenges in their isolation and characterization, currently available guidelines describe and establish methods with rigor and standardization, facilitating reproducible studies across different laboratories worldwide.

Regarding the composition of helminth EVs, the majority of the studies have focused on their protein content, with subsequent reports on their small RNA content, and still limited publications addressing their lipid and glycan composition.

EVs participate in helminth–host communication, regulating and modulating the host immune system both in vitro and in vivo. These characteristics position them as promising candidates for controlling helminthiases. Different approaches have demonstrated their potential as biomarkers for diagnosis, vaccination in animal models, and therapeutic tools for various diseases, not limited to helminthiases but also encompassing diseases associated with the immune response. This review also discusses potential future research directions in the study of helminth EVs.

Keywords

Helminth · Extracellular vesicles · Host–parasite interaction · Excretory/secretory products

A. Marcilla (✉) · C. M. Sánchez-López
Área de Parasitología, Departament de Farmàcia i Tecnologia Farmacèutica i Parasitologia, Universitat de València, Burjassot, Valencia, Spain

Joint Research Unit on Endocrinology, Nutrition and Clinical Dietetics, Health Research IIS La Fe-Universitat de València, Valencia, Spain
e-mail: antonio.marcilla@uv.es

A. González-Arce · A. Cortés
Área de Parasitología, Departament de Farmàcia i Tecnologia Farmacèutica i Parasitologia, Universitat de València, Burjassot, Valencia, Spain

D. Bernal
Departament de Bioquímica i Biologia Molecular, Facultat de Ciències Biològiques, Universitat de València, Burjassot, Valencia, Spain

J. D. Ramírez González (ed.), *Recent Advances in Parasitomics*,
https://doi.org/10.1007/978-3-031-70591-5_13

1 Introduction

It is estimated that nearly a third of the human population is affected by helminths, which are extraordinarily prevalent in low-income countries, with more than 2000 million people infected [1]. Parasitic worms cause substantial morbidity and an enormous negative impact on the socio-economical development of affected individuals and their communities. In spite of this, and the fact that the number of people affected by helminthiases is increasing, these diseases are classified as "neglected" by the World Health Organization, meaning that they are placed at the very bottom of the global health agenda [2]. In the livestock sector, helminth infections are also responsible for important economic losses linked to direct and indirect costs, such as reduced animal performance and repeated anthelmintic treatments, respectively [3].

Interestingly, besides direct interactions of the parasite surface with host tissues, the pathology of helminthiases is related to parasite excretory/secretory products (ESPs), which interact with the host immune system in multiple ways (e.g., binding to or degrading selected molecules, or acting directly on host cells) [4]. Indeed, helminths establish long-term infections associated with a marked decrease and regulation of the host immune response, which, in addition, can interfere with the response to other infectious agents, resulting in aggravated disease outcomes [5]. ESPs include a variety of molecules, such as proteins, nucleic acids, lipids, carbohydrates, and small metabolites that worms release during feeding and migration. These products could therefore play a central role in establishing a beneficial niche for the parasite, although most of the interaction pathways between these and the host are still under investigation [6]. Moreover, ESPs may represent excellent candidates for vaccination against and/or diagnosis of helminth infections, whereas their immunoregulatory properties have led to exploring their use for treating autoimmune diseases [6–9].

Back in 2012, our group discovered that ESPs contain extracellular vesicles (EVs)—*membrane-surrounded, nanosized particles, secreted by cells into the extracellular space*—loaded with "moonlighting proteins" [10]. Interestingly, these proteins can perform unrelated functions depending on their cellular or developmental context, without compartmentalizing them into different protein domains [11]. Structures similar to EVs had been already visualized by transmission electron microscopy in the 60s at the surface of *Schistosoma mansoni* and *Fasciola hepatica* [12–14] and they were even named EVs in *Echinostoma caproni* by Andresen and coworkers in 1989 [15]. All helminth species studied up to now have been shown to produce and release EVs with key roles in their interaction with the mammalian host [16–22].

In this chapter, we describe the origin and types of helminth EVs, as well as current methods for their purification from parasite culture media or host samples. An optimal isolation of EVs is essential for their molecular and functional characterization, two aspects that are also thoroughly reviewed herein. Finally, we explore the potential applications of helminth EVs in the control of helminthiases and treatment of chronic inflammatory disorders and discuss current challenges and opportunities to advance the understanding of EV's immunobiology and their eventual translation into clinical practice.

2 Origin of EVs in Helminths

EVs are small nanostructures surrounded by a lipid bilayer that are released by all cell types, from prokaryotes to higher eukaryotic organisms [23]. They can be classified into diverse subtypes based on their biogenesis, biophysical and biochemical properties, and content. Currently, two major groups of EVs are recognized according to their size and biogenesis: ectosomes and exosomes. Briefly, the first ones are derived from the cell plasma membrane and include microvesicles and apoptotic bodies, among others. In contrast, exosomes have an endosomal origin; they are enclosed in multivesicular bodies formed in the cytoplasm and released outside the cell via the fusion of the limiting membrane of such multivesicular bodies with the plasma membrane [24].

In helminths, flatworms and nematodes exhibit marked differences in their external surface and that has a direct impact on the quantity of EVs produced by each group. While flatworms have an active tegument structure that participates in the secretion of EVs, roundworms present a rigid cuticular exoskeleton, mainly composed of collagen and other extracellular matrix proteins [25]. As a result, the secretion of EVs in nematodes occurs through the anterior apertures and/or the posterior end of the worm, besides some apertures in the cuticle known as excretory/secretory pores [26].

To date, the routes for EV secretion in helminths have not been entirely elucidated, although current investigations point toward the parasite intestine as the main source of EVs [19]. Most of these investigations have been carried out using *F. hepatica* as a model, whose EVs are released by the tegument, and also through the intestine, the oral sucker, the protonephridia system, and the gastrodermis [27].

3 Purification of Helminth EVs

Research on EVs is affected by the utilization of different and sometimes poorly reproducible isolation methodologies, diverse nomenclature, and the absence of standardized protocols for data acquisition and analysis. These factors limit the interpretation of results obtained from published research [28–30].

The International Society of Extracellular Vesicles supports the use of a new standardized nomenclature, suggesting the use of the term "extracellular vesicles" unless researchers can reliably demonstrate the origin and correct purification of their preparations. The Minimal Information for Studies of Extracellular Vesicles (MISEV) guidelines recommend the use of general nomenclature to describe EV preparations. Furthermore, they recommend characterizing EVs based on size (small, <200 nm, or medium/large >200 nm), biochemical composition (e.g., CD81+/CD63+), density (low, medium, high), or producing organism/cell [31–33].

While the study of mammalian EVs has largely benefited from various guidelines, such as MISEV [31, 32], the study of helminth EVs has faced additional challenges due to the lack of sufficient quantities of EVs, specific EV markers, and standardized methodologies [34]. Furthermore, the specific isolation of a particular subpopulation is difficult because of several reasons, including: (i) size overlap among the different EV populations; (ii) the occurrence of shared proteins in their biogenesis (such as those of the ESCRT complex); and (iii) the presence of widely used exosome markers also in ectosomes, although in varying degrees of abundance [24, 30, 35]. Moreover flatworms' surface can release higher amounts of EVs in comparison to nematodes, requiring fewer parasites and less time of incubation to obtain similar quantities [36]. It is important to mention that prolonged incubations of the parasites can lead to loss of viability depending on the worm species [36].

Recently, two "road maps" have been introduced to assist the helminth community in enhancing rigor and reproducibility in helminth EV studies [28, 30].

Variability across studies is further increased by the several methods described for EV obtention. EVs have been isolated using very diverse techniques, including differential ultracentrifugation (dUC), size exclusion chromatography (SEC), precipitation techniques based on polyethylene glycol, immunocapture, or the use of microfluidics [37]. Each of these methods has its own advantages and disadvantages, providing varying performance in terms of purity, preservation of physicochemical properties of EVs, or heterogeneity of the EV population obtained [38]. The two most commonly used techniques for isolating helminth EVs are dUC and SEC.

3.1 Differential Ultracentrifugation

To date, classical UC remains the most widely used approach for isolation of EVs [39, 40]. This technique is based on the sequential separation of particles based on their size and density, achieved through a series of centrifugal forces applied over a given time period [38]. The effectiveness of UC depends on several factors, including the sedimentation rate, type of rotor, viscosity of the sample, and the centrifugation time (parameters that must be specified according to current guidelines, as they have a significant impact on the population of isolated EVs) [33].

Despite its widespread use, UC has several limitations. It may lead to the co-isolation of proteins not present in EVs and particles of similar size or density and can also cause the formation of vesicle aggregates. Consequently, the popularity and use of this technique have declined, probably due to technological advances in EV isolation methods that are less demanding in time and resources and that are applicable to a high number of clinical samples [40].

Variations of UC, such as density gradient UC using sucrose or iodixanol, can improve the purity and yield of the isolated EVs [38]. However, this method comes with drawbacks, including significant loss of EVs, laborious protocols requiring highly qualified personnel, and a considerable time investment (up to 2 days) [41]. Overall, all these advantages and disadvantages make density gradient UC an interesting isolation protocol for validation tests in research laboratories, but not very useful in clinical settings.

3.2 Size Exclusion Chromatography

This technique, used in pioneering studies of EV isolation, uses the biological fluid to be analyzed as a mobile phase and a porous filtration polymer as the stationary phase [42]. The differential elution of molecules in the fluid is facilitated by the nature of the stationary phase and the chosen molecular exclusion limit: larger particles, including EVs, elute first, followed by proteins, other metabolites, and non-membrane-bound soluble substances. The stationary phase can consist of various polymers, such as agarose (sepharose) or dextrans (Sephadex™), among others [38].

The size of the particle determines the number of pores of the polymer it can pass through, affecting its elution time. Therefore, large particles will elute faster than small molecules. [43]. Additionally, the shape and hydration layer of EVs impact their elution time, with non-spherical EVs eluting differently from spherical ones of equal weight. Flow rate also plays a crucial role in the isolation of EVs, with low speeds providing higher resolution, specificity, and integrity of isolated EVs. It is important to note that SEC does not differentiate between exosomes and ectosomes of the same size; thus, it needs its combination with immunocapture techniques to fully discern between both populations [38].

SEC has been shown to outperform conventional UC and other techniques, such as ultrafiltration, for the purity of the obtained EVs, largely eliminating contamination with soluble proteins [41]. Moreover, it is efficient, with an average processing time of 20 min per sample, compared with techniques like gradient UC, which require longer isolation times [43]. SEC also preserves the integrity of EVs to a greater extent compared with conventional UC, where EVs may suffer damage or aggregation from high pressure [38].

SEC has a lower EV recovery rate than conventional UC, although greater than density gradient UC. Moreover, its speed, scalability, and minimal equipment requirements have led to its increasing use [43, 44]. Normally, it is used in combination with other techniques, such as initial centrifugation steps to remove cellular debris and/or ultrafiltration to concentrate the sample before it is loaded into the column. Many studies combine density gradient UC and SEC to obtain high-purity samples and eliminate sucrose or iodixanol from the buffer during the SEC [38].

Despite advancements, the isolation and characterization of EVs remain important scientific challenges, particularly in areas like Parasitology. Recent guidelines emphasize the need for standardization and the search for specific EV markers, as the existing ones for EVs from human cells, highlighting ongoing efforts to address these challenges [28, 30].

4 Composition of Helminth EVs

Pioneering studies in the field of helminth EVs focused on identifying proteins present in the vesicular component of parasite ESPs [10, 45, 46]. In addition to proteins, the presence of small RNAs [namely, microRNAs (miRNAs)] in helminth EVs was reported for the first time in the trematode *Dicrocoelium dendriticum* [47]. Soon after this publication, these nucleic acids were also described in nematode EVs and linked to host gene regulation and suppression of innate type 2 responses in vivo [45]. In contrast, compositional and functional profiles of other important biological molecules found in EVs, such as lipids and carbohydrates, have been less explored thus far. In the following sections, current knowledge of the molecular composition of helminth EVs is reviewed (Fig. 1) with emphasis on their protein cargo, as this is the central topic of this book.

4.1 Nucleic Acids

Although EVs are known to contain a wide variety of nucleic acids, including DNA, messenger RNA, and small RNAs (Fig. 1), thus far, studies of helminth EVs have focused on the latter and, in particular, in miRNAs [48–53]. In fact, miRNAs have been proposed as key players in the immunomodulation mediated by helminths [21], as many parasite miRNAs in EVs exhibit homology to those of the host, and some of them can target host genes involved in the immune response [54].

Two families of miRNAs, mir-10 and let-7, have been found in EVs of all analyzed helminth species, while three additional families are shared

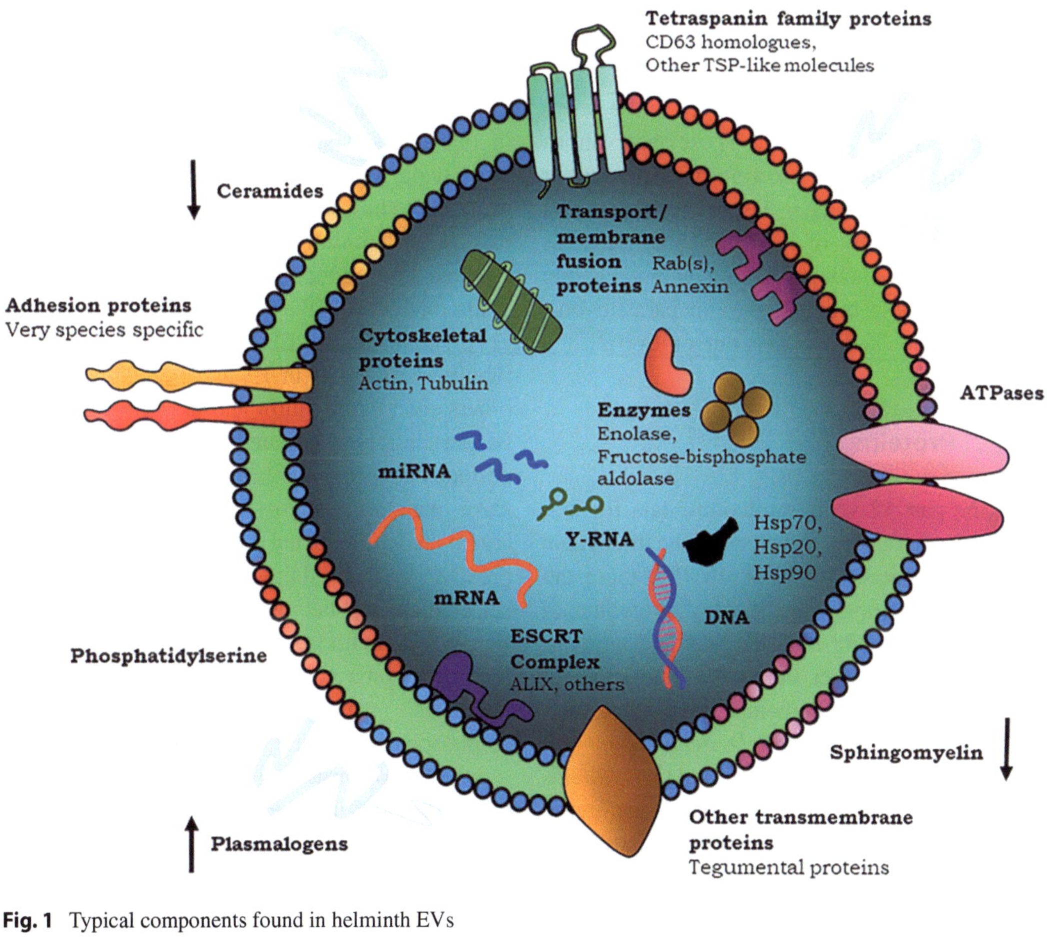

Fig. 1 Typical components found in helminth EVs

among trematodes (mir-71, mir-190, and mir-2) [21]. The role(s) of all identified miRNAs in the biology of the parasite and its interaction with the host have not been fully analyzed. However, it has been demonstrated that miRNA profiles are significantly different in EVs derived from adult worms compared with juvenile phases, suggesting their controlled packaging to perform specific functions in the host [55]. Altogether, this evidence has fostered the study of helminth miRNAs and their potential use in theranostics—*treatment strategies combining therapeutics and diagnostics* [21, 48, 50, 56–61].

As for other classes of small RNAs present in helminth EVs, little is yet known about their profiles and biological functions. Nonetheless, a seminal study observed that secondary small interfering RNAs (siRNAs) are overrepresented in EVs of the parasitic nematode *H. polygyrus,* compared with miRNAs [62]. Moreover, the profile of exported secondary siRNAs differed from that observed in adult worms, a finding linked to the identification of an extracellular worm Argonaute (i.e., exWAGO) involved in the selective export of siRNAs in EVs. These observations support the existence of sorting mechanisms for selective RNA loading in EVs and suggest that extending the study of EV-linked nucleic acids beyond miRNAs may provide novel insights into the role of these biomolecules in parasite–host communication and infection outcomes [62].

4.2 Proteins

Over the past 12 years, proteomic data for helminth EVs have been gathered for the three main lineages of parasitic worms. Briefly, there are several studies analyzing the EV proteome of trematodes infecting both humans and animals, including: (i) the liver flukes *Fasciola* spp., for which different developmental stages have been studied [10, 27, 46, 63–69], *D. dendriticum* [47, 66], *Clonorchis sinensis* [70, 71] and *Opisthorchis viverrini* [72]; (ii) blood flukes of the genus *Schistosoma* [52, 53, 73–77]; (iii) the gastrointestinal trematodes *E. caproni* [10] and *Calicophoron daubyeni* [78]; and, more recently, (iv) the lung fluke *Paragonimus kellicotti* [79]. In cestodes, proteomic descriptions are available for EVs released by *Echinococcus* spp. [57, 80, 81], *Taenia* spp. [82, 83], *Hymenolepis diminuta* [84], and *Anoplocephala perfoliata* [85], whereas in nematodes, the EV protein cargo has been described for gastrointestinal species, such as *Heligmosomoides polygyrus* [45] *Trichuris muris* [86–88], *Toxocara canis* [89], *Trichinella spiralis* [90–92], *Ascaris suum* [59], *Anisakis pegreffi* [93], *Haemonchus contortus* [94–98], and *Teladorsagia circumcincta* [99], as well as in the filarial worm, *Brugia malayi* [100].

The protein composition of helminth EVs has been recently reviewed by several authors [16, 21, 101] and the reader is directed to those works to gain further insights into this topic. However, in general, comparative analyses of EV protein cargoes across helminth species have revealed an enormous variability among parasites. EVs appear to contain a very specific cargo according to the helminth species studied and, in fact, when compared using the Pfam database, the majority of proteins found in helminth EVs are only found in one parasite species [16, 21]. Nonetheless, large-scale analysis of available helminth EV proteomes showed that, while a universal protein marker for helminth EVs does not seem to exist, selected Pfam domains and protein families are common to EVs of the different helminth lineages [21]. For instance, "EF-hand" type proteins and tubulins are found in EVs of flatworms, while others such as ubiquitin and annexins appear to be particularly associated with trematodes and cestodes, respectively. In nematode EVs, in contrast, metallopeptidases are common [21]. A recent study conducted an orthology-based comparative analysis between EV proteomes of selected species belonging to the three helminth lineages (trematoda, cestoda, and nematoda) and identified some common proteins among them, including actin, HSP90, and fructose-bisphosphate aldolase [16]. Moreover, varied repertoires of structural and moonlighting proteins, proteins involved in both canonical and non-canonical EV biogenesis, and RNA-binding proteins, among others with unknown function, were detected in each lineage [16] (Fig. 1).

Considering that EVs are important mediators in host–parasite communication, it seems plausible that differences across helminth species are related, at least partially, to the specificity of this interaction [66]. Additional biological and technical sources of variation have been suggested, including the origin of each EV population (e.g., gastrodermis vs. tegument), the developmental stage studied, the method for EV purification, and the availability of datasets for protein annotation [19]. Studies conducted on the liver flukes *F. hepatica* and *D. dendriticum* may serve as good examples to illustrate these statements.

Briefly, proteomic studies of *F. hepatica* EVs obtained through different isolation methods (differential UC, SEC, and gravity flow) led to the identification of both overlapping and unique proteins, thus suggesting a possible impact of the purification method on reported EV proteomic profiles [10, 27, 46, 63–66, 68, 69]. Comparative analysis of the proteomes of EVs released by adult specimens of *F. hepatica* and *D. dendriticum* (both isolated by SEC) revealed quantitative and qualitative differences between them [66]. A higher number of proteins were identified in *F. hepatica* EVs compared with those from *D. dendriticum*. This outcome could be attributed to the absence of specific datasets for *D. dendriticum* proteins, a critical issue in proteomics when working with non-model organisms [102, 103]. Supporting this hypothesis, EV protein markers such as TSG101 and CD63rec were detected using immuno-gold labeling on EVs of both *F. hepatica* and *D. dendriticum*, although proteomic analyses only identified these proteins in *F. hepatica* EVs [66].

More than half of the proteins identified in *D. dendriticum* EVs were also detected in EVs from *F. hepatica*, indicating a high degree of homology between the protein content of EVs of both species, probably due to the presence of many similar structural proteins and enzymes participating in EV biogenesis [34]. However, the set of proteins identified only in EVs from *D. dendriticum* (30 proteins) or *F. hepatica* (181 proteins) may be responsible for the different roles of these EVs [66]. Indeed, functional classification of both proteomes also showed differences between both EV isolates (e.g., the absence of proteins with antioxidant and enzyme regulator activities in *D. dendriticum*) [66]. Furthermore, in vitro assays demonstrated that *D. dendriticum* and *F. hepatica* EVs induce a distinct phenotype in human macrophages [66] and establish specific interactions with hepatic cells in culture [104] (see below).

The study of EV proteomes from immature stages of *F. hepatica*, namely, egg/miracidia, and newly excysted metacercariae and juveniles showed that their protein cargo evolves across the parasite life cycle [67, 69]. In juveniles, the greatest variety of identified proteins, including those involved in membrane trafficking and cellular physiology, may reflect their intense migratory activity, driven by locomotion based on wave movements that require a precise reorganization of their cytoskeleton [105]. Strikingly, compared with proteins from adult EVs, no cysteine proteases (i.e., cathepsins) have been identified neither in EVs from eggs/miracidia nor NEJs/juveniles in culture [69].

4.3 Lipids

Lipids are also an important group of components of EVs, although, thus far, they have not been as extensively studied as proteins or small RNAs. These biological compounds not only constitute key components of EVs, forming their membrane and protecting the internal content, but also many lipids can perform specific functions, such as contributing to EV uptake or modulating the host immune response [106]. To date, the only studies available on the lipid content of helminth EVs are those carried out by Simbari and co-workers and Wang and co-workers on EVs of *H. polygyrus* and *H. contortus,* respectively [107]. The former identified differences in the lipid profile of parasite EVs and that of murine cell lines, which may have a significant impact on their stiffness [107]. Wang et al. observed that the lipid profile of *H. contortus* EVs differed from that published for *H. polygyrus*, with *Haemonchus* displaying a lower abundance of plasmalogens in its vesicles [94–97].

4.4 Carbohydrates

Carbohydrates constitute another group of molecules that are yet understudied in helminth EVs. Both proteins and lipids can be glycosylated, thus affecting their subsequent subcellular localization and functionality. In fact, it has been confirmed that carbohydrates are key regulators of the cellular tropism and uptake of EVs [59]. To date, the only studies about carbohydrates present in helminth EVs have been performed in *F. hepatica* [64, 65] and *S. mansoni* [108–110]. In the first case, it was confirmed that the carbohydrate content of EVs was very different from that of the tegument and, additionally, the elimination of glycans from the EV surface blocked their internalization by macrophages in vitro [64, 65]. In *S. mansoni*, differences in the carbohydrate profile were found depending on the larval phase of the parasite [108–110]. In addition, it was shown that EV-associated glycans act as ligands for immune cell receptors such as CD209 (DC-SIGN) in dendritic cells [108–110]. Binding to this pattern recognition receptor is involved in *Schistosoma* EV uptake by dendritic cells and affects the activation of immunostimulatory and immunoregulatory pathways in the latter [108–110]. Therefore, although more studies are needed, carbohydrates appear to be important in the interaction with host cells and the modulation of subsequent immune responses, as well as in promoting the recruitment of EVs themselves.

5 Regulatory Effects of Helminth EVs

The evidence summarized above provides strong support to the fact that helminth EVs play significant roles in modulating the host immune response. Several studies, conducted both in vitro and in vivo, have demonstrated the extent of this interaction.

5.1 Regulatory Effects of Helminth EVs Studied In Vitro

There is increasing evidence of the internalization of helminth EVs by cells of the host innate immune system, mainly macrophages. Both helminths and their ESPs can activate selected signaling pathways in host cells, leading to the regulation of immune phenotypes [111, 112]. However, it is important to note that the effects of purified and concentrated EVs may differ from those observed with complete parasite ESPs, which include both EVs and soluble components in different proportions. A summary of currently available studies focusing on the functionality of helminth EVs in vitro is presented in Tables 1 and 2, where key molecules, such as specific proteins, miRNAs, and other upregulated or downregulated molecules involved in the interaction with host cells are indicated. Overall, these studies provide a wealth of evidence of the crucial role of helminth-derived EVs in modulating the host immunity.

Previous reports have investigated the role of *F. hepatica* EVs in stimulating immune cells, such as dendritic cells derived from mouse bone marrow [65]. In this study, the internalization of EVs induced the secretion of TNF-α and the expression of membrane co-stimulatory markers (CD80, CD86, CD40, and OX40 L) as well as intracellular markers (SOCS1 and SOCS3) in dendritic cells. However, the effects described in these cells were not sufficient to fully elucidate all the immunomodulatory actions of this parasite during infection, suggesting that other soluble components within the ESPs may play significant roles [65].

Recently, a functional study of *F. hepatica* and *D. dendriticum* EVs on human macrophages in culture was reported [66]. Interestingly, *Fasciola* EVs reduced the migratory capacity of monocytes and induced a mixed M1/M2 response, demonstrating anti-inflammatory properties in LPS-activated macrophages. In contrast, EVs of

Table 1 Effects described for nematode EVs in in vitro studies (adapted and updated from the publication in [36])

Nematode	Receptor cell	Function	Effect/mechanism	References
Brugia malayi	Murine J774A.1M1 macrophages	M1-type macrophage activation	Significant increases in G-CSF, MCP-1, IL-6 and MIP-2 levels	[113]
Brugia malayi	THP-1-derived monocytes and dendritic cells	Immune metabolism regulation	Decreased mTOR phosphorylation	[114]
Brugia malayi	Aag2 mosquito clonal cells	Immunomodulatory properties	Changes in mRNA expression and miRs implicated in melanization	[115]
Heligmosomoides polygyrus	BMDM (C57BL/6, BALB/c, and ST2−/− mice)	BMDM activation suppression	Decrease in the levels of type 1 molecules (IL-6 and TNF-α), type 2 molecules (Relmα and Ym1), and the ST2 subunit of the IL-33 receptor	[18]
Heligmosomoides polygyrus	Murine MODE-K epithelial cell line	Immunomodulatory properties	Decreased levels of Dusp1 and IL-33R	[45]
Trichinella spiralis	Human PBMCs	Immunomodulatory properties	Significant increases in IL-10, IL-6, and IL-17a	[91]
Trichinella spiralis	BMDM (BALB/c mice); human kidney fibroblasts (HKF)	M2b type macrophage activation; inhibition of fibroblast activation	Increased Arg-1 protein expression and CCL-1 mRNA; decreased iNOS, TNF-α, IL-1β, IL-6, and IL-23 in macrophages; decreased α-SMA, Collagen I, IV, and VI protein levels	[116]
Trichinella spiralis	Porcine intestinal epithelial cells (IPEC-J2)	Inflammatory properties, promotion of apoptosis	Increased mRNA expression of IL-1β; decreased mRNA expression of IL-10, TGF-β, TLR-5, and MUC-1	[117, 118]
Trichinella spiralis	Porcine intestinal epithelial cells	Promotion of apoptosis	Decreased levels of Bcl2 by its miR-153	[94–97]
Trichuris muris	Enteric organoids (C57BL/6 N mice)	Immunomodulatory properties	Decreased levels of genes implicated in type I IFN signaling	[119]

PBMC, peripheral blood mononuclear cells; *BMDM*, bone marrow-derived macrophages

D. dendriticum did not affect monocyte migration and exhibited proinflammatory properties. These findings may be reflective of differences in the life cycle of each parasite species, which would result in different host immune responses. In particular, only *F. hepatica* migrates to the bile duct through the liver parenchyma, which may require eliciting an immune response to heal deep erosions.

In addition to monocytes/macrophages, the effects of these EVs on hepatic stellate cells (HSC) and hepatocytes, both of which are in

Table 2 Effects described for plathelminth EVs in in vitro studies (adapted and updated from the publication in [36])

Helminth	Receptor cell	Function	Effect/mechanism	References
Schistosoma japonicum	RAW264.7 murine macrophage cell line	Immunomodulation; promotion of proliferation	Changes in RNA expression profile; increases in TNF-α protein levels	[49]
Schistosoma japonicum	NCTC clon 1469 murine hepatic cell line	Immunomodulation	Decrease in Gins4, Tysnd1, and Utp3 expression via their miRs	[52, 53]
Schistosoma mansoni	Th cells (ICR mice)	Immunomodulation	Changes in gene expression via miRs	[120]
Schistosoma mansoni	HUVECs and THP-1 monocytes	Vascular remodeling, immunomodulation	Alterations in gene expression	[74]
Schistosoma mansoni	Human monocyte-derived dendritic cells	Immunomodulation	Ves glycoconjugates favor their internalization; increases in secretion of IL-6, IL-10, IL-12, and PD-L1	[110]
Opisthorchis viverrini	H69 human cholangiocyte cell line	Wound-healing properties; immunomodulation; tumorigenic	Promotion of cell proliferation	[121]
Fasciola gigantica	Human intrahepatic epithelial cells	Increased oxidative stress	Increased levels of reactive oxygen species and phosphorylation of H2A.X and Chk.1	[122]
Fasciola hepatica	BMDCs (BALB/c and OT-II mice)	Immunomodulation	Increased mRNA levels of TNF-α, CD80, CD86, CD40, OX40L; decreased ICAM-1 levels	[65]
Echinococcus multilocularis	RAW264.7 murine macrophage cell line	Immunomodulation; decrease in NO production	Increases in gene expression involved in LPS/TLR4 pathway; suppression in mRNA levels of IL-1α, IL-1β, and iNOS	[123]
Echinococcus multilocularis	HUVECs	Promotion of angiogenesis	Increased proliferation and tube formation	[124]
Echinococcus granulosus	Murine BMDCs (CF-1 mice)	Immunomodulation; BMDCs activation	Increases in mRNA levels of CD86; decreases in MHC class II molecule expression	[81]
Echinococcus granulosus	Murine PBMCs (BALB/c mice)	Immunomodulation; inhibition of lymphocyte proliferation	Decrease in secretion of IL-10, IL-6, IL-17a, and IFN-γ; increases in IL-2 and IL-4	[125]
Clonorchis sinensis	Murine bile epithelial cells (C56BL/6 mice)	Proinflammatory properties	Increased phosphorylation of ERK, p38, AKT, and p65; increased levels of TNF-α and IL-6	[94–97]
Echinococcus granulosus	Murine BMDCs (BALB/c mice)	Proinflammatory properties	Increased NF-κB p65/p50 ratio and levels of IL-6, TNF-α, IFN-γ, IL-1b, and IL-12 proteins	[70, 71]
Echinococcus granulosus (treated with metformin or albenzadol sulfoxide)	Murine BMDCs (CF-1 mice)	Immunomodulation; BMDCs activation	Increases in mRNA levels of IL-12, IL-6, and TNF-α in cells treated with Ves from vermes exposed to metformin; increases in IL-12 mRNA levels in those exposed to albendazole	[80]
Spirometra erinaceieuropeai	RAW264.7 murine macrophage cell line stimulated with LPS	Anti-inflammatory properties	Decrease in mRNA levels of IL-6, IL-1β, and TNF-α	[126]

(continued)

Table 2 (continued)

Helminth	Receptor cell	Function	Effect/mechanism	References
Taenia pisiformis	Peritoneal macrophages (rabbits from New Zealand)	Immunomodulation; decrease in macrophage proliferation	Differential expression of miRs and proteins upon uptake, involved in the regulation of IL-1β and NF-κB2 expression	[94–97]
Taenia pisiformis	RAW264.7 murine macrophage cell line	Immunomodulation; polarization to M2 phenotype	Increases in mRNA levels of Arg-1, IL-4, IL-6, IL-10, and IL-13; decreases in iNOS, IL-12, and IFN-γ	[83]

BMDCs, bone marrow-derived dendritic cells; *PBMC*, peripheral blood mononuclear cell; *HUVECs*, human umbilical vein endothelial cells

direct contact with the worms and constitute the main cells involved in the development of liver pathologies caused by these parasites, have been recently analyzed [104]. These studies have revealed that EVs establish unique and specific interactions with HSC and hepatocytes, suggesting different functions for EVs derived from each parasite, which might be linked to the different migration routes toward their final niche in the bile ducts. *Fasciola* EVs have a cytostatic effect on HSC but induce extracellular matrix secretion and elicit anti-inflammatory responses in hepatocytes. On the contrary, EVs of *D. dendriticum* have a more potent antiproliferative effect than *Fasciola* EVs and trigger a global inflammatory response, with increased levels of NF-κB and other inflammatory mediators in both cell types. Therefore, EV–host cell interactions may significantly influence the progression of the disease, contributing to generating conditions that may favor the establishment of the helminths in the host [104].

5.2 Regulatory Effects of Helminth EVs Studied In Vivo

The effects of EVs from both nematodes and flatworms on their corresponding hosts have also been analyzed in vivo, using several experimental models. These studies, as summarized in Tables 3 and 4, have focused on the immunomodulatory properties of EVs, as well as their potential application in anti-helminth vaccines and therapeutic formulations to ameliorate chronic inflammatory conditions, such as colitis [36]. The rationale behind such anticipated clinical applications is described in the following sections, where the current needs for EV-based products to be translatable into clinical practice are thoroughly discussed.

Table 3 Effects described for nematode EVs in in vivo studies (adapted and updated from the publication in [36])

Nematode	Animal model	Administration	Effect	References
Heligmosomoides polygyrus	BALB/c mice	Intranasal administration of EVs	Suppression of eosinophilia and type 2 innate response; inhibition of IL-33R and DUSP1 production	[45]
Heligmosomoides polygyrus	BALB/c and ST2−/− mice	Intraperitoneal vaccination with VEs + adjuvant	Protective immunity only in wild type mice	[18]
Trichuris muris	C57BL/6 mice	Subcutaneous vaccination with EVs	Protective immunity	[87]
Trichuris muris	BALB/c mice (TNBS-induced colitis)	Intraperitoneal injection of EVs	No protective effect in colitis	[86]
Nippostrongylus brasiliensis	BALB/c mice (TNBS-induced colitis)	Intraperitoneal injection of EVs	Improvement of colitis; decrease in IL-1β, IL-6, IL-17a, and IFN-γ levels; increase in IL-10 levels	[127]
Trichinella spiralis	BALB/c mice (TNBS-induced colitis)	Intraperitoneal injection of EVs	Improvement of colitis; decrease in IL-1β, TNF-α, IL-17a, and IFN-γ levels; increase in IL-10, IL-4, IL-13, and TGF-β levels; reduction of Th1 and Th17 cell population; increase in Th2 cell population	[128]
Trichinella spiralis	BALB/c mice	Subcutaneous vaccination with EVs	23.4% reduction in adult parasite burden and 43.7% reduction in larval burden. Induction of IgG, IgE, and IgM. Increase in serum levels of IL-4, IL-10, IL-12, and IFN-γ (Th1/Th2 response)	[90]
Trichinella spiralis	Swiss albino mice	Subcutaneous vaccination with EVs	78.5% reduction in adult parasite burden and 97% reduction in larval burden. Increase in serum levels of IFN-γ and induction of IgG	[129]

TNBS, 2,4,6-trinitrobenzene sulfonic acid

Table 4 Effects described for plathelminth EVs in in vivo studies (adapted and updated from the publication in [36])

Helminth	Animal model	Administration	Effect	References
Echinostoma caproni	BALB/c mice	Subcutaneous vaccination with EVs	Induction of Th2/Treg response; decrease in egg production; increase in levels of IL-4, TGF-β, and IFN-γ	[130]
Opisthorchis viverrini	Syrian hamster	Intraperitoneal vaccination with EVs + adjuvant	Protective immunity; decrease in egg production and worm burden	[121]
Fasciola hepatica	C57BL6 and RAG1 −/− mice (DSS-induced colitis)	Subcutaneous injection of EVs	Improvement of colitis; decrease in mRNA levels of TNF-α, IL-6, and IL-17a; suppression of MAPK/NF-κB signaling pathway; reduced neutrophil infiltration	[131]
Schistosoma mansoni (egg)	Swiss albino mice	Subcutaneous injection of EVs (with or without adjuvant)	46.58% reduction in adults and 93% reduction in eggs; increase in serum levels of IFN-γ and induction of IgG	[132]
Fasciola gigantica	Female ICR mice	Intraperitoneal vaccination with EVs + adjuvant (15 k EVs and 100 k EVs)	60.38% and 37.73% reduction in adults for vaccination with 15K and 100K VEs, respectively; induction of IgG	[68]
Fasciola hepatica	BALB/c mice	Intraperitoneal injection of EVs + adjuvant	Induction of mixed Th1/Th2 response; increased levels of IFN-γ and IL-2	[65]

DSS, dextran sodium sulfate

6 Use of Helminth EVs in Vaccines

The numerous proteomic studies that followed the discovery of helminth EVs have led to the identification of multiple vaccine candidates within these nanoparticles. This has provided the rationale for their utilization in the development of vaccines against parasitic worms affecting both animals and humans, for which effective prevention remains a significant challenge [16, 133]. As predicted, experimental immunization with parasite-derived EVs has resulted in the development of specific antibody responses (Tables 3 and 4), confirming that these vesicles transport antigenic proteins, which can be successfully internalized and presented to activate the host's adaptive immunity. Furthermore, EV inoculation has induced protection against subsequent infection in several models of helminthiasis (Tables 3 and 4), thus validating their potential utility as vaccine targets.

Despite these promising results, EV-based immunoprophylaxis is still far from translation into practical applications for helminth control. Thus far, the level of immunity achieved through this strategy varies considerably among studies, with better outcomes usually observed against nematodes [18, 87, 90, 129] than platyhelminths [68, 121, 130, 132]. For instance, compared with control mice, immunization with EVs in alum adjuvant resulted in a significant reduction in the establishment of adult *H. polygyrus* following exposure to infective larvae of this gastrointestinal nematode [18]. In contrast, injection with purified EVs (in PBS) of the intestinal trematode *E. caproni* did not affect worm burdens but impaired parasite development and reduced disease severity [130].

In addition to differences in the immunobiology of each helminth infection, variations in vaccine formulations and immunization regimes across studies may have also influenced the results. For instance, The use of adjuvants to enhance antibody responses, although not strictly necessary [87, 130], may improve vaccination efficacy. The mode of action of EV-based vaccines remains to be fully elucidated. Nevertheless, it has been suggested that antibodies likely play important roles in protection [26]. Antibodies binding to EVs can reduce their immunosuppressive functions, either neutralizing their uptake by host cells [121] or, if eventually internalized, directing them to the lysosomes for degradation [18]. Moreover, exposure to EV antigens primes the host immune system to detect the parasite upon forthcoming colonization, although potential variations in vesicles' composition across developmental stages might influence vaccine efficacy against specific life stages [26].

Among EV-associated proteins, tetraspanins have been repeatedly studied as potential targets for vaccine development—even before the discovery of helminth EVs—and have yielded promising results against both nematodes [134] and platyhelminths [121, 135, 136]. Unlike the many vaccine candidates identified as part of the internal EV cargo [16, 21], tetraspanins are membrane proteins located on the vesicular surface and with presumed functions in EV biogenesis, cargo selection, and interaction with host cells [137]. However, as these molecules are also present on the worms' surface, whether protection results from antibody-mediated neutralization of EV function or the activation of effector mechanisms directly on the parasites' body is difficult to ascertain with current data [121].

Identifying key target antigens and unveiling the exact mechanisms through which they induce protection against infection is pivotal to fully exploit the potential of parasite EVs, or rather their molecules, for immunoprophylactic purposes. However, the use of native helminth EVs in vaccine formulations is hindered by the lack of high throughput systems to produce them in sufficient amounts and with high-quality standards. This process entails challenges linked not only to EV isolation and purification but also to the culture of the parasites themselves [36]. Moreover, the use of complete parasite EVs for vaccination might give rise to safety concerns. For instance, it has been speculated that responses to conserved EV antigens displaying substantial similarity to host molecules might result in undesired autoimmune reactions [26]. In our own experience, the administration of complete EVs from adult *F.*

hepatica to mice prior to experimental infection not only failed to protect the animals against parasite colonization but it worsened their health status following the administration of metacercariae (Marcilla et al., unpublished results). Future studies are still necessary to elucidate the immunopathological mechanisms underlying this outcome.

The identification of individual antigens that could be specifically targeted to block EV functions is expected to increase both production efficiency and safety of EV-based vaccines. The use of discrete antigens would also facilitate the evaluation of the efficacy of the antibody responses elicited by the vaccine, as well as the identification of polymorphic variation across parasite isolates that may hamper the utility of certain vaccine candidates [138]. Future research in the field of EV-based vaccination should also ascertain the type of response the vaccine has to induce to maximize its efficacy [139]. Collectively, this knowledge is essential to optimize vaccine formulations through, for instance, the rational combination of several antigens or the use of the most appropriate adjuvants and delivery systems. Moreover, this will aid in defining vaccination regimes that ensure the greatest protection against infection, while guaranteeing safeness in vaccine use.

7 Use of Helminth EVS as Biomarkers

The stability and accessibility of EVs in several body fluids such as blood, urine, and saliva, along with their rich and diverse composition of proteins and nucleic acids, among other biomolecules, have motivated the biomedical community to use them as diagnostic biomarkers for various diseases, including infections [140, 141]. In the field of helminthology, growing evidence demonstrates that parasitic worms release EVs into the host body fluids, which may represent a source of biomarkers for novel and improved diagnostic tools [50].

Currently, "gold standard" diagnostic assays for helminth infections involve microscopic examination of potentially hazardous biological materials [142]. While highly specific, microscopy-based methods require trained specialists to conduct the analyses and may produce false negatives linked to low infection intensities or intermittent presence of parasite forms in test samples. Furthermore, these techniques cannot be applied during the pre-patent period of the infection, that is, before parasite eggs, or larvae, are produced and become detectable [143]. To overcome these limitations, immunological and molecular diagnostic methods have been developed, which are not exempt from disadvantages. Serology, for example, cannot distinguish between present and past infections, while antigen detection may be hindered by cross-reactivity among helminth species [142, 144]. Molecular techniques can be highly specific and sensitive, and negativize upon parasite elimination. However, they are often subjected to technical challenges that are difficult to overcome in field settings [142, 143] and, as immunodiagnostics, do not provide information regarding the staging of the parasitic disease. In this context, EV-based diagnostic approaches hold promise for improving the accessibility and detectability of parasite molecules in host samples, thus enhancing the sensibility of the analyses and enabling the detection of early and/or low-burden infections. Furthermore, molecules transported in either parasite- or host-derived EVs might provide valuable information regarding disease progression [6, 140, 141].

In principle, whole helminth EVs, as well as their associated molecules, can be detected in host samples through immunodiagnostics, proteomics, PCR, or NGS technologies [145]. This hypothesis is supported by several proof-of-concept studies published to date. For instance, specific parasite proteins have been identified in blood-derived EVs from individuals naturally infected by filarial nematodes [146] and patients affected by abdominal cystic echinococcosis [147] via mass spectrometry. Moreover, in the latter study, some cestode proteins found in circulating EVs were specifically associated with either active or inactive disease and represent promising candidate biomarkers to differentiate cyst staging, evaluate treatment efficacy, and monitor patients' progression [147].

Amplification of parasite miRNAs in serum-derived EV-enriched fractions by quantitative reverse transcription PCR was suggested to improve the detection of human schistosomiasis and follow-up of individuals with low infection burdens [148]. However, the sensitivity and specificity of the analyses decreased considerably when applied to whole serum samples, highlighting the dependence of the method on the application of EV enrichment procedures prior to the molecular analysis [148]. EV-based diagnostic methods are often hindered by difficulties in selectively capturing and detecting the vesicles of interest [149] and that may affect the outcome of the diagnosis [145]. Hence, the development of optimized and standardized methods for the purification/enrichment of helminth EVs in host biofluids would expand their potential for diagnostic purposes, particularly when an increased sensibility is pursued [50].

Profiling of host EVs may also offer insights into the stage or progression of helminth diseases, although this area remains largely underexplored. Mammalian EVs have been recognized as potential prognostic biomarkers for various conditions, including cancer, autoimmune disorders like systemic lupus erythematosus, and acute inflammatory diseases [118, 150–152]. Moreover, alterations of host-cell-derived EVs have been reported during infections by intracellular parasitic protozoa [153–157] and bacteria [158, 159]. In contrast, the impact of helminths on EV production by host cells, including both quantitative and compositional alterations, remains largely underexplored.

Recent studies have investigated the molecular composition of host EVs, including miRNAs and proteins, in helminth-infected versus naïve mice. For example, Li and co-workers found differences in the miRNA profiles of plasma EVs isolated from mice experimentally infected with *Schistosoma japonicum,* compared with uninfected animals [160]. Similarly, the protein cargo of plasma EVs changed over the course of experimental murine infection with *Echinococcus granulosus* [161]. In humans, Fratini et al. [147] compared the proteomic profiles of EV-enriched plasma fractions between patients with active and inactive cystic echinococcosis, as well as uninfected individuals [147]. The functional profiles characterizing pooled samples from infected subjects—with either active or inactive disease—reflected the ongoing immune response against the parasite. Furthermore, a subset of proteins was identified that were specifically linked to each stage of the disease (active or inactive), suggesting that host EVs might represent novel sources for prognostic biomarkers in helminthiases. Notwithstanding, extensive studies are yet necessary to identify EV-associated diagnostic and prognostic markers for helminth infections and to validate them in large cohorts of infected and uninfected individuals. Additionally, the development of diagnostic tools for helminth infections must consider factors, such as cost and usability, to ensure their translation into practical applications that improve the management of helminth infections of humans and livestock in real-world scenarios [142, 162].

Nanotechnological advances are being made toward the development of simple and rapid strategies to analyze EV populations and their cargoes, without requiring complex sample pretreatments and sophisticated systems for signal detection [163, 164]. Lab-on-a-chip devices implementing microfluidics technologies and label-free methods have enabled the differentiation of host and parasite (i.e., *A. suum*) EVs in "mock" samples, prepared by mixing purified vesicles from each organism, based on compositional differences between them—i.e., the specific presence of CD63 in the surface of host-derived EVs [165]. However, simultaneous detection of multiple EV markers would be necessary to detect/separate diverse EV populations in complex biological samples like blood or urine [165]. The discovery of specific markers for helminth EVs would greatly facilitate this task [30]. Upon detection, the protein and/or RNA content of captured EVs could be analyzed on the same chip, thus enabling the identification and/or quantification of selected molecules of interest, such as biomarkers of disease progression [163]. Despite its enormous biomedical potential, research in this field of nanotechnology is still in its early stages. To translate these advancements into point-of-care devices for the diagnosis of helminthiases, parallel efforts are needed to advance the characterization of helminth (and host) EV composition along the course of the infection. This comprehensive approach will be essential for the development of effective diagnostic tools for helminth infections.

8 Helminth EVs as Therapeutic Tools

Helminth infections commonly lead to prototypical Th2 responses, followed by the activation of complex regulatory networks that are orchestrated by the parasite to downregulate host's immunity and pathology, and thus, promote their own survival [166, 167]. The discovery of the mechanisms underpinning the immunomodulatory abilities of helminths supported the hypothesis that controlled infections by parasitic worms could be employed with medical purposes to ameliorate chronic inflammatory disorders, such as inflammatory bowel disease, rheumatoid arthritis, psoriasis, multiple sclerosis, or allergies, among others [167, 168]. Since the 1990s, the so-called helminth therapy has been evaluated in numerous preclinical and clinical studies with promising results [167, 169]. Nevertheless, the use of worms as therapeutic agents raises a series of ethical and practical concerns that continue to be a subject of debate, limiting the widespread application of these practices [169]. Alternatively, the use of individual excreted/secreted molecules mediating helminths' modulatory properties has been proposed as a suitable means to overcome issues related to the use of live parasites [170, 171]. In this context, EVs emerged as promising therapeutic candidates, as they transport several regulatory biomolecules as part of their native cargo and are capable of transferring them directly to host cells [172].

However, thus far, only a handful of studies have explored the potential of helminth EVs to ameliorate inflammation in vivo, employing, in most cases, murine models of chemically induced colitis (see Table 3). It has been demonstrated that subcutaneous administration of *F. hepatica* EVs protected mice against DSS-induced acute colitis [131]. Along with improved histopathological manifestations, reduced levels of proinflammatory cytokines (namely, TNF-α, IL-6, and IL-17A) and diminished activation of the NF-κB and MAPK pathways were observed in the colon of wild-type mice exposed to parasite EVs prior to colitis induction. Remarkably, *Fasciola* EVs ameliorated disease manifestations also in *Rag1*$^{-/-}$ mice, indicating that lymphocytes did not participate in the observed protection, which, instead, could be mediated by alternatively activated macrophages [131]. Similarly, reduced expression of inflammatory cytokines (IL-1β, TNF-α, IL-6, IL-17A, and/or IFN-γ) and upregulation of immunosuppressive IL-10 were associated with improved colitis symptoms in mice intraperitoneally injected with EVs from the gastrointestinal nematodes *Nippostrongylus brasiliensis* [127] and *T. spiralis* [128, 173] before exposure to the colitogenic agent, TNBS. In contrast, EVs from the related nematode *T. muris* did not ameliorate disease manifestations when administered to mice under the same conditions as *Nippostrongylus* EVs [86]. *In silico* predictions suggested that differences in host gens putatively targeted by EV-linked miRNAs might contribute to these variations between worm species, although this hypothesis still needs experimental validation [127].

miRNAs carried in *H. polygyrus* exosomes may also contribute to the ability of these vesicles to suppress innate type 2 responses in the airways of mice [45]. Intranasal administration of EVs from this gastrointestinal nematode reduced the expression of the IL-33 receptor (ST2) in lung innate lymphoid cells of mice exposed to the allergenic fungus *Alternaria* and suppressed the expression of IL-5 and IL-13 in these cells. This phenotype was associated with reduced infiltration of eosinophils in the lungs of allergen-exposed mice, indicating that helminth EVs have the potential to suppress pathogenic immune responses outside the gastrointestinal tract. Therefore, they could also be explored for the treatment of non-intestinal inflammatory conditions [45].

9 Future Directions

Since their "re-discovery" [10], helminth EVs have aroused the interest among helminthologists and immunologists due to their key roles in host–parasite communication. As discussed earlier, disrupting EV-mediated interactions between the worm and its host has emerged as an innovative approach to fight against, treat, reduce helminthiases that is worth investigating. Additionally, it has been speculated that the immunosuppressive properties of helminth-derived EVs could be harnessed to develop novel therapies for ameliorating chronic inflammatory conditions. Progress in either of these directions requires a comprehensive understanding of the composition of the different populations of helminth-derived EVs produced by all the intra-mammalian developmental stages (including vesicles secreted and non-secreted out of the worms' body [27]), as well as of all aspects of their interaction with host cells [19, 36].

This, in turn, involves overcoming technical and methodological challenges currently associated with the study of these biological nanoparticles [30]. For instance, tracking cellular uptake, intracellular trafficking, and the fate of helminth EVs in host cells are pivotal to investigate their biological functions [18, 64, 109, 110]. However, fluorescent labeling methods commonly used for this purpose can introduce artifacts that compromise the validity of the results and may lead to erroneous conclusions [174]. It is strongly recommended to apply purification methods to remove the unbound dye following EV staining and to quantify the background signal remaining after cleanup. However, the most appropriate methods for this may vary depending on the labeling strategy and no clear criteria exist to choose them [30, 175]. Recently, a method for in vivo EV labeling in whole worms has been described, which reduces the amount of off-target labeling, thereby minimizing issues associated with unspecific staining [176]. Furthermore, to confirm the specificity of the internalization process, labeled synthetic liposomes can be employed as a negative control, as these artificial particles are devoid of membrane proteins, glycans, and lipid domains that participate in EV uptake [104].

In vitro studies of EV-cell interactions and subsequent responses in target cells have been invaluable in grasping the importance of EVs in host–parasite relationships [19, 36]. Nonetheless, the question remains as to what extent the experimental conditions set to obtain parasite EVs, as well as to investigate their interaction with host cells, mimic the environment inside the host and may affect EV cargo and function. Zhou et al. [125] observed remarkable differences in the protein profiles of EVs released by protoscolex of *E. granulosus* in culture compared with those isolated from patients' hydatid cyst fluid, including variations in the number and identity of proteins found in each EV population [125]. Future studies comparing the cargoes of vesicles released by cultured worms versus parasite EVs purified from host samples (e.g., blood, urine, or feces) are therefore necessary to assess whether these variations occur in other helminth species and, if so, their potential impact on EV-mediated functions. Assessing the biodistribution and bioavailability of helminth-derived EVs in vivo, in contrast, relies on the eventual identification of universal biomarkers for helminth EVs, or the generation of worms capable of producing "trackable" EVs within the host body (e.g., via genome editing or by incorporating the dye into the worms before infection) [30].

In addition to host cells, investigating the interaction of helminth EVs with other colonizers of the host body, namely, the microbiota, will expand our understanding of EV-mediated inter-kingdom communication and provide novel insights into the exact mechanisms through which these parasites modulate the composition of the host gut flora. Moreover, this emerging field of study may pave the way for the discovery of novel antimicrobial products [177].

Indeed, owing to their cargo and biological properties, EVs hold enormous potential for developing a range of innovative biomedical applications for disease prevention and control [178]. As stated above, expanding the knowledge of helminth EVs can substantially contribute to these goals. Nevertheless, the medical use of helminth-derived EVs, for instance, as vaccines or immunosuppressants, faces a major limitation that will likely restrain their direct translation into clinical practice. Producing these vesicles on an industrial scale and in compliance with good manufacturing practices seems unfeasible. Parasitic worms cannot be propagated out of their hosts and the amount of EVs produced by each specimen in culture is limited. Hence, repeated animal infections and subsequent euthanizing would be necessary to obtain the helminths to produce EVs. This procedure is as unethical as it is inefficient and forces the search for alternative means to harness the biomedical potential of helminth EVs.

Over the past years, accumulating evidence has pointed toward bacterial EVs and plant-derived vesicle-like nanoparticles (PDVLNs) as promising vehicles for the delivery of both chemical drugs and active biomolecules [179–182]. Compared with synthetic liposomes, the inherent biocompatibility of natural vesicles, along with their ability to cross biological barriers, tropism to target tissues/cells, and optimal cellular uptake, generally confer these nanoparticles an enhanced efficiency for drug delivery (see [183]). Moreover, several methodological strategies are being developed to modify the cargo of these and other natural vesicles following purification and to decorate them with diverse surface moieties to actively target specific tissues and cells [183]. Therefore, it seems plausible that selected recombinant parasite proteins and/or mimetic peptides and miRNAs could be incorporated into plant or bacterial vesicles that serve as nanocarriers for delivering these molecules to host cells.

Identification of individual helminth biomolecules mediating EV-linked, desired effects will be pivotal to this end, as will exploring potential synergies resulting from the combination of these molecules. Studies of the intrinsic biological activity, biodistribution, safety, and stability of carrier vesicles will also be essential for an optimal selection of the delivery system depending on the purpose of the formulation and/or the administration route. For instance, it has been recently shown that PDVLNs obtained from pomegranate juice display anti-inflammatory properties [184], which might be harnessed in helminth-based immunosuppressant formulations to potentiate their effects. In contrast, bacterial EVs might contribute to the nonspecific stimulation of the host immune system in vaccine applications [185]. Inevitably, this knowledge must be accompanied by technical and methodological developments in the design and manufacturing of EV-based delivery systems, including but not limited to their large-scale production, optimal drug loading, and active cellular targeting. The implementation of regulatory proceedings for the evaluation and authorization of these products by competent drug agencies will be equally necessary [186, 187]. Hence, while the translation of experimental findings to clinical applications may not be imminent, prospects in this field are certainly exciting.

Acknowledgments Dr. Lynne Yenush is acknowledged for critically reading the manuscript. This work was supported by the Agencia Estatal de Investigación, Ministerio de Ciencia e Innovación, Spain (Grant PID2019-105713GBI00/AEI/10.13039/501100011033), and Conselleria de Educación, Universidades y Empleo, Generalitat Valenciana, Valencia, Spain (Grant PROMETEO/2020/071). C.M.S-L was supported by a predoctoral fellowship (PRE2020-092458) funded by AEI/10.13039/501100011033. A.G-A was supported by a predoctoral fellowship (CPI-20-353) funded by Conselleria de Educación, Universidades y Empleo, Generalitat Valenciana, Valencia, Spain (PROMETEO/2020/071). Research Group member of GEIVEX and "Red Traslacional para la Aplicación Clínica de Vesículas Extracelulares, Tentacles" (RED2018-102411-T, Agencia Estatal de Investigación, Spain).

References

1. Jourdan PM, Lamberton PHL, Fenwick A, Addiss DG (2018) Soil-transmitted helminth infections. Lancet 391:252–265
2. Weatherhead JE, Hotez PJ, Mejia R (2017) The global state of helminth control and elimination in children. Pediatr Clin N Am 64:867–877
3. Charlier J, Rinaldi L, Musella V, Ploeger HW, Chartier C, Vineer HR, Hinney B, von Samson-Himmelstjerna G, Băcescu B, Mickiewicz M, Mateus TL, Martinez-Valladares M, Quealy S, Azaizeh H, Sekovska B, Akkari H, Petkevicius S, Hektoen L, Höglund J, Morgan ER, Bartley DJ, Claerebout E (2020) Initial assessment of the eco-

nomic burden of major parasitic helminth infections to the ruminant livestock industry in europe. Prev Vet Med 182:105103
4. Lightowlers MW, Rickard MD (1988) Excretory-secretory products of helminth parasites: effects on host immune responses. Parasitology 96(Suppl):123
5. Simon GG (2016) Impacts of neglected tropical disease on incidence and progression of HIV/AIDS, tuberculosis, and malaria: scientific links. Int J Infect Dis 42:54–57
6. Zakeri A, Hansen EP, Andersen SD, Williams AR, Nejsum P (2018) Immunomodulation by helminths: intracellular pathways and extracellular vesicles. Front Immunol 9:2349
7. Hewitson JP, Grainger JR, Maizels RM (2009) Helminth immunoregulation: the role of parasite secreted proteins in modulating host immunity. Mol Biochem Parasitol 167:1–11
8. Maizels RM, Smits HH, McSorley HJ (2018) Modulation of host immunity by helminths: the expanding repertoire of parasite effector molecules. Immunity 49:801–818
9. Shepherd C, Navarro S, Wangchuk P, Wilson D, Daly NL, Loukas A (2015) Identifying the immunomodulatory components of helminths. Parasite Immunol 37:293–303
10. Marcilla A, Trelis M, Cortés A, Sotillo J, Cantalapiedra F, Minguez MT, Valero ML, Sánchez del Pino MM, Muñoz-Antoli C, Toledo R, Bernal D (2012) Extracellular vesicles from parasitic helminths contain specific excretory/secretory proteins and are internalized in intestinal host cells. PLoS One 7:e45974
11. Singh N, Bhalla N (2020) Moonlighting proteins. Annu Rev Genet 54:265–285
12. Senft AW, Philpott DE, Pelofsky AH (1961) Electron microscope observations of the integument, flame cells, and gut of *Schistosoma mansoni*. J Parasitol 47:217–229
13. Threadgold LT (1963) The ultrastructure of the "cuticle" of *Fasciola hepatica*. Exp Cell Res 30:238–242
14. Threadgold LT (1967) Electron-microscope studies of *Fasciola hepatica*. 3. Further observations on the tegument and associated structures. Parasitology 57:633–637
15. Andresen K, Simonsen PE, Andersen BJ, Birch-Andersen A (1989) *Echinostoma caproni* in mice: shedding of antigens from the surface of an intestinal trematode. Int J Parasitol 19:111–118
16. Ancarola ME, Maldonado LL, García LCA, Franchini GR, Mourglia-Ettlin G, Kamenetzky L, Cucher MA (2023) A comparative analysis of the protein cargo of extracellular vesicles from helminth parasites. Life (Basel) 13:2286
17. Coakley G, Maizels RM, Buck AH (2015) Exosomes and other extracellular vesicles: the new communicators in parasite infections. Trends Parasitol 31:477–489
18. Coakley G, McCaskill JL, Borger JG, Simbari F, Robertson E, Millar M, Harcus Y, McSorley HJ, Maizels RM, Buck AH (2017) Extracellular vesicles from a helminth parasite suppress macrophage activation and constitute an effective vaccine for protective immunity. Cell Rep 19:1545–1557
19. Drurey C, Maizels RM (2021) Helminth extracellular vesicles: interactions with the host immune system. Mol Immunol 137:124–133
20. Hoffmann KF, Hokke CH, Loukas A, Buck AH (2020) Helminth extracellular vesicles: great balls of wonder. Int J Parasitol 50:621–622
21. Sotillo J, Robinson MW, Kimber MJ, Cucher M, Ancarola ME, Nejsum P, Marcilla A, Eichenberger RM, Tritten L (2020) The protein and microRNA cargo of extracellular vesicles from parasitic helminths - current status and research priorities. Int J Parasitol 50:635–645
22. Tritten L, Geary TG (2018) Helminth extracellular vesicles in host-parasite interactions. Curr Opin Microbiol 46:73–79
23. Yáñez-Mó M, Siljander PR, Andreu Z, Zavec AB, Borràs FE, Buzas EI, Buzas K, Casal E, Cappello F, Carvalho J, Colás E, Cordeiro-da Silva A, Fais S, Falcon-Perez JM, Ghobrial IM, Giebel B, Gimona M, Graner M, Gursel I, Gursel M, Heegaard NHH, Hendrix A, Kierulf P, Kokubun K, Kosanovic M, Kralj-Iglic V, Krämer-Albers E, Laitinen S, Lässer C, Lener T, Ligeti E, Linē A, Lipps G, Llorente A, Lötvall J, Manček-Keber M, Marcilla A, Mittelbrunn M, Nazarenko I, Nolte-'t Hoen ENM, Nyman TA, O'Driscoll L, Olivan M, Oliveira C, Pállinger É, Del Portillo HA, Reventós J, Rigau M, Rohde E, Sammar M, Sánchez-Madrid F, Santarém N, Schallmoser K, Ostenfeld MS, Stoorvogel W, Stukelj R, Van der Grein SG, Vasconcelos MH, Wauben MHM, De Wever O (2015) Biological properties of extracellular vesicles and their physiological functions. J Extracell Vesicles 4:27066
24. Buzas EI (2023) The roles of extracellular vesicles in the immune system. Nat Rev Immunol 23:236–250
25. de la Torre-Escudero E, Bennett APS, Clarke A, Brennan GP, Robinson MW (2016) Extracellular vesicle biogenesis in helminths: more than one route to the surface? Trends Parasitol 32:921–929
26. Drurey C, Coakley G, Maizels RM (2020) Extracellular vesicles: new targets for vaccines against helminth parasites. Int J Parasitol 50:623–633
27. Bennett APS, de la Torre-Escudero E, Oliver NAM, Huson KM, Robinson MW (2020) The cellular and molecular origins of extracellular vesicles released by the helminth pathogen, *Fasciola hepatica*. Int J Parasitol 50:671–683
28. Fernández-Becerra C, Xander P, Alfandari D, Dong G, Aparici-Herraiz I et al (2023) Guidelines for the purification and characterization of extracellular vesicles of parasites. J Extracell Biol 2:e117
29. Liu C, Su C (2019) Design strategies and application progress of therapeutic exosomes. Theranostics 9:1015–1028

30. White R, Sotillo J, Ancarola ME, Borup A, Boysen AT, Brindley PJ, Buzás EI, Cavallero S, Chaiyadet S, Chalmers IW, Cucher MA, Dagenais M, Davis CN, Devaney E, Duque-Correa MA, Eichenberger RM, Fontenla S, Gasan TA, Hokke CH, Kosanovic M, Kuipers ME, Laha T, Loukas A, Maizels RM, Marcilla A, Mazanec H, Morphew RM, Neophytou K, Nguyen LT, Nolte-'t Hoen E, Povelones M, Robinson MW, Rojas A, Schabussova I, Smits HH, Sungpradit S, Tritten L, Whitehead B, Zakeri A, Nejsum P, Buck AH, Hoffmann KF (2023) Special considerations for studies of extracellular vesicles from parasitic helminths: a community-led roadmap to increase rigour and reproducibility. J Extracell Vesicles 12:e12298
31. Théry C, Witwer KW, Aikawa E et al (2018) Minimal information for studies of extracellular vesicles 2018 (MISEV2018): a position statement of the international society for extracellular vesicles and update of the MISEV2014 guidelines. J Extracell Vesicles 7:1535750
32. Welsh JA, Goberdhan DCI, O'Driscoll L et al (2024) Minimal information for studies of extracellular vesicles (MISEV2023): from basic to advanced approaches. J Extracell Vesicles 13:e12404
33. Witwer KW, Goberdhan DC, O'Driscoll L, Théry C, Welsh JA, Blenkiron C, Buzás EI, Di Vizio D, Erdbrügger U, Falcón-Pérez JM, Fu Q, Hill AF, Lenassi M, Lötvall J, Nieuwland R, Ochiya T, Rome S, Sahoo S, Zheng L (2021) Updating MISEV: evolving the minimal requirements for studies of extracellular vesicles. J Extracell Vesicles 10:e12182
34. Cwiklinski K, Dalton JP (2022) Exploiting comparative omics to understand the pathogenic and virulence-associated protease: anti-protease relationships in the zoonotic parasites *Fasciola hepatica* and *Fasciola gigantica*. Genes (Basel) 13:1854
35. van Niel G, D'Angelo G, Raposo G (2018) Shedding light on the cell biology of extracellular vesicles. Nat Rev Mol Cell Biol 19:213–228
36. Sánchez-López CM, Trelis M, Bernal D, Marcilla A (2021) Overview of the interaction of helminth extracellular vesicles with the host and their potential functions and biological applications. Mol Immunol 134:228–235
37. Akbar A, Malekian F, Baghban N, Kodam SP, Ullah M (2022) Methodologies to isolate and purify clinical grade extracellular vesicles for medical applications. Cells 11:186
38. Sidhom K, Obi PO, Saleem A (2020) A review of exosomal isolation methods: is size exclusion chromatography the best option? Int J Mol Sci 21:6466
39. Galiano A, Minguez MT, Sánchez CM, Marcilla A (2020) Isolation and analysis of *Fasciola hepatica* extracellular vesicles. Methods Mol Biol 2137:37–50
40. Williams S, Jalal AR, Lewis MP, Davies OG (2023) A survey to evaluate parameters governing the selection and application of extracellular vesicle isolation methods. J Tissue Eng 14:20417314231155114
41. Konoshenko MY, Lekchnov EA, Vlassov AV, Laktionov PP (2018) Isolation of extracellular vesicles: general methodologies and latest trends. Biomed Res Int 2018:8545347
42. Pan BT, Johnstone RM (1983) Fate of the transferrin receptor during maturation of sheep reticulocytes *in vitro*: selective externalization of the receptor. Cell 33:967–978
43. Monguió-Tortajada M, Gálvez-Montón C, Bayes-Genis A, Roura S, Borràs FE (2019) Extracellular vesicle isolation methods: rising impact of size-exclusion chromatography. Cell Mol Life Sci 76:2369–2382
44. Clos-Sansalvador M, Monguió-Tortajada M, Roura S, Franquesa M, Borràs FE (2022) Commonly used methods for extracellular vesicles' enrichment: implications in downstream analyses and use. Eur J Cell Biol 101:151227
45. Buck AH, Coakley G, Simbari F, McSorley HJ, Quintana JF, Le Bihan T, Kumar S, Abreu-Goodger C, Lear M, Harcus Y, Ceroni A, Babayan SA, Blaxter M, Ivens A, Maizels RM (2014) Exosomes secreted by nematode parasites transfer small RNAs to mammalian cells and modulate innate immunity. Nat Commun 5:5488
46. Cwiklinski K, de la Torre-Escudero E, Trelis M, Bernal D, Dufresne PJ, Brennan GP, O'Neill S, Tort J, Paterson S, Marcilla A, Dalton JP, Robinson MW (2015) The extracellular vesicles of the helminth pathogen, *Fasciola hepatica*: biogenesis pathways and cargo molecules involved in parasite pathogenesis. Mol Cell Proteomics 14:3258–3273
47. Bernal D, Trelis M, Montaner S, Cantalapiedra F, Galiano A, Hackenberg M, Marcilla A (2014) Surface analysis of *Dicrocoelium dendriticum*. The molecular characterization of exosomes reveals the presence of miRNAs. J Proteome 105:232–241
48. Cucher MA, Ancarola ME, Kamenetzky L (2021) The challenging world of extracellular RNAs of helminth parasites. Mol Immunol 134:150–160
49. Liu J, Zhu L, Wang J, Qiu L, Chen Y, Davis RE, Cheng G (2019) *Schistosoma japonicum* extracellular vesicle miRNA cargo regulates host macrophage functions facilitating parasitism. PLoS Pathog 15:e1007817
50. Mu Y, McManus DP, Gordon CA, Cai P (2021) Parasitic helminth-derived microRNAs and extracellular vesicle cargos as biomarkers for helminthic infections. Front Cell Infect Microbiol 11:708952
51. Quintana JF, Babayan SA, Buck AH (2017) Small RNAs and extracellular vesicles in filarial nematodes: from nematode development to diagnostics. Parasite Immunol 39
52. Zhu L, Liu J, Dao J, Lu K, Li H, Gu H, Liu J, Feng X, Cheng G (2016) Molecular characterization of *S. japonicum* exosome-like vesicles reveals their regulatory roles in parasite-host interactions. Sci Rep 6:25885
53. Zhu S, Wang S, Lin Y, Jiang P, Cui X, Wang X, Zhang Y, Pan W (2016) Release of extracellular vesicles containing small RNAs from the eggs of *Schistosoma japonicum*. Parasit Vectors 9:574–572

54. Fontenla S, Langleib M, de la Torre-Escudero E, Domínguez MF, Robinson MW, Tort J (2022) Role of *Fasciola hepatica* small RNAs in the interaction with the mammalian host. Front Cell Infect Microbiol 11:812141
55. Fromm B, Ovchinnikov V, Høye E, Bernal D, Hackenberg M, Marcilla A (2017) On the presence and immunoregulatory functions of extracellular microRNAs in the trematode *Fasciola hepatica*. Parasite Immunol 39
56. Alizadeh Z, Mahami-Oskouei M, Spotin A, Ahmadpour E, Cai P, Sandoghchian Shotorbani S, Pashazadeh F, Ansari F, Mohammadi H (2022) MicroRNAs in helminth parasites: a systematic review. Curr Mol Med 22:779–808
57. Ancarola ME, Marcilla A, Herz M, Macchiaroli N, Pérez M, Asurmendi S, Brehm K, Poncini C, Rosenzvit M, Cucher M (2017) Cestode parasites release extracellular vesicles with microRNAs and immunodiagnostic protein cargo. Int J Parasitol 47:675–686
58. Cai Q, He B, Weiberg A, Buck AH, Jin H (2019) Small RNAs and extracellular vesicles: new mechanisms of cross-species communication and innovative tools for disease control. PLoS Pathog 15:e1008090
59. Hansen EP, Fromm B, Andersen SD, Marcilla A, Andersen KL, Borup A, Williams AR, Jex AR, Gasser RB, Young ND, Hall RS, Stensballe A, Ovchinnikov V, Yan Y, Fredholm M, Thamsborg SM, Nejsum P (2019) Exploration of extracellular vesicles from *Ascaris suum* provides evidence of parasite-host cross talk. J Extracell Vesicles 8:1578116
60. Soekmadji C, Li B, Huang Y, Wang H, An T, Liu C, Pan W, Chen J, Cheung L, Falcon-Perez JM, Gho YS, Holthofer HB, Le MTN, Marcilla A, O'Driscoll L, Shekari F, Shen TL, Torrecilhas AC, Yan X, Yang F, Yin H, Xiao Y, Zhao Z, Zou X, Wang Q, Zheng L (2020) The future of extracellular vesicles as theranostics - an ISEV meeting report. J Extracell Vesicles 9:1809766
61. Zhou C, Tuersong W, Liu L, Di W, He L, Li F, Wang C, Hu M (2024) Non-coding RNA in the gut of the blood-feeding parasitic worm, *Haemonchus contortus*. Vet Res 55:1
62. Chow FW, Koutsovoulos G, Ovando-Vázquez C, Neophytou K, Bermúdez-Barrientos JR, Laetsch DR, Robertson E, Kumar S, Claycomb JM, Blaxter M, Abreu-Goodger C, Buck AH (2019) Secretion of an argonaute protein by a parasitic nematode and the evolution of its siRNA guides. Nucleic Acids Res 47:3594–3606
63. Davis CN, Phillips H, Tomes JJ, Swain MT, Wilkinson TJ, Brophy PM, Morphew RM (2019) The importance of extracellular vesicle purification for downstream analysis: a comparison of differential centrifugation and size exclusion chromatography for helminth pathogens. PLoS Negl Trop Dis 13:e0007191
64. de la Torre-Escudero E, Gerlach JQ, Bennett APS, Cwiklinski K, Jewhurst HL, Huson KM, Joshi L, Kilcoyne M, O'Neill S, Dalton JP, Robinson MW (2019) Surface molecules of extracellular vesicles secreted by the helminth pathogen *Fasciola hepatica* direct their internalisation by host cells. PLoS Negl Trop Dis 13:e0007087
65. Murphy A, Cwiklinski K, Lalor R, O''Connell B, Robinson MW, Gerlach J, Joshi L, Kilcoyne M, Dalton JP, O'Neill SM (2020) *Fasciola hepatica* extracellular vesicles isolated from excretory-secretory products using a gravity flow method modulate dendritic cell phenotype and activity. PLoS Negl Trop Dis 14:e0008626
66. Sánchez-López CM, González-Arce A, Soler C, Ramírez-Toledo V, Trelis M, Bernal D, Marcilla A (2023) Extracellular vesicles from the trematodes *Fasciola hepatica* and *Dicrocoelium dendriticum* trigger different responses in human THP-1 macrophages. J Extracell Vesicles 12:e12317
67. Sánchez-López CM, Trelis M, Jara L, Cantalapiedra F, Marcilla A, Bernal D (2020) Diversity of extracellular vesicles from different developmental stages of *Fasciola hepatica*. Int J Parasitol 50:663–669
68. Sheng Z, Wu C, Wang D, Zhong S, Yang X, Rao G, Peng H, Feng S, Li J, Huang W, Luo H (2023) Proteomic analysis of exosome-like vesicles from *Fasciola gigantica* adult worm provides support for new vaccine targets against fascioliasis. Parasit Vectors 16:62–67
69. Trelis M, Sánchez-López CM, Sánchez-Palencia LF, Ramírez-Toledo V, Marcilla A, Bernal D (2022) Proteomic analysis of extracellular vesicles from *Fasciola hepatica* hatching eggs and juveniles in culture. Front Cell Infect Microbiol 12:903602
70. Zhang X, Gong W, Duan C, Cai H, Shen Y, Cao J (2022) *Echinococcus granulosus* protoscoleces-derived exosome-like vesicles and egr-miR-277a-3p promote dendritic cell maturation and differentiation. Cells 11:3220
71. Zhang X, Duan S, Li X, Ding J, Zuo L, Sun B, Zhang X, Jiang X, Gao Y, Hu X, Han S (2022) Differences in the secretory exosomes of *Clonorchis sinensis* adults at different incubation times. Acta Trop 234:106604
72. Chaiyadet S, Sotillo J, Smout M, Cantacessi C, Jones MK, Johnson MS, Turnbull L, Whitchurch CB, Potriquet J, Laohaviroj M, Mulvenna J, Brindley PJ, Bethony JM, Laha T, Sripa B, Loukas A (2015) Carcinogenic liver fluke secretes extracellular vesicles that promote cholangiocytes to adopt a tumorigenic phenotype. J Infect Dis 212:1636–1645
73. Du P, Giri BR, Liu J, Xia T, Grevelding CG, Cheng G (2020) Proteomic and deep sequencing analysis of extracellular vesicles isolated from adult male and female *Schistosoma japonicum*. PLoS Negl Trop Dis 14:e0008618
74. Kifle DW, Pearson MS, Becker L, Pickering D, Loukas A, Sotillo J (2020) Proteomic analysis of two populations of *Schistosoma mansoni*-derived extra-

cellular vesicles: 15k pellet and 120k pellet vesicles. Mol Biochem Parasitol 236:111264
75. Nowacki FC, Swain MT, Klychnikov OI, Niazi U, Ivens A, Quintana JF, Hensbergen PJ, Hokke CH, Buck AH, Hoffmann KF (2015) Protein and small non-coding RNA-enriched extracellular vesicles are released by the pathogenic blood fluke *Schistosoma mansoni*. J Extracell Vesicles 4:28665
76. Samoil V, Dagenais M, Ganapathy V, Aldridge J, Glebov A, Jardim A, Ribeiro P (2018) Vesicle-based secretion in schistosomes: analysis of protein and microRNA (miRNA) content of exosome-like vesicles derived from *Schistosoma mansoni*. Sci Rep 8:3286–3284
77. Sotillo J, Pearson M, Potriquet J, Becker L, Pickering D, Mulvenna J, Loukas A (2016) Extracellular vesicles secreted by *Schistosoma mansoni* contain protein vaccine candidates. Int J Parasitol 46:1–5
78. Allen NR, Taylor-Mew AR, Wilkinson TJ, Huws S, Phillips H, Morphew RM, Brophy PM (2021) Modulation of rumen microbes through extracellular vesicle released by the rumen fluke *Calicophoron daubneyi*. Front Cell Infect Microbiol 11:661830
79. Di Maggio LS, Fischer K, Yates D, Curtis KC, Rosa BA, Martin J, Erdmann-Gilmore P, Sprung RSW, Mitreva M, Townsend RR, Weil GJ, Fischer PU (2023) The proteome of extracellular vesicles of the lung fluke *Paragonimus kellicotti* produced *in vitro* and in the lung cyst. Sci Rep 13:13726–1372x
80. Nicolao MC, Rodrigues CR, Coccimiglio MB, Ledo C, Docena GH, Cumino AC (2023) Characterization of protein cargo of *Echinococcus granulosus* extracellular vesicles in drug response and its influence on immune response. Parasit Vectors 16:255–256
81. Nicolao MC, Rodriguez Rodrigues C, Cumino AC (2019) Extracellular vesicles from *Echinococcus granulosus* larval stage: isolation, characterization and uptake by dendritic cells. PLoS Negl Trop Dis 13:e0007032
82. Liang P, Mao L, Zhang S, Guo X, Liu G, Wang L, Hou J, Zheng Y, Luo X (2019) Identification and molecular characterization of exosome-like vesicles derived from the *Taenia asiatica* adult worm. Acta Trop 198:105036
83. Wang L, Liu T, Liang P, Zhang S, Li T, Li Y, Liu G, Mao L, Luo X (2020) Characterization of exosome-like vesicles derived from *Taenia pisiformis* cysticercus and their immunoregulatory role on macrophages. Parasit Vectors 13:318
84. Mazanec H, Koník P, Gardian Z, Kuchta R (2021) Extracellular vesicles secreted by model tapeworm *Hymenolepis diminuta*: biogenesis, ultrastructure and protein composition. Int J Parasitol 51:327–332
85. Wititkornkul B, Hulme BJ, Tomes JJ, Allen NR, Davis CN, Davey SD, Cookson AR, Phillips HC, Hegarty MJ, Swain MT, Brophy PM, Wonfor RE, Morphew RM (2021) Evidence of immune modulators in the secretome of the equine tapeworm *Anoplocephala perfoliata*. Pathogens 10:912
86. Eichenberger RM, Talukder MH, Field MA, Wangchuk P, Giacomin P, Loukas A, Sotillo J (2018) Characterization of *Trichuris muris* secreted proteins and extracellular vesicles provides new insights into host-parasite communication. J Extracell Vesicles 7:1428004
87. Shears RK, Bancroft AJ, Hughes GW, Grencis RK, Thornton DJ (2018) Extracellular vesicles induce protective immunity against *Trichuris muris*. Parasite Immunol 40:e12536
88. Tritten L, Tam M, Vargas M, Jardim A, Stevenson MM, Keiser J, Geary TG (2017) Excretory/secretory products from the gastrointestinal nematode *Trichuris muris*. Exp Parasitol 178:30–36
89. Sperotto RL, Kremer FS, Aires Berne ME, Costa de Avila LF, da Silva Pinto L, Monteiro KM, Caumo KS, Ferreira HB, Berne N, Borsuk S (2017) Proteomic analysis of *Toxocara canis* excretory and secretory (TES) proteins. Mol Biochem Parasitol 211:39–47
90. Gao X, Yang Y, Liu X, Xu F, Wang Y, Liu L, Yang Y, Liu M, Bai X (2022) Extracellular vesicles from *Trichinella spiralis*: proteomic analysis and protective immunity. PLoS Negl Trop Dis 16:e0010528
91. Kosanović M, Cvetković J, Gruden-Movsesijan A, Vasilev S, Svetlana M, Ilić N, Sofronić-Milosavljević L (2019) *Trichinella spiralis* muscle larvae release extracellular vesicles with immunomodulatory properties. Parasite Immunol 41:e12665
92. Li C, Li C, Xu F, Wang H, Jin X, Zhang Y, Liu X, Wang R, You X, Liu M, Bai X, Yang Y (2023) Identification of antigens in the *Trichinella spiralis* extracellular vesicles for serological detection of early stage infection in swine. Parasit Vectors 16:387–387
93. Palomba M, Rughetti A, Mignogna G, Castrignanò T, Rahimi H, Masuelli L, Napoletano C, Pinna V, Giorgi A, Santoro M, Schininà ME, Maras B, Mattiucci S (2023) Proteomic characterization of extracellular vesicles released by third stage larvae of the zoonotic parasite *Anisakis pegreffii* (nematoda: Anisakidae). Front Cell Infect Microbiol 13:1079991
94. Wang L, Liu T, Pu G, Chen G, Li H, Zhang S, Li Y, Luo X (2023) Global profiling of miRNA and protein expression patterns in rabbit peritoneal macrophages treated with exosomes derived from *Taenia pisiformis* cysticercus. Genomics 115:110690
95. Wang R, Lin L, Han Y, Li Z, Zhen J, Zhang Y, Sun F, Lu Y (2023) Exosome-delivered miR-153 from *Trichinella spiralis* promotes apoptosis of intestinal epithelial cells by downregulating Bcl2. Vet Res 54:52–56
96. Wang T, Koukoulis TF, Vella LJ, Su H, Purnianto A, Nie S, Ang C, Ma G, Korhonen PK, Taki AC, Williamson NA, Reid GE, Gasser RB (2023) The proteome and lipidome of extracellular vesicles from *Haemonchus contortus* to underpin explorations of host-parasite cross-talk. Int J Mol Sci 24:10955

97. Wang Y, Wang X, Zhang N, Yu Y, Bao P, Ma Y, Zhang H, Zhang X, Zhang X, Gong P, Li X, Li J (2023) Extracellular vesicles of *Clonorchis sinensis* promote IL-6 and TNF-α secretion via the toll-like receptor 9-mediated ERK pathway in biliary epithelial cells. Dev Comp Immunol 139:104555
98. Wu F, Chen X, Du Z, Chen Y, Tong D, Zhang J, Yang Y, Ma G, Du A (2024) Proteomic differences between extracellular vesicles and extracellular vesicle-depleted excretory/secretory products of barber's pole worm. Parasit Vectors 17:17–16
99. Tzelos T, Matthews JB, Buck AH, Simbari F, Frew D, Inglis NF, McLean K, Nisbet AJ, Whitelaw CBA, Knox DP, McNeilly TN (2016) A preliminary proteomic characterisation of extracellular vesicles released by the ovine parasitic nematode, *Teladorsagia circumcincta*. Vet Parasitol 221:84–92
100. Harischandra H, Yuan W, Loghry HJ, Zamanian M, Kimber MJ (2018) Profiling extracellular vesicle release by the filarial nematode *Brugia malayi* reveals sex-specific differences in cargo and a sensitivity to ivermectin. PLoS Negl Trop Dis 12:e0006438
101. Montaño KJ, Loukas A, Sotillo J (2021) Proteomic approaches to drive advances in helminth extracellular vesicle research. Mol Immunol 131:1–5
102. Heck M, Neely BA (2020) Proteomics in non-model organisms: a new analytical frontier. J Proteome Res 19:3595–3606
103. Toledo R, Bernal MD, Marcilla A (2011) Proteomics of foodborne trematodes. J Proteome 74:1485–1503
104. Sánchez-López CM, González-Arce A, Ramírez-Toledo V, Bernal D, Marcilla A (2024) Unraveling new players in helminth pathology: extracellular vesicles from *Fasciola hepatica* and *Dicrocoelium dendriticum* exert different effects on hepatic stellate cells and hepatocytes. Int J Parasitol 24
105. González-Miguel J, Becerro-Recio D, Siles-Lucas M (2021) Insights into *Fasciola hepatica* juveniles: crossing the fasciolosis rubicon. Trends Parasitol 37:35–47
106. Whitehead B, Boysen AT, Mardahl M, Nejsum P (2020) Unique glycan and lipid composition of helminth-derived extracellular vesicles may reveal novel roles in host-parasite interactions. Int J Parasitol 50:647–654
107. Simbari F, McCaskill J, Coakley G, Millar M, Maizels RM, Fabriás G, Casas J, Buck AH (2016) Plasmalogen enrichment in exosomes secreted by a nematode parasite versus those derived from its mouse host: implications for exosome stability and biology. J Extracell Vesicles 5:30741
108. Dagenais M, Gerlach JQ, Wendt GR, Collins JJ, Atkinson LE, Mousley A, Geary TG, Long T (2021) Analysis of *Schistosoma mansoni* extracellular vesicles surface glycans reveals potential immune evasion mechanism and new insights on their origins of biogenesis. Pathogens 10:1401
109. Kuipers ME, Nguyen DL, van Diepen A, Mes L, Bos E, Koning RI, Nolte-'t Hoen ENM, Smits HH, Hokke CH (2023) Life stage-specific glycosylation of extracellular vesicles from *Schistosoma mansoni* schistosomula and adult worms drives differential interaction with C-type lectin receptors DC-SIGN and MGL. Front Mol Biosci 10:1125438
110. Kuipers ME, Nolte-'t Hoen ENM, van der Ham AJ, Ozir-Fazalalikhan A, Nguyen DL, de Korne CM, Koning RI, Tomes JJ, Hoffmann KF, Smits HH, Hokke CH (2020) DC-SIGN mediated internalisation of glycosylated extracellular vesicles from *Schistosoma mansoni* increases activation of monocyte-derived dendritic cells. J Extracell Vesicles 9:1753420
111. Gazzinelli-Guimaraes PH, Nutman TB (2018) Helminth parasites and immune regulation. F1000Res 7. eCollection 2018
112. Joshi P, Mishra PKK (2022) Functional diversity of the excretory/secretory proteins of nematode parasites. Acta Parasitol 67:619–627
113. Zamanian M, Fraser LM, Agbedanu PN, Harischandra H, Moorhead AR, Day TA, Bartholomay LC, Kimber MJ (2015) Release of small RNA-containing exosome-like vesicles from the human filarial parasite *Brugia malayi*. PLoS Negl Trop Dis 9:e0004069
114. Ricciardi A, Bennuru S, Tariq S, Kaur S, Wu W, Elkahloun AG, Arakelyan A, Shaik J, Dorward DW, Nutman TB, Tolouei Semnani R (2021) Extracellular vesicles released from the filarial parasite *Brugia malayi* downregulate the host mTOR pathway. PLoS Negl Trop Dis 15:e0008884
115. Loghry HJ, Kwon H, Smith RC, Sondjaja NA, Minkler SJ, Young S, Wheeler NJ, Zamanian M, Bartholomay LC, Kimber MJ (2023) Extracellular vesicles secreted by *Brugia malayi* microfilariae modulate the melanization pathway in the mosquito host. Sci Rep 13:8778–8779
116. Wu J, Liao Y, Li D, Zhu Z, Zhang L, Wu Z, He P, Wang L (2022) Extracellular vesicles derived from *Trichinella spiralis* larvae promote the polarization of macrophages to M2b type and inhibit the activation of fibroblasts. Front Immunol 13:974332
117. Wang R, Zhang Y, Zhen J, Zhang J, Pang Z, Song X, Lin L, Sun F, Lu Y (2022) Effects of exosomes derived from *Trichinella spiralis* infective larvae on intestinal epithelial barrier function. Vet Res 53:87
118. Wang X, Tian L, Lu J, Ng IO (2022) Exosomes and cancer - diagnostic and prognostic biomarkers and therapeutic vehicle. Oncogenesis 11:54–55
119. Duque-Correa MA, Schreiber F, Rodgers FH, Goulding D, Forrest S, White R, Buck A, Grencis RK, Berriman M (2020) Development of caecaloids to study host-pathogen interactions: new insights into immunoregulatory functions of *Trichuris muris* extracellular vesicles in the caecum. Int J Parasitol 50:707–718
120. Meningher T, Barsheshet Y, Ofir-Birin Y, Gold D, Brant B, Dekel E, Sidi Y, Schwartz E, Regev-Rudzki N, Avni O, Avni D (2020) Schistosomal extracellu-

lar vesicle-enclosed miRNAs modulate host T helper cell differentiation. EMBO Rep 21:e47882
121. Chaiyadet S, Sotillo J, Krueajampa W, Thongsen S, Brindley PJ, Sripa B, Loukas A, Laha T (2019) Vaccination of hamsters with *Opisthorchis viverrini* extracellular vesicles and vesicle-derived recombinant tetraspanins induces antibodies that block vesicle uptake by cholangiocytes and reduce parasite burden after challenge infection. PLoS Negl Trop Dis 13:e0007450
122. Guo A, Wang L, Meng X, Zhang S, Sheng Z, Luo X, Huang W, Wang S, Cai X (2021) Extracellular vesicles from *Fasciola gigantica* induce cellular response to stress of host cells. Exp Parasitol 231:108173
123. Zheng Y, Guo X, Su M, Guo A, Ding J, Yang J, Xiang H, Cao X, Zhang S, Ayaz M, Luo X (2017) Regulatory effects of *Echinococcus multilocularis* extracellular vesicles on RAW264.7 macrophages. Vet Parasitol 235:29–36
124. Liu C, Cao J, Zhang H, Field MC, Yin J (2023) Extracellular vesicles secreted by *Echinococcus multilocularis*: important players in angiogenesis promotion. Microbes Infect 25:105147
125. Zhou X, Wang W, Cui F, Shi C, Ma Y, Yu Y, Zhao W, Zhao J (2019) Extracellular vesicles derived from *Echinococcus granulosus* hydatid cyst fluid from patients: isolation, characterization and evaluation of immunomodulatory functions on T cells. Int J Parasitol 49:1029–1037
126. Kondo Y, Ito D, Taniguchi R, Tademoto S, Horie T, Otsuki H (2023) Extracellular vesicles derived from *Spirometra erinaceieuropaei* plerocercoids inhibit activation of murine macrophage RAW264.7 cells. Parasitol Int 95:102742
127. Eichenberger RM, Ryan S, Jones L, Buitrago G, Polster R, Montes de Oca M, Zuvelek J, Giacomin PR, Dent LA, Engwerda CR, Field MA, Sotillo J, Loukas A (2018) Hookworm secreted extracellular vesicles interact with host cells and prevent inducible colitis in mice. Front Immunol 9:850
128. Yang Y, Liu L, Liu X, Zhang Y, Shi H, Jia W, Zhu H, Jia H, Liu M, Bai X (2020) Extracellular vesicles derived from *Trichinella spiralis* muscle larvae ameliorate TNBS-induced colitis in mice. Front Immunol 11:1174
129. Ashour DS, Ibrahim FMK, Elshamy AM, Zoghroban HS (2022) *Trichinella spiralis*-derived extracellular vesicles induce a protective immunity against larval challenge in mice. Pathog Dis 80:ftac040
130. Trelis M, Galiano A, Bolado A, Toledo R, Marcilla A, Bernal D (2016) Subcutaneous injection of exosomes reduces symptom severity and mortality induced by *Echinostoma caproni* infection in BALB/c mice. Int J Parasitol 46:799–808
131. Roig J, Saiz ML, Galiano A, Trelis M, Cantalapiedra F, Monteagudo C, Giner E, Giner RM, Recio MC, Bernal D, Sánchez-Madrid F, Marcilla A (2018) Extracellular vesicles from the helminth *Fasciola hepatica* prevent DSS-induced acute ulcerative colitis in a T-lymphocyte independent mode. Front Microbiol 9:1036
132. Mossallam SF, Abou-El-Naga IF, Abdel Bary A, Elmorsy EA, Diab RG (2021) *Schistosoma mansoni* egg-derived extracellular vesicles: a promising vaccine candidate against murine schistosomiasis. PLoS Negl Trop Dis 15:e0009866
133. Mekonnen GG, Pearson M, Loukas A, Sotillo J (2018) Extracellular vesicles from parasitic helminths and their potential utility as vaccines. Expert Rev Vaccines 17:197–205
134. Dakshinamoorthy G, Munirathinam G, Stoicescu K, Reddy MV, Kalyanasundaram R (2013) Large extracellular loop of tetraspanin as a potential vaccine candidate for filariasis. PLoS One 8:e77394
135. Curti E, Kwityn C, Zhan B, Gillespie P, Brelsford J, Deumic V, Plieskatt J, Rezende WC, Tsao E, Kalampanayil B, Hotez PJ, Bottazzi ME (2013) Expression at a 20L scale and purification of the extracellular domain of the *Schistosoma mansoni* TSP-2 recombinant protein: a vaccine candidate for human intestinal schistosomiasis. Hum Vaccin Immunother 9:2342–2350
136. Dang Z, Yagi K, Oku Y, Kouguchi H, Kajino K, Matsumoto J, Nakao R, Wakaguri H, Toyoda A, Yin H, Sugimoto C (2012) A pilot study on developing mucosal vaccine against alveolar echinococcosis (AE) using recombinant tetraspanin 3: vaccine efficacy and immunology. PLoS Negl Trop Dis 6:e1570
137. Andreu Z, Yáñez-Mó M (2014) Tetraspanins in extracellular vesicle formation and function. Front Immunol 5:442
138. Maizels RM, Kurniawan-Atmadja A (2002) Variation and polymorphism in helminth parasites. Parasitology 125(Suppl):25
139. McManus DP (2020) Recent progress in the development of liver fluke and blood fluke vaccines. Vaccines (Basel) 8:553
140. Carrière J, Barnich N, Nguyen HTT (2016) Exosomes: from functions in host-pathogen interactions and immunity to diagnostic and therapeutic opportunities. Rev Physiol Biochem Pharmacol 172:39–75
141. Fuhrmann G, Neuer AL, Herrmann IK (2017) Extracellular vesicles - a promising avenue for the detection and treatment of infectious diseases? Eur J Pharm Biopharm 118:56–61
142. Archer J, LaCourse JE, Webster BL, Stothard JR (2020) An update on non-invasive urine diagnostics for human-infecting parasitic helminths: what more could be done and how? Parasitology 147:873–888
143. Miswan N, Singham GV, Othman N (2022) Advantages and limitations of microscopy and molecular detections for diagnosis of soil-transmitted helminths: an overview. Helminthologia 59:321–340
144. Romasanta A, Romero JL, Arias M, Sánchez-Andrade R, López C, Suárez JL, Díaz P, Díez-Baños P, Morrondo P, Paz-Silva A (2003) Diagnosis of parasitic zoonoses by immunoenzymatic assays--analy-

sis of cross-reactivity among the excretory/secretory antigens of *Fasciola hepatica*, *Toxocara canis*, and *Ascaris suum*. Immunol Investig 32:131–142
145. Coumans FAW, Brisson AR, Buzas EI, Dignat-George F, Drees EEE, El-Andaloussi S, Emanueli C, Gasecka A, Hendrix A, Hill AF, Lacroix R, Lee Y, van Leeuwen TG, Mackman N, Mäger I, Nolan JP, van der Pol E, Pegtel DM, Sahoo S, Siljander PRM, Sturk G, de Wever O, Nieuwland R (2017) Methodological guidelines to study extracellular vesicles. Circ Res 120:1632–1648
146. Yates D, Di Maggio LS, Rosa BA, Sprung RW, Erdmann-Gilmore P, Townsend RR, Budge PJ, Kamgno J, Mitreva M, Weil GJ, Fischer U (2023) Identification of biomarker candidates for filarial parasite infections by analysis of extracellular vesicles. Front Parasitol 2
147. Fratini F, Tamarozzi F, Macchia G, Bertuccini L, Mariconti M, Birago C, Iriarte A, Brunetti E, Cretu CM, Akhan O, Siles-Lucas M, Díaz A, Casulli A (2020) Proteomic analysis of plasma exosomes from cystic echinococcosis patients provides *in vivo* support for distinct immune response profiles in active vs inactive infection and suggests potential biomarkers. PLoS Negl Trop Dis 14:e0008586
148. Meningher T, Lerman G, Regev-Rudzki N, Gold D, Ben-Dov IZ, Sidi Y, Avni D, Schwartz E (2017) Schistosomal microRNAs isolated from extracellular vesicles in sera of infected patients: a new tool for diagnosis and follow-up of human schistosomiasis. J Infect Dis. 215(3):378–386
149. Kim C, Lee K (2019) Polydiacetylene (PDA) liposome-based immunosensor for the detection of exosomes. Biomacromolecules 20:3392–3398
150. Deng Y, Zou Y, Song X, Jiang A, Wang M, Qin Q, Song Y, Yue C, Yang D, Yu B, Lu H, Zheng Y (2023) Potential of extracellular vesicles for early prediction of severity and potential risk stratification in critical inflammatory diseases. J Cell Commun Signal 17:1283–1292
151. Revenfeld ALS, Bæk R, Nielsen MH, Stensballe A, Varming K, Jørgensen M (2014) Diagnostic and prognostic potential of extracellular vesicles in peripheral blood. Clin Ther 36:830–846
152. Xu K, Liu Q, Wu K, Liu L, Zhao M, Yang H, Wang X, Wang W (2020) Extracellular vesicles as potential biomarkers and therapeutic approaches in autoimmune diseases. J Transl Med 18:432
153. Esteves S, Lima C, Costa I, Osório H, Fernández-Becerra C, Santarém N, Cordeiro-da-Silva A (2023) Characterization and proteomic analysis of plasma EVs recovered from healthy and diseased dogs with canine leishmaniosis. Int J Mol Sci 24:5490
154. Gioseffi A, Hamerly T, Van K, Zhang N, Dinglasan RR, Yates PA, Kima PE (2020) *Leishmania*-infected macrophages release extracellular vesicles that can promote lesion development. Life Sci Alliance 3:e202000742
155. Madeira RP, Dal'Mas Romera LM, de Cássia Buck P, Mady C, Ianni BM, Torrecilhas AC (2021) New biomarker in Chagas disease: extracellular vesicles isolated from peripheral blood in chronic Chagas disease patients modulate the human immune response. J Immunol Res 2021:6650670
156. Olajide JS, Qu Z, Yang S, Yang B, Xu X, Wang J, Cai J (2023) Quantitative proteomic analysis of local and systemic extracellular vesicles during *Eimeria falciformis* infectious cycle in the host. Parasit Vectors 16:339–33x
157. Tedford E, Badya NB, Laing C, Asaoka N, Kaneko S, Filippi BM, McConkey GA (2023) Infection-induced extracellular vesicles evoke neuronal transcriptional and epigenetic changes. Sci Rep 13:6913–6912
158. Emerson LE, Mosby CA, Enslow S, Hui WW, Jones MK, Ferraro MJ (2024) Changes in lipid composition of host-derived extracellular vesicles following *Salmonella* infection. Microbiol Spectr 12:e0279623. Epub 2023 Dec 11
159. Hui WW, Emerson LE, Clapp B, Sheppe AE, Sharma J, Del Castillo J, Ou M, Maegawa GHB, Hoffman C, Larkin Iii J, Pascual DW, Ferraro MJ (2021) Antigen-encapsulating host extracellular vesicles derived from *Salmonella*-infected cells stimulate pathogen-specific Th1-type responses *in vivo*. PLoS Pathog 17:e1009465
160. Li S, Giri BR, Liu J, He X, Cai P, Jing Z, Cheng G (2022) Characterization of microRNA cargo of extracellular vesicles isolated from the plasma of *Schistosoma japonicum*-infected mice. Front Cell Infect Microbiol 12:803242
161. Shi C, Zhou X, Yang W, Wu J, Bai M, Zhang Y, Zhao W, Yang H, Nagai A, Yin M, Gao X, Ding S, Zhao J (2022) Proteomic analysis of plasma-derived extracellular vesicles from mice with *Echinococcus granulosus* at different infection stages and their immunomodulatory functions. Front Cell Infect Microbiol 12:805010
162. Charlier J, Williams DJ, Ravinet N, Claerebout E (2023) To treat or not to treat: diagnostic thresholds in subclinical helminth infections of cattle. Trends Parasitol 39:139–151
163. Guo S, Tao S, Dawn H (2018) Microfluidics-based on-a-chip systems for isolating and analysing extracellula vesicles. J Extracell Vesicles 7(1):15082711
164. Leggio L, Paternò G, Vivarelli S, Bonasera A, Pignataro B, Iraci N, Arrabito G (2023) Label-free approaches for extracellular vesicle detection. iScience 26:108105
165. Wang Y, Yuan W, Kimber M, Lu M, Dong L (2018) Rapid differentiation of host and parasitic exosome vesicles using microfluidic photonic crystal biosensor. ACS Sens 3:1616–1621
166. Maizels RM, McSorley HJ (2016) Regulation of the host immune system by helminth parasites. J Allergy Clin Immunol 138:666–675
167. Wammes LJ, Mpairwe H, Elliott AM, Yazdanbakhsh M (2014) Helminth therapy or elimination: epidemiological, immunological, and clinical considerations. Lancet Infect Dis 14:1150–1162

168. Helmby H (2015) Human helminth therapy to treat inflammatory disorders - where do we stand? BMC Immunol 16:12–13
169. Sobotková K, Parker W, Levá J, Růžková J, Lukeš J, Jirků Pomajbíková K (2019) Helminth therapy - from the parasite perspective. Trends Parasitol 35:501–515
170. Chakraborty P, Aravindhan V, Mukherjee S (2023) Helminth-derived biomacromolecules as therapeutic agents for treating inflammatory and infectious diseases: what lessons do we get from recent findings? Int J Biol Macromol 241:124649
171. Ni Y, Xiong R, Zhu Y, Luan N, Yu C, Yang K, Wang H, Xu X, Yang Y, Sun S, Shi L, Padde JR, Chen L, Chen L, Hou M, Xu Z, Lai R, Ji M (2023) A target-based discovery from a parasitic helminth as a novel therapeutic approach for autoimmune diseases. EBioMedicine 95:104751
172. Montaner S, Galiano A, Trelis M, Martin-Jaular L, Del Portillo HA, Bernal D, Marcilla A (2014) The role of extracellular vesicles in modulating the host immune response during parasitic infections. Front Immunol 5:433
173. Gao X, Yang Y, Liu X, Wang Y, Yang Y, Boireau P, Liu M, Bai X (2021) Extracellular vesicles derived from Trichinella spiralis prevent colitis by inhibiting M1 macrophage polarization. Acta Trop 213:105761. Epub 2020 Nov 19. PMID: 33221281.
174. Simonsen JB (2019) Pitfalls associated with lipophilic fluorophore staining of extracellular vesicles for uptake studies. J Extracell Vesicles 8:1582237
175. Rautaniemi K, Zini J, Löfman E, Saari H, Haapalehto I, Laukka J, Vesamäki S, Efimov A, Yliperttula M, Laaksonen T, Vuorimaa-Laukkanen E, Lisitsyna ES (2021) Addressing challenges in the removal of unbound dye from passively labelled extracellular vesicles. Nanoscale Adv 4:226–240
176. Boysen AT, Whitehead B, Stensballe A, Carnerup A, Nylander T, Nejsum P (2020) Fluorescent labeling of helminth extracellular vesicles using an *in vivo* whole organism approach. Biomedicines 8:213
177. Rooney J, Northcote HM, Williams TL, Cortés A, Cantacessi C, Morphew RM (2022) Parasitic helminths and the host microbiome - a missing 'extracellular vesicle-sized' link? Trends Parasitol 38:737–747
178. Zhang Y, Dou Y, Liu Y, Di M, Bian H, Sun X, Yang Q (2023) Advances in therapeutic applications of extracellular vesicles. Int J Nanomedicine 18:3285–3307
179. Feng J, Xiu Q, Huang Y, Troyer Z, Li B, Zheng L (2023) Plant-derived vesicle-like nanoparticles as promising biotherapeutic tools: present and future. Adv Mater 35:e2207826
180. Gurunathan S, Kim J (2023) Bacterial extracellular vesicles: emerging nanoplatforms for biomedical applications. Microb Pathog 183:106308
181. Mu N, Li J, Zeng L, You J, Li R, Qin A, Liu X, Yan F, Zhou Z (2023) Plant-derived exosome-like nanovesicles: current progress and prospects. Int J Nanomedicine 18:4987–5009
182. Xie J, Li Q, Haesebrouck F, Van Hoecke L, Vandenbroucke RE (2022) The tremendous biomedical potential of bacterial extracellular vesicles. Trends Biotechnol 40:1173–1194
183. van der Koog L, Gandek TB, Nagelkerke A (2022) Liposomes and extracellular vesicles as drug delivery systems: a comparison of composition, pharmacokinetics, and functionalization. Adv Healthc Mater 11:e2100639
184. Sánchez-López CM, Manzaneque-López MC, Pérez-Bermúdez P, Soler C, Marcilla A (2022) Characterization and bioactivity of extracellular vesicles isolated from pomegranate. Food Funct 13:12870–12882
185. Morishita M, Horita M, Higuchi A, Marui M, Katsumi H, Yamamoto A (2021) Characterizing different probiotic-derived extracellular vesicles as a novel adjuvant for immunotherapy. Mol Pharm 18:1080–1092
186. Cheng K, Kalluri R (2023) Guidelines for clinical translation and commercialization of extracellular vesicles and exosomes based therapeutics. Extracell Vesicle 2:100029
187. Ghodasara A, Raza A, Wolfram J, Salomon C, Popat A (2023) Clinical translation of extracellular vesicles. Adv Healthc Mater 12:e2301010

Recent Advances in Parasitomics: Omics in the Study of Ascariasis

Nina L. Tang and Jill E. Weatherhead

Abstract

Ascaris is the most common helminth infection in the world, leading to significant global morbidity. Despite the high burden of disease globally, ascariasis remains widely understudied and underfunded leading to a large knowledge gap in basic biological processes. "Omic" technological advancements have significantly accelerated our comprehension of the biological underpinnings of numerous diseases of profound human impact. But adoption of these technologies in *Ascaris* research has been limited. Moving forward, application of omic technology could provide the first steps to addressing critical knowledge gaps in *Ascaris* research. Genomic techniques have provided unique insights into the epidemiology and evolution of ascariasis across the world. The application of cutting-edge omic technology in proteomics, metabolomics, lipidomics, and microbiome studies has quickly provided insights into essential *Ascaris* biological pathways. Detailed characterization of *Ascaris* biology is a key step in the development of novel interventions including vaccines that are desperately needed to prevent *Ascaris*-induced morbidity and essential for global elimination efforts.

Keywords

Ascaris · Metagenomics · Proteomics · Transcriptomics · Vaccine development

1 Introduction

Ascaris is a parasitic nematode of significant global importance. Its infection of the intestinal lumen results in ascariasis in humans and pigs and is responsible for enduring morbidity and mortality across many regions of the world [1–4]. In recent decades, the use of mass drug administrations and improvement of living conditions have resulted in reductions in human ascariasis. However, the global burden of ascariasis in humans persists with an estimated global prevalence of about 732 million [4]. There is considerable variability in the burden of *Ascaris* infection

N. L. Tang (✉)
Department of Pediatrics, Division of Tropical Medicine, Baylor College of Medicine, Houston, TX, USA
e-mail: nina.tang@bcm.edu

J. E. Weatherhead
Department of Pediatrics, Division of Tropical Medicine, Baylor College of Medicine, Houston, TX, USA

Department of Medicine, Division of Infectious Diseases, Baylor College of Medicine, Houston, TX, USA

National School of Tropical Medicine, Baylor College of Medicine, Houston, TX, USA

J. D. Ramírez González (ed.), *Recent Advances in Parasitomics*,
https://doi.org/10.1007/978-3-031-70591-5_14

between geographic regions with the highest rates of infection in Southern Asia, Southeastern Asia, South America, Eastern Africa, and Western Africa. Prevalence also varies with age with children experiencing a higher global prevalence of 13.5% compared to adults for whom the prevalence is 8.85% [4].

Ascariasis primarily affects humans (*Ascaris lumbricoides*) and pigs (*Ascaris suum*), although it has also been recorded in other species [5]. Regardless, the establishment of *Ascaris* infection is identical in these organisms and inter-species infection occurs [6]. Ascariasis is acquired through fecal–oral transmission. *Ascaris* eggs are ingested inadvertently through contaminated food or water, and infectious stage-3 larvae (L3) hatch in the gastrointestinal system of the host [7]. These larvae are carried to the cecum and proximal colon where they penetrate the intestinal mucosa and reach the portal blood [8]. They continue on to the liver and eventually the alveoli of the lungs [9]. From the alveoli, they migrate to the pharynx where they are swallowed and return to the small intestine as L4 [10]. The parasites remain in the small intestine where the L4 mature into adult worms, reaching up to 35 cm in length, and can reside in the host for approximately 1–2 years. Adult female worms produce up to 200,000 eggs per day that are excreted in the host's feces [11].

Morbidity and mortality associated with ascariasis are typically related to the worm burden in an individual. The most common acute and severe complications result from large *Ascaris* adult worm burden in the intestinal lumen including intestinal obstruction—accounting for 72% of all *Ascaris*-associated complications [12]—as well as biliary obstruction, intestinal colic, acute pancreatitis, intestinal perforation, and appendicitis [13–15]. High worm burden has also been directly correlated to the severity of malnutrition in children, and it is known that anthelmintic treatment can lead to recoveries in pediatric height and weight [16]. The intensity of infection in children has also been correlated to various types of cognitive impairment, though the strength of this relationship has been contested [17–21].

Ascariasis is a neglected tropical disease [22] and, as such, has historically received disproportionately less research attention than its significant global prevalence warrants. In this chapter, we will review the contributions of studies using genomics and transcriptomics, proteomics, and lipidomics that have advanced our understanding of *Ascaris* and highlight the need for ongoing metagenomic research.

2 Genomic and Transcriptomic Sequencing in *Ascaris* Biology and Epidemiology

2.1 The role of *Caenorhabditis elegans*

The evolutionary proximity of *Ascaris* to the free-living nematode, *Caenorhabditis elegans,* has facilitated many advances in our understanding of *Ascaris*. Early sequencing of the mitochondrial genomes of *A. suum* and *C. elegans* provided clues regarding the evolutionary divergence of *Ascaris* species from other nematodes [23]. And, comparative analysis of cDNA enriched in L3 larva of *A. suum* with that of *C. elegans* enabled the identification of *Ascaris* genes associated with essential processes like embryonic and larval development, cellular processing and signaling, and metabolism [24]. Subsequent sequencing studies included other parasitic nematodes such as *Haemonchus contortus* in addition to *A. suum* and *C. elegans,* allowing for the identification of a conserved family of 241 intestinal proteins that composed ~20% of each nematode's intestinal transcriptome [25]. Therefore, comparative analyses with *C. elegans* were integral to the initial progress made in understanding *Ascaris* biology.

Despite *Ascaris* research benefiting from the larger body of literature on *C. elegans*, the greater size of *Ascaris* worms enabled researchers to conduct work that was less accessible in the smaller organisms of *C. elegans*. One such study dissected *A. suum* and used RNA sequencing to study the tissue-specific transcriptomes within the head, pharynx, and intestine of male and female *A. suum* as well as the testis and seminal vesicle and the uterus and ovary in male and female worms, respectively [26]. This allowed for an appreciation of differential expression across tissues within a single organism.

2.2 Sex-Specific Biology

In the final infective stages of ascariasis, the host organism is infected with both male and female adult worms in the small intestine. Genomic tools have enabled us to better understand sex-dependent transcriptomic disparities, allowing us to better appreciate the biology of these organisms and more effectively identify anthelmintic targets. This work has revealed female *Ascaris* transcriptomes enriched with *C. elegans* orthologs of molecules associated with reproduction and development while male transcriptomes were more heavily weighted toward the expression of major sperm proteins, kinases, phosphatases, actins, and myosins [27]. A later study of microRNA (miRNA) profiles of male and female adult *A. suum* expanded on these findings by discovering that there were more sex-specific miRNAs than shared ones, including targets such as elongation factors, heat shock proteins, and growth factors. Importantly, bioinformatic analyses of these sex-specific miRNAs found that they spanned broad functional categories such as cellular components, molecular functions, and biological processes [28]. This deeper understanding of sex-specific phenotypes enables better prediction of anthelmintic targets that efficiently interrupt physiological processes essential for both male and female *Ascaris*.

2.3 Epidemiology

Arguably, the most important impact of genomic techniques on the study of *Ascaris* is the assembly of genomes and enabling of genotyping and haplotyping for epidemiological purposes. In 2003, a study of nucleotide variation in nuclear ribosomal DNA spanning the first internal transcribed spacer (ITS-1) from *Ascaris* collected from humans and pigs across multiple regions in China discovered the existence of five genotypes: three found in both human and pig *Ascaris* (G1-G3) and two unique to human *Ascaris* (G4-G5) [29]. Of these, G1 was the most common in humans and G3 was the most common in pigs. Following this, there have been continuous discoveries of more genotypes of human ascariasis including the discovery of G6 in Brazil [30] and 8 new sequence variants in India [31].

In addition to ITS-1, mitochondrial cytochrome *c* oxidase subunit 1 (*cox*1) and NADH dehydrogenase subunit 1 (*nad*1) have been used for haplotyping of human and pig *Ascaris* [32–36]. These works have revealed higher genetic diversity in pig *Ascaris* than human *Ascaris* and high gene flow between worms infecting different host species [34]. It has also been observed that some haplotypes are found in vastly different geographical areas. For example, haplotypes found in China have also been identified in southwestern Uganda and Brazil [35, 36]. Therefore, the increasing accessibility of sequencing has allowed for more detailed epidemiological descriptions of *Ascaris* across geographic regions and host species.

2.4 The Species Question

One example that especially highlights the impact of genomic tools in the study of *Ascaris* is the historical debate of *Ascaris* taxonomy. Ascariasis has long been observed in humans and pigs, and the organisms responsible for their infection were believed to be of two different species: *Ascaris lumbricoides* in humans and *Ascaris suum* in pigs. In 1999, analyses of human and pig *Ascaris* found important early evidence that suggested that these parasitic organisms belonged to one species [37]. They found identical 5.8S sequences and second internal transcribed spacer (ITS-2) sequences between the organisms, and only 1.3% difference in the first internal transcribed spacer (ITS-1). These findings did not definitively argue in favor of their being the same species or not since there are examples of multiple species with identical ITS-2 sequences, and small differences in ITS-1 could represent either population variation within a species or population variation between species. However, this study marked one of the earliest impacts of genomic sequencing in what continues to be a contentious debate.

Additional studies using modern technologies have continued to add to this debate in the hopes of bringing it to rest. In 2003, a study in China using automated cycle sequencing to identify genotypes of *Ascaris* determined that three genotypes were shared among worms infecting humans and pigs [29]. Other microsatellite sequence analyses have also found high similarity, heterozygote deficiency, weak population differentiation, and high gene flow between *Ascaris* populations [34, 38, 39]—together suggesting that *A. suum* and *A. lumbricoides* are, indeed, one species.

Analyses of mitochondrial *cox*1 sequences determined that human *Ascaris* infections in Europe were derived from pigs [40]. Additional studies identified cross-infections in multiple provinces in China [41] and Africa [40], and multiple analyses of mitochondrial genomes from pig and human *Ascaris* have found high similarity between organisms [42, 43] and evidence of genetic mosaics [44]. Sequencing analyses have also led to the identification of hybrid *Ascaris* species capable of infecting both host species in China and Guatemala [41, 45]. And, additional zoonotic sequencing studies of *Ascaris* in chimpanzees and gibbons have found that non-human primates are also infected by a single species of *Ascaris* that has genetic similarity to *A. lumbricoides* [5].

At present, many consider the body of existing evidence to show that *A. lumbricoides* and *A. suum* constitute a single species, collectively called *Ascaris*. Some posit that *A. lumbricoides* has taxonomic priority, so what was formally known as *A. suum* should be understood as *A. lumbricoides* moving forward [46]. With the publication of full *Ascaris* genomes [44, 47, 48] and the increasing availability of sequencing technologies, the body of research on this topic continues to grow. Regardless of one's individual analysis of the existing data, the application of continually advancing genomic sequencing techniques in the study of ascariasis has given rise to the wealth of data that continues to fuel this debate.

3 Proteomic Clues into Host–Parasite Biology

3.1 Excretory-Secretory Products

Ascaris excretory-secretory (ES) product is a collection of more than 250 heterogeneous proteins and miRNAs secreted by *Ascaris* worms during the larval migration phase (L3- and L4-stage) [49, 50] and the adult intestinal phase of the life cycle [49]. Proteins secreted in the *Ascaris* ES product have important roles in biological functions including worm development and survival [51]. Characterization of the *Ascaris* ES secretome using mass spectrometry has revealed that protein composition of the ES product is not only distinct from other *Ascaris* fluid compartments but is distinct for each *Ascaris* developmental stage [52].

The dynamic nature of the ES product reflects the varying needs of the organism at different life stages. For larvae, ES proteins aid in development and migration across host tissue. Motor activity proteins such as myosin-4, paramyosin, and tropomyosin are more commonly identified from pre-migratory L3, whereas structural proteins like cuticlin-1 and cuticle collagen 12 and 13 are more prominently found in migrating L3 in the lungs. L4 also have a distinct secretome, secreting glycosyl hydrolases that degrade complex carbohydrates required for energy metabolism [53]. While the majority of the *Ascaris* secretome is specific to larval development stages, some proteins like 14-3-3-like proteins and serpin-like protein are conserved across all larval developmental stages [53]. Larval stage ES proteins are also immunogenic. Proteins such as enolase, pyruvate dehydrogenase, fructose bisphosphate aldolase, GAPDH, phosphoglycerate kinase, and triose phosphate isomerase, used in metabolic, cellular repair and antioxidant processes, are known potent inducers of the host immune response during larval migration [54].

However, immune responses to specific *Ascaris* proteins during larval migration are not uniform among host immune compartments. For example, larval 42 kDa antigen is recognized at 7 days post-infection in liver lymph nodes, but not at 14 days post-infection. The same 42 kDa antigen is detected in the lungs at 14 days post-infection reflecting the unique interaction between *Ascaris* ES product and the host immune response during larval migration [55].

Adult worm ES product supports energy metabolism and shapes the intestinal immune cell repertoire. Similar to L4, adult worms which are anaerobic utilize fermentative metabolism through the upregulation of glycosyl hydrolases [56]. In contrast to the immunogenic properties of larval ES product, adult worm ES product is largely involved in immune evasion or immune regulation. Adult worm ES product can suppress plasmacytoid dendritic subtypes of conventional dendritic cells and natural killer cells regulating type 1 cytokine responses [49]. Characterization of adult ES product-derived OV17 protein demonstrated the protein is an epitope for CD4 T helper cells [49]. Other adult worm-derived ES proteins such as protein 1 (PAS-1) inhibit pro-inflammatory cytokine immune responses and upregulate immunomodulatory cytokines like IL-10 [57]. Beyond proteins, miRNA can also be detected in adult ES product and may also contribute to the modulation of the host immune response through gene regulation. However, there remains a considerable lack of research, and thus understanding, pertaining to the function of *Ascaris* ES product-derived miRNA [50, 58].

Despite the work done to date evaluating the composition and function of *Ascaris* ES product, there remain significant limitations to studying the *Ascaris* ES product secretome. Generally, the secretome is evaluated in ex vivo assays; however, ex vivo studies, which are different environmental conditions compared to in vivo, may influence protein composition resulting in non-biologically relevant data. Additionally, little is known about the secretory pathways involved in the release of the ES product leaving gaps in knowledge about the biologic importance of these proteins [52]. Despite these limitations, the prospect of preventing *Ascaris* development, migration, and host immune evasion by neutralizing or inhibiting proteins within *Ascaris* ES product remains an intriguing approach to identifying novel targets for vaccine and therapeutic development [52, 55].

3.2 Extracellular Vesicles

Extracellular vesicles (EVs), membrane-enclosed nanoparticles, have been identified in *Ascaris*-secreted fluid including ES products [50] and adult body fluid and tissue. Transmission electron microscopy provides direct evidence that EVs are released from *Ascaris* worms at each developmental stage and range in size from 80 to 200 nm. Similar to ES product, EV protein composition is also distinct at each development stage. Composition of EVs includes proteases, peptidases, oxidases, reductases, kinases, heat shock proteins, and other proteins related to cell–cell communication and cellular structure [50, 52]. In comparison to other compartments such as adult body fluid, adult ES product EVs have the greatest diversity in protein composition with 167 unique proteins identified [50]. Furthermore, miRNAs, which are thought to have a role in immunity and/or inflammation, have been detected in high abundance within EVs from fluid secreted from L3, L4, and adult worms [50]. Among the developmental stages, 24 unique miRNAs are conserved within the EVs. Co-culture of *A. suum* EVs with monocytic cell lines or primary human PBMCs followed by challenge with LPS demonstrated reduction in pro-inflammatory cytokine TNF-α concentrations compared to cultures without *A. suum* EVs. Conversely, *Ascaris* EVs had no effect on the release of IL-1β [59]. Additional investigation into the biologic role of EVs from each *Ascaris* compartment, the origin of EVs within the parasite, and the mechanisms regulating the release of EVs are critical for understanding host–parasite interactions and evaluating the therapeutic potential of *Ascaris* EVs [58].

4 Lipidomics

Lipidomics is a rapidly evolving technology that has provided discrete insight into the significant biologic influence of lipids within the host–parasite relationship [60]. However, to date, limited knowledge exists regarding the *Ascaris*-specific lipidome. Using untargeted lipidomic processes with LC-MS/MS, more than 500 lipids, within 18 lipid classes, have been identified across *Ascaris* fluid compartments and developmental stages. Glycolipids, glycerophospholipids, and sphingolipids were characterized at each developmental stage, with the greatest abundance of lipids found in L3 stage. The variation in lipid content at each developmental stage further reflects the unique metabolic needs of the organisms, from energy storage to structural integrity and environmental adaptation [60]. While further research is required, higher abundance of triglycerides from L3 and decreasing abundance of triglycerides during development into adult stage worms provide the first insight into the involvement of the lipidome in *Ascaris* physiology [61]. Other lipids such as carboxylic acids (acetic acid) have also been identified in adult *Ascaris* worms [62], but the importance of acetic acid and other carboxylic acids remains unknown.

5 Future Directions

5.1 Diagnostic Tools

For parasitoses like ascariasis, there is a persistent tension between technological advancements and cost of implementation. Microscopy has historically been the mainstay of diagnosis in field settings due to its simplicity and low cost [63]. However, advances in polymerase chain reaction techniques (PCR)—enabled by improvements in sequencing for primer/probe design [64]—have increased our capacity to detect *Ascaris* DNA at lower concentrations [63–68]. Specifically, quantitative PCR (qPCR) has been tested by many different groups in various field conditions in both multiplexed and multi-parallel designs [63–65, 67–70]. These data demonstrate consistently more "positive" results by qPCR than with traditional microscopy techniques. However, two primary concerns remain: (1) the absence of a gold standard and (2) the increased cost of qPCR compared to microscopy.

Without a gold standard, it's unclear whether the increased detection of positive samples by qPCR compared to microscopy is the result of increased sensitivity or reduced specificity [63, 66, 68, 69]. To reduce cross-reactivity, it is recommended that primers/probes are tested with genomic DNA of co-endemic and likely cross-reactive species [69]. Groups also strategically select species-specific targets such as high copy-number repetitive sequences that, by design, are less susceptible to cross-reactivity [64, 66]. Therefore, despite the definitive assurance that a gold standard would provide, it is more likely that a higher detection rate with qPCR is the result of advances in sensitivity over weaknesses in specificity.

In *Ascaris* endemic regions, widespread implementation of qPCR may not be feasible due to the need for specialized equipment, training, and labor [63]. However, recent advances in loop-mediated amplification (LAMP) assays for diagnosis of ascariasis may resolve these financial barriers associated with molecular diagnostics. LAMP leverages the amplification benefit of PCR but allows for this to occur isothermally, reducing the labor and run-time [71, 72]. The colorimetric read-out of LAMP assays also allows for interpretation of results with the naked eye. LAMP has been lauded for its potential to satisfy the WHO's ASSURED criteria for point-of-care diagnostic tools [73] and thus may soon become the pre-eminent application of omics in field-friendly diagnostic tools for ascariasis.

5.2 Anthelmintic Resistance

Resistance to multiple classes of anthelmintics has been observed in zoonotic parasites, including helminths similar to *Ascaris* [74]. Anthelminthic resistance is measured by low fecal egg count reduction—implying reduced treatment efficiency—and decreases in intervals between egg reappearance—suggesting failure to achieve durable cure following treatment. With widespread use of anthelmintics in both veterinary and human helminth disease, the potential for anthelmintic resistance in human ascariasis has become an escalating topic of concern [75, 76].

Thus far, the data regarding the prevalence of anthelmintic resistance in *Ascaris* has been mixed. In a cohort of school children in Northwest Ethiopia with extensive exposure to anthelmintics, there was reduced efficacy of mebendazole and albendazole against *Ascaris*, suggesting an emergence of resistance in this area [77]. However, in a large study of ascariasis in children in Brazil, Cameroon, Cambodia, Ethiopia, India, Tanzania, and Vietnam, a single 400 mg dose of albendazole was very effective with the lowest cure rate of all sites being 95.2% [78]. Therefore, the phenotype of anthelmintic resistance has not been consistently observed.

Molecular techniques such as PCR, SmartAmp2, Sanger sequencing, and pyrosequencing have enabled genotypic screening for resistance-associated single-nucleotide polymorphisms (SNPs). In equine helminths, the primary SNPs associated with benzimidazole resistance are found in beta-tubulin genes at codons 167 [79], 198 [80], and 200 [81]. Therefore, screening for anthelmintic resistance in *Ascaris* has centered on detection of SNPs at these target sites. Most studies across various regions have found no signs of resistance-associated SNPs [82–86]. However, there is documentation of resistance-associated SNPs at codon 167 in *Ascaris* collected from independent cohorts in Haiti and Kenya [87]. Importantly, this study showed that following anthelmintic treatment, the proportions of heterozygotes containing a resistant beta-tubulin allele increased while the proportion of homozygous susceptible populations decreased. These findings suggest that SNPs at codon 167 confer anthelmintic resistance with anthelmintic treatment through exertion of selective pressure for these resistant alleles. Another study of *Ascaris* collected from 7 sites in Brazil found that 0.5% of eggs contained SNPs at codon 200 in beta-tubulin isotype-1 [88]. However, this study did not report any associated health history, so the relationship between the resistance allele and any associated phenotype is less clear. Importantly, a recent study described the existence of additional beta-tubulin genes in *Ascaris* and revealed differential expression of these genes in specific life stages and tissues [89]. This development suggests that previous studies focusing on screening resistance-associated SNPs in a single beta-tubulin gene may have missed additional beta-tubulin genes housing other resistant alleles.

Overall, the incorporation of molecular genomic techniques has enabled these large-scale epidemiologic evaluations of resistance genotypes [90, 91]. Although the current arsenal of molecular techniques enables surveillance for known resistance-associated SNPs, all continue to face obstacles regarding implementation and/or efficiency. There is also a need for more global assessments of resistance-associated variants across the *Ascaris* genome—not just in known beta-tubulin genes [76]. Hopefully, the increasing accessibility of whole genome sequencing will enable future unbiased descriptions of the effect of persistent anthelmintic use to provide additional insights into the development of resistance as well as inform the design of novel anthelmintics that would be unaffected by existing resistance mechanisms [76].

5.3 Ascariasis and the Microbiome

Insufficient data exist regarding the effect of *Ascaris* larval migration and worm development on the host microbiome, particularly in key viscera such as the liver and the lungs. However, microbiome shifts in the intestines and stool have been briefly explored with varying results. Using 16S rRNA gene sequencing, adult *Ascaris* infection in pigs, a native host of ascariasis, resulted in decreased microbial diversity at the location of infection within the small intestines [92, 93]. Enrichment of *Lactobacillus*, *Megasphaera*, *Prevotella*, and *Succinivibrio* species was identified in the small intestines of *Ascaris*-infected pigs. However, alterations in microbial diversity in the host small intestine did not correlate with worm burden. The impact of ascariasis on the colon and stool microbiome remains uncertain, with one study suggesting decreased diversity in the colon [93] and another suggesting increased diversity in the colon [94], although these differences may reflect variations in timing of analysis post-infection and host diet. Microbiota disruption in the stool and colon was also linked to altered biological pathways, including carbohydrate, amino acid metabolism, DNA repair, and fatty acid biosynthesis [93], although the pathological significance of these changes remains unclear. The underlying factors causing microbial disruption are also not yet understood but may be driven by the parasite secretome. *Ascaris* ES product contains antimicrobial properties that interfere with bacterial growth and biofilm formation [93, 95]. Advancement of sequencing techniques, whole genome sequencing, and metabolomic analysis will provide more detailed insight into the host microbiome changes and aid in assessing their influence on host pathophysiology [96]. Furthermore, more expansive investigation beyond the host intestines to include the liver and lungs will elucidate a more comprehensive understanding of host–parasite interplay.

Lastly, *Ascaris* is a multi-cellular organism and contains its own unique microbiome. The intestinal microbiome of the *Ascaris* worm reflects the host intestinal microbiome, albeit with less robust diversity [92]. Overabundance of *Clostridium* sensu stricto, *Escherichia-Shigella*, *Lactobacillus*, *Prevotella*, *Romboutsia*, and *Streptococcus* as well as *Staphylococcus* and *Pseudomonas* [92] has been recorded. The interactions between the the *Ascaris* microbiome and the host microbiome remain poorly understood.

6 Conclusions

While the "omics" field continues to evolve at a rapid pace, the use of omic technology in *Ascaris* research remains at the starting line. Despite the high burden of disease globally, ascariasis remains widely understudied and underfunded leading to a large knowledge gap. Further exploration of the *Ascaris* proteome, lipidome, and the influence of these parasite-derived molecules on the host microbiome using advanced omic technology will significantly advance the field of ascariasis. Understanding the biological processes of ascariasis is a key feature in the development of novel interventions including vaccines that are desperately needed to prevent *Ascaris*-induced morbidity and essential to global elimination efforts.

References

1. Walsh JA, Warren KS (1979) Selective primary health care. N Engl J Med 301(18):967–974. https://doi.org/10.1056/NEJM197911013011804
2. de Silva NR, Chan MS, Bundy DA (1997) Morbidity and mortality due to ascariasis: re-estimation and sensitivity analysis of global numbers at risk. Trop Med Int Health 2(6):519–528. https://doi.org/10.1046/j.1365-3156.1997.d01-320.x
3. Pullan RL, Smith JL, Jasrasaria R, Brooker SJ (2014) Global numbers of infection and disease burden of soil transmitted helminth infections in 2010. Parasit Vectors 7(1):37. https://doi.org/10.1186/1756-3305-7-37
4. Holland C, Sepidarkish M, Deslyper G et al (2022) Global prevalence of Ascaris infection in humans (2010–2021): a systematic review and meta-analysis. Infect Dis Poverty 11(1):113. https://doi.org/10.1186/s40249-022-01038-z
5. Xie Y, Niu L, Zhao B et al (2013) Complete mitochondrial genomes of chimpanzee- and gibbon-derived Ascaris isolated from a zoological garden in Southwest China. PLoS One 8(12):e82795. https://doi.org/10.1371/journal.pone.0082795
6. CDC - DPDx - Ascariasis. Published July 19, 2019. https://www.cdc.gov/dpdx/ascariasis/index.html. Accessed 31 December 2023
7. Geenen PL, Bresciani J, Boes J et al (1999) The morphogenesis of Ascaris suum to the infective third-stage larvae within the egg. J Parasitol 85(4):616–622. https://doi.org/10.2307/3285733
8. Murrell KD, Eriksen L, Nansen P, Slotved HC, Rasmussen T (1997) Ascaris suum: a revision of its early migratory path and implications for human ascariasis. J Parasitol 83(2):255–260. https://doi.org/10.2307/3284450
9. Roepstorff A, Eriksen L, Slotved HC, Nansen P (1997) Experimental Ascaris suum infection in the pig: worm population kinetics following single inoculations with three doses of infective eggs. Parasitology 115(4):443–452. https://doi.org/10.1017/S0031182097001480

10. Dold C, Holland CV (2011) Ascaris and ascariasis. Microbes Infect 13(7):632–637. https://doi.org/10.1016/j.micinf.2010.09.012
11. Olsen LS, Kelley GW, Sen HG (1958) Longevity and egg-production of Ascaris suum. Trans Am Microsc Soc 77(4):380–383. https://doi.org/10.2307/3223859
12. de Silva NR, Guyatt HL, Bundy DAP (1997) Morbidity and mortality due to Ascaris-induced intestinal obstruction. Trans R Soc Trop Med Hyg 91(1):31–36. https://doi.org/10.1016/S0035-9203(97)90384-9
13. Louw JH (1966) Abdominal complications of ascaris lumbricoides infestation in children. Br J Surg 53(6):510–521. https://doi.org/10.1002/bjs.1800530606
14. Thein-Hlaing (1987) A profile of ascariasis morbidity in Rangoon Children's Hospital, Burma. J Trop Med Hyg 90(4):165–169
15. Chrungoo RK, Hangloo VK, Faroqui MM, Khan M (1992) Surgical manifestations and management of ascariasis in Kashmir. J Indian Med Assoc 90(7):171–174
16. Thein-Hlaing T-T, Than-Saw M-L-K, Myint-Lwin (1991) A controlled chemotherapeutic intervention trial on the relationship between Ascaris lumbricoides infection and malnutrition in children. Trans R Soc Trop Med Hyg 85(4):523–528. https://doi.org/10.1016/0035-9203(91)90242-Q
17. Levav M, Mirsky AF, Schantz PM, Castro S, Cruz ME (1995) Parasitic infection in malnourished school children: effects on behaviour and EEG. Parasitology 110(Pt 1):103–111. https://doi.org/10.1017/s0031182000081105
18. Jardim-Botelho A, Raff S, DeÁvila RR et al (2008) Hookworm, Ascaris lumbricoides infection and polyparasitism associated with poor cognitive performance in Brazilian schoolchildren. Trop Med Int Health 13(8):994–1004. https://doi.org/10.1111/j.1365-3156.2008.02103.x
19. Taylor-Robinson DC, Maayan N, Soares-Weiser K, Donegan S, Garner P (2012) Deworming drugs for soil-transmitted intestinal worms in children: effects on nutritional indicators, haemoglobin and school performance. Cochrane Database Syst Rev 11. https://doi.org/10.1002/14651858.CD000371.pub5
20. Liu C, Luo R, Yi H et al (2015) Soil-transmitted helminths in southwestern China: a cross-sectional study of links to cognitive ability, nutrition, and school performance among children. PLoS Negl Trop Dis 9(6):e0003877. https://doi.org/10.1371/journal.pntd.0003877
21. Blouin B, Casapia M, Joseph L, Gyorkos TW (2018) A longitudinal cohort study of soil-transmitted helminth infections during the second year of life and associations with reduced long-term cognitive and verbal abilities. PLoS Negl Trop Dis 12(7):e0006688. https://doi.org/10.1371/journal.pntd.0006688
22. WHO Neglected tropical diseases -- global. https://www.who.int/health-topics/neglected-tropical-diseases. Accessed 30 December 2023
23. Okimoto R, Macfarlane JL, Clary DO, Wolstenholme DR (1992) The mitochondrial genomes of two nematodes, Caenorhabditis elegans and Ascaris suum. Genetics 130(3):471–498. https://doi.org/10.1093/genetics/130.3.471
24. Huang CQ, Gasser RB, Cantacessi C et al (2008) Genomic-bioinformatic analysis of transcripts enriched in the third-stage larva of the parasitic nematode Ascaris suum. PLoS Negl Trop Dis 2(6):e246. https://doi.org/10.1371/journal.pntd.0000246
25. Yin Y, Martin J, Abubucker S et al (2008) Intestinal transcriptomes of nematodes: comparison of the parasites Ascaris suum and Haemonchus contortus with the free-living Caenorhabditis elegans. PLoS Negl Trop Dis 2(8):e269. https://doi.org/10.1371/journal.pntd.0000269
26. Rosa BA, Jasmer DP, Mitreva M (2014) Genome-wide tissue-specific gene expression, co-expression and regulation of co-expressed genes in adult nematode Ascaris suum. PLoS Negl Trop Dis 8(2):e2678. https://doi.org/10.1371/journal.pntd.0002678
27. Cantacessi C, Zou FC, Hall RS et al (2009) Bioinformatic analysis of abundant, gender-enriched transcripts of adult Ascaris suum (Nematoda) using a semi-automated workflow platform. Mol Cell Probes 23(5):205–217. https://doi.org/10.1016/j.mcp.2009.03.003
28. Xu MJ, Fu JH, Nisbet AJ et al (2013) Comparative profiling of microRNAs in male and female adults of Ascaris suum. Parasitol Res 112(3):1189–1195. https://doi.org/10.1007/s00436-012-3250-x
29. Peng W, Yuan K, Zhou X, Hu M, Abs El-Osta YG, Gasser RB (2003) Molecular epidemiological investigation of Ascaris genotypes in China based on single-strand conformation polymorphism analysis of ribosomal DNA. Electrophoresis 24(14):2308–2315. https://doi.org/10.1002/elps.200305455
30. Leles D, Araújo A, Vicente ACP, Iñiguez AM (2009) Molecular diagnosis of ascariasis from human feces and description of a new Ascaris sp. genotype in Brazil. Vet Parasitol 163(1):167–170. https://doi.org/10.1016/j.vetpar.2009.03.050
31. Das K, Chowdhury P, Ganguly S (2015) Internal transcribed spacer 1 (ITS1) based sequence typing reveals phylogenetically distinct Ascaris population. Comput Struct Biotechnol J 13:478–483. https://doi.org/10.1016/j.csbj.2015.08.006
32. Peng W, Yuan K, Hu M, Zhou X, Gasser RB (2005) Mutation scanning-coupled analysis of haplotypic variability in mitochondrial DNA regions reveals low gene flow between human and porcine Ascaris in endemic regions of China. Electrophoresis 26(22):4317–4326. https://doi.org/10.1002/elps.200500276
33. Betson M, Halstead FD, Nejsum P et al (2011) A molecular epidemiological investigation of Ascaris on Unguja, Zanzibar using isoenyzme analysis, DNA barcoding and microsatellite DNA profiling. Trans R Soc Trop Med Hyg 105(7):370–379. https://doi.org/10.1016/j.trstmh.2011.04.009

34. Zhou C, Li M, Yuan K, Hu N, Peng W (2011) Phylogeography of Ascaris lumbricoides and A. suum from China. Parasitol Res 109(2):329–338. https://doi.org/10.1007/s00436-011-2260-4
35. Betson M, Nejsum P, Llewellyn-Hughes J et al (2012) Genetic diversity of Ascaris in southwestern Uganda. Trans R Soc Trop Med Hyg 106(2):75–83. https://doi.org/10.1016/j.trstmh.2011.10.011
36. Iñiguez AM, Leles D, Jaeger LH, Carvalho-Costa FA, Araújo A, Amazonas Research Group (2012) Genetic characterisation and molecular epidemiology of Ascaris spp. from humans and pigs in Brazil. Trans R Soc Trop Med Hyg 106(10):604–612. https://doi.org/10.1016/j.trstmh.2012.06.009
37. Zhu X, Chilton NB, Jacobs DE, Boes J, Gasser RB (1999) Characterisation of Ascaris from human and pig hosts by nuclear ribosomal DNA sequences. Int J Parasitol 29(3):469–478. https://doi.org/10.1016/S0020-7519(98)00226-4
38. Shao CC, Xu MJ, Alasaad S et al (2014) Comparative analysis of microRNA profiles between adult Ascaris lumbricoides and Ascaris suum. BMC Vet Res 10:99. https://doi.org/10.1186/1746-6148-10-99
39. Zhou C, Jian S, Peng W, Li M (2018) Genetic diversity of Ascaris in China assessed using simple sequence repeat markers. Korean J Parasitol 56(2):175–181. https://doi.org/10.3347/kjp.2018.56.2.175
40. Betson M, Nejsum P, Bendall RP, Deb RM, Stothard JR (2014) Molecular epidemiology of ascariasis: a global perspective on the transmission dynamics of Ascaris in people and pigs. J Infect Dis 210(6):932–941. https://doi.org/10.1093/infdis/jiu193
41. Zhou C, Li M, Yuan K, Deng S, Peng W (2012) Pig Ascaris: an important source of human ascariasis in China. Infect Genet Evol 12(6):1172–1177. https://doi.org/10.1016/j.meegid.2012.04.016
42. Liu GH, Wu CY, Song HQ et al (2012) Comparative analyses of the complete mitochondrial genomes of Ascaris lumbricoides and Ascaris suum from humans and pigs. Gene 492(1):110–116. https://doi.org/10.1016/j.gene.2011.10.043
43. Nejsum P, Hawash MBF, Betson M, Russell SJ, Gasser RB, Andersen LO (2017) Ascaris phylogeny based on multiple whole mtDNA genomes. Infect Genet Evol 48:4–9. https://doi.org/10.1016/j.meegid.2016.12.003
44. Easton A, Gao S, Lawton SP et al (2020) Molecular evidence of hybridization between pig and human Ascaris indicates an interbred species complex infecting humans. eLife 9:e61562. https://doi.org/10.7554/eLife.61562
45. Criscione CD, Anderson JD, Sudimack D et al (2007) Disentangling hybridization and host colonization in parasitic roundworms of humans and pigs. Proc R Soc B Biol Sci 274(1626):2669–2677. https://doi.org/10.1098/rspb.2007.0877
46. Leles D, Gardner SL, Reinhard K, Iñiguez A, Araujo A (2012) Are Ascaris lumbricoides and Ascaris suum a single species? Parasit Vectors 5(1):42. https://doi.org/10.1186/1756-3305-5-42
47. Jex AR, Liu S, Li B et al (2011) Ascaris suum draft genome. Nature 479(7374):529–533. https://doi.org/10.1038/nature10553
48. Wang J, Veronezi GMB, Kang Y, Zagoskin M, O'Toole ET, Davis RE (2020) Comprehensive chromosome end remodeling during programmed DNA elimination. Curr Biol 30(17):3397–3413.e4. https://doi.org/10.1016/j.cub.2020.06.058
49. Ebner F, Morrison E, Bertazzon M et al (2020) CD4+ Th immunogenicity of the Ascaris spp. secreted products. NPJ Vaccines 5:25. https://doi.org/10.1038/s41541-020-0171-z
50. Hansen EP, Fromm B, Andersen SD et al (2019) Exploration of extracellular vesicles from Ascaris suum provides evidence of parasite-host cross talk. J Extracell Vesicles 8(1):1578116. https://doi.org/10.1080/20013078.2019.1578116
51. Joshi P, Mishra PKK (2022) Functional diversity of the excretory/secretory proteins of nematode parasites. Acta Parasitol 67(2):619–627. https://doi.org/10.1007/s11686-022-00523-7
52. Chehayeb JF, Robertson AP, Martin RJ, Geary TG (2014) Proteomic analysis of adult Ascaris suum fluid compartments and secretory products. PLoS Negl Trop Dis 8(6):e2939. https://doi.org/10.1371/journal.pntd.0002939
53. Wang T, Van Steendam K, Dhaenens M et al (2013) Proteomic analysis of the excretory-secretory products from larval stages of Ascaris suum reveals high abundance of glycosyl hydrolases. PLoS Negl Trop Dis 7(10):e2467. https://doi.org/10.1371/journal.pntd.0002467
54. González-Miguel J, Morchón R, Gussoni S et al (2014) Immunoproteomic approach for identification of Ascaris suum proteins recognized by pigs with porcine ascariasis. Vet Parasitol 203(3):343–348. https://doi.org/10.1016/j.vetpar.2014.03.031
55. Jungersen G, Eriksen L, Nansen P, Lind P, Rasmussen T, Meeusen ENT (2001) Regional immune responses with stage-specific antigen recognition profiles develop in lymph nodes of pigs following Ascaris suum larval migration. Parasite Immunol 23(4):185–194. https://doi.org/10.1046/j.1365-3024.2001.00373.x
56. Islam MK, Miyoshi T, Yokomizo Y, Tsuji N (2004) The proteome expression patterns in adult Ascaris suum under exposure to aerobic/anaerobic environments analyzed by two-dimensional electrophoresis. Parasitol Res 93(2):96–101. https://doi.org/10.1007/s00436-004-1101-0
57. Antunes MFP, Titz TO, Batista IFC et al (2015) Immunosuppressive PAS-1 is an excretory/secretory protein released by larval and adult worms of the ascarid nematode Ascaris suum. J Helminthol 89(3):367–374. https://doi.org/10.1017/S0022149X14000200
58. Gazzinelli-Guimaraes PH, Nutman TB (2018) Helminth parasites and immune regulation. F1000Res. https://doi.org/10.12688/f1000research.15596.1
59. Borup A, Boysen AT, Ridolfi A et al (2022) Comparison of separation methods for immuno-

modulatory extracellular vesicles from helminths. J Extracell Biol 1(5):e41. https://doi.org/10.1002/jex2.41
60. Wang T, Nie S, Reid GE, Gasser RB (2021) Helminth lipidomics: technical aspects and future prospects. Curr Res Parasitol Vector Borne Dis 1:100018. https://doi.org/10.1016/j.crpvbd.2021.100018
61. Wang T, Nie S, Ma G et al (2020) Quantitative lipidomic analysis of Ascaris suum. PLoS Negl Trop Dis 14(12):e0008848. https://doi.org/10.1371/journal.pntd.0008848
62. Wangchuk P, Yeshi K, Loukas A (2023) Metabolomics and lipidomics studies of parasitic helminths: molecular diversity and identification levels achieved by using different characterisation tools. Metabolomics 19(7):63. https://doi.org/10.1007/s11306-023-02019-5
63. Khurana S, Singh S, Mewara A (2021) Diagnostic techniques for soil-transmitted helminths - recent advances. Res Rep Trop Med 12:181–196. https://doi.org/10.2147/RRTM.S278140
64. Pilotte N, Maasch JRMA, Easton AV, Dahlstrom E, Nutman TB, Williams SA (2019) Targeting a highly repeated germline DNA sequence for improved real-time PCR-based detection of Ascaris infection in human stool. PLoS Negl Trop Dis 13(7):e0007593. https://doi.org/10.1371/journal.pntd.0007593
65. Mejia R, Vicuña Y, Broncano N et al (2013) A novel, multi-parallel, real-time polymerase chain reaction approach for eight gastrointestinal parasites provides improved diagnostic capabilities to resource-limited at-risk populations. Am J Trop Med Hyg 88(6):1041–1047. https://doi.org/10.4269/ajtmh.12-0726
66. Pilotte N, Papaiakovou M, Grant JR et al (2016) Improved PCR-based detection of soil transmitted helminth infections using a next-generation sequencing approach to assay design. PLoS Negl Trop Dis 10(3):e0004578. https://doi.org/10.1371/journal.pntd.0004578
67. Meurs L, Polderman AM, Melchers NVSV et al (2017) Diagnosing polyparasitism in a high-prevalence setting in Beira, Mozambique: detection of intestinal parasites in fecal samples by microscopy and real-time PCR. PLoS Negl Trop Dis 11(1):e0005310. https://doi.org/10.1371/journal.pntd.0005310
68. Basuni M, Mohamed Z, Ahmad M, Zakaria NZ, Noordin R (2012) Detection of selected intestinal helminths and protozoa at Hospital Universiti Sains Malaysia using multiplex real-time PCR. Trop Biomed 29(3):434–442
69. O'Connell EM, Nutman TB (2016) Molecular diagnostics for soil-transmitted helminths. Am J Trop Med Hyg 95(3):508–513. https://doi.org/10.4269/ajtmh.16-0266
70. Easton AV, Oliveira RG, O'Connell EM et al (2016) Multi-parallel qPCR provides increased sensitivity and diagnostic breadth for gastrointestinal parasites of humans: field-based inferences on the impact of mass deworming. Parasit Vectors 9(1):38. https://doi.org/10.1186/s13071-016-1314-y
71. Shiraho EA, Eric AL, Mwangi IN et al (2016) Development of a loop mediated isothermal amplification for diagnosis of *Ascaris lumbricoides* in fecal samples. J Parasitol Res 2016:e7376207. https://doi.org/10.1155/2016/7376207
72. Rashwan N, Diawara A, Scott ME, Prichard RK (2017) Isothermal diagnostic assays for the detection of soil-transmitted helminths based on the SmartAmp2 method. Parasit Vectors 10(1):496. https://doi.org/10.1186/s13071-017-2420-1
73. Deng MH, Zhong LY, Kamolnetr O, Limpanont Y, Lv ZY (2019) Detection of helminths by loop-mediated isothermal amplification assay: a review of updated technology and future outlook. Infect Dis Poverty 8(1):20. https://doi.org/10.1186/s40249-019-0530-z
74. Matthews JB (2014) Anthelmintic resistance in equine nematodes. Int J Parasitol Drugs Drug Resist 4(3):310–315. https://doi.org/10.1016/j.ijpddr.2014.10.003
75. Geerts S, Gryseels B (2001) Anthelmintic resistance in human helminths: a review. Trop Med Int Health 6(11):915–921. https://doi.org/10.1046/j.1365-3156.2001.00774.x
76. Vercruysse J, Albonico M, Behnke JM et al (2011) Is anthelmintic resistance a concern for the control of human soil-transmitted helminths? Int J Parasitol Drugs Drug Resist 1(1):14–27. https://doi.org/10.1016/j.ijpddr.2011.09.002
77. Adugna S, Kebede Y, Moges F, Tiruneh M (2007) Efficacy of mebendazole and albendazole for Ascaris lumbricoides and hookworm infections in an area with long time exposure for antihelminthes, Northwest Ethiopia. Ethiop Med J 45(3):301–306
78. Vercruysse J, Behnke JM, Albonico M et al (2011) Assessment of the anthelmintic efficacy of Albendazole in school children in seven countries where soil-transmitted helminths are endemic. PLoS Negl Trop Dis 5(3):e948. https://doi.org/10.1371/journal.pntd.0000948
79. Silvestre A, Cabaret J (2002) Mutation in position 167 of isotype 1 β-tubulin gene of Trichostrongylid nematodes: role in benzimidazole resistance? Mol Biochem Parasitol 120(2):297–300. https://doi.org/10.1016/S0166-6851(01)00455-8
80. Ghisi M, Kaminsky R, Mäser P (2007) Phenotyping and genotyping of *Haemonchus contortus* isolates reveals a new putative candidate mutation for benzimidazole resistance in nematodes. Vet Parasitol 144(3):313–320. https://doi.org/10.1016/j.vetpar.2006.10.003
81. Kwa MSG, Veenstra JG, Roos MH (1994) Benzimidazole resistance in *Haemonchus contortus* is correlated with a conserved mutation at amino acid 200 in β-tubulin isotype 1. Mol Biochem Parasitol 63(2):299–303. https://doi.org/10.1016/0166-6851(94)90066-3
82. Diawara A, Drake LJ, Suswillo RR et al (2009) Assays to detect beta-tubulin codon 200 polymorphism in Trichuris trichiura and Ascaris lumbricoides. PLoS Negl Trop Dis 3(3):e397. https://doi.org/10.1371/journal.pntd.0000397

83. Krücken J, Fraundorfer K, Mugisha JC et al (2017) Reduced efficacy of albendazole against Ascaris lumbricoides in Rwandan schoolchildren. Int J Parasitol Drugs Drug Resist 7(3):262–271. https://doi.org/10.1016/j.ijpddr.2017.06.001
84. Zuccherato LW, Furtado LF, da Silva Medeiros C, da Silva Pinheiro C, Rabelo ÉM (2018) PCR-RFLP screening of polymorphisms associated with benzimidazole resistance in Necator americanus and Ascaris lumbricoides from different geographical regions in Brazil. PLoS Negl Trop Dis 12(9):e0006766. https://doi.org/10.1371/journal.pntd.0006766
85. Matamoros G, Rueda MM, Rodríguez C et al (2019) High endemicity of soil-transmitted helminths in a population frequently exposed to Albendazole but no evidence of antiparasitic resistance. Trop Med Infect Dis 4(2):73. https://doi.org/10.3390/tropicalmed4020073
86. Palma A, Matamoros G, Escobar D, Sánchez AL, Fontecha G (2020) Absence of mutations associated with resistance to benzimidazole in the beta-tubulin gene of Ascaris suum. Rev Soc Bras Med Trop 53:e20190155. https://doi.org/10.1590/0037-8682-0155-2019
87. Diawara A, Halpenny CM, Churcher TS et al (2013) Association between response to albendazole treatment and β-tubulin genotype frequencies in soil-transmitted helminths. PLoS Negl Trop Dis 7(5):e2247. https://doi.org/10.1371/journal.pntd.0002247
88. Furtado LFV, da Silva Medeiros C, Zuccherato LW et al (2019) First identification of the benzimidazole resistance-associated F200Y SNP in the beta-tubulin gene in Ascaris lumbricoides. PLoS One 14(10):e0224108. https://doi.org/10.1371/journal.pone.0224108
89. Roose S, Avramenko RW, Pollo SMJ et al (2021) Characterization of the β-tubulin gene family in Ascaris lumbricoides and Ascaris suum and its implication for the molecular detection of benzimidazole resistance. PLoS Negl Trop Dis 15(9):e0009777. https://doi.org/10.1371/journal.pntd.0009777
90. Wang J (2021) Genomics of the parasitic nematode Ascaris and its relatives. Genes 12(4):493. https://doi.org/10.3390/genes12040493
91. Gandasegui J, Martínez-Valladares M, Grau-Pujol B et al (2020) Role of DNA-detection-based tools for monitoring the soil-transmitted helminth treatment response in drug-efficacy trials. PLoS Negl Trop Dis 14(2):e0007931. https://doi.org/10.1371/journal.pntd.0007931
92. Midha A, Jarquín-Díaz VH, Ebner F et al (2022) Guts within guts: the microbiome of the intestinal helminth parasite Ascaris suum is derived but distinct from its host. Microbiome 10(1):229. https://doi.org/10.1186/s40168-022-01399-5
93. Wang Y, Liu F, Urban JF, Paerewijck O, Geldhof P, Li RW (2019) Ascaris suum infection was associated with a worm-independent reduction in microbial diversity and altered metabolic potential in the porcine gut microbiome. Int J Parasitol 49(3):247–256. https://doi.org/10.1016/j.ijpara.2018.10.007
94. Williams AR, Krych L, Ahmad HF et al (2017) A polyphenol-enriched diet and Ascaris suum infection modulate mucosal immune responses and gut microbiota composition in pigs. PLoS One 12(10):e0186546. https://doi.org/10.1371/journal.pone.0186546
95. Midha A, Ebner F, Schlosser-Brandenburg J, Rausch S, Hartmann S (2021) Trilateral relationship: Ascaris, microbiota, and host cells. Trends Parasitol 37(3):251–262. https://doi.org/10.1016/j.pt.2020.09.002
96. Ramírez-Carrillo E, Gaona O, Nieto J et al (2020) Disturbance in human gut microbiota networks by parasites and its implications in the incidence of depression. Sci Rep 10(1):3680. https://doi.org/10.1038/s41598-020-60562-w

Part V

Vectors

Genomics of Triatominae, the Chagas Disease Vectors

Antonella Bacigalupo and Sebastián Pita

Abstract

This chapter aims to introduce the reader to the importance of bugs from the Triatominae subfamily, which transmit the parasite that causes Chagas disease. It discusses the role of genomics in advancing our understanding of triatomines' biology, and how this knowledge could be involved in the improvement of vector control strategies, toward ending the vectorial transmission of this neglected disease. Genomics of Triatominae includes insights from their mitochondrial genomes—mitogenomes—highlighting their structure and genetic variations, as a means to explore triatomine diversity and evolutionary history. The chapter further examines whole genome sequencing and analyses of Triatominae, demonstrating how comprehensive genomic data can illuminate aspects such as their physiology, biochemistry, and phylogenetic relations. By reviewing the available data, we discuss the quality of genome assemblies and the technologies used. The functional genomics section addresses key areas such as behavior, digestion, metabolism, immunity, and insecticide resistance, emphasizing the genetic basis of these crucial traits. Additionally, the chapter discusses the repetitive DNA elements of the genome—the repeatome—and its role in genome structure. Finally, the phylogenomics section integrates genomic data to resolve evolutionary relationships within Triatominae, enhancing our understanding of the taxonomy and evolution of these insects.

Keywords

Triatominae · Insect genomics · Whole genome · Mitogenome · Repeatome · Phylogenomics

1 Introduction to Triatominae and Genomics

The Triatominae insect subfamily (Hemiptera: Reduviidae) is composed of at least 160 species: 157 extant and three fossils [1–3]. These reduviids have evolved to feed on vertebrates, unlike the rest of the family Reduviidae, which predates insects [4]. Some of their adaptations include a thin piercing-sucking mouthpart that folds below the head [5] (Fig. 1). Notwithstanding, a remarkable recently paper, re-described the Triatominae

A. Bacigalupo (✉)
School of Biodiversity, One Health and Veterinary Medicine, University of Glasgow, Glasgow, Scotland, UK
e-mail: a.bacigalupo.1@research.gla.ac.uk

S. Pita
Sección Genética Evolutiva, Facultad de Ciencias, Universidad de la República, Montevideo, Uruguay
e-mail: spita@fcien.edu.uy

J. D. Ramírez González (ed.), *Recent Advances in Parasitomics*,
https://doi.org/10.1007/978-3-031-70591-5_15

Fig. 1 *Triatoma infestans*. This triatomine is one of the main vectors of *Trypanosoma cruzi*, the parasite that causes Chagas disease. Credits: Fernando Moya

subfamily, to include three new tribes, that involve several genera: Zeluroides, Neivacoris, Zelurus and Opisthacidius [6]. Their relevance relates to their role as vectors of *Trypanosoma cruzi*, which causes Chagas disease in humans. Vectorial transmission occurs when the infected insect, which carries the protozoa in its digestive tract, releases its dejections that contain the infective form of the parasite. *Trypanosoma cruzi* can enter the body directly through mucous membranes or by the puncture site. Then, it multiplies and spreads to several organs. A naïve vector gets infected when it feeds on an infected host, and the parasite multiplies again in the insect's gut; another form of transmission that involves triatomines is oral, by consumption of contaminated food or beverages [7, 8]. Chagas disease is a neglected tropical disease, in part for its silent progression, where cardiac and digestive muscle tissues get increasingly damaged, until they are no longer able to perform their function, requiring surgical interventions to survive. Sudden death is also a possibility if untreated. The second part of the problem is that most of the population that are exposed to the vectors live in rural or low-income urban areas, so they do not have proper access to information and treatment. Regarding the drugs available, they are just a few, which can cause adverse reactions, and their efficacy is higher in children [8]. Triatominae control is mainly based on residual insecticide spraying in houses, but this approach is unsustainable in low resource settings, with reports of resistance in some triatomine populations [9–11].

Genomics have emerged as a tool to explore the genome; that is, the whole of the DNA contained in species cells, both in the nucleus and in the mitochondria. This could elucidate the mechanisms of adaptation of triatomines to their particular feeding source: blood; it could be used to evaluate within-taxa relationships, and in general, to explore genome-wide particularities of these insect vectors.

2 Mitogenomes

The mitochondrial genome is a circular molecule of DNA, located in the mitochondria of eukaryotic cells. Due to several characteristics such as their presence in multiple copies within the cell, the utilization of mitochondrial markers has gained widespread popularity and serves as an invaluable resource for studying the evolutionary history of organisms. High copy number facilitates the extraction of mitochondrial DNA and the subsequent PCR reaction, even from samples with limited amounts of starting material, such as museum or degraded specimens.

The mitochondrial genome has been widely utilized as a mitochondrial marker in investigating the phylogenetics of Triatominae. Since the late 1990s [12, 13], an extensive body of research has leveraged mitochondrial DNA markers to explore the genetic variability within and among Triatominae species. Mitochondrial markers have been employed to assess the diversity and relationships among tribes, complexes, subcomplexes, and populations within a particular species, along with the genetic structure and population dynamics of Triatominae, offering valuable information for understanding their evolutionary history and distribution patterns. By analyzing the sequence variations in mitochondrial genes, researchers have delineated distinct lineages, identified genetic boundaries, and uncovered cryptic species within the Triatominae subfamily (reviewed in [14–18]). A milestone in Triatominae molecular systematics is the Hypsa and colleagues' paper where a great number of species were analyzed—the most extensive taxon sampling at the time. Among several results, the work evidenced that Old World triatomines are related to *Triatoma* species; hence *Triatoma* genus is paraphyletic [19]. Furthermore, mitochondrial markers have proven instrumental in elucidating the dispersal patterns and colonization events of Triatominae species across different geographical regions. By comparing mitochondrial DNA sequences from populations in various locations, researchers have been able to infer historical migration routes, assess the impact of human activities on gene flow, and investigate the processes driving population differentiation ([20], reviewed in [14]).

As soon as 2001, Dotson and Beard published the first complete mitochondrial genome for a Triatominae—even the first for Hemiptera as a whole. This was the *Triatoma dimidiata* mitogenome, a major vector from Central and South America [21]. However, it was not until 2017 that the second mitogenome was published. In this case was the most important vector species in the Southern Cone, *T. infestans* [22]. The methodology used in this study was the extraction of mitochondrial reads from next-generation sequencing (NGS) and their subsequent alignment, method that would be used in subsequent research to obtain mitogenomes. In recent years, the advent of advanced sequencing technologies and the development of bioinformatic tools have significantly enhanced our ability to obtain complete mitochondrial genome sequences. Nowadays, mitogenomes from 27 species are available for Triatominae. Successive papers contributed with new mitogenomes: *Triatoma rubrofasciata* [23]; *Triatoma migrans*, *Panstrongylus rufotuberculatus*, and *Rhodnius pictipes* [24]; and *Triatoma boliviana* [25], but the most relevant contribution came with 20 new species reported by Aguilera-Uribe et al. [18]. Although not all were fully complete, all 13 protein-coding genes (PCGs), 22 tRNAs, and two rRNAs were obtained, except for *Linshcosteus costalis* and *P. lignarius*. In addition, the paper includes several mitogenomes for *T. dimidiata* [18]. In all Triatominae species the gene order was conserved. This is not surprising since within Insecta, the order of the mitochondrial genes is highly conserved and has led to the proposal of an ancestral gene order. The majority of hemipteran families that have been sequenced possess this ancestral insect arrangement, including triatomines (reviewed in [26]).

The length of the mitogenomes varies between 15,699 (*Triatoma mexicana*) and 17,323 bp (*T. migrans*). However, these differences are mostly due to variability in the repetitive regions, mainly in the control region. This region could be split into four different parts. A first conserved GC-rich region (i.e., 445 bp in *T. boliviana*) is located after the long-rRNA gene and ends in a sequence of several guanines; second, an AT-rich region (i.e., 237 bp in *T. boliviana*); third, the most variable region that contains a repetitive DNA motif tandemly organized with a high similarity between the different copies of the repeated sequence. Each species presents its own repetitive motif, being mainly responsible for mitogenome size variation. At last, a short conserved region (i.e., 87 bp in *T. boliviana*) contains two inverted repeats of 10 bp with the potential to form a stem and loop structure, which has been related to the replication mechanism [27].

One of the main uses of the mitogenomes is the phylogenetic reconstruction of triatomines species. Gene-based analysis—leaving the repetitive regions aside—has proved to obtain robust phylogenies [25]. The most extended methodology includes a partitioned analysis of each of the 13 PCGs, and both rRNAs, to calculate a phylogenetic tree, by maximum likelihood (ML) or Bayesian inference. In Fig. 2, an ML tree is depicted, showing the two main tribes of Triatominae: Rhodniini and Triatomini; the latter is divided into three main lineages: Dispar, North, and South American Triatomini (sensu [14]). These results are in concordance with a recent molecular phylogeny using nuclear ultraconserved elements [28]. As seems to be the consensus, the differentiation of Triatomini and Rhodniini is the most ancestral splitting within the subfamily Triatominae (Fig. 2). Within Triatomini, there are three main lineages: Dispar, North, and South American Triatomini (sensu [14]).

The analysis of the mitochondrial genome has played a pivotal role in modern genomics,

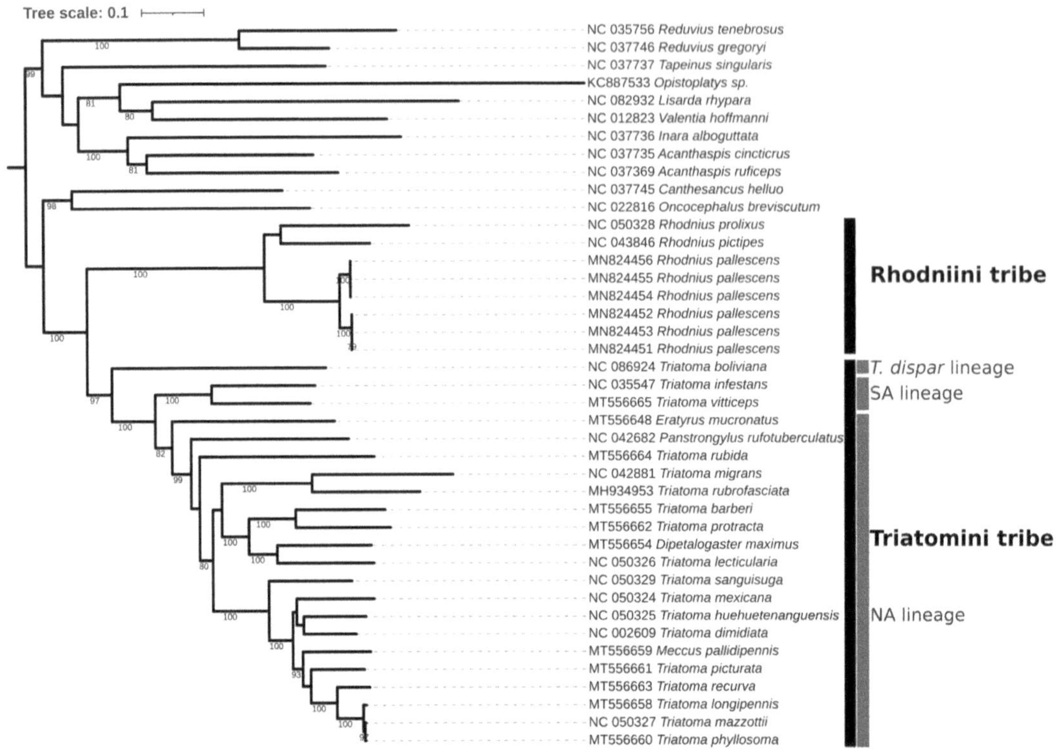

Fig. 2 Maximum likelihood phylogenetic tree calculated using a partitioned analysis of 13 PCGs and two rRNAs. Node support was calculated with 1000 bootstrap pseudo-replicates and is depicted in the tree when greater than 80. NCBI accession numbers of mitogenomes used in the analysis are indicated for each terminal. As outgroups, all mitogenomes available from Harpactorinae, Reduviinae, and Stenopodainae were included. NA: North American; SA: South American

increasing our understanding of evolution and biological diversity. In the case of Triatominae, these comprehensive sequences have shed light on the structural characteristics and gene organization of triatomine mitogenomes. Their complete sequencing has provided important evidence about genetic variation, population dynamics, and evolutionary insights.

3 Triatominae Whole Genomes

To date, there are whole genome assemblies available only for six species of triatomines. The first reduviid genome to be sequenced was of *Rhodnius prolixus* [15]. Remarkably, this first published assembly was thoroughly evaluated by functional experiments performed by a multitude of researchers from more than 30 institutions that provided their particular expertise to annotate this genome. The first *R. prolixus* assembly is highly fragmented (Fig. 3), but nevertheless comprises more than 15,000 protein-coding genes and over 700 RNA genes [15]. It included interspersed repeats of DNA as transposable elements (TEs), which were mainly *mariner* TEs—short transposons with one open reading frame (ORF) encoding a transposase—comprised 5.6% of this genome [15, 29, 30]. The *R. prolixus* genome also showed horizontal gene transfers from bacteria of the genus *Wolbachia* (Rickettsiales: Anaplasmataceae) [15]. The RNAi pathway was described, which allows for gene silencing; the immune deficiency (IMD), Jack-STAT, and Toll immune pathways' components were almost all present, with the exception of some parts of the IMD pathway [15]. There were expansions of

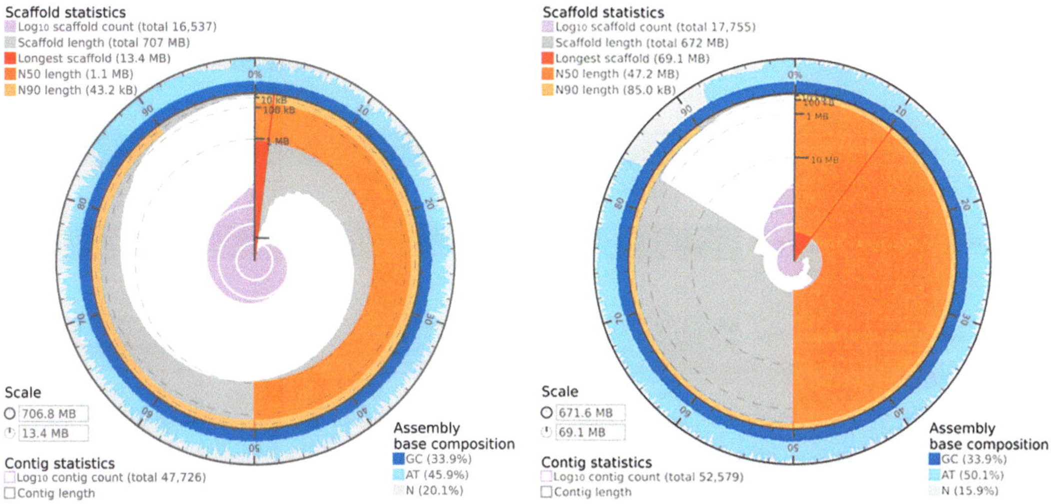

Fig. 3 Graphical representation of the *Rhodnius prolixus* scaffold-level assembly (left) and chromosomal-level assembly (right), generated with the assembly-stats software (Copyright (c) 2015–2016 Richard Challis). The outer circle indicates the nucleotide base composition (A/T in light blue, C/G in dark blue, and N in gray). In red, the proportion of the longest scaffold regarding the whole genome assembled is shown; in dark orange, the N50; and in light orange, the N90. The cumulative number of scaffolds is depicted in purple, and the inner arc in gray is the length of the scaffolds

genes coding for detoxification enzymes, including carboxylesterases and cytochrome P450. Most of the components of the development and signaling pathways were described, with reductions in the tyrosine kinome and expansions in transcription factors and in cuticle development, when compared to other insect species. Unexpectedly, the genes for enzymes to produce the amino acid selenocysteine were found in the *R. prolixus* genome, as to that date all insects reported had non-selenium, cysteine-based enzymes [15].

Regarding hematophagy, the main characteristic of this subfamily that separates them from predatory reduviids, they present expansions related to chemoreception to detect hosts, including gustatory and olfactory receptors, and chemosensory and odorant-binding proteins. Some of the salivary and digestive proteins also showed expansions: lipocalins, nitrophorins, and aspartic peptidases were among them. In turn, less genes coding for amiloride-sensitive sodium channels were found, which may attain the elimination of the excess of liquids involved in a bloodmeal without losing minerals [15]. Many other genes involved in processes related to hematophagy were described, including humidity and temperature detection, behavioral control, locomotion and vision, mechanoreception, memory, neuropeptides and hormone precursors, circadian clock control, heme/iron binding, transport, metabolism, and energy pathways. *Rhodnius prolixus* seemed to lack some pathways for amino acid biosynthesis, which presumably are obtained from the bloodmeal [15].

Molecules involved in the lipid metabolism were searched in the *R. prolixus* genome, and genes coding for putative lipases were annotated [31]. Expression of mRNA of some lipases was found in digestive tissue, fat body, flight muscles, gonads, or nervous system; a few were expressed in all these tissues, while the mRNA of one of the tested lipases was not detected in any of them [31]. Microsomal and cytosolic fatty acid synthases were also annotated, along with secreted and calcium-independent phospholipases, fatty acid elongases, membrane-bound fatty acid desaturases, fatty acid reductases—which showed an expansion—acyl-coenzyme-A synthetases, mitochondrial acyl-CoA dehydrogenases, components of the mitochondrial fatty acid β-oxidation, and the carnitine shuttle, among others [31]. The latest *R. prolixus* assembly is available in public repositories: GenBank

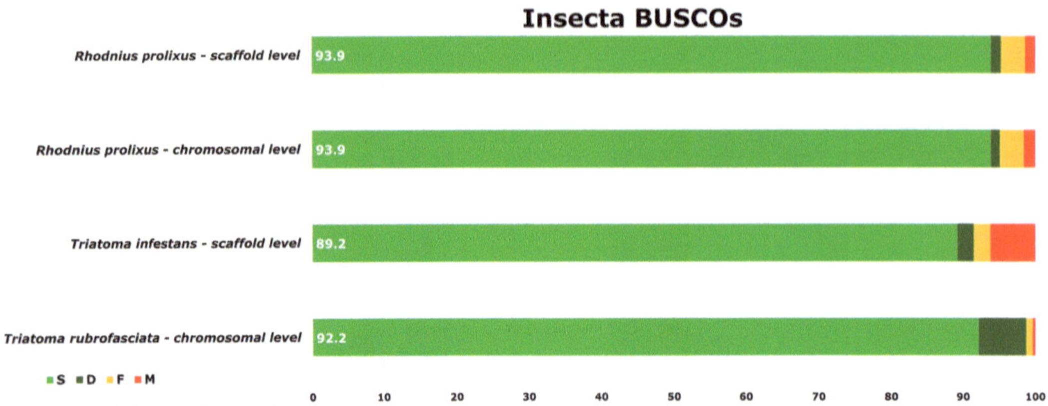

	Triatoma rubrofasciata - chromosomal level	Triatoma infestans - scaffold level	Rhodnius prolixus - chromosomal level	Rhodnius prolixus - scaffold level
S	92.2	89.2	93.9	93.9
D	6.5	2.3	1.2	1.3
F	0.9	2.3	3.3	3.4
M	0.4	6.2	1.6	1.4

Fig. 4 Insecta BUSCOs: The assemblies from *Rhodnius prolixus*, *Triatoma infestans*, and *Triatoma rubrofasciata* were searched for 1367 near-universal single copy benchmark orthologous genes (BUSCOs) available in the insecta_odb10 database. The bars represent the percentage corresponding to single (light green), duplicated (dark green), fragmented (yellow), and missing Insecta BUSCO genes

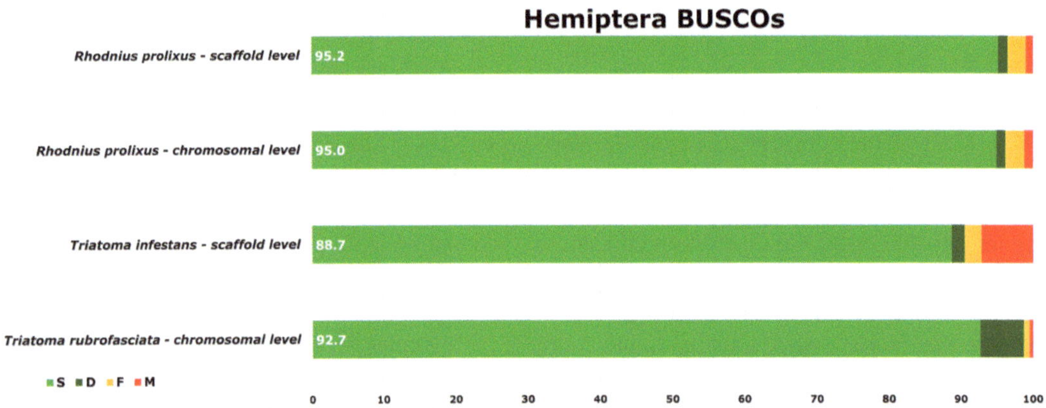

	Triatoma rubrofasciata - chromosomal level	Triatoma infestans - scaffold level	Rhodnius prolixus - chromosomal level	Rhodnius prolixus - scaffold level
S	92.7	88.7	95.0	95.2
D	6.0	1.8	1.2	1.3
F	0.8	2.4	2.6	2.5
M	0.5	7.1	1.2	1.0

Fig. 5 Hemiptera BUSCOs: The assemblies from *Rhodnius prolixus*, *Triatoma infestans*, and *Triatoma rubrofasciata* were searched for 2510 near-universal single copy benchmark orthologous genes (BUSCOs) available in the hemiptera_odb10 database. The bars represent the percentage corresponding to single (light green), duplicated (dark green), fragmented (yellow), and missing Hemiptera BUSCO genes

[32] assembly accession: GCA_000181055.3 (Fig. 3, left panel) and also an improved structural annotation (RproC3.5) can be obtained in VectorBase [33]. The benchmarking near-universal single copy orthologs (BUSCOs) [34]—metrics for the assembly—are available in Figs. 4 and 5.

Lately, a chromosomal-level assembly was produced for *R. prolixus* [35, 36],[1] using the contigs available from the first assembly [15] and

[1]Unpublished genome assemblies and sequencing data for *Rhodnius prolixus* are used with permission from the DNA Zoo Consortium (dnazoo.org).

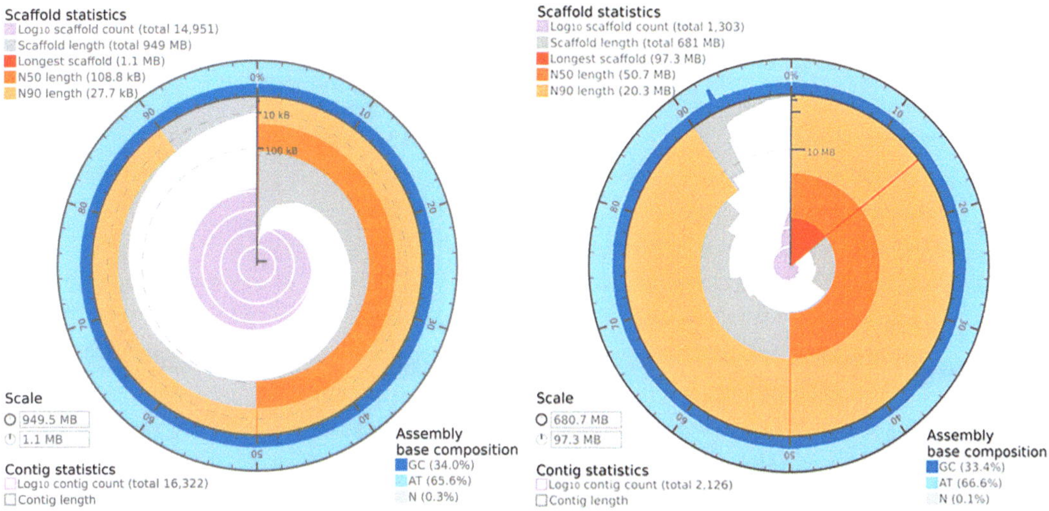

Fig. 6 Graphical representation of the *Triatoma infestans* scaffold-level assembly (left) and the *Triatoma rubrofasciata* (right), generated with the assembly-stats software (Copyright (c) 2015–2016 Richard Challis). The outer circle indicates the nucleotide base composition (A/T in light blue, C/G in dark blue, and N in gray). In red, the proportion of the longest scaffold regarding the whole genome assembled is shown; in dark orange, the N50; and in light orange, the N90. The cumulative number of scaffolds is depicted in purple, and the inner arc in gray is the length of the scaffolds

others produced with the HiC technique, which evaluates fixed DNA fragments to discern which contigs belong to the same chromosome, i.e., chromosome conformation [37]. This chromosomal assembly improved its contiguity and reduced the unassigned bases (Fig. 3, right panel), but it did not improve its completeness, measured as BUSCOs (Figs. 4 and 5). There was no associated annotation for this assembly. The first study that used this chromosomal assembly analyzed gene conflict between autosomes and sex chromosomes of Reduvioidea, showing faster evolution rate of X chromosome linked loci. This would impact phylogenetic reconstructions, depending on the loci involved. For *R. prolixus*, chromosome 10 of the chromosomal assembly would be the X chromosome [38].

There is a second species with a whole genome assembly available in a public repository: *Triatoma infestans* (GCA_011037195.1) [39]; however, there is no scientific publication about it yet. This scaffold-level assembly is also very fragmented (Fig. 6, left panel). It only includes the annotation of the ORFs. The completeness of this assembly regarding BUSCOs is shown in Figs. 4 and 5. Some putative genes were identified by sequencing cDNA libraries of expressed sequence tags (ESTs) before the release of the *T. infestans* assembly, with many related to development and reproduction, including the juvenile hormone esterase, hexamerins, and vitellogenin—genes that probably participate in copulation, embryonic development, and wing formation [40].

The first triatomine chromosomal-level assembly was obtained for *Triatoma rubrofasciata* [41], a species distributed in tropical areas worldwide [42]. This whole genome assembly was constructed by sequencing with Illumina and PacBio, and the chromosomal assembly used the HiC technique (Fig. 6, right panel). Its estimated genome size of 757 Mb [41] is unexpected, given that many species of the same genus have genome sizes above 1 Gb [16, 43]; however, to our knowledge the C-value for this species is not available. More than 50% of the genome was composed of repeats [41]. The chromosomal assembly size is 680.73 Mb, so roughly 92% of the contigs available were anchored into 13 chromosomes [41], and this corresponded with previous karyotype studies [44, 45]. The completeness of this assembly regarding BUSCOs can be found in Figs. 4

Table 1 Contiguity and composition statistics of the triatomine assemblies available

Statistic per assembly	*R. prolixus*—scaffold	*R. prolixus*—chromosomal	*T. infestans*—scaffold	*T. rubrofasciata*—chromosomal
Assembly span (bp)	706,824,083	671,559,592	949,486,157	680,726,098
N (%)	20.12	15.92	0.34	0.06
GC (%)	33.94	33.94	34.02	33.38
AT (%)	66.06	66.06	65.98	66.62
Scaffold count	16,537	17,755	14,951	1303
Longest scaffold (bp)	13,425,595	69,096,158	1,054,224	97,329,580
Scaffold N50 length	1,088,772	47,232,656	108,830	50,700,875
Scaffold N50 count	170	6	2457	6
Scaffold N90 length	43,238	85,000	27,682	20,341,594
Scaffold N90 count	1250	397	9200	12
Contig count	47,726	52,579	16,322	2126
Contig N50 length	35,751	33,994	95,318	2,722,109
Contig N50 count	4418	4565	2767	76
Contig N90 length	5522	4573	25,620	338,786
Contig N90 count	18,563	20,111	10,298	301

and 5. The annotation available includes over 12,000 genes; however, the protein annotation does not reach the completeness level of the genomic one. This assembly and its corresponding annotation are not currently on GenBank but are available in a public repository (GigaScience repository, GigaDB[2]). Statistics of the previously mentioned assemblies can be found in Table 1. During the writing process of this chapter, three new triatomine species have had assemblies recently released to GenBank: *Panstrongylus geniculatus* (GCA_964188295.1; GCA_964188385.1), *Triatoma dimidiata* (GCA_964198125.1; GCA_964198115.1)—each species with one chromosomal-level and one scaffold-level assembly—and *Triatoma sanguisuga*—with a contig-level assembly (GCA_041753995.1). As these were unavailable during the review for this chapter, their details were not included.

[2]GigaDB Liu Q, Guo YH, Zhang Y et al. (2019) Supporting data for "A chromosomal-level genome assembly for the insect vector for Chagas disease, *Triatoma rubrofasciata*" GigaScience Database. https://doi.org/10.5524/100614

4 Functional Genomics in Triatomines

4.1 Behavior

Transcriptomic studies have unveiled the molecular machinery in charge of processing signals from the environment. Among them, olfaction is crucial for finding food, mates, refuge, etc. Neuromodulatory, neuropeptide precursors and neuroendocrine-related genes are expressed in the triatomine antennae [46]. The antennal sensory neurons project to the central nervous system, particularly to the antennal lobe, and to abdominal ganglia [47], to process the olfactory information. The brain of *Rhodnius prolixus* expresses neuromodulatory genes, including neuropeptide precursors and their enzymes, neurohormones, and their receptors, and biogenic amine enzymes, neurotransmitter receptors, nuclear receptors, clock genes, sensory receptors, and takeout genes [48]. Using peptidomics, Traverso et al. [49] unveiled that some neuropeptide precursors are highly conserved and specific for the Triatominae subfamily, and others differentiate among the different genera analyzed.

Proteomics were used to identify odorant-binding proteins (OBPs), chemosensory proteins (CSPs), odorant receptors (ORs), transient receptor potential channels (TRPs), and gustatory receptors (GRs) in the antennae of *R. prolixus*, along with other proteins related to general cellular functions [50]. The odorant receptors are composed of a binding unit and an odorant receptor coreceptor (Orco). By silencing the *Orco* gene, which is highly expressed in the *R. prolixus* antennae, Franco et al. [51] observed impaired host detection, reduced blood ingestion, ecdysis and oviposition, and increased mortality. Olfactory and ionotropic receptors (IRs) have also been found in triatomine legs and genitalia [52], and while some OBPs are found in the antennae, proboscis, heads, and legs, others are found only in the antennae of *R. prolixus* [53]. Comparative full-body transcriptomes revealed that OBPs were higher in *T. dimidiata* and *T. infestans*; CSPs were also abundant [54].

There are four triatomine species whose antennal transcriptomes have been investigated: *Triatoma brasiliensis* [55, 56], *R. prolixus* [51, 52, 57–60], *Rhodnius robustus* [60, 61], and *T. infestans* [62]. In *T. brasiliensis*, gene content was estimated in ~25,000; from these, gene ontology indicated that the majority of annotated contigs were involved in binding and catalytic molecular functions [55]. When analyzing pools of field-captured *T. brasiliensis* in sylvatic, peridomestic, and domestic settings, CSPs and OBPs seemed to be differentially expressed in higher proportion in peridomestic settings; also several takeout proteins and just a few ORs, TRPs, and one sensory neuron membrane protein (SNMP) were under-expressed in domiciliary individuals, but with some cuticle protein transcripts overexpressed [56]. Head, antennae, and rostrum transcriptome of *R. robustus* sylvatic specimens presented overexpression compared to laboratory-reared ones of CSPs, an OBP, ORs, IRs, TRPs, and one GR [60].

Glutathione transferases (GSTs) were highly expressed in antennae of *R. prolixus*, where they may degrade pheromones or other odors, along with cytochromes P450 (CYPs), which might be related to odors/pheromones processing [63]. The expression of antennal genes varied according to ontology and feeding. In *R. prolixus*, wide changes in the expression of multiple antennal genes occur shortly after molting, especially ORs, CSPs, OBPs, and odorant degrading enzymes (ODEs) [58]. Ontogeny also modifies the expression of olfactory and neural genes: *R. prolixus* adults presented higher expression than nymphs [46, 57, 64], while sex shows differential expression in adults of *R. prolixus* and *R. robustus*, with some genes expressed only in one sex [60]; however, another report indicates that the difference between sexes is minimal [64]. In general, males presented upregulation of chemosensory genes compared to females [60]. Silencing an OBP gene caused male *R. prolixus* to reduce the time spent with females, so that particular protein is likely to bind pheromones [53].

The antennal transcriptome after blood ingestion causes a reduced response to hosts, depending on the triatomine level of starvation; more than 50% of the genes modified their expression during the first week after feeding in *R. prolixus*: most of the olfactory genes were downregulated after a bloodmeal in *R. prolixus* and *T. infestans*, with just a few upregulated [57, 58, 62]. Some olfactory coreceptors were also transcribed in rostri, tarsi, tibial pads, and genitalia [52].

Genomic data available (e.g., [15]) has been used to assist in the assembly of transcriptomes, with both de novo and reference-based approaches being applied [59]. The increased use of RNAseq has allowed the annotation of more transcripts in the genome of *R. prolixus*; when compared with *R. robustus*, most of the studied transcripts were conserved among these closely related species. Another study evaluated the head and salivary gland transcriptome of *R. montenegrensis* and compared it with *R. robustus* (actually *R. prolixus*, see later), showing many fixed differences among these, confirming *R. montenegrensis* species status [61].

Infection can modulate the expression of genes related to behavior: in *R. prolixus*, infection by *Trypanosoma rangeli* decreased the expression of the *foraging* gene (*Rpfor*), which would be related to increased locomotion [65]; however, the expression of this gene has different effects according to sex in fifth-stage nymphs [66].

4.2 Digestion and Metabolism

Early studies of the saliva composition of triatomines recurred to cDNA random amplification and high-performance liquid chromatography (HPLC) to evaluate the salivary gland expressed proteins, unveiling an expansion in the lipocalin protein family, and new proteins were described in *R. prolixus*, including nitrophorins [67]; the reddish color of their salivary gland is caused by these proteins, which constitute about 50% of the salivary gland content [68]. In salivary glands of *Rhodnius brethesi* and *R. robustus*, nitrophorins and lipocalins were identified, with salivary platelet aggregation inhibitors and biogenic amine binding proteins in lower proportions, among others [69]. In *T. brasiliensis*, the majority of the secreted proteins were lipocalins, which prevent coagulation, platelet aggregation, and vasoconstriction; also previously unreported serine proteases and some Kazal peptides, which would be involved in vasodilation [70]. For *T. infestans*, cDNA amplification and reverse-phase liquid chromatography (RPLC), 2D electrophoresis and mass spectrometry, and liquid chromatography–tandem mass spectrometry (LC–MS/MS) were used to identify the proteins expressed in the salivary gland; the most frequently expressed proteins were lipocalins, followed by the pore-forming proteins trialysins, along with hemolysins, defensins, and other immunity-related peptides, among others [71, 72]. In *T. dimidiata*, lipocalins were also the main transcripts found, with a procalin homolog as the most frequent one [73]. The salivary proteome found in *Dipetalogaster maximus* was similar, with lipocalins as the most expressed proteins, followed by the antigen 5 family, which in other insects has allergen characteristics [74]. In *Panstrongylus megistus*, as in the other members of the Triatomini tribe, lipocalins were the most frequent salivary proteins, with some Kazal-type proteinase inhibitor, tryalisin, and triatin [75]; *Triatoma matogrossensis* followed the pattern, with lipocalins as the most frequently secreted proteins, followed by trialysins, hemolysins, Kazal-type, antigen 5, and metalloproteases, among others [76]. In *Triatoma rubida*, lipocalins were again the most frequent proteins, but followed by the antigen 5 family, with others in lower proportion [77]. *Panstrongylus chinai* also presented lipocalins as the main secreted protein family, followed by Kazal peptides and hemolysins, phosphatases, and antigen 5 proteins, with others in lower proportion [78].

The RNA interference technique (RNAi) was applied to the salivary nitrophorin 2 of *R. prolixus*, protein that transports nitric oxide, which in turn reduces platelet aggregation and causes vasodilatation, achieving a reduction of the anticoagulant activity of the saliva after two injections of double strand RNA (dsRNA) and molt into the next stage, and also when the dsRNA was provided by ingestion, which validated the use of this knockdown technique in triatomines [79]; later, nitrophorins 1–4 were simultaneously knocked down, showing that these specimens required more time to feed compared to controls, due to more bites and longer probing time to find a vessel, and a reduced contraction frequency of their cibarial pump [80]. A thrombin inhibitor named brasiliensin, another anticoagulant, was evaluated by RNAi in *T. brasiliensis*, showing that its expression was vastly reduced in the anterior midgut of knockdown specimens, and caused a reduction in the amount of blood that these consumed compared to controls, showing that anticoagulant activity is required by triatomines to feed effectively [81]. The activity of an insulin-like peptide (ILP) was evaluated in *R. prolixus* by injection with dsRNA of the molecule, which was also quantified by qPCR and visualized with in situ fluorescent RNA hybridization: the ILP was highly expressed in the central nervous system, particularly in the dorsal central part of the brain; after knockdown, carbohydrates and lipids were increased in the hemolymph, so its role probably relates to the assimilation of these nutrients [82].

Later on, next-generation sequencing techniques were employed to evaluate the transcriptome of triatomine species. In *R. prolixus*, the differential expression in different compartments of the digestive tract was assessed by cDNA 454 sequencing, which was complemented with proteomic evaluation by LC-MS/MS [83]. Highly expressed proteins included peritrophins; mucines; transporters; digestive enzymes involved in carbohydrate digestion including glycosidases, hexosaminidases, glycoside hydrolase, α-amylase, and an α-fucosidase; protein digestion included several aspartyl and cysteinyl proteases with some cathepsin and trypsin-like serine proteases also involved. In turn, protease inhibitors from the Kazal and pacifastin family were also found, mainly related to anticoagulant properties in the anterior midgut. Lipocalins were also highly expressed in the digestive tract, along with some chemosensory-binding, juvenile hormone-binding, odorant-binding, and takeout proteins. Protein kinases and protein phosphatases were found, part of the signaling pathways; and RNA processing, translation, and protein secretion were enhanced compared to the whole-body transcripts, as well as detoxification enzymes, like cytochrome P450s and glutathione transferases (GSTs). Iron and heme metabolism molecules were detected, while lipid and amino acid metabolism proteins were highly expressed [83].

Illumina sequencing provided an expanded sialotranscriptome of *T. infestans*, increasing the number of different enzymes found in the salivary gland, along with new protease inhibitors, lipocalins, odorant-binding proteins, antigen 5 family proteins, antimicrobial peptides, and lysozymes, finding differences in the expression of some proteins according to developmental stage and geographical origin [84]. The same technique was also applied to study the sialotranscriptome of *P. megistus* that, until then, had only one salivary protein sequence publicly available, where the most expressed secreted proteins were lipocalins, followed by Kazal domain peptides, inositol phosphatase, hemolysins and trialysins, serine proteases, apyrases, antigen 5 family, and metalloproteases, among several others in lower proportion [85]. In *Triatoma pallidipennis*, a general correspondence of the transcriptome with the saliva proteome was found, with triabins as the most represented in both, followed by several unknown proteins, COestearase, and trypsin in the transcriptome, with a new angiotensin-converting enzyme found in the saliva and transcripts [86]. *Panstrongylus lignarius* presented lipocalins as the major group of salivary gland transcripts, followed by juvenile hormone-binding proteins, Kazal type peptides, and glycine-rich proteins, among many others; in the fat body, many hemolymph peptides are synthesized, along with vitellogenin and transferrin [87]. *Triatoma rubrofasciata* showed mainly secreted proteins in the sialotranscriptome, composed by lipocalins as the most frequent compound family, with antigen 5, Kazal-type serine protease inhibitor, inositol polyphosphate 5-phosphatase, and apyrase 5-nucleotidase, to name a few [88].

Applying transcriptomic techniques to evaluate speciation in triatomines has had promising results when comparing *R. montenegrensis* and *R. robustus*; the head and salivary gland transcriptome showed many single-nucleotide polymorphisms (SNPs) fixed between both species, some of which seemed to have evolved under positive selection, so they might be used as molecular markers to differentiate these two species. These genes included some that participate in host interaction and in cell metabolism [61]. Comparing whole-body transcriptomes of three species of *Triatoma* from different geographical origins showed enhanced mitochondrial transcription, and nearly half of the predicted proteome was similar to the *R. prolixus* proteins [54]. Pacifastin and cystatin—protease inhibitors—were enriched, along with calycins, which bind lipophilic compounds, including triabin/procalin proteins. The odorant-binding proteins were enriched in *T. infestans* and *T. pallidipennis*, showing higher divergence in metabolism, including chemosensory proteins, odorant, ionotropic, gustatory, pickpocket, and mechanoreceptors. *Triatoma infestans* showed higher numbers of non-LTR retrotransposons [54].

4.3 Immunity

The first studies of triatomine immunity used degenerated primers to find defensins and lysozymes. In *R. prolixus*, the hemolymph presented defensins, lysozymes, pathogen-recognition proteins, an immunoglobulin, and antimicrobial peptides, along with peptidase inhibitors and oxidant detoxifier proteins, among which the heme-binding proteins were more abundant [89, 90]. Defensins were highly expressed after blood ingestion first in the fat body and later in the midgut [89]. In situ hybridization of specific probes allowed the location of the digestive sites of expression of their genes, with qPCR to evaluate differential expression according to time before or after feeding; in general, it caused an increase in the cardia and stomach of *T. brasiliensis* [91].

Regarding the expression of immunity molecules on the triatomine digestive tract, pathogen-recognition molecules like lectins and peptidoglycan recognition proteins were identified, along with several involved in the Toll pathway, one regulator from the IMD pathway, and proteins with dsRNA binding domain, with lysozymes and defensins, among others [83]. Using a shotgun proteomic approach (nano-LC/MS/MS), the immunity-related proteins in the midgut and rectum of *T. infestans*, *R. prolixus*, *P. megistus*, and *D. maxima* were analyzed, finding proteins from the Toll, IMD, and Jackstat canonical immune pathways, along with defensins, heat shock proteins, and annexins, among others [92]. The Toll pathway is activated by pathogen-associated molecular patterns (PAMPs) present in fungi and Gram-positive bacteria [93], while Gram-negative bacteria prompt the IMD pathway [94, 95]; viruses are the PAMPs of the Jackstat and RNAi pathways [96, 97]. The first publication of the genome of *R. prolixus* seemed to lack parts of the IMD pathway described in dipterans [15]; however, by both bioinformatic and in vivo experiments exposing the triatomine to Gram-negative bacteria and gene silencing of one of its reported components by RNAi, most of these missing genes were found, and the IMD pathway was shown to be active [98, 99].

Regarding immunity against infection, expressed sequence tags (ESTs) in the fat body and the intestine were evaluated in *R. prolixus* after infection by injection in the thorax of bacteria, *T. cruzi*, or sterile media, respectively. Most of the matches involved metabolism, defense, stress response, receptor, ribosomal and transcriptional control [100]. Later, in experiments feeding *T. cruzi* to *R. prolixus*, it was shown that different clones of the parasite elicited differential immune responses, with one reducing the microbiota, nitrites, and nitrates and enhancing parasite infection and phenoloxidase activity, and the other showing some changes that were not sufficient to produce successful infection of the triatomine [101]. Midgut infection with *T. rangeli*, on the other hand, altered microbiota and immune response of this triatomine species, with increased antimicrobial activity against *Serratia marcescens* and some under- and overexpression of different antimicrobial peptides and lysozymes, and reduction of phenoloxidase activity [102]. The Nuclear Factor Kappa B transcription factors (NF-κB TF), including *relish*, *dorsal* and *cactus*, are proteins that trigger the transcription of antimicrobial peptides (AMPs) like defensins and prolixicin, which was confirmed by suppressing the translocation of NF-κB TF to the nucleus, showing a reduction of AMPs expression in *R. prolixus* and lower antibacterial activity in the midgut [103]. In the hemolymph, ingestion of *T. cruzi* caused over- and under-expression of different proteins, with increases of a secreted cathepsin, endopeptidase, and superoxide dismutase; early post-feeding, defensins increase, but later they were diminished, along with several other proteins, while there was under-expression of heat shock proteins, a transglutaminase, a peptidase, and a putative vitellogenin, among others; only a putative hemolymph juvenile hormone-binding protein and a chemosensory protein were increased a day after feeding, and several other proteins were decreased [90].

In terms of immune memory, a recent study compared survival and ecdysis in insects with previous exposure to dead *T. cruzi* from two strains and found that, depending on the strain, this exposure increased *R. prolixus* survival or triggered an earlier molting, which would be related to the strain virulence [104].

4.4 Insecticide Resistance

Insecticide resistance—that is, the failure to achieve death of insects exposed to an insecticide—can be mediated by several mechanisms. Among them, metabolic resistance is mediated by the action of several enzymes over the insecticide compounds, increasing their detoxification [105]. ATP-binding cassette (ABC) transporters, carboxylesterases, cytochrome P450 monooxygenases (P450s), and glutathione-S-transferases (GSTs) are among the main enzyme families involved in this resistance [105, 106]. Point mutations cause an increase in enzyme activity and/or overexpression, which in turn increases the detoxification of the insecticide [105]. Furthermore, the sequestering of pyrethroids is another property particularly of esterases and glutathione-S-transferases that is related to insecticide resistance [107].

In the genome of *R. prolixus*, many carboxylesterases were predicted, which participate in several processes, including development, digestion, detoxification, pheromone and hormone processing, and neural functions [105]. Regarding cytochrome P450s, they participate in synthesis, metabolism, and degradation of many compounds. The cytochrome P450 genes (CYPs) are expanded in *R. prolixus*, with clusters in tandem of CYP2, CYP3, CYP4, and mitochondrial CYP; this expansion is more pronounced in the CYP3 clan [105]. Glutathione-S-transferases participate in detoxification and protection from oxidative stress; most of the GSTs found in *R. prolixus* were cytosolic [105].

The transcriptomes of *R. prolixus*, *T. dimidiata*, *T. infestans*, and *T. pallidipennis* were also inspected for enzymes involved in metabolic resistance [63]. They showed higher levels of CYP transcripts than other model insects, with CYP3 and CYP4 clades composing the majority of them and mitochondrial and CYP2 clades present in lower levels. Triatomines also attained comparatively higher expression of pheromone/hormone processing class of carboxylcholinesterases (CCEs) of the E clade and of cytosolic Sigma GSTs. When comparing a small set of genes between resistant and susceptible *T. infestans* individuals, there was an overexpression of a CYP4 gene in the resistant ones, which could be related to the capacity of these insects to metabolize pyrethroids [63].

An endogenous clock regulation has been observed in *T. infestans* for some P450s and for the nicotinamide adenine dinucleotide phosphate cytochrome P450 reductase (CPR), which supplies the electrons required by P450s reactions, differing its expression pattern according to stage and sex and also depending on the light regimen [108].

Analysis of the transcriptome from the integument—cuticle and its epidermis—of *T. infestans* showed many CYP transcripts, with some of CYP3 clan overexpressed in resistant individuals, while only one CCE of the detoxification group was significantly upregulated in these resistant specimens [104]. ABC transporters from subfamilies that have been associated with insecticide resistance in other insect species were also described [104]. Later on, further bioinformatic analyses included more CYP4 members in *T. infestans* transcriptome, with higher theoretical docking capacity to deltamethrin than the ones from *R. prolixus*; some of these CYP4 were overexpressed in resistant individuals [105].

Another mechanism related to insecticide resistance is through point mutations occurring in the site of action of an insecticide, reducing its affinity, which would in turn reduce its knockdown effect, a process known as *kdr* resistance. In triatomines, *kdr* resistance in pyrethroids' target site, the voltage-gated sodium channel, was found by cloning and sequencing; some *T. infestans* pyrethroid-resistant populations presented a non-synonymous mutation turning leucine into phenylalanine, and other populations showed another mutation converting leucine into isoleucine [111–113].

The third insecticide resistance mechanism involves the physicochemical properties of the cuticle, where its width and composition have been proven significantly different in resistant versus susceptible *T. infestans* individuals [114]. In the integument's transcriptome of resistant *T. infestans*, a couple of structural cuticle protein genes with an RR1 motif were significantly overexpressed, compared to the susceptible individuals [109]. Moreover, some CYP4 are involved in the production of cuticular hydrocarbons, which reduce insecticide penetration [105].

Other functional genomic studies have been performed in triatomines, encompassing the reproductive system. Transcriptomic and RNAi experiments have uncovered regulatory pathways of the reproduction, which include transcripts produced not only in the ovaries, but in the fat body, and some in the central nervous system, and the relation of reproduction with the feeding status of the females [115–118]. Furthermore, embryo development has also been studied, unveiling genes showing regulatory functions [119].

Next-generation sequencing techniques have increased exponentially the transcript data that can be obtained from specific tissues, allowing a detailed analysis of over- and under-expression according to ontogeny, physiological status, and experimental treatments, promoting the verification of these findings using RNAi, to confirm the intricate mechanisms involved in genome expression. This brief overview of the functional genomics of Triatominae shows that the main tribes—Triatomini and Rhodniini—present relevant differences, but when new species are analyzed, more within-tribe variation can be detected, adding up to manifest the diversity of these vector species.

5 Repeatomes

DNA repeats, as the name suggests, are segments of DNA that recur within a genome. Therefore, a DNA repeat is a sequence of nucleotides that appears in more than one locus in a genome. These repeats could include several types of sequences, depending on the structure, length, location, origin, and mode of multiplication. The distribution of repetitive elements within the genome may manifest as either directly adjacent arrays known as tandem repeats or as dispersed repetitions throughout the genome referred to as interspersed repeats. Some repetitive sequences encode essential genetic information, such as families of genes crucial for various biological functions, such as immune responses or sensory perception. In addition, central functions for the cells, including DNA folding, depend on histone genes, which are also repetitive DNA families. On the other hand, non-coding repetitive sequences also occupy a significant portion of the genome. Examples include the ribosomal RNA genes (rRNA), which play a fundamental role in protein synthesis, and various classes of small non-coding RNAs involved in regulatory processes. Beyond these, certain repetitive elements have been linked to the maintenance of chromosomal integrity and stability, exemplifying the multifaceted nature of DNA repeats in the intricate dance of genomic organization.

Notwithstanding, within DNA repeats two prominent types of sequences emerge as the most frequent components: transposable elements (TEs) and satellite DNA (satDNA). Formerly dubbed as “selfish DNA”, TEs exhibit a remarkable ability to move within the genome, dictating their own replication and propagation. Nowadays, TEs are seen as key players in organism evolution, shaping their genomes [120, 121]. On a parallel note, satDNA, despite their seemingly redundant nature—non-coding, head-to-tail arrays of tandemly repeated DNA sequences—could also propagate through their own mechanisms, such as recombination slippage or extrachromosomal circular DNA [122]. Beyond the well-established roles of centromeric or telomeric repeats, transcripts originating from satDNA have been documented in various species, including triatomines [123], suggesting a potential involvement of satellite non-coding RNAs (ncRNAs) in gene expression regulation [124, 125]. Furthermore, the rapid turnover of satDNA has been demonstrated to promote genome plasticity, serving as a catalyst for genome evolution, occasionally contributing to the process of speciation [126–129].

The first attempt to determine the repetitive landscape of a Triatominae genome came with the analysis of the first genome assembly of a triatomine species: *R. prolixus*. According to the annotation of *R. prolixus* assembly, the repetitive part of the genome reached 5.6% of the genome.

The main repetitive sequences were class II TEs being the 3% of the genome [15]. However, as will be explained later, this value was highly conservative. One of the limitations at that time was the limited access to genome sequencing, together with the bioinformatic tools available capable of studying the repetitive DNA. With the massification of NGS data, several tools were developed, which had the power to determine the complete amount of repetitive DNA in a genome [130]. The term "repeatome" emerged to include the set of all repetitive DNA sequences composing the genomes [131]. Within Triatominae, the first repeatome determination was in 2017, when the *T. infestans* repeatome was published [132]. The decision was based not only on the epidemiologic importance of *T. infestans* in the South Cone, but due to genomic variation existing within the species. This variation involves, among other characteristics, its genome size. *Triatoma infestans* is divided into two main lineages: Andean and Non-Andean, with significant differences in their genome size: 1.936 Gbp and 1.487 Gbp per haploid genome, respectively [16, 43]. Prior to the repeatome determination, there was no information about the type of sequences responsible for this variation. At first glance, the *T. infestans* genome was revealed to be highly repetitive, contrasting with the *R. prolixus* genome. The repetitive fraction reached up to 44% and 34% for Andean and Non-Andean lineages respectively. In addition, the satDNA sequences were the biggest fraction within the repeatome of both lineages, 33% of the Andean genome and 25% of the Non-Andean. In fact, several families of satDNA were described and mapped to the chromosomes, revealing that satDNA sequences were not only present in heterochromatic regions, but also dispersed in euchromatic regions [132]. The satDNA fraction emerged as the major player in *T. infestans* genome size polymorphism. In accordance with cytogenetic data on constitutive heterochromatin amount, while non-repetitive DNA remained similar, the repetitive DNA content on the Andean lineage was greater than in the Non-Andean. Specifically, the Andean lineage presented almost double of satDNA content than the Non-Andean lineage (630.78 Mbp and 376.05 Mbp per haploid genome, respectively). In detail, just a few satDNA families located mainly in the heterochromatin revealed an extraordinary variation directly related to the genome content.

The trend of highly repetitive genome, populated with satDNA sequences, was then reinforced with the elucidation of the *Triatoma delpontei* repeatome. In addition to analyzing the largest heteropteran genome, *T. delpontei* is a sister species to *T. infestans*. Therefore, this brought the opportunity to extend the analysis of how the repeatome impacts genome size variation among closely related species. The *T. delpontei* genome size scales up to 2.7 pg (C-value) [43], and half of it is composed by satDNA sequences; precisely, 59.6% of the genome is repetitive DNA, being 51.1% satDNA families. The mechanism by which the genome varies the repetitive DNA content is the same as in *T. infestans*: just a few satDNA families, located mainly in heterochromatic regions, are accumulated in the genome. In the *T. delpontei* genome, four satDNA families represent more than 48% of the genome. Remarkably, the great majority of families are shared with *T. infestans*, which supports the so-called "library hypothesis" that describes the evolution of satDNA families, stating that related species share a pool of ancestral satDNA families. Therefore, it is expected that closely related species share a larger proportion of those ancestral satDNA families than evolutionary distant species [133, 134]. Comparison with *R. prolixus* also supports this hypothesis, showing fewer shared families (see later).

Constitutive heterochromatin is a great source of karyotype variation within Triatominae. In spite of being conserved in chromosome number, Triatominae karyotypes possess variability in the amount and distribution of C-heterochromatic blocks, especially in species within the Triatomini tribe [135]. Therefore, the process in which heterochromatin is formed is a key player in karyotypic evolution. *Triatoma infestans* and *T. delpontei* genomes showed that heterochromatic regions are formed by just a few satDNA families. The fact that heterochromatic blocks were present in the telomeric region [132], and, in parallel, all satDNA families presented dispersed copies along all the genome [127, 133], led to the hypothesis that repeats reaching subtelomeres are prone to be involved in an expansion process. Ectopic recombination between non-homologous subtelomeres could enhance this process, spreading the repeats. Future analyses of other distant species are needed to confirm the hypothesis [133].

The abundance of repeats in *Rhodnius* species genomes seems to be greatly different. As mentioned earlier, *R. prolixus* genome assembly was reported to bear 5.6% of repetitive sequences [24]. However, using novel bioinformatic tools to analyze repetitive sequences using unassembled NGS reads, Castro and collaborators demonstrate that *R. prolixus* repetitive DNA actually ranged between 19% and 23% of the genome. This result was supported also by the evaluation of other sibling species, *R. montenegrensis* and *R. marabaensis* (formerly *Rhodnius robustus* II and III, respectively), with similar results [134]. *Rhodnius* genomes are characterized by highly abundant Class II elements. However, data on satDNA sequences were completely missing [134]. Further studies, including satDNA determination, reported a similar value for total repetitive DNA (24% of the genome), being 8.6% satDNA sequences [118]. Not only the *Rhodnius* and *Triatoma* species analyzed differed on the most abundant type of sequences (Class II TEs versus satDNA, respectively), but their genomes seem to possess different features. *Triatoma infestans* and *T. delpontei* bear few extremely amplified families which represent the great majority of the entire satDNA content, and more than 10% of the genome, while *R. prolixus* satDNA families descend gradually, with the top family being 2.13%, and just nine families above 0.1%. The fact that the *R. prolixus* genome is devoid of C-heterochromatin—excepting the Y chromosome—could be responsible for this scenario. *Triatoma* genus species lacking C-heterochromatin, such as *T. rubrovaria*, must be analyzed to further discuss this issue. However, this also demonstrates the protagonism of heterochromatin on genome architecture during evolution.

The role of transcripts derived from satDNA has been questioned in the past, but recent evidence has changed the scenario. Nowadays, it is accepted that these non-coding RNAs might have functions in different cellular contexts, such as cancer, stress response, development, or cell proliferation [135–137]. Evidence on satDNA transcripts was found in *R. prolixus* transcriptomes from antennae and gonads. Although transcription level appears to be significantly correlated with satDNA abundance in the genome, each family is transcribed under different patterns. Therefore, transcription could have a specific role in those tissue environments. As cited before, several examples are reported on satDNA-derived non-coding RNAs; however, research on triatomines is pending.

In summary, the remarkable diversity in the abundance and composition of repetitive elements within the genomes of Triatominae species unravels a complex tapestry of genomic evolution. Notably, constitutive heterochromatin, predominantly integrated by satDNA, emerges as a key player in shaping karyotypic variations. While Triatomini genomes, exemplified by *T. infestans* and *T. delpontei*, showcase a dominance of a few highly amplified satellite DNA families, Rhodniini species exhibit a distinct landscape characterized by the prevalence of Class II TEs and a gradual descent in the abundance of satDNA families. Future analyses, encompassing diverse species within both genera, are imperative to unveil the full spectrum of mechanisms through which these repetitive elements contribute to the dynamic evolution of Triatominae genomes. In the light of this picture, it is becoming evident that repetitive elements are not mere bystanders but influential protagonists on Triatominae genome architecture.

6 Phylogenomics of Triatominae

As defined by Steenwyk et al. [138] phylogenetics 'aims to reconstruct the evolutionary histories of organisms, genes, traits or other biological features by examining the distribution of inherited characters in descendant lineages and tracing them back in time to identify ones that shared a common ancestor". Therefore, a phylogenetic tree attempts to represent the evolutionary history of the analyzed organisms. It must be taken into account that trees are a representation of the best-supported hypotheses: the statistically most probable path rather than the absolute truth. This representation requires the identification of homologous characters that are common to different organisms and the inference of phylogenetic trees from the comparison of these characters using mathematical methods [139]. Nowadays, phylogenies are mostly inferred using DNA or protein sequences. Commonly, trees are calculated using one or few genes or gene fragments, such as the popular mitochondrial markers. Usually, each gene tree retrieves a different topology [140]. Reasons behind this could be due to sampling error or discordance between gene trees. Increasing the number of genes and total sequence length can reduce the sampling error and yield more robust phylogenies [141].

Phylogenetic studies in triatomines, like in many other biological groups, have seen significant advances over time. Traditionally, the phylogenetic analysis involved few popular molecular markers, such as the mitochondrial genes *cytochrome c oxidase subunit I* (*coI*), *cytochrome b* (*cytb*), ribosomal 12S and 16S, or the nuclear marker ITS-2 (reviewed in [13], and [142]).

Phylogenomics is the comprehensive study of evolutionary relationships through the analysis of complete genomic information. Unlike conventional phylogenetics, which relies on specific gene sequences, phylogenomics delves into the entirety of an organism's genetic material. This holistic approach harnesses the power of massive datasets, encompassing not only coding regions but also non-coding elements, intergenic spaces, and structural variations, thereby offering a more accurate and detailed portrayal of evolutionary dynamics.

The advent of phylogenomics has marked a transformative era. While traditional phylogenetics has long been the cornerstone for reconstructing evolutionary trees based on select genetic markers, phylogenomics expands the horizon by leveraging entire genomic datasets. This shift from a gene-centric to a genome-wide perspective provides a richer, more nuanced understanding of evolutionary histories. The main issue for the gene-centric approach is, as mentioned earlier, the incongruence between two phylogenies. This could be due to several reasons. Orthology determination could be a serious problem for a phylogenetic reconstruction. Mechanisms such as gene duplication, horizontal gene transfer, or lineage sorting must be taken into account [143]. However, sometimes it could be that the dataset is not best suited for the organisms involved in the analysis, in terms of the gene that is being analyzed or the length of the dataset [144]. Therefore, the strength of the genome-wide approach is that the analysis could potentially solve issues such as horizontal gene transfer, gene duplications, convergent evolution, or incomplete lineage sorting, providing a more robust framework for resolving intricate evolutionary relationships. This minimizes the impact of singular gene histories, unveiling a more reliable depiction of species divergence and the dynamics of genomic evolution over time.

Rhodnius phylogeny is constantly under revision. As occurs with all triatomines, *Rhodnius* species classification is largely based on morphological and chromatic features. However, the *Rhodnius* genus presents several complexities, since morphological variation is limited, resulting in species complexes or cryptic species. Recently, phylogeny reconstruction using genomic data has appeared. The first article using an omics approach was from de Carvalho et al. [56]. In this initial study, authors conducted large-scale RNA-sequencing data generation for *R. montenegrensis* and *R. robustus* to enable a comparative analysis aimed at determining whether they were distinct species or the same. The findings presented by the researchers were based on the 13,695 SNPs detected, distributed across 2196 distinct contigs. Among these SNPs, 8350 were found to be fixed in *R. montenegrensis*, while 7888 were fixed in *R. robustus*. Subsequent SNP calling, utilizing the more comprehensive *R. robustus* transcriptome assembly, unveiled 3055 fixed interspecific variations alongside the identification of 216 transcripts exhibiting substantial divergence, solely harboring fixed differences between the two species. Consequently, the authors concluded that indeed *R. montenegrensis* represents a separate species from *R. robustus*. It should be noted that *R. robustus* is actually a cryptic species complex, consisting of 5 lineages and *R. prolixus*. Indeed, *R. montenegrensis* is widely confirmed as one of these lineages within the complex: the so-called *R. robustus* II [145]. Another significant conclusion achieved using transcriptomic data was the confirmation of hybridization within insectary colonies [145]. Indeed, Brito et al. [145] noted that the comparison between species that de Carvalho et al. [56] performed was actually between *R. montenegrensis* (*R. robustus* II) and *R. prolixus*.

A phylogenomic approach that is gaining prominence in the study of triatomines is restriction-site associated DNA sequencing (RAD-seq). RAD-seq is a DNA sequencing technique based on identifying and sequencing DNA regions near specific restriction sites. This technique enables high-resolution genomic data collection and is particularly useful for phylogenomic studies by providing genetic markers distributed across the genome, facilitating the reconstruction of phylogenetic relationships among species. Additionally, RAD-seq can be employed to compare populations, making it applicable at various evolutionary scales. It is worth noting that there are several variants of the technique depending on the enzyme or enzymes used. The first article to introduce this technique in triatomines was by Hernandez-Castro et al. [146]. In this study, they aimed to evaluate the effectiveness of three distinct 2b restriction site-associated DNA (2b-RAD) genotyping strategies in *R. ecuadoriensis* to achieve adequate genomic resolution for dissecting micro-evolutionary processes. They concluded that 2b-RAD genotyping represents a cost-effective and methodologically straightforward approach for generating high-resolution genomic data for triatomines, capable of distinguishing between different vector populations at epidemiologically relevant scales. The main results of the work involved the identification of two distinct populations, one located in the Manabí province (Western Ecuador) and the other in the Loja province (Southwestern Ecuador). These results are in accordance with the recent data on molecular markers and morphometrics, which defines certain differentiation between Manabí and Loja populations [147]. Further research is much needed to better describe the demographic and evolutionary history of *R. ecuadoriensis*. A spin-off of this work focused on the Loja province, aimed at a better understanding of migration behavior of triatomines and genomic adaptations to domiciliation [148]. The authors confirmed *R. ecuadoriensis* does frequently invade houses from the forest in Southern Ecuador. In addition, genome scans of SNPs comparing specimens' pairs from domestic/sylvatic environments lead to the finding of some outlier loci that are mapped to the *R. prolixus* genome, related to salivary protein precursors and development of the embryo in the available *R. prolixus* annotation. Lastly, they propose that orography could act as the major barrier to genetic connectivity between populations [148].

Another example of the complexity of Rhodniini phylogeny was addressed by Kieran et al. [149]. *Rhodnius pallescens* is an important Chagas disease vector in Colombia and Panama, where morphometric, genetic, and cytogenetic variability has been described [150, 151]. When a dark chromatic morph was discovered in the highlands of Veraguas Province, the authors questioned whether this represented an isolated population. This time, a 3RAD approach was employed to generate a 2216 SNP dataset, which was also compared with several complete mitochondrial genomes. The results using both the RAD-seq and mtDNA data displayed highly differentiated clades with essentially no gene flow between the dark and light chromatic morphs from Veraguas and central Panama, respectively. The authors suggest that this dark morph could represent a new species, but further research is needed [152]. Furthermore, a comparison with populations outside Panama would have a significant impact.

A similar scenario, where several cryptic species or a species complex was described, occurs within the Triatomini tribe, with *T. dimidiata*. Orantes et al. [155] developed a RAD-seq-based analysis pipeline to study mixed-species DNA extracted from *T. dimidiata* abdomens. In this way, the analysis covered not only the vector but also *T. cruzi* genetic variation—when present—microbial diversity, and the bloodmeal history. The analysis was conducted in an epidemiologically relevant area, encompassing Guatemala, Belize, El Salvador, and Nicaragua. One of the authors' main findings regarding the vector was the ability to perform population genomics analysis without a reference genome, which is actually one of the strengths of RAD-seq. Furthermore, the chapter reinforces the importance of genome-wide analyses, since the SNP dataset enables population analyses across different spatial scales: within populations and between geographically distant ones [155]. In fact, further investigation was focused on the consequences of insecticide spraying. It is well known that the efficacy of spraying campaigns decays through time, and insect reinfestation is one of the main issues in Chagas disease management. Is not clear whether this reinfestation is caused by an incomplete elimination from infested houses, within-village dispersal, or influx from other villages or sylvatic environments. Cahan et al. [156] studied an area in a time lapse that involved a spraying campaign and employed RAD-seq to test this question. Triatomines collected from reinfested houses post-spraying were most closely related to pre-spray specimens from the same house, leading to the hypothesis that the spraying was not able to completely eliminate the insects. However, house colonization by unrelated insects was also detected after 1 year of spraying, but it was mainly from neighboring houses. Therefore, authors concluded that house structural improvements are critical for long-term reduction of *T. dimidiata* infestation. However, it must be noted that this situation could change in different localities or depending on the vector species [156].

A wider study dealing with the *T. dimidiata* species complex, using RAD-seq, was able to disentangle the issue [157]. *Triatoma dimidiata* s.l. is distributed across an extensive geographic region, from Mexico to Ecuador. It has been largely described that *T. dimidiata* exhibits considerable morphological diversity, which has led to multiple splitting and merging of the species (reviewed in [14, 158, 159]). Using SNPs data, authors determine at least three well-defined, monophyletic lineages: group 3, nowadays named *Triatoma huehuetenanguensis* [160], appeared as the earliest branching clade; group 4, now described as *Triatoma mopan* [161], was recovered as sister clade of *T. dimidiata s.s.* (which comprises groups 1 and 2) plus *Triatoma hegneri*. The analysis was not able to fully resolve this last clade, involving *T. dimidiata s.s.* [157]. These results allow further discussion on nuclear and mitochondrial phylogeny since the different approaches showed different results. Authors suggest that controversies could arise due to ancestral hybridization between the lineages and mitochondrial introgression.

Population genomics has been also applied in *Triatoma brasiliensis*, a major vector of Chagas disease in Northeast Brazil. In order to improve vector control campaigns, Viana et al. [162] performed a ddRAD-seq approach to better understand the population dynamics of vectors. In a scenario very common for triatomine species, peridomiciliary reinfestations by *T. brasiliensis*, coupled with elevated rates of *T. cruzi* infection, are recurrent in the state of Rio Grande do Norte, Brazil. The results based on 3002 neutral loci evidenced low genetic structuration, indicating local panmixia. This was in disagreement with mitochondrial markers data, hence depicting the importance of exploring genome-wide approaches. The authors were also able to detect 11 loci putatively under selection. From these, most of them were similar to hypothetical proteins from species of other taxa. Only one of them was annotated as a mobile element from the lineage Jockey [162].

There are several variations in the RAD-seq implementation to perform phylogenomics. One of them, called DArTseq Genotyping-by-Sequencing, was employed in a much discussed topic, the phylogeny within the *Mepraia* genus. In a very fancy analysis, San Juan et al. [163] tested if the center–periphery hypothesis (CPH) could be fitted to *Mepraia spinolai*. This hypothesis states that populations near the distribution limit of a species present lower genetic diversity and higher differentiation than populations located in the center [164]. The obtained results partially support the CPH and identified biotic (abdomen area) and abiotic (winter/summer temperatures) factors that would relate to genetic diversity in this restricted-dispersal species of epidemiological relevance. The authors did not observe an increase in genetic differentiation of peripheral populations expected with CPH. Hybridization with other *Mepraia* species could be influencing this. In fact, as said earlier, *Mepraia* phylogeny is far from being resolved. A phylogenomic approach using transcriptomic data was performed, but results suggested several alternative topologies are supported. Hence, the question remains unanswered, but suggests that events such as hybridization or incomplete lineage sorting are taking place within the genus [165].

Another approach using genome-wide markers deals with phylogenies at a deeper scale. Ultraconserved elements (UCE) are highly conserved DNA sequences found across a wide range of organisms. Kieran et al. [28] employed this methodology to test the evolutionary origin of triatomines, as this is a matter of great debate. This discussion also involves the origin of their hematophagous feeding habit. All possible hypotheses were suggested for the origin of the Triatominae subfamily: monophyly, polyphyly, and paraphyly (reviewed by [14] and by [166]); the latter theory was drawn since the genus *Opisthacidius* appeared as a sister of the Rhodniini tribe [167]. However, UCE analysis strongly supports Triatominae as monophyletic, with *Opisthacidius* genus as sister taxa [28]. What is more, results depicted that the Triatominae subfamily is basally split into two major clades: one combining Cavernicolini, Rhodniini, and the *Microtriatoma* genus from the Bolboderini tribe and a second clade composed by Triatomini and the *Belminus* genus, also from Bolboderini. Therefore, it is interesting to note support for the placement of Cavernicolini as more related to Rhodniini, but highly important is the paraphyly of the Bolboderini tribe, with *Microtriatoma* close to Rhodniini—as previously suggested [168]—and the novelty of *Belminus* close to Triatomini [28].

Furthermore, several questions within Triatominae were addressed in this study: (i) One-step Old world colonization? (ii) How is the relationship between the three lineages within Rhodniini? (iii) Does the *Mepraia* genus belongs to the South or North lineage within Triatomini? As for the first question, UCE analysis suggests that the Old World was colonized only once by Triatominae species. Therefore, the *T. rubrofasciata* placement in America would be a secondary colonization. Regarding the Rhodniini tribe, UCE phylogenomics results indicated that the prolixus (including *Psammolestes* genus) and pallescens groups are sister taxa, with the pictipes group as their closest relative. One long-lasting question is the placement of the *Mepraia* genus. UCE data place *Mepraia* within the North lineage of Triatomini [28]. As is depicted here, the utilization of UCE for elucidating the Triatominae phylogeny proved to be extremely useful.

Several questions remain to be answered, such as the evolution of the rest of untested species and tribes such as Alberproseniini. Increased genomic sampling from the family Reduviidae would allow us to test further the origin of hematophagy within Triatominae. We hope that further implementation of genomics in endemic countries provides relevant answers to Chagas disease prevention and control, contributing to end its neglected status.

Acknowledgments AB is a scholar supported by the Agencia Nacional de Investigación y Desarrollo (ANID) Programa Becas—Doctorado Becas Chile 2019, grant number 72200391. SP is an investigator of the "Programa de Desarrollo de las Ciencias Básicas" (PEDECIBA) and member of the SNI from the "Agencia Nacional de Investigación e Innovación" (ANII).

We thank Carezza Botto-Mahan for providing the means to obtain a picture of a wild specimen of *Triatoma infestans*, and Fernando Moya for capturing and donating the picture; Martin S. Llewellyn, for allowing AB to spend time in the making of this chapter; and Juan David Ramírez, for giving us the opportunity to condense and disseminate the current status of Chagas disease vectors' genomics. Finally, we would like to highlight all the researchers listed in the references—and also those who were not included—who worked hard not only to satisfy their curiosity but to improve the future of the people living in endemic areas, who would benefit from breakthroughs in Chagas disease control.

References

1. Alevi KCC, de Oliveira J, da Silva RD et al (2021) Trends in taxonomy of Chagas disease vectors (Hemiptera, Reduviidae, Triatominae): from Linnaean to integrative taxonomy. Pathogens 10:1627. https://doi.org/10.3390/pathogens1012162
2. Gil-Santana HR, Chavez T, Pita S et al (2022) *Panstrongylus noireaui*, a remarkable new species of Triatominae (Hemiptera, Reduviidae) from Bolivia. ZooKeys 1104:203–225. https://doi.org/10.3897/zookeys.1104.81879
3. Zhao Y, Fan M, Li H et al (2023) Review of kissing bugs (Hemiptera: Reduviidae: Triatominae) from China with descriptions of two new species. Insects 14:450. https://doi.org/10.3390/insects14050450
4. Panfilio KA, Angelini DR (2018) By land, air, and sea: hemipteran diversity through the genomic lens. Curr Opin Insect Sci 25:106–115. https://doi.org/10.1016/j.cois.2017.12.005
5. Otálora-Luna F, Pérez-Sánchez AJ, Sandoval C et al (2015) Evolution of hematophagous habit in Triatominae (Heteroptera: Reduviidae). Rev Chil Hist Nat 88(1):4. https://doi.org/10.1186/s40693-014-0032-0
6. Masonick PK, Knyshov A, Gordon ERL et al (2024) A revised classification of the assassin bugs (Hemiptera: Heteroptera: Reduviidae) based on combined analysis of phylogenomic and morphological data. Syst Entomol:1–37. https://doi.org/10.1111/syen.12646
7. Guarneri A, Lorenzo M (eds) (2021) Triatominae - the biology of Chagas disease vectors. Springer. https://doi.org/10.1007/978-3-030-64548-9
8. Tarleton RL, Gurtler RE, Urbina JA et al (2014) Chagas disease and the London declaration on neglected tropical diseases. PLoS Negl Trop Dis 8(10):e3219. https://doi.org/10.1371/journal.pntd.0003219
9. Echeverria JE, Bustamante Gomez MB, Pessoa G et al (2018) Resistance to deltamethrin by domestic and wild *Triatoma infestans* populations in the municipality of Toro Toro, Potosi, Bolivia. Parasit Vectors 11(1):92. https://doi.org/10.1186/s13071-018-2663-5
10. Gomez MB, D'Avila GCP, Orellana ALG et al (2014) Susceptibility to deltamethrin of wild and domestic populations of *Triatoma infestans* of the Gran Chaco and the Inter-Andean Valleys of Bolivia. Parasit Vectors 7:497. https://doi.org/10.1186/s13071-014-0497-3
11. Grijalva MJ, Villacís AG, Ocaña-Mayorga S et al (2022) Evaluation of the effectiveness of chemical control for Chagas disease vectors in Loja Province, Ecuador. Vector borne Zoonotic Dis 22(9):449–458. https://doi.org/10.1089/vbz.2021.0089
12. García BA, Powell JR (1998) Phylogeny of species of *Triatoma* (Hemiptera: Reduviidae) based on mitochondrial DNA sequences. J Med Entomol 35(3):232–238. https://doi.org/10.1093/jmedent/35.3.232
13. Stothard JR, Yamamoto Y, Cherchi A et al (1998) A preliminary survey of mitochondrial sequence variation in Triatominae (Hemiptera: Reduviidae) using polymerase chain reaction-based single strand conformational polymorphism (SSCP) analysis and direct sequencing. Bull Entomol Res 88(5):553–560. https://doi.org/10.1017/S0007485300026079
14. Monteiro FA, Weirauch C, Felix M et al (2018) Evolution, systematics, and biogeography of the Triatominae, vectors of Chagas disease. Adv Parasitol 99:265–344. https://doi.org/10.1016/BS.APAR.2017.12.002
15. Mesquita RD, Vionette-Amaral R, Lowenberger C et al (2015) Genome of *Rhodnius prolixus*, an insect vector of Chagas disease, reveals unique adaptations to hematophagy and parasite infection. Proc Natl Acad Sci USA 112(48):14936–14941. https://doi.org/10.1073/pnas.1506226112
16. Panzera F, Pérez R, Panzera Y et al (2010) Cytogenetics and genome evolution in the subfamily Triatominae (Hemiptera, Reduviidae). Cytogenet Genome Res 128:77–87. https://doi.org/10.1159/000298824
17. Gómez-Palacio A, Pita S, Abad-Franch F et al (2023) Molecular and cytogenetic evidence for sibling species in the Chagas disease vector *Triatoma maculata*. Med Vet Entomol 37(2):316–329. https://doi.org/10.1111/mve.12633
18. Aguilera-Uribe M, Meza-Lázaro RN, Kieran TJ et al (2020) Phylogeny of the North-Central American clade of blood-sucking reduviid bugs of the tribe Triatomini (Hemiptera: Triatominae) based on the mitochondrial genome. Infect Genet Evol 84:104373. https://doi.org/10.1016/j.meegid.2020.104373
19. Hypsa V, Tietz DF, Zrzavý J et al (2002) Phylogeny and biogeography of Triatominae (Hemiptera: Reduviidae): molecular evidence of a New World origin of the Asiatic clade. Mol Phylog Evol 23(3):447–457. https://doi.org/10.1016/S1055-7903(02)00023-4
20. Panzera F, Ferreiro MJ, Pita S et al (2014) Evolutionary and dispersal history of *Triatoma infestans*, main vector of Chagas disease, by chromosomal markers. Infect Genet Evol 27:105–113. https://doi.org/10.1016/j.meegid.2014.07.006
21. Dotson EM, Beard CB (2001) Sequence and organization of the mitochondrial genome of the Chagas disease vector, *Triatoma dimidiata*. Insect Mol Biol 10:205–215. https://doi.org/10.1046/j.1365-2583.2001.00258.x

22. Pita S, Panzera F, Vela J et al (2017) Complete mitochondrial genome of *Triatoma infestans* (Hemiptera, Reduviidae, Triatominae), main vector of Chagas disease. Infect Genet Evol 54:158–163. https://doi.org/10.1016/j.meegid.2017.06.003
23. Dong L, Ma X, Wang M et al (2018) Complete mitochondrial genome of the Chagas disease vector, *Triatoma rubrofasciata*. Korean J Parasitol 56:515–519. https://doi.org/10.3347/kjp.2018.56.5.515
24. Zhao Y, Jiang M, Wu Y et al (2019) Mitochondrial genomes of three kissing bugs (Reduviidae: Triatominae) and their phylogenetic implications. Int J Biol Macromol 134:36–42. https://doi.org/10.1016/j.ijbiomac.2019.05.020
25. Pita S, Mora P, Rojas-Cortez M et al (2023) The complete nucleotide sequence and gene organization of the mitochondrial genome of *Triatoma boliviana* (Hemiptera, Reduviidae, Triatominae) and phylogenetic comparisons. Arthropoda 1:3–11. https://doi.org/10.3390/entomology1010002
26. Wang Y, Chen J, Jiang L et al (2015) Hemipteran mitochondrial genomes: features, structures and implications for phylogeny. Int J Mol Sci 16:11834–14290. https://doi.org/10.3390/ijms160612382
27. Zhang DX, Szymura JM, Hewitt GM (1995) Evolution and structural conservation of the control region of insect mitochondrial DNA. J Mol Evol 40:382–391. https://doi.org/10.1007/BF00164024
28. Kieran TJ, Gordon ERL, Zaldívar-Riverón A et al (2021) Ultraconserved elements reconstruct the evolution of Chagas disease-vectoring kissing bugs (Reduviidae: Triatominae). Syst Entomol 46:725–740. https://doi.org/10.1111/syen.12485
29. Dupeyron M, Baril T, Bass C et al (2020) Phylogenetic analysis of the *Tc1/mariner* superfamily reveals the unexplored diversity of *pogo*-like elements. Mob DNA 11:21. https://doi.org/10.1186/s13100-020-00212-0
30. Filée J, Rouault JD, Harry M et al (2015) Mariner transposons are sailing in the genome of the blood-sucking bug *Rhodnius prolixus*. BMC Genomics 16:1061. https://doi.org/10.1186/s12864-015-2060-9
31. Majerowicz D, Calderón-Fernández GM, Alves-Bezerra M et al (2017) Lipid metabolism in *Rhodnius prolixus*: lessons from the genome. Gene 596:27–44. https://doi.org/10.1016/j.gene.2016.09.045
32. Sayers EW, Bolton EE, Brister JR et al (2022) Database resources of the national center for biotechnology information. Nucleic Acids Res 50(D1):D20–D26. https://doi.org/10.1093/nar/gkab1112
33. Giraldo-Calderón GI, Harb OS, Kelly SA et al (2022) VectorBase.org updates: bioinformatic resources for invertebrate vectors of human pathogens and related organisms. Curr Opin Insect Sci 50:100860. https://doi.org/10.1016/j.cois.2021.11.008
34. Manni M, Berkeley MR, Seppey M et al (2021) BUSCO update: novel and streamlined workflows along with broader and deeper phylogenetic coverage for scoring of eukaryotic, prokaryotic, and viral genomes. Mol Biol Evol 38(10):4647–4654. https://doi.org/10.1093/molbev/msab199
35. Dudchenko O, Batra SS, Omer AD et al (2017) De novo assembly of the *Aedes aegypti* genome using Hi-C yields chromosome-length scaffolds. Science 356(6333):92–95. https://doi.org/10.1126/science.aal3327
36. Dudchenko O, Shamim MS, Batra S et al (2018) The Juicebox assembly tools module facilitates de novo assembly of mammalian genomes with chromosome-length scaffolds for under $1000. bioRxiv. https://doi.org/10.1101/254797
37. van Berkum NL, Lieberman-Aiden E, Williams et al (2010) Hi-C: a method to study the three-dimensional architecture of genomes. J Vis Exp 39:e1869. https://doi.org/10.3791/1869
38. Knyshov A, Gordon ERL, Masonick PK et al (2023) Chromosome-aware phylogenomics of assassin bugs (Hemiptera: Reduvioidea) elucidates ancient gene conflict. Mol Biol Evol 40(8):msad168. https://doi.org/10.1093/molbev/msad168
39. Justi SA, Dorn PL, Cahan SH et al (2020) The genome of *Triatoma infestans* (Hemiptera: Reduviidae): the most important Chagas disease vector. https://www.ncbi.nlm.nih.gov/datasets/genome/GCA_011037195.1/
40. Avila ML, Tekiel V, Moretti G et al (2011) Gene discovery in *Triatoma infestans*. Parasit Vectors 4:39. https://doi.org/10.1186/1756-3305-4-39
41. Liu Q, Guo Y, Zhang Y et al (2019) A chromosomal-level genome assembly for the insect vector for Chagas disease, *Triatoma rubrofasciata*. Gigascience 8(8):giz089. https://doi.org/10.1093/gigascience/giz089
42. Patterson JS, Schofield CJ, Dujardin JP et al (2001) Population morphometric analysis of the tropicopolitan bug *Triatoma rubrofasciata* and relationships with Old World species of *Triatoma*: evidence of New World ancestry. Med Vet Entomol 15:443–451. https://doi.org/10.1046/j.0269-283x.2001.00333.x
43. Panzera F, Ferrandis I, Ramsey JM et al (2007) Genome size determination in chagas disease transmitting bugs (Hemiptera-Triatominae) by flow cytometry. Am J Trop Med Hyg 76:516–521. https://doi.org/10.4269/ajtmh.2007.76.516
44. Hieu HV, Do LT, Pita S et al (2019) Biological attributes of the kissing bug *Triatoma rubrofasciata* from Vietnam. Parasit Vectors 12:585. https://doi.org/10.1186/s13071-019-3844-6
45. Panzera F, Pérez R, Hornos S et al (1996) Chromosome numbers in the Triatominae (Hemiptera-Reduviidae): a review. Mem Inst Oswaldo Cruz 91(4):515–518. https://doi.org/10.1590/s0074-02761996000400021
46. Latorre-Estivalis JM, Sterkel M, Ons S et al (2020) Transcriptomics supports local sensory regulation in the antenna of the kissing-bug *Rhodnius prolixus*.

BMC Genomics 21:101. https://doi.org/10.1186/s12864-020-6514-3
47. Barrozo RB, Couton L, Lazzari CR et al (2009) Antennal pathways in the central nervous system of a blood-sucking bug, *Rhodnius prolixus*. Arthropod Struct Dev 38(2):101–110. https://doi.org/10.1016/j.asd.2008.08.004
48. Nevoa JC, Latorre-Estivalis JM, Pais FSM et al (2023) Global characterization of gene expression in the brain of starved immature *Rhodnius prolixus*. PLoS One 18(3):e0282490. https://doi.org/10.1371/journal.pone.0282490
49. Traverso L, Sierra I, Sterkel M et al (2016) Neuropeptidomics in *Triatoma infestans*. Comparative transcriptomic analysis among triatomines. J Physiol Paris 110(3 Part A):83–98. https://doi.org/10.1016/j.jphysparis.2016.12.005
50. Oliveira DS, Brito NF, Nogueira FCS et al (2017) Proteomic analysis of the kissing bug *Rhodnius prolixus* antenna. J Insect Physiol 100:108–118. https://doi.org/10.1016/j.jinsphys.2017.06.004
51. Franco TA, Oliveira DS, Moreira MF et al (2016) Silencing the odorant receptor co-receptor RproOrco affects the physiology and behavior of the Chagas disease vector *Rhodnius prolixus*. Insect Biochem Mol Biol 69:82–90. https://doi.org/10.1016/j.ibmb.2015.02.012
52. Latorre-Estivalis JM, de Oliveira ES, Esteves BB et al (2016) Patterns of expression of odorant receptor genes in a Chagas disease vector. Insect Biochem Mol Biol 69:71–81. https://doi.org/10.1016/j.ibmb.2015.05.002
53. Oliveira DS, Brito NF, Franco TA et al (2018) Functional characterization of odorant binding protein 27 (RproOBP27) from *Rhodnius prolixus* antennae. Front Physiol 9:1175. https://doi.org/10.3389/fphys.2018.01175
54. Martínez-Barnetche J, Lavore A, Beliera M et al (2018) Adaptations in energy metabolism and gene family expansions revealed by comparative transcriptomics of three Chagas disease triatomine vectors. BMC Genomics 19:296. https://doi.org/10.1186/s12864-018-4696-8
55. Marchant A, Mougel F, Almeida C et al (2015) De novo transcriptome assembly for a non-model species, the blood-sucking bug *Triatoma brasiliensis*, a vector of Chagas disease. Genetica 143:225–239. https://doi.org/10.1007/s10709-014-9790-5
56. Marchant A, Mougel F, Jacquin-Joly E et al (2016) Under-expression of chemosensory genes in domiciliary bugs of the Chagas disease vector *Triatoma brasiliensis*. PLoS Negl Trop Dis 10(10):e0005067. https://doi.org/10.1371/journal.pntd.0005067
57. Latorre-Estivalis JM, Omondi BA, DeSouza O et al (2015) Molecular basis of peripheral olfactory plasticity in *Rhodnius prolixus*, a Chagas disease vector. Front Ecol Evol 3:74. https://doi.org/10.3389/fevo.2015.00074
58. Latorre-Estivalis JM, Große-Wilde E, Fernandes GR et al (2022) Changes in antennal gene expression underlying sensory system maturation in *Rhodnius prolixus*. Insect Biochem Mol Biol 140:103704. https://doi.org/10.1016/j.ibmb.2021.103704
59. Marchant A, Mougel F, Mendonça V et al (2016) Comparing de novo and reference-based transcriptome assembly strategies by applying them to the blood-sucking bug *Rhodnius prolixus*. Insect Biochem Mol Biol 69:25–33. https://doi.org/10.1016/j.ibmb.2015.05.009
60. Marchant A, Mougel F, Jacquin-Joly E et al (2021) Chemosensory gene expression for two closely relative species *Rhodnius robustus* and *R. Prolixus* (Hemiptera, Reduviidade, Triatominae) vectors of Chagas disease. Front Ecol Evol 9:725504. https://doi.org/10.3389/fevo.2021.725504
61. de Carvalho DB, Congrains C, Chahad-Ehlers S et al (2017) Differential transcriptome analysis supports *Rhodnius montenegrensis* and *Rhodnius robustus* (Hemiptera, Reduviidae, Triatominae) as distinct species. PLoS One 12(4):e0174997. https://doi.org/10.1371/journal.pone.0174997
62. Latorre Estivalis JM, Traverso L, Pontes G et al (2022) The antennal transcriptome of *Triatoma infestans* reveals substantial expression changes triggered by a blood meal. BMC Genomics 23:861. https://doi.org/10.1186/s12864-022-09059-6
63. Traverso L, Lavore A, Sierra I et al (2017) Comparative and functional triatomine genomics reveals reductions and expansions in insecticide resistance-related gene families. PLoS Negl Trop Dis 11(2):e0005313. https://doi.org/10.1371/journal.pntd.0005313
64. Latorre-Estivalis J, Robertson H, Walden K et al (2017) The molecular sensory machinery of a Chagas disease vector: expression changes through imaginal moult and sexually dimorphic features. Sci Rep 7:40049. https://doi.org/10.1038/srep40049
65. Marliére NP, Latorre-Estivalis JM, Lorenzo MG et al (2015) Trypanosomes modify the behavior of their insect hosts: effects on locomotion and on the expression of a related gene. PLoS Negl Trop Dis 9(8):e0003973. https://doi.org/10.1371/journal.pntd.0003973
66. Marliére NP, Lorenzo MG, Villegas LEM et al (2020) Co-existing locomotory activity and gene expression profiles in a kissing-bug vector of Chagas disease. J Insect Physiol 122:104021. https://doi.org/10.1016/j.jinsphys.2020.104021
67. Ribeiro JM, Andersen J, Silva-Neto MA et al (2004) Exploring the sialome of the blood-sucking bug *Rhodnius prolixus*. Insect Biochem Mol Biol 34(1):61–79. https://doi.org/10.1016/j.ibmb.2003.09.004
68. Champagne DE, Nussenzveig RH, Ribeiro JM (1995) Purification, partial characterization, and cloning of nitric oxide-carrying heme proteins (nitrophorins) from salivary glands of the blood-sucking insect *Rhodnius prolixus*. J Biol Chem 270(15):8691–8695. https://doi.org/10.1074/jbc.270.15.8691

69. Bussacos AC, Nakayasu ES, Hecht MM et al (2011) Diversity of anti-haemostatic proteins in the salivary glands of *Rhodnius* species transmitters of Chagas disease in the greater Amazon. J Proteomics 74(9):1664–1672. https://doi.org/10.1016/j.jprot.2011.06.022
70. Santos A, Ribeiro JM, Lehane MJ et al (2007) The sialotranscriptome of the blood-sucking bug *Triatoma brasiliensis* (Hemiptera, Triatominae). Insect Biochem Mol Biol 37(7):702–712. https://doi.org/10.1016/j.ibmb.2007.04.004
71. Assumpção TC, Francischetti IM, Andersen JF et al (2008) An insight into the sialome of the blood-sucking bug *Triatoma infestans*, a vector of Chagas' disease. Insect Biochem Mol Biol 38(2):213–232. https://doi.org/10.1016/j.ibmb.2007.11.001
72. Charneau S, Junqueira M, Costa CM et al (2007) The saliva proteome of the blood-feeding insect *Triatoma infestans* is rich in platelet-aggregation inhibitors. Int J Mass Spectrom 268(2–3):265–276. https://doi.org/10.1016/j.ijms.2007.05.004
73. Kato H, Jochim RC, Gomez EA et al (2010) A repertoire of the dominant transcripts from the salivary glands of the blood-sucking bug, *Triatoma dimidiata*, a vector of Chagas disease. Infect Genet Evol 10(2):184–191. https://doi.org/10.1016/j.meegid.2009.10.012
74. Assumpção TC, Charneau S, Santiago PB et al (2011) Insight into the salivary transcriptome and proteome of *Dipetalogaster maxima*. J Proteome Res 10(2):669–679. https://doi.org/10.1021/pr100866h
75. Bussacos AC, Nakayasu ES, Hecht MM et al (2011) Redundancy of proteins in the salivary glands of *Panstrongylus megistus* secures prolonged procurement for blood meals. J Proteomics 74(9):1693–1700. https://doi.org/10.1016/j.jprot.2011.04.028
76. Assumpção TCF, Eaton DP, Pham VM et al (2012) An insight into the Sialotranscriptome of *Triatoma matogrossensis*, a kissing bug associated with Fogo Selvagem in South America. Am J Trop Med Hyg 86(6):1005–1014. https://doi.org/10.4269/ajtmh.2012.11-0690
77. Ribeiro JMC, Assumpção TCF, Pham MV (2012) An insight into the Sialotranscriptome of *Triatoma rubida* (Hemiptera: Heteroptera). J Med Entomol 49(3):563–572. https://doi.org/10.1603/ME11243
78. Kato H, Jochim RC, Gomez EA et al (2017) Salivary gland transcripts of the kissing bug, *Panstrongylus chinai*, a vector of Chagas disease. Acta Trop 174:122–129. https://doi.org/10.1016/j.actatropica.2017.06.022
79. Araujo RN, Santos A, Pinto FS (2006) RNA interference of the salivary gland nitrophorin 2 in the triatomine bug *Rhodnius prolixus* (Hemiptera: Reduviidae) by dsRNA ingestion or injection. Insect Biochem Mol Biol 36(9):683–693. https://doi.org/10.1016/j.ibmb.2006.05.012
80. Araujo RN, Soares AC, Paim RM (2009) The role of salivary nitrophorins in the ingestion of blood by the triatomine bug *Rhodnius prolixus* (Reduviidae: Triatominae). Insect Biochem Mol Biol 39(2):83–89. https://doi.org/10.1016/j.ibmb.2008.10.002
81. Araujo RN, Campos IT, Tanaka AS et al (2007) Brasiliensin: a novel intestinal thrombin inhibitor from *Triatoma brasiliensis* (Hemiptera: Reduviidae) with an important role in blood intake. Int J Parasitol 37(12):1351–1358. https://doi.org/10.1016/j.ijpara.2007.04.017
82. Defferrari MS, Orchard I, Lange AB (2016) Identification of the first insulin-like peptide in the disease vector *Rhodnius prolixus*: involvement in metabolic homeostasis of lipids and carbohydrates. Insect Biochem Mol Biol 70:148–159. https://doi.org/10.1016/j.ibmb.2015.12.009
83. Ribeiro JMC, Genta FA, Sorgine MHF et al (2014) An insight into the transcriptome of the digestive tract of the bloodsucking bug, *Rhodnius prolixus*. PLoS Negl Trop Dis 8(1):e2594. https://doi.org/10.1371/journal.pntd.0002594
84. Schwarz A, Medrano-Mercado N, Schaub GA et al (2014) An updated insight into the Sialotranscriptome of *Triatoma infestans*: developmental stage and geographic variations. PLoS Negl Trop Dis 8(12):e3372. https://doi.org/10.1371/journal.pntd.0003372
85. Ribeiro JMC, Schwarz A, Francischetti IMB (2015) A deep insight into the Sialotranscriptome of the Chagas disease vector, *Panstrongylus megistus* (Hemiptera: Heteroptera). J Med Entomol 52(3):351–358. https://doi.org/10.1093/jme/tjv023
86. Hernández-Vargas MJ, Gil J, Lozano L et al (2017) Proteomic and transcriptomic analysis of saliva components from the hematophagous reduviid *Triatoma pallidipennis*. J Proteome 162:30–39. https://doi.org/10.1016/j.jprot.2017.04.022
87. Nevoa JC, Mendes MT, da Silva MV et al (2018) An insight into the salivary gland and fat body transcriptome of *Panstrongylus lignarius* (Hemiptera: Heteroptera), the main vector of Chagas disease in Peru. PLoS Negl Trop Dis 12(2):e0006243. https://doi.org/10.1371/journal.pntd.0006243
88. Mizushima D, Tabbabi A, Yamamoto DS et al (2020) Salivary gland transcriptome of the Asiatic *Triatoma rubrofasciata*. Acta Trop 210:105473. https://doi.org/10.1016/j.actatropica.2020.105473
89. Lopez L, Morales G, Ursic R et al (2003) Isolation and characterization of a novel insect defensin from *Rhodnius prolixus*, a vector of Chagas disease. Insect Biochem Mol Biol 33(4):439–447. https://doi.org/10.1016/s0965-1748(03)00008-0
90. Ouali R, Vieira LR, Salmon D et al (2022) *Rhodnius prolixus* hemolymph immuno-physiology: deciphering the systemic immune response triggered by *Trypanosoma cruzi* establishment in the vector using quantitative proteomics. Cells 11:1449. https://doi.org/10.3390/cells11091449
91. Araújo CAC, Waniek PJ, Stock P et al (2006) Sequence characterization and expression patterns of defensin and lysozyme encoding genes from the gut of the reduviid bug *Triatoma brasiliensis*. Insect

Biochem Mol Biol 36(7):547–560. https://doi.org/10.1016/j.ibmb.2006.04.003
92. Gumiel M, Mattos DP, Vieira CS et al (2020) Proteome of the Triatomine digestive tract: from catalytic to immune pathways; focusing on Annexin expression. Front Mol Biosci 7:589435. https://doi.org/10.3389/fmolb.2020.589435
93. Valanne S, Wang JH, Rämet M (2011) The *Drosophila* toll signaling pathway. J Immunol 186(2):649–656. https://doi.org/10.4049/jimmunol.1002302
94. Kleino A, Silverman N (2014) The *Drosophila* IMD pathway in the activation of the humoral immune response. Dev Comp Immunol 42(1):25–35. https://doi.org/10.1016/j.dci.2013.05.014
95. Myllymäki H, Valanne S, Rämet M (2014) The *Drosophila* Imd signaling pathway. J Immunol 192(8):3455–3462. https://doi.org/10.4049/jimmunol.1303309
96. Morin-Poulard I, Vincent A, Crozatier M (2013) The *Drosophila* JAK-STAT pathway in blood cell formation and immunity. JAKSTAT 2:3. https://doi.org/10.4161/jkst.25700
97. Wang XH, Aliyari R, Li WX et al (2006) RNA interference directs innate immunity against viruses in adult *Drosophila*. Science 312(5772):452–454. https://doi.org/10.1126/science.1125694
98. Salcedo-Porras N, Guarneri A, Oliveira PL et al (2019) *Rhodnius prolixus*: identification of missing components of the IMD immune signaling pathway and functional characterization of its role in eliminating bacteria. PLoS One 14(4):e0214794. https://doi.org/10.1371/journal.pone.0214794
99. Zumaya-Estrada FA, Martínez-Barnetche J, Lavore A et al (2018) Comparative genomics analysis of triatomines reveals common first line and inducible immunity-related genes and the absence of Imd canonical components among hemimetabolous arthropods. Parasit Vectors 11:48. https://doi.org/10.1186/s13071-017-2561-2
100. Ursic-Bedoya RJ, Lowenberger CA (2007) *Rhodnius prolixus*: identification of immune-related genes up-regulated in response to pathogens and parasites using suppressive subtractive hybridization. Dev Comp Immunol 31(2):109–120. https://doi.org/10.1016/j.dci.2006.05.008
101. Castro DP, Moraes CS, Gonzalez MS et al (2012) *Trypanosoma cruzi* immune response modulation decreases microbiota in *Rhodnius prolixus* gut and is crucial for parasite survival and development. PLoS One 7(5):e36591. https://doi.org/10.1371/journal.pone.0036591
102. Vieira CS, Mattos DP, Waniek PJ et al (2015) *Rhodnius prolixus* interaction with *Trypanosoma rangeli*: modulation of the immune system and microbiota population. Parasit Vectors 8:135. https://doi.org/10.1186/s13071-015-0736-2
103. Vieira CS, Moreira OC, Batista KKS et al (2018) The NF-κB inhibitor, IMD-0354, affects immune gene expression, bacterial microbiota and *Trypanosoma cruzi* infection in *Rhodnius prolixus* midgut. Front Physiol 9:1189. https://doi.org/10.3389/fphys.2018.01189
104. Carmona-Peña SP, Vázquez-Chagoyán JC, Castro DP et al (2022) Benefits and costs of immune memory in *Rhodnius prolixus* against *Trypanosoma cruzi*. Microb Pathog 165:105505. https://doi.org/10.1016/j.micpath.2022.105505
105. Schama R, Pedrini N, Juárez MP et al (2016) *Rhodnius prolixus* supergene families of enzymes potentially associated with insecticide resistance. Insect Biochem Mol Biol 69:91–104. https://doi.org/10.1016/j.ibmb.2015.06.005
106. Dermauw W, Van Leeuwen T (2014) The ABC gene family in arthropods: comparative genomics and role in insecticide transport and resistance. Insect Biochem Mol Biol 45:89–110. https://doi.org/10.1016/j.ibmb.2013.11.001
107. Hemingway J (2000) The molecular basis of two contrasting metabolic mechanisms of insecticide resistance. Insect Biochem Mol Biol 30(11):1009–1015. https://doi.org/10.1016/s0965-1748(00)00079-5
108. Varela GM, Stroppa MM, García BA (2019) Daily variations in the expression of genes related to insecticide resistance in the Chagas disease vector *Triatoma infestans* (Hemiptera: Reduviidae). Am J Trop Med Hyg 100(6):1482–1485. https://doi.org/10.4269/ajtmh.19-0155
109. Calderón-Fernández GM, Moriconi DE, Dulbecco AB et al (2017) Transcriptome analysis of the *Triatoma infestans* (Hemiptera: Reduviidae) integument. J Med Entomol 54(6):1531–1542. https://doi.org/10.1093/jme/tjx151
110. Dulbecco AB, Moriconi DE, Calderón-Fernández GM et al (2018) Integument CYP genes of the largest genome-wide cytochrome P450 expansions in triatomines participate in detoxification in deltamethrin-resistant *Triatoma infestans*. Sci Rep 8:10177. https://doi.org/10.1038/s41598-018-28475-x
111. Capriotti N, Mougabure-Cueto G, Rivera-Pomar R et al (2014) L925I mutation in the Para-type Sodium Channel is associated with Pyrethroid resistance in *Triatoma infestans* from the Gran Chaco region. PLoS Negl Trop Dis 8(1):e2659. https://doi.org/10.1371/journal.pntd.0002659
112. Fabro J, Sterkel M, Capriotti N et al (2012) Identification of a point mutation associated with pyrethroid resistance in the para-type sodium channel of *Triatoma infestans*, a vector of Chagas' disease. Infect Genet Evol 12(2):487–491. https://doi.org/10.1016/j.meegid.2011.12.006
113. Sierra I, Capriotti N, Fronza G et al (2016) Kdr mutations in *Triatoma infestans* from the Gran Chaco are distributed in two differentiated foci: implications for pyrethroid resistance management. Acta Trop 158:208–213. https://doi.org/10.1016/j.actatropica.2016.03.014
114. Pedrini N, Mijailovsky SJ, Girotti JR et al (2009) Control of Pyrethroid-resistant Chagas disease vectors with entomopathogenic fungi. PLoS Negl

Trop Dis 3(5):e434. https://doi.org/10.1371/journal.pntd.0000434
115. Brito T, Julio A, Berni M et al (2018) Transcriptomic and functional analyses of the piRNA pathway in the Chagas disease vector *Rhodnius prolixus*. PLoS Negl Trop Dis 12(10):e0006760. https://doi.org/10.1371/journal.pntd.0006760
116. Coelho VL, de Brito TF, de Abreu Brito IA et al (2021) Analysis of ovarian transcriptomes reveals thousands of novel genes in the insect vector *Rhodnius prolixus*. Sci Rep 11:1918. https://doi.org/10.1038/s41598-021-81387-1
117. Leyria J, Orchard I, Lange AB (2020) What happens after a blood meal? A transcriptome analysis of the main tissues involved in egg production in *Rhodnius prolixus*, an insect vector of Chagas disease. PLoS Negl Trop Dis 14(10):e0008516. https://doi.org/10.1371/journal.pntd.0008516
118. Leyria J, Orchard I, Lange AB (2020) Transcriptomic analysis of regulatory pathways involved in female reproductive physiology of *Rhodnius prolixus* under different nutritional states. Sci Rep 10:11431. https://doi.org/10.1038/s41598-020-67932-4
119. Lavore A, Pascual A, Salinas FM et al (2015) Comparative analysis of zygotic developmental genes in *Rhodnius prolixus* genome shows conserved features on the tracheal developmental pathway. Insect Biochem Mol Biol 64:32–43. https://doi.org/10.1016/j.ibmb.2015.06.012
120. Mérel V, Boulesteix M, Fablet M et al (2020) Transposable elements in *Drosophila*. Mob DNA 11:23. https://doi.org/10.1186/s13100-020-00213-z
121. Wells JN, Feschotte C (2020) A field guide to eukaryotic transposable elements. Ann Rev Genet 54:539–561. https://doi.org/10.1146/annurev-genet-040620-022145
122. Sneideman MPL, Meller VH (2021) *Drosophila* satellite repeats at the intersection of chromatin, gene regulation and evolution. In: Ugarković Đ (ed) Satellite DNAs in physiology and evolution, Progress in molecular and subcellular biology, vol 60. Springer, Cham. https://doi.org/10.1007/978-3-030-74889-0_1
123. Montiel EE, Panzera F, Palomeque T et al (2021) Satellitome analysis of *Rhodnius prolixus*, one of the main Chagas disease vector species. Int J Mol Sci 22:6052. https://doi.org/10.3390/ijms22116052
124. Ferreira D, Meles S, Escudeiro A et al (2015) Satellite non-coding RNAs: the emerging players in cells, cellular pathways and cancer. Chromosom Res 23:479–493. https://doi.org/10.1007/s10577-015-9482-8
125. Pezer Ž, Brajković J, Feliciello I et al (2011) Transcription of satellite DNAs in insects. In: Ugarkovic D (ed) Long non-coding RNAs, Progress in molecular and subcellular biology, vol 51. Springer, Berlin. https://doi.org/10.1007/978-3-642-16502-3_8
126. de Lima LG, Ruiz-Ruano FJ (2022) In-depth satellitome analyses of 37 *Drosophila* species illuminate repetitive DNA evolution in the *Drosophila* genus. Genome Biol Evol 14:evac064. https://doi.org/10.1093/gbe/evac064
127. Ferree PM, Prasad S (2012) How can satellite DNA divergence cause reproductive isolation? Let us count the chromosomal ways. Genet Res Int 2012:e430136. https://doi.org/10.1155/2012/430136
128. Hughes SE, Hawley RS (2009) Heterochromatin: a rapidly evolving species barrier. PLoS Biol 7:e1000233. https://doi.org/10.1371/journal.pbio.1000233
129. Jagannathan M, Yamashita YM (2021) Defective satellite DNA clustering into chromocenters underlies hybrid incompatibility in *Drosophila*. Mol Biol Evol 38:4977–4986. https://doi.org/10.1093/molbev/msab221
130. Satovic E, Cvitanic MT, Plohl M (2020) Tools and databases for solving problems in detection and identification of repetitive DNA sequences. Period Biol. https://doi.org/10.18054/pb.v121-122i1-2.10571
131. Maumus F, Quesneville H (2014) Deep investigation of *Arabidopsis thaliana* junk DNA reveals a continuum between repetitive elements and genomic dark matter. PLoS One 9:e94101. https://doi.org/10.1371/journal.pone.0094101
132. Pita S, Panzera F, Mora P et al (2017) Comparative repeatome analysis on *Triatoma infestans* Andean and non-Andean lineages, main vector of Chagas disease. PLoS One 12:e0181635. https://doi.org/10.1371/journal.pone.0181635
133. Fry K, Salser W (1977) Nucleotide sequences of HS-alpha satellite DNA from kangaroo rat *Dipodomys ordii* and characterization of similar sequences in other rodents. Cell 12:1069–1084. https://doi.org/10.1016/0092-8674(77)90170-2
134. Mestrović N, Plohl M, Mravinac B, Ugarković Đ (1998) Evolution of satellite DNAs from the genus *Palorus*DOUBLEHYPHENexperimental evidence for the "library" hypothesis. Mol Biol Evol 15:1062–1068. https://doi.org/10.1093/oxfordjournals.molbev.a026005
135. Panzera F, Pita S, Lorite P et al (2021) Chromosome structure and evolution of Triatominae: a review. In: Guarneri A, Lorenzo M (eds) Triatominae - the biology of Chagas disease vectors, Entomology in focus, vol 5. Springer, Cham. https://doi.org/10.1007/978-3-030-64548-9_4
136. Mora P, Pita S, Montiel EE et al (2023) Making the genome huge: the case of *Triatoma delpontei*, a Triatominae species with more than 50% of its genome full of satellite DNA. Genes 14:371. https://doi.org/10.3390/genes14020371
137. Castro MRJ, Goubert C, Monteiro FA et al (2020) Homology-free detection of transposable elements unveils their dynamics in three ecologically distinct *Rhodnius* species. Genes 11:170. https://doi.org/10.3390/genes11020170
138. Ferreira D, Escudeiro A, Adega F et al (2019) DNA methylation patterns of a satellite non-coding sequence—FA-SAT in cancer cells: its expression cannot be explained solely by DNA methyla-

tion. Front Genet 10:101. https://doi.org/10.3389/fgene.2019.00101
139. Halbach R, Miesen P, Joosten J et al (2020) A satellite repeat-derived piRNA controls embryonic development of *Aedes*. Nature 580:274–277. https://doi.org/10.1038/s41586-020-2159-2
140. Louzada S, Lopes M, Ferreira D et al (2020) Decoding the role of satellite DNA in genome architecture and plasticity-an evolutionary and clinical affair. Genes 11:72. https://doi.org/10.3390/genes11010072
141. Steenwyk JL, Li Y, Zhou X et al (2023) Incongruence in the phylogenomics era. Nat Rev Genet 24:834–850. https://doi.org/10.1038/s41576-023-00620-x
142. Delsuc F, Brinkmann H, Philippe H (2005) Phylogenomics and the reconstruction of the tree of life. Nat Rev Genet 6:361–375. https://doi.org/10.1038/nrg1603
143. Rokas A, Williams B, King N et al (2003) Genome-scale approaches to resolving incongruence in molecular phylogenies. Nature 425:798–804. https://doi.org/10.1038/nature02053
144. Cheon S, Zhang J, Park C (2020) Is Phylotranscriptomics as reliable as phylogenomics? Mol Biol Evol 37(12):3672–3683. https://doi.org/10.1093/molbev/msaa181
145. Pavan MG, Lazoski C, Monteiro FA (2021) Speciation processes in Triatominae. In: Guarneri A, Lorenzo M (eds) Triatominae - the biology of Chagas disease vectors, Entomology in focus, vol 5. Springer, Cham. https://doi.org/10.1007/978-3-030-64548-9_3
146. Rokas A, Carroll SB (2005) More genes or more taxa? The relative contribution of gene number and taxon number to phylogenetic accuracy. Mol Biol Evol 22(5):1337–1344. https://doi.org/10.1093/molbev/msi121
147. Jeffroy O, Brinkmann H, Delsuc F (2006) Phylogenomics: the beginning of incongruence? Trends Genet 22(4):225–231. https://doi.org/10.1016/j.tig.2006.02.003
148. Brito RN, Geraldo JA, Monteiro FA et al (2019) Transcriptome-based molecular systematics: *Rhodnius montenegrensis* (Triatominae) and its position within the *Rhodnius prolixus–Rhodnius robustus* cryptic–species complex. Parasit Vectors 12:305. https://doi.org/10.1186/s13071-019-3558-9
149. Hernandez-Castro LE, Paterno M, Villacís AG, Andersson B, Costales JA et al (2017) 2b-RAD genotyping for population genomic studies of Chagas disease vectors: *Rhodnius ecuadoriensis* in Ecuador. PLoS Negl Trop Dis 11(7):e0005710. https://doi.org/10.1371/journal.pntd.0005710
150. Abad-Franch F, Monteiro FA, Pavan MG et al (2021) Under pressure: phenotypic divergence and convergence associated with microhabitat adaptations in Triatominae. Parasit Vectors 14:195. https://doi.org/10.1186/s13071-021-04647-z
151. Hernandez-Castro LE, Villacís AG, Jacobs A et al (2022) Population genomics and geographic dispersal in Chagas disease vectors: landscape drivers and evidence of possible adaptation to the domestic setting. PLoS Genet 18(2):e1010019. https://doi.org/10.1371/journal.pgen.1010019
152. Kieran TJ, Bayona-Vásquez NJ, Varian CP (2020) Population genetics of two chromatic morphs of the Chagas disease vector *Rhodnius pallescens* Barber, 1932 in Panamá. Infect Genet Evol 84:104369. https://doi.org/10.1016/j.meegid.2020.104369
153. Gómez-Palacio A, Jaramillo-Ocampo N, Triana-Chávez O et al (2008) Chromosome variability in the Chagas disease vector *Rhodnius pallescens* (Hemiptera, Reduviidae, Rhodniini). Mem Inst Oswaldo Cruz 103(2):160–164. https://doi.org/10.1590/s0074-02762008000200006
154. Gómez-Palacio A, Jaramillo-O N, Caro-Riaño H et al (2012) Morphometric and molecular evidence of intraspecific biogeographical differentiation of *Rhodnius pallescens* (HEMIPTERA: REDUVIIDAE: RHODNIINI) from Colombia and Panama. Infect Genet Evol 12(8):1975–1983. https://doi.org/10.1016/j.meegid.2012.04.003
155. Orantes LC, Monroy C, Dorn PL et al (2018) Uncovering vector, parasite, blood meal and microbiome patterns from mixed-DNA specimens of the Chagas disease vector *Triatoma dimidiata*. PLoS Negl Trop Dis 12(10):e0006730. https://doi.org/10.1371/journal.pntd.0006730
156. Cahan SH, Orantes LC, Wallin KF et al (2019) Residual survival and local dispersal drive reinfestation by *Triatoma dimidiata* following insecticide application in Guatemala. Infect Genet Evol 74:104000. https://doi.org/10.1016/j.meegid.2019.104000
157. Justi SA, Cahan S, Stevens L et al (2018) Vectors of diversity: genome wide diversity across the geographic range of the Chagas disease vector *Triatoma dimidiata* sensu lato (Hemiptera: Reduviidae). Mol Phylogenet Evol 120:144–150. https://doi.org/10.1016/j.ympev.2017.12.016
158. Dorn PL, Monroy C, Curtis A (2007) *Triatoma dimidiata* (Latreille, 1811): a review of its diversity across its geographic range and the relationship among populations. Infect Genet Evol 7(2):343–352. https://doi.org/10.1016/j.meegid.2006.10.001
159. Dorn PL, de la Rúa NM, Axen H et al (2016) Hypothesis testing clarifies the systematics of the main central American Chagas disease vector, *Triatoma dimidiata* (Latreille, 1811), across its geographic range. Infect Genet Evol 44:431–443. https://doi.org/10.1016/j.meegid.2016.07.046
160. Lima-Cordón RA, Monroy MC, Stevens L et al (2019) Description of *Triatoma huehuetenanguensis* sp. n., a potential Chagas disease vector (Hemiptera, Reduviidae, Triatominae). ZooKeys 820:51–70. https://doi.org/10.3897/zookeys.820.27258
161. Dorn PL, Justi SA, Dale C et al (2018) Description of *Triatoma mopan* sp. n. from a cave in Belize (Hemiptera, Reduviidae, Triatominae). ZooKeys 775:69–95. https://doi.org/10.3897/zookeys.775.22553

162. Viana MC, Alves-Pereira A, Oliveira MAP et al (2024) Population genetics and genomics of *Triatoma brasiliensis* (Hemiptera, Reduviidae) in an area of high pressure of domiciliary infestation in northeastern Brazil. Acta Trop 252:107144. https://doi.org/10.1016/j.actatropica.2024.107144
163. San Juan E, Araya-Donoso R, Véliz D et al (2021) Genetic diversity in a restricted-dispersal kissing bug: the Centre–periphery hypothesis halfway. Mol Ecol 30:4660–4672. https://doi.org/10.1111/mec.16093
164. Eckert CG, Samis KE, Lougheed SC (2008) Genetic variation across species' geographical ranges: the central–marginal hypothesis and beyond. Mol Ecol 17:1170–1188. https://doi.org/10.1111/j.1365-294X.2007.03659.x
165. Belintani T, Congrains C, Pinotti H et al (2023) Transcriptome-based phylogenomic analysis reveals evidence of ancient hybridization events in the *Mepraia* genus (Hemiptera: Reduviidae: Triatominae). Front Ecol Evol 11:1215319. https://doi.org/10.3389/fevo.2023.1215319
166. Weirauch C (2021) Origin and evolution of Triatominae. In: Guarneri A, Lorenzo M (eds) Triatominae - the biology of Chagas disease vectors, Entomology in focus, vol 5. Springer, Cham. https://doi.org/10.1007/978-3-030-64548-9_1
167. Hwang WS, Weirauch C (2012) Evolutionary history of assassin bugs (Insecta: Hemiptera: Reduviidae): insights from divergence dating and ancestral state reconstruction. PLoS One 7(9):e45523. https://doi.org/10.1371/journal.pone.0045523
168. Georgieva AY, Gordon ERL, Weirauch C (2017) Sylvatic host associations of Triatominae and implications for Chagas disease reservoirs: a review and new host records based on archival specimens. PeerJ 5:e3826. https://doi.org/10.7717/peerj.3826

Studying the Interactions Between Microbiomes and *Leishmania* Parasites in Sand Flies: A Source of New Targets for Pathogen Control

Rafael José Vivero-Gomez,
Daniel Fernando Largo, Gloria Cadavid-Restrepo,
Daniela Duque-Granda,
and Claudia Ximena Moreno-Herrera

Abstract

The sand fly gut microbiota has recently emerged as an encouraging field to be explored for vector-based disease control. Previous studies reported that variation in the microbiota residing in the insect gut might be mainly explained by the effects of host habitat, diet, developmental stage, and phylogeny, all contributing to the structure of insect gut microbiota. However, a better understanding of vector–microbiota–pathogen interactions is vital, as it can lead to the discovery of new tools to block disease transmission. Diverse data suggest that the sand fly midgut microbiome is a critical factor for *Leishmania* growth and differentiation to its infective state prior to disease transmission. The review focuses on the background of microbiota and *Leishmania*—sand fly interactions, which have been studied for some vectors such as *Lutzomyia, Pintomyia*, and *Phlebotomus*, among others. On the other hand, we also consider information about how endosymbionts (such as *Wolbachia, Cardinium, Microsporidia, Arsenophonus, Spiroplasma, Rickettsia*) could influence the establishment of parasites and other pathogens such as arboviruses in sand flies. The lack of systematic studies regarding the detection, diversity, endosymbiont phylogeny, and evaluation of their impact on pathogens in phlebotomine sand flies needs further exploration. The information currently generated has allowed the development of in vitro and in vivo studies of co-infections with different species of parasites, bacteria, and sand flies, as well as the impact of antibiotic treatment, which enhances the development of paratransgenesis studies for the control of the disease transmission. Finally, the microbiota also plays a role in environmental effects caused by variables of climate change (e.g., thermotolerance) and other variables such as insecticide resistance that influence the distribution and survival of sand flies and can have a significant effect on the development of parasite promastigotes of *Leishmania*, thus favoring their transmission in new territories. Deepening and increasing knowledge about the sand flies–*Leishmania*–

R. J. Vivero-Gomez (✉) · D. F. Largo ·
G. Cadavid-Restrepo · D. Duque-Granda ·
C. X. Moreno-Herrera
Grupo de Microbiodiversidad y Bioprospección,
Laboratorio de Biología Celular y Molecular,
Universidad Nacional de Colombia sede Medellín,
Medellín, Colombia
e-mail: rjviverog@unal.edu.co

J. D. Ramírez González (ed.), *Recent Advances in Parasitomics*,
https://doi.org/10.1007/978-3-031-70591-5_16

microbiome and human host interactions, as well as the influence of climate change and the dynamics of blood intake, could help us to have a better understanding and to impact key objectives of the interaction, allowing the development of sustainable and cost-effective strategies to interrupt the transmission of leishmaniasis.

Keywords

Sand fly · Gut microbiota · *Leishmania* · Endosymbionts · In vitro and in vivo co-infections · Paratransgenesis · Vector competence

1 Background

Leishmaniasis is caused by a protozoan parasite from over 20 *Leishmania* species, with approximately 70 distinct domestic and sylvatic mammal species acting as potential reservoirs of the infection [1, 2]. The incidence of leishmaniasis can be affected by changes in urbanization, deforestation, or human incursion into forested areas [3, 4]. Climate change further exacerbates the spread of leishmaniasis through alterations in temperature and rainfall, which impact the size and geographic distribution of sand fly populations [5]. Currently, leishmaniasis transmission is reported in 98 countries and 3 territories across 5 continents, with over 1 billion people living in endemic areas at risk of infection [1]. An estimated 30,000 new cases of visceral leishmaniasis (VL) and more than 1 million new cases of cutaneous leishmaniasis (CL) occur annually [2]. Despite these alarming figures, there is currently no vaccine available, and treatment with antiparasitic drugs is mainly injectable, which is often unsuccessful and complex and should be administered by highly experienced health personnel.

The principal vectors of *Leishmania* (*Le.*) parasites are phlebotomine sand flies (Diptera: Phlebotominae). Within the vector, *Leishmania* undergoes a complex developmental process through several morphologically and functionally distinct forms [6–9]. However, of the almost 900 described phlebotomine species, only about 10% (98 species) are listed as proven or suspected vectors of *Leishmania*, with 42 *Phlebotomus* (*Ph.*) species in the Old World and 56 *Lutzomyia* (*Lu.*) species in the New World [9]. The temporal and spatial distribution of sand flies depends on local geographical and climatic factors, and while they are primarily recognized as vectors for *Leishmania*, they may also transmit other pathogens such as bacteria and viruses [10].

The vector competence of sand flies depends on several factors, including their preference for feeding on humans (anthropophilic) or other relevant reservoir hosts (zoonotic agents), and their ability to become infected with the same *Leishmania* species occurring in mammals from the same geographical area [11, 12]. Studies have demonstrated that species-specific differences in vectorial competence are directly correlated with the ability of promastigotes to attach to the sand fly midgut [10, 11, 13]. The ability of the parasite to resist or inhibit killing by proteolytic enzymes released into the midgut soon after blood feeding, and its capacity to maintain infection in the midgut during excretion of the digested blood, can determine the specificity of parasite–sand fly interactions [13]. Some sand fly species are considered restrictive or specific vectors, transmitting only particular species of *Leishmania*, such as *Ph. papatasi* and *Le. major* [14]. Other sand fly species are considered permissive or nonspecific, harboring experimental infections of several *Leishmania* species (e.g., *Lu. longipalpis* with *Le. infantum chagasi* or *Le. mexicana*) [12, 14]. Moreover, vector permissiveness suggests broader or even nonspecific binding processes [10].

Leishmania is not alone in the sand fly midgut; it interacts with a rich and diverse microbiome consisting of viruses, bacteria, fungi, and genes, with bacterial predominance (microbiota or bacteriome), which naturally colonizes the gut. Similar to other insect vectors, molecules produced by specific bacteria or resulting from interactions between bacterial consortiums may directly or indirectly interfere with *Leishmania* development within the sand fly. Some studies also indicate that host genetic factors can influence microbial composition to some extent, potentially by changing nutrient availability through dietary choices, by host metabolism, or by triggering immune factors [15, 16].

The natural gut microbiota of adult sand flies is acquired from various sources, including plants sugars and blood from a wide range of hosts, or through re-colonization of the gut by microbes ingested by terrestrial-dwelling larval stages [17–19]. The microbiota sand fly also varies through their life stages and can affect their biology and fitness parameters, such as longevity, survival, and behavior. Additionally, the intestinal microbiota may play a role in thermotolerance and potentially in insecticide resistance [20, 21]. Finally, the gut microbiota of sand flies significantly impacts vector competence and disease transmission. Recent studies have shown a connection between bacteria deposited during *Leishmania*-infected sand fly bites and the clinical outcomes of leishmaniasis, suggesting that the gut microbes of *Lu. longipalpis* are egested into the host skin alongside *Leishmania* triggering neutrophil infiltration and facilitating parasitic establishment [16, 22, 23].

The role of gut microorganisms in controlling vector-borne diseases is highly relevant and offers several perspectives or targets [24]. These include anti-microbiota vaccines as a tool to manipulate vector microbiota and interfere with pathogen development within their vectors [25], populations transformed (genetically modified or not) using intracellular symbionts like *Wolbachia* or Microsporidia [26, 27] to assess how microbiota affect phenotypes impacting vectorial capacity, and the evaluation of bacteria with enzymatic and antiparasitic bioactivities or toxic metabolites that can be used in vitro and in vivo for the design of paratransgenesis strategies to control *Leishmania* transmission [28–30], in addition to the discovery of molecules with insecticide (biopesticides) activity from cultivable bacteria of sand flies that can provide information on their mechanisms of action. Biotechnological approaches using microorganisms have significant potential to control insect vector populations and reduce their vector competence, serving as alternatives to synthetic insecticides [28].

A better understanding of vector–microbiota–pathogen interactions, particularly regarding the impact of host physiology, vector density, biting rate, survival, insecticide resistance, immune responses, and vector competence, is essential for discovering new tools to block disease transmission and provide critical information for the development of intervention strategies for sand fly control. Advances in omics technologies can help unravel this complexity, but microorganisms (culturable or intracellular bacterial) can also be selected based on known forms of transmission or inheritance (vertical, transstadial, venereal, transovarial) and their evolutionary relationships (intraspecific and interspecific).

In this chapter, we highlight some of the most important findings on the bacteria composition and diversity in the gut of sand flies, the presence of bacterial endosymbionts, the role of gut microbiota during the development of *Leishmania* using in vitro and in vivo models, and the relationship of the microbiota and its possible role in response to environmental variables. Additional themes related to current and future strategies using the microbiome to prevent transmission of *Leishmania* parasites by targeting the vector are also discussed. A systematic and in-depth literature search was conducted by screening several online databases (National Center for Biotechnology Information, Web of Science, Science Direct, PubMed, Scopus, Wiley Online Library, and Google Scholar). The search involved combining "sand flies" (in singular and plural form) with terms such as "microbiota," "microbiome," "paratransgenesis," "vector control," "biotechnological approaches," "*Leishmania* interactions," "co-infection," "antibiotic treatment," "skin microbiome," "virome," "skin," "climate change variables," "insecticides," "insecticide resistance," "endosymbiont," "biological control," "reproductive manipulator bacteria," "endosymbiont *Leishmania* interaction," "thermotolerance."

2 Microbiota: Composition and Diversity in the Sand Fly Gut

Microbiota plays a crucial role in vector competence and in the development of infective *Leishmania* species in the gut of the sand fly, also influencing pathogen resistance, nutrition, oviposition, and larval development [31–34]. Furthermore, sand flies are capable of mounting general and specific humoral immune responses to the presence of bacteria [35, 36].

This is the primary reason why the association between the microbiota and sand flies has been studied, employing both culture-dependent and culture-independent methodologies. However, the culture-dependent method is inherently limited due to the restricted range of microorganisms that can be cultivated in a laboratory setting. Despite this limitation, obtaining isolates or their metabolite extracts for assays directed against the parasite in vitro is important. Recent advances in molecular methods in microbial ecology and sequencing platforms have significantly expanded our understanding of sand flies' microbiome, offering a more comprehensive view [31, 37]. It is important to synergize the benefits of both methodologies to develop alternatives needed for biological control and to progress toward the development of a paratransgenic approach. Table 1 provides a summary of studies that characterize microbiota in relevant sand fly species which are further discussed in this chapter.

Table 1 Summary of studies characterizing main bacterial taxa by sand fly species and their taxonomic assignment methods

Sand fly species	Characteristic taxa	Methods	References
Ph. duboscqi	*Rhizobiaceae, Ochrobactrum, Wolbachia*	16S rRNA gene sequencing (V3–V4 regions) and ITS fungal identification	[16]
	Rahnella, Ochrobactrum, Bradyrhizobium, Leifsonia, Yersinia	16S rRNA gene sequencing (V1–V4 region)	[38]
	Microbacterium, Chloroflexi	16S rRNA gene sequencing (V6–V8 region) and TGGE	[39]
	Ochrobactrum, Serratia, Stenotrophomonas	Bacteriological identification and RAPD	[40]
Ph. papatasi	*Rhizobiaceae, Serratia, Wolbachia*	16S rRNA gene sequencing (V3–V4 region) and ITS fungal identification	[16]
	Spiroplasma, Wolbachia	16S rRNA gene sequencing (V4 region)	[41]
	Microbacteriaceae, Staphylococcaceae, Xanthomonadaceae	16S rRNA gene sequencing (fragment of 1.5 kbp) and gyrB	[42]
	Bacillus, Brevibacillus, Solibacillus, Paenibacillus, Terribacillus	16S rRNA gene sequencing	[43]
	Pseudomonas, Bacillus, Serratia, Micrococcus, Enterobacter	Bacteriological identification	[44]
	Microbacterium	16S rRNA gene sequencing (V6-V8 region) and TGGE	[39]
	Enterobacter, Aeromonas, Erwinia, Pseudomonas	Bacteriological identification	[45]
Ph. tobbi	*Wolbachia* *Cutibacterium, Rhizobiaceae, Pseudomonas, Staphylococcus*	16S rRNA gene sequencing (V4 region)	[41]
Ph. neglectus	*Wolbachia* *Staphylococcus*	16S rRNA gene sequencing (V4 region)	[41]
Ph. similis	*Cutibacterium, Rhizobiaceae, Pseudomonas, Staphylococcus*	16S rRNA gene sequencing (V4 region)	[41]

(continued)

Table 1 (continued)

Sand fly species	Characteristic taxa	Methods	References
Ph. perniciosus	*Stenotrophomonas, Staphylococcus, Bacillus, Serratia, Enterococcus, Veillonella*	16S rRNA gene (V3–V5 regions) and DGGE	[46]
Ph. sergenti	*Enterobacter, Acinetobacter*	Bacteriological identification	[44]
Ph. kandelakii	*Escherichia*	Bacteriological identification	[44]
Ph. perfiliewi	*Enterobacter, Escherichia*	Bacteriological identification	[44]
Ph. halepensis	*Klebsiella, Micrococcus*	Bacteriological identification	[44]
Lu. longipalpis	*Achromobacter, Tsukamurella, Staphylococcus, and Stenotrophomonas.*	16S rRNA gene sequencing (V3–V4 regions) and ITS fungal identification	[16]
	Bacillus, Enterococcus, Lysinibacillus, Staphylococcus, Erwinia, Enterobacter, Escherichia and Providencia	16S rRNA gene sequencing (V1–V9 regions)	[47]
	Serratia, Enterobacter, Pseudomonas, Stenotrophomonas, Enhydrobacter, Chryseobacterium.	16S rRNA gene (V3–V4 regions)	[48]
	Enterobacteriaceae, Pseudomonadaceae, Acetobacteraceae	16S rRNA gene (V4 region)	[49]
	Ralstonia, Erwinia, Bradyrhizobium	16S rRNA gene sequencing and DGGE	[50]
	Serratia, Enterobacter, Pseudomonas, Stenotrophomonas, Yokenella, Acinetobacter	Bacteriological identification	[51]
	Acinetobacter, Stenotrophomonas, Flavimonas, Enterobacter, Klebsiella, Bacillus, Staphylococcus	Bacteriological identification	[52]
Pi. evansi	*Pseudomonas, Acinetobacter, Enterobacter, Ochrobactrum, Microbacterium, Streptomyces, Bacillus, Lysobacter, Rummeliibacillus*	16S rRNA gene sequencing and DGGE (V3–V6 regions), TGGE	[18]
	Methylobacterium, Staphylococcus, Oxalobacteriaceae, Pseudomonadaceae	16S rRNA gene sequencing (V4 region)	[53]
Lu. ayacuchensis	*Ochrobactrum, Micrococcus, Corynebacterium, Lawsonella, Cutibacterium*	16S rRNA gene sequencing (V3–V4 regions)	[54]
Lu. cruzi	*Ralstonia, Saccharibacter, Staphylococcus, Chryseobacterium*	16S rRNA gene sequencing and DGGE	[50]
Ny. intermedia	*Ochrobactrum, Bradyrhizobium, Pseudomonas*	16S rRNA gene sequencing (V1–V3 regions)	[17]
Pi. evansi	*Ralstonia, Bacillus, Burkholderia, Paraburkholderia, Vibrionimonas*	16S rRNA gene sequencing (V4 region)	[34, 55]

Culture-dependent methods were the first ones used to identify bacteria present in the gut of sand flies [29]. These seminal studies combined the processing of guts in sterile conditions with the isolation in culture media or variations intended to facilitate the growth of this kind of bacteria [44]. Efforts to widen the spectrum of isolates allowed the identification of characteristic genera and species in epidemiologically important sand flies [46]. For instance, in *Ph. papatasi,* culture-dependent analysis of its gut microbiota was performed using standard biochemical bacteriological methods [45]. From 17 bacterial species isolated, *Enterobacter cloacae* was the dominant species and the Enterobacteriaceae family the most abundant. In the case of wild-captured *Lu. longipalpis* from Brazil, the culture-dependent studies have shown the predominance of 8 species belonging to the Enterobacteriaceae family: *Acinetobacter lwoffii, Stenotrophomonas maltophilia, Pseudomonas putida, Flavimonas oryzihabitans, Enterobacter cloacae, Klebsiella ozaenae, Bacillus thuringiensis,* and *Staphylococcus* spp. [52]. These studies were limited by the biochemical and morphological information available and could have biased the ideal identification of the bacteria present in the sand fly gut. Other studies with similar methodologies are found for colony-reared *Lu. longipalpis* fed with blood and sucrose [56], *Ph. papatasi, Ph. sergenti, Ph. kandelakii, Ph. Perfiliewi*, and *Ph. halepensis* [44], and *Ph. duboscqi* [40]. These studies are of high importance nowadays because control via paratransgenesis depends on the availability of these bacteria in axenic conditions, as will be discussed later in this chapter.

Later, analysis based on 16S rDNA amplification by PCR followed by typing, footprinting, or sequencing allowed us to characterize bacterial communities in sand flies in more depth. Molecular methods such as denaturation gradient gel electrophoresis (PCR-DGGE), temperature gradient gel electrophoresis (TGGE), and random amplified polymorphic DNA (RAPD), which hinge on the whole genome to assign a unique pattern, have been used to explore sand fly microbiota composition [39, 46, 50]. For instance, fingerprinting methods allow a rapid differentiation between developmental stages of two different *Ph. duboscqi* and *Ph. papatasi* [39]. Nevertheless, it is also important to state that DGGE has the limitation of sharing similar migration of bands for different species [57], showing different migration patterns associated with heterogeneous DNA for the same sample, as well as polymorphisms associated with the 16S rRNA gene [58].

With the advancement of next-generation sequencing (NGS) technologies, it has been easier and more accessible to rely on sequence information to characterize microbiota in sand flies. Now, it is possible to readily use different combinations of hypervariable regions of the 16S rRNA gene widening the resolution of the species that can be identified. That is why many recent studies in the field utilize this method as the primary means to assign taxonomic groups. The use of different hypervariable region combinations of this gene also influences the taxonomic assignment of the bacterial species found (Table 1). For example, profiling using the V3-V4 hypervariable region of the 16S rRNA gene allowed the identification of a core group of six genera common between different feeding conditions in wild-captured *Lu. longipalpis: Enterobacter, Serratia, Stenotrophomonas, Enhydrobacter, Pseudomonas*, and *Chryseobacterium* [48]. Similarly, using *Lu. longipalpis* Jacobina strain and sequencing of the V4 region of 16S rRNA gene, a dynamical evaluation of the microbiota composition showed the flexibility of this bacterial community [49]. Upon feeding with sucrose, blood, or *Le. Infantum*-infected blood, linear discrimination analysis identified Phyllobacteriaceae (family), Actinobacteria (phylum), and *Trabulsiella* (genera), respectively, as the distinguished taxa for every treatment. Both studies highlight the advantages of metataxonomics, as it determines relative abundances to the species level and allows massive examination of samples from different conditions [59].

Despite the vast number of methods available, some variations are found between the same species from different origins and conditions. This is exemplified in studies conducted using three laboratory-reared sand flies: *Ph. papatasi, Ph. duboscqi,* and *Lu. longipalpis* [16]. It was determined that Proteobacteria was the main Phylum in all samples, in accordance with results from a meta-analysis based on culture-dependent and culture-independent studies [46]. There were differences between the two *Phlebotomus* species, with distinct composition linked to Rhizobiaceae and Yersiniaceae, while in the case of *Lu. longipalpis, Achromobacter, Tsukamurella, Staphylococcus,* and *Stenotrophomonas* were the main bacterial genera. Interestingly, when the experiments were repeated with slight changes in the food batch, a different proportion of those genera appeared, demonstrating the associated role of the feeding source in bacterial composition.

Other components of the microbiota that deserve attention are archaea, viruses, and fungi. Usually, the characterization of fungal species has been done using morphological identification

[44], ITS sequencing [16, 54], or metagenomics [60]. For *Lu. ayacuchensis, Papiliotrema* was the most abundant genus, among others such as *Penicillium, Exserohilum, Cladosporium, Sarocladium, Malassezia,* and *Debaryomyces* [54]. In the case of *Ph. papatasi, Ph. duboscqi,* and *Lu. longipalpis, Meyerozyma* was the main fungal genus in the adult stage, while *Chaetomium* and *Microascus* were the main fungal genera in the larval stage [16]. The archaeal component has not been explored widely yet, but available data show this domain is not present in rearing substrates treated with larvae of *Ph. Papatasi* [61] and their contribution to gut microbiota in *Phlebotomus* species is minimal [41]. Despite the apparent secondary role of these microbiota members in sand flies, they may possess significant ecological functions and relationships that remain undiscovered. Utilizing advanced techniques such as metagenomics or metatranscriptomics [60, 62] may be crucial in elucidating these potential complexities. Naturally, the selection of the methodological approach will be dependent on the characteristics of the sample and the hypothesis under scrutiny. In the end, to fully comprehend the role of microbiota in *Leishmania* infection and other physiological traits in sand flies, it is imperative to explore these groups in depth.

3 Bacterial Endosymbionts Identified from Sand Flies: Relationship with their Host and Effects on Pathogens

Endosymbionts form diverse associations with their host, and insects are no exception to such interactions. Bacteria hosted in different tissues form long-term relationships that shape their evolution while providing benefits that can go from nutrition to defense against pathogens [63, 64]. Endosymbionts can be classified as obligate or facultative, with mutualist, pathogenic, or commensal interactions [65]. However, different perspectives can be considered in such classifications, depending on whether the interaction is dominated by the host or the symbiont. Traditionally, the survival of one depends on the other and vice versa in an "obligate mutualist" interaction, which differs from those depicted as "facultative" where the host can survive without the symbiont, but from an insect-microorganism perspective, it does not always consider whether the microbe survives outside this relationship [66]. Moreover, those that are in an obligate partnership with their host experience genome reduction and share a co-evolutionary history, coexisting in intracellular as well as extracellular environments [67]. On the other hand, facultative symbionts are not always present in all individuals of the same species; they are known for being reproductive manipulators and can be involved in mutualist interactions that provide traits of ecological importance [68].

In insect vectors, interactions with endosymbionts may be influenced by the presence of pathogens, as their fitness interest may collide, leading to changes in traits related to reproduction or behavior that ultimately can impact disease transmission [69]. Similarly, the establishment of pathogens such as arboviruses and parasites in insect vectors is also influenced by the gut microbiota [70]. However, the lack of systematic studies regarding the detection, diversity, endosymbiont phylogeny, and evaluation of their impact on pathogens (viruses and parasites) in insect vectors, such as the phlebotomine sand flies of the Old and New World, needs further exploration. This field may provide new insights into co-evolution, virome, and bacteriome modulation, behavior alterations, and the survival of the sand flies, contributing to the development of new alternatives for biological control strategies.

An early approach to understanding the sand fly–endosymbiont interaction, focusing on areas from Central and South America, Africa, and Asia, was conducted by Ono et al. [71]. They found that 27% of the surveyed sand flies corresponding to species of *Ph. papatasi, Ph. perniciosus, Lu. shannoni,* and *Lu. whitmani* (see Fig. 1) were infected with different *Wolbachia* strains. This bacterium, which is maternally transmitted and an obligate intracellular endosymbiont, belonged to the groups A and B according to their phylogeny. More recent studies have detected the presence of *Wolbachia* in sand flies collected in Colombia, including *Pi. evansi, Lu. gomezi, Ev. dubitans, Mi. micropyga,* and *Br. hamata,* alongside *Cardinium* in *Tr. triramula, Mi. cayennensis,* and *Ps. shannoni* [34, 53].

Similarly, Lozano-Sardaneta et al. [78] identified *Wolbachia* of supergroup A in *Ps. shannoni* in Mexico, while another survey by Lozano-Sardaneta et al. [79] revealed the presence of supergroup A and B strains of the endosymbiont in *Ps. corossoniensis, Lu. cruciata, Pi. ovallesi, Ps. panamensis,* and *Ny. ylephiletor*, as well as *Bartonella* infection in one specimen of *Pi. ovallesi*.

The endosymbiont is not exclusive to phlebotomine sand flies of the New World, as it has also been reported in sand flies collected in Europe and Asia (Fig. 1); for instance, *Wolbachia* of the supergroup B have been detected in insects of *Sergentomyia minuta* and *Ph. perniciosus,* haplotypes of group A have been identified in *Ph. perfiliewi transcaucasicus, Ph. kandelakii, Ph. papatasi,* and *Ph. sergenti,* and the strain *w*Ri has been reported in *P. chinensis* [72, 75, 76, 80, 81].

The significance of detecting natural occurrences of endosymbionts like *Wolbachia* in sand flies lies in evidence suggesting their potential to activate the immune system of the host, particularly the Th1 pro-inflammatory response against *Leishmania* [82] and in their potential of exhibiting antiviral properties interfering with virus transmission, particularly by reducing viral accumulation, thus being useful for biological control strategies, such as the control of dengue transmission by *Ae. aegypti* [83]. Furthermore, findings indicate a positive correlation between the symbiont and susceptibility to insecticides such as Deltamethrin in sand flies of *Ph. papatasi* [20]. Despite the challenges in obtaining cultures of the endosymbiont, some researchers have successfully established cell lines of strains such as *w*Pap and *w*Pip; although these were ineffective, they serve as valuable tools for further studies [84].

Other endosymbionts of the genera *Arsenophonus, Cardinium, Microsporidia, Asaia, Rickettsia,* and *Spiroplasma* are also present in sand flies from both the Old and New Worlds. For instance, *Microsporidia* and *Rickettsia* have been found in *Pi. pia* [34], while the latter has been reported in *P. chinensis* and *Ps. aclydifera* [76, 77]. *Arsenophonus* has been detected in sand flies of *Pi. evansi, Ps. panamensis,* and *Lu. longipalpis* [73], and has been found associated with insect temperature preferences in a positive correlation with *Bacillus* sp. in *Lu. longipalpis* [21]. Additionally, *Spiroplasma* has been identified in specimens of *Phlebotomus* sp. and *Ph. chinensis* [74, 76].

The discovery of diverse endosymbionts among sand fly populations highlights the intricate ecological relationships shaping vector biology across geographical regions. While research has predominantly focused on *Wolbachia* and its role as a reproductive manipulator, ultimately impacting the insect population densities, it is noteworthy that other symbionts also exhibit sim-

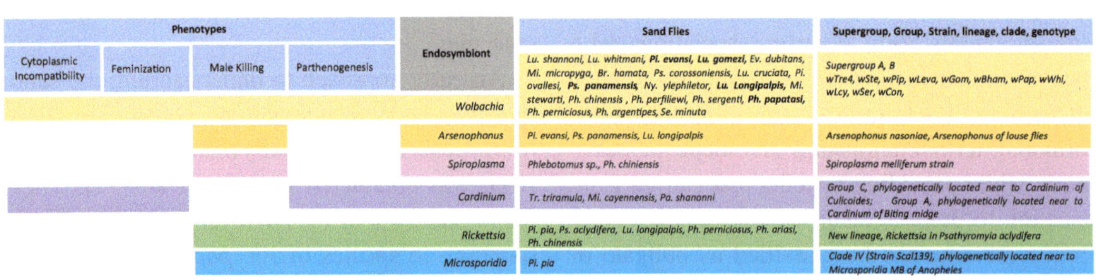

Fig. 1 Phenotypes expressed in the reproductive system of insects associated with endosymbionts and the different strains reported in sand flies from around the world [34, 53, 71–81]. Further research is needed to better understand the distribution of *Wolbachia* in sand flies, ecoregions, and its potential implications for disease transmission and vector control strategies. The specific strains of *Wolbachia* present in sand flies may vary depending on the geographical location and the species of sand fly involved

ilar reproductive manipulation traits. Their respective phenotypes are depicted in Fig. 1.

In addition to the effects these bacteria have over the reproductive system of their hosts, symbionts such as *Spiroplasma* are known for providing defense against nematode parasites through ribosome inactivating proteins (RIP) that have toxic activity against parasitic nematodes and wasps [85]. Similarly, *Microsporidia* has shown the potential to impair the full development of parasites of *Plasmodium* in *Anopheles arabiensis* mosquitoes by preventing the parasite from reaching the salivary glands [26], while *Asaia* bacteria are thought to activate the immune system of mosquitoes, thus affecting parasite transmission [86]. Regarding the mechanisms in which endosymbionts act, Massey and Newton [87] investigated *Wolbachia, Arsenophonus, Cardinium, Rickettsia,* and *Spiroplasma* toxin families and found 16 that have the potential to affect the biology of the host, mostly related to their participation in male killing, cytoplasmic incompatibility, and as a virulence factor associated with RIP, while other toxins are thought to participate in tissue colonization and transmission of the bacteria to the next host. However, the roles these endosymbionts play in phlebotomine sand flies are yet to be elucidated. Overall, the presence of diverse endosymbionts highlights the complexity of sand fly microbiomes and their potential roles in vector-borne disease dynamics. This broader understanding of symbiotic associations opens avenues for exploring novel biological control strategies and elucidating the interplay between symbionts and sand fly ecology.

4 Effects of Gut Microbiota on *Leishmania* Interactions

The recognition of the importance of bacterial interactions with *Leishmania* and the sand fly vector has been proven by several published studies on the sand fly gut microbiota. These studies have involved experimental trials using different species of *Leishmania*, varied feeding states, antibiotic treatments, co-infections with bacteria, and analyses of diversity at different post-infection time points until the development of infective stage metacyclic promastigotes [8, 12, 19, 31, 38, 49].

The *Leishmania* parasite develops entirely inside the female gut of the sand fly and is therefore exposed to the sand fly gut microbiota. Microbes may be acquired during feeding on plants or animals, as well as being vertically transmitted via larvae and pupal stages [19]. In the sand fly's midgut, amastigotes transform into promastigotes, which migrate to the foregut. Here, the promastigotes transform into free-swimming infective metacyclic promastigotes, ready for regurgitation when the sand fly takes another blood meal [8, 88]. *Leishmania* parasites can face different mechanisms of action or regulatory strategies by the gut microbiota, such as binding, inactivation, and direct lysis of bacterial microbiota, binding and lysis of midgut cells, formation of biofilms, activation of immune response signaling pathways, release of antipathogenic components, induction of antimicrobial peptides (AMP), melanization, and the production of reactive oxygen species (ROS), among others [8, 12, 33, 89].

These interactions have a significant impact on *Leishmania*–microbiota interactions, suggesting that sand fly vector competition can be modulated [90]. Then, can bacterial communities in the gut promote or block the development of *Leishmania*? Some studies have shown that symbiotic bacteria of the genera *Serratia, Lysinibacillus,* and *Pseudocitrobacter* inhibit the growth of *Leishmania* parasites (*Le.i. chagasi, Le. major, and Le. amazonensis*) in the midgut of *Lu. Longipalpis* during in vivo co-infection assays [47] (see Fig. 2). Additionally, when females of *Lu. longipalpis* were fed with *Ochrobactrum intermedium, Asaia* sp., and *Pseudozyma* sp. the number of sand flies infected with *Le. mexicana* was lower. However, the study also reported that *Le. mexicana* can protect *Lu. longipalpis* against the bacterial infection by *Serratia* [91] (Fig. 2), suggesting a mutually beneficial relationship between the host and the *Leishmania*.

In *Pi. evansi,* another vector recognized in the New World, the genera *Bacillus* and *Aeromonas* are the most abundant in the midgut of females infected with *Le. infantum* [34, 55] (Fig. 2). In contrast, *Ralstonia* seemed to have a higher abundance in uninfected females of *Pi. evansi* [55]. Meanwhile, Kelly et al. [49] identified that bacterial richness progressively decreased (mainly *Pseudomonadaceae* sp.) in *Le. infantum*-infected females of *Lu. longipalpis*, while *Acetobacteraceae* sp. was dominant (Fig. 2).

Recently, Vivero et al. (2024, in press) evaluated the effect of the gut microbiota of *Lu. longipalpis* on its susceptibility to infection with *Le. infantum* and *Le. braziliensis,* finding that the group of females exposed to *Le. infantum* and uninfected had a high relative abundance of *Pseudomonas* (75–85%), in contrast to the guts positive for *Le. infantum* infection (30–55%) under experimental infection or naturally infected with the same species of the parasite (35%). Additionally, ASVs that moderately increased in guts infected with *Le. infantum* were Enterobacteriaceae (25%), *Enterobacter* (10%), and *Klebsiella* (5%) (Fig. 2). On the other hand, the group positive for *Le. braziliensis* under experimental infection showed a higher relative abundance of *Pseudomonas* (95%) and less proportion of *Enterobacteriaceae* compared to the guts with *Le. infantum* (Fig. 2). Finally, several studies have shown that sucrose-fed sand flies contained a highly diverse and stable midgut microbiome, while blood feeding caused a decrease in microbial richness that eventually recovered.

Louradour et al. [38] used a colonized population of *Ph. duboscqi* to conduct co-infection assays with *Leishmania major*, as well as antibiotic and sucrose treatments, aiming to evaluate the vector competence of this species. Their findings indicated that sucrose utilization by the microbiota is essential to create the appropriate osmotic conditions required for the survival of infective stage promastigotes in vivo. Additionally, this research identified a dominance of *Leifsonia* sp., *Serratia rubidaea, Gordonia,* and *Ranhella* in the midguts of the antibiotic-treated flies, suggesting that the growth and differentiation of the parasite might be directly inhibited by the proliferation of this bacterial species (see Fig. 2). *Enterobacter* was found to be especially beneficial to the parasite, with the majority of flies harboring infections that were equivalent to controls [38].

Further in vivo co-infection studies with both wild-caught and laboratory-reared sand flies are needed to better understand *Leishmania*–microbiota–sand fly interactions. Gut bacterial diversity and composition vary depending on factors such as species, locality, food source, and infection status [15, 31, 34, 43]. Collectively, these diverse datasets suggest that the sand fly midgut microbiome plays a crucial role in *Leishmania* growth and differentiation into its infective state before disease transmission [8, 49, 91]. All this information can be used to develop new tools aimed at blocking disease transmission and provide critical insights for the development of intervention strategies for sand fly control [25, 92].

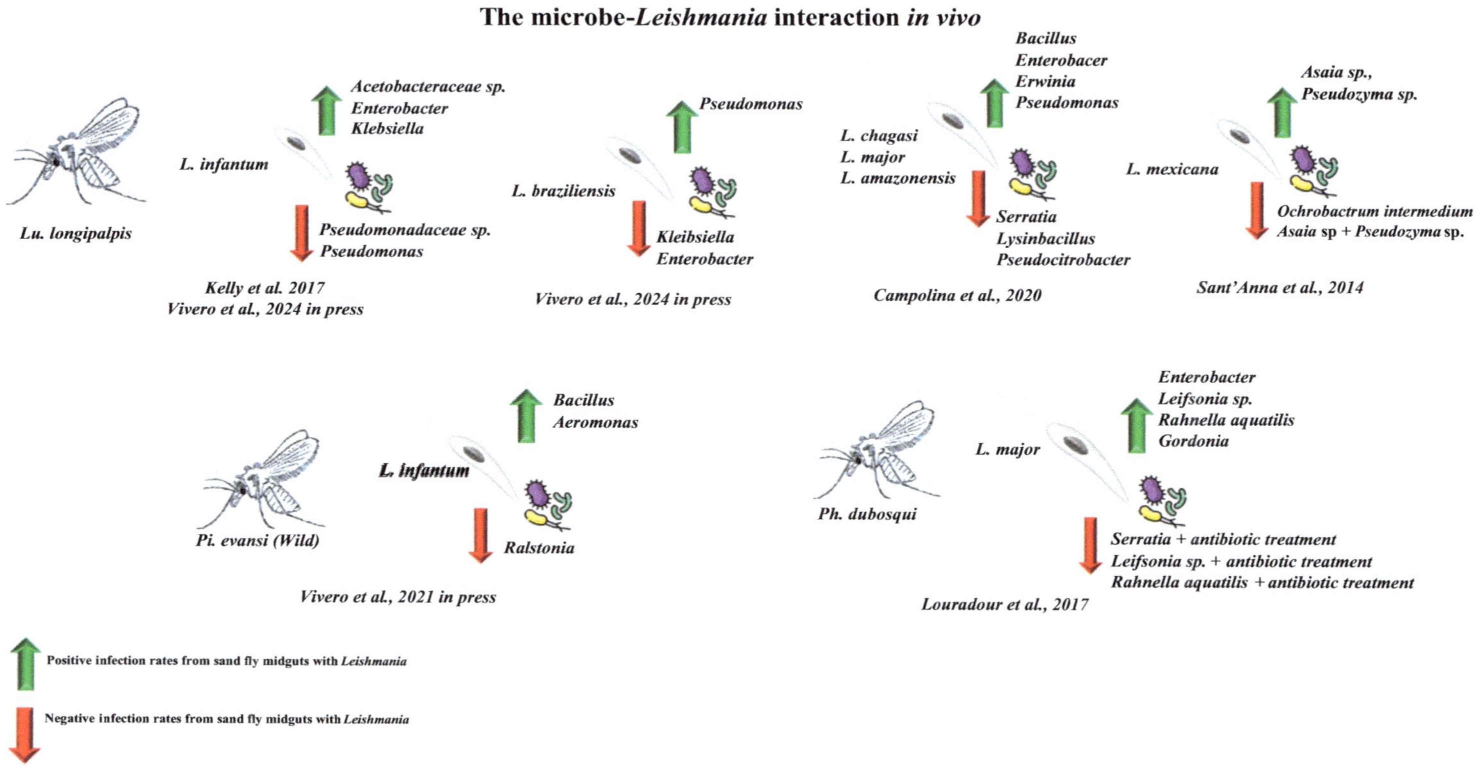

Fig. 2 The effect of in vivo co-infection between several species of *Leishmania* and gut microbiota. Most of the studies involve female sand flies with a six-day post-infection period, as well as the co-infection with bacteria previously isolated in cultures (following oral exposure)

5 In Vitro Co-infection Using Diverse Species of *Leishmania* and Bacterial Species

Understanding the range of effector molecules and mechanisms involved in the interaction between microbiota and *Leishmania* species highlights the complexity of processes occurring within the sand fly gut. The microbiota inhabiting this organ in insects plays crucial roles, including nutrition, immune regulation and maturation, reproduction, and host development [15, 93–96]. Moreover, it has been established that the composition of sand fly microbiota is dynamic and influenced by various factors. This intricate microenvironment is what *Leishmania* parasites encounter upon entering the gut, and the reason to conduct research using specific bacterial species and their metabolic products in vitro. Typically, this involves axenic *Leishmania* species cell culture, individual bacterial or fungal isolates, and sometimes sand fly cell lines capable of mimicking insect vector behavior [38, 91, 97]. This research is complemented by transcriptomics, viability assays with supernatants, or cell extracts evaluation [98, 99], significantly contributing to reduce the knowledge gap in this area.

The functional role of bacterial members within the microbial community is crucial for understanding their contribution to *Leishmania* infection progression. Several studies have examined the role of individual or combined bacterial isolates obtained from the gut of Phlebotominae species. For instance, *Serratia marcescens* strain SM365 demonstrated the ability to lyse *Le. chagasi* promastigotes, in contrast with the unpigmented DB11 strain, which does not synthesize prodigiosin, a molecule associated with antimicrobial properties through plasma membrane damage [100, 101]. The interaction was shown to be mediated by *Leishmania* mannose-dependent surface binding in the flagellar pocket, the same place where *S. marcescens* SM365 was found forming filamentous biofilm-like structures. In this case, the interaction was avoided after mannose addition, possibly by the binding of this molecule to *S. marcescens* fimbriae, allowing escape of *Le.i. chagasi* of the deleterious effect [100].

Similarly, challenging *Le. braziliensis* with the same two strains of *S. marcescens* confirmed the lytic effect [102]. Also, prodigiosin was discarded as the responsible metabolite for this effect when blocking the synthesis of this molecule at 37 °C [103], and still observing a decrease in the number of parasites [102]. The interaction between bacteria and *Leishmania* is further supported by studies of *Le. mexicana* challenged with *S. marcescens* cells and filtered used media [90], supporting the hypothesis of a secreted molecule as the mediator of the interaction. While studying this interaction, from the perspective of microbial ecology, bacteria do not always exhibit a function exclusively on their own [104, 105]. Therefore, the co-culture of more than one bacterial isolate, which has shown positive results against *Leishmania* infection in vivo [91], should be applied in more in vitro-based research.

Other bacteria–*Leishmania* interactions show that the presence of live bacteria is essential to affect *Leishmania* development. During in vitro incubation of fungal *Pseudozyma* sp. or bacterial *Asaia* sp. and *O. intermedium* cultures in conjunction with *Le. mexicana*, a reduction in promastigote abundance is observed. However, this phenomenon is not evident when utilizing a filtrate of the bacterial media instead [91]. This result was supported by similar trends observed in vivo, highlighting the validity of the in vitro outcome [91]. Additionally, *Pantoea ananatis*, *E. cloacae*, and *Ochrobactrum anthropi* cells isolated from *Lu. evansi* inhibited procyclic-like but not metacyclic-like promastigotes of *Le. infantum* [106]. A plausible explanation lies in the less specialized traits of procyclic promastigotes versus metacyclics, the infective stage that is more adapted to harsh conditions [107].

Additionally, in vitro co-cultures of *Lysinibacillus* and *Serratia* registered a harmful effect over time against *Le. i. chagasi, Le. major, Le. amazonensis,* and *Le. braziliensis* represented as a decrease in the number of parasites [47]. Metacyclic promastigotes represented only 1% in this study, indicating the direct incidence of bacteria in the life cycle of *Leishmania* species. Variables affecting both microbiota and parasite metacyclogenesis in the sand fly gut, such as pH, temperature [108], immune response [15, 109], metabolic activity [40, 106], culture phase, and media composition [110], might determine this effect; however, more studies to cover this issue are necessary. These studies have limitations in establishing the relationship with the life cycle of *Leishmania* species, and future investigations should differentiate between different states. Table 2 summarizes bacteria–*Leishmania* co-infection studies in vitro.

The combination of the in vitro assays with the functional characterization of microbiota in the sand fly gut enables the evaluation of *Leishmania* control strategies. Furthermore, *E. cloacae, S. rubidaea, or Rahnella aquatilis* were utilized to assess the role of microbiota in sand fly vector competence for *Le. major* [38]. Significant lethality of promastigotes in vitro was observed when the sucrose concentration ranged from 30% to 15%, with similar effects observed at concentrations lower than 7.5%. These results imply that fine-tuning osmotic conditions is necessary to achieve homeostasis in the gut microenvironment, particularly for sucrose, an essential substrate. The optimal nutrient concentration permits the survival and development of *Leishmania* during the metacyclogenesis process [111]. Identifying the microbiota components that modulate this parameter could offer another avenue for biological control of *Leishmania.* Additionally, 12 Gram-negative bacterial isolates from *Pi. evansi,* evaluated for their proteolytic, lipolytic, hemolytic, cellulolytic, amylolytic, and nitrogen metabolism functions, demonstrate the multiple roles they can play in the interaction with the insect host and *Leishmania* [30]. These isolates also show potential as sources of biotechnological and pharmaceutical activities. Among these isolates, only *Enterobacter hormaechei* showed leishmanicidal activity against *Le. braziliensis* [30].

Insect cell lines have facilitated the evaluation of both bacterial isolates and *Leishmania* infections under closely controlled conditions. This method has enabled the establishment of stable cell lines of *Lu. longipalpis* with *Wolbachia* *w*MelPop-CLA strain [97]. These bacteria have previously shown natural infection of *P. papatasi* cell lines [84]. The introduction was difficult and selective to only Lulo and not in the LL5 cell line,

Table 2 In vitro studies challenging *Leishmania* species with different bacterial isolates

Leishmania species	Bacterial species	Sand fly species	Developmental stage of Leishmania	References
Le. i. chagasi	*Serratia marcescens* (SM365 and DB11 strains)	Not evaluated	4 days old promastigotes	[100]
Le. braziliensis	*Serratia marcescens* (SM365 and DB11 strains)	Not evaluated		[102]
Le. mexicana	*Pseudozyma* sp. *Asaia* sp. and *Ochrobactrum intermedium*	*Lu. longipalpis.*	Promastigotes	[91]
Le. infantum	*Pantoea ananatis, Enterobacter cloacae*, and *Ochrobactrum anthropi*. (Sand flies were challenged with Serratia)	*Pi. evansi* (Bacterial isolates source)	Promastigotes, (metacyclic or procyclic)	[106]
Le. major	*Enterobacter cloacae, Serratia rubidaea*, or *Rahnella aquatilis*, isolated from a laboratory-reared colony of *P. duboscqi*	*P. duboscqi*	Promastigotes (metacyclic or non-metacyclic)	[38]
Le. infantum	*Wolbachia* (*w*MEL and *w*MelPop-CLA)	*Lu. longipalpis* (LL5 and LULO cell lines)	Promastigotes	[97]
Le. i. chagasi, Le. major, Le. amazonensis, and *Le. braziliensis*	13 distinct bacterial genera (*Bacillus, Enterococcus, Erwinia, Enterobacter, Escherichia, Klebsiella, Lysinibacillus, Pseudocitrobacter, Providencia, Pseudomonas, Serratia, Staphylococcus, and Solibacillus*)	*Lu. longipalpis*	Promastigotes	[47]

demonstrating that not only the genetic background but also the cytoplasmic composition is an important factor of this interaction [112]. Despite previous reports of virus and parasite reduction effects of *Wolbachia* in other insects [113–115], levels of *Leishmania*-infected cells did not change with the presence of *Wolbachia* at different times post-coinfection [97]. This suggests that, at least in vitro, this strain of *Wolbachia* does not have a negative effect on *Leishmania* infection. Despite this result, infecting an insect cell line with *Wolbachia* represents the first step toward inoculation of the adult insect host [116]. Consequently, in vivo assays should be performed with other endosymbionts reported to have parasite reduction effects in insects.

In addition to the direct cellular interaction of microbiota with *Leishmania,* the extracts and specific metabolites possibly responsible for the bioactivity have been tested in vitro. Using cell-free culture supernatants of *Xenorhabdus* and *Photorhabdus* bacteria at concentrations ranging from 1.25 to 10%, antiprotozoal activity against different parasitic protozoa, including *Le. tropica,* was measured [117]. Initially, only *Xenorhabdus* species appeared to have a negative effect on *Le. tropica*. Subsequently, to identify antiparasitic cellular products, the easyPACId method creates mutant strains expressing one main protein product at a time under the control of an L-arabinose inducible promoter [118]. Fabclavine, a hybrid peptide/polyketide [119], and Xenocoumacin, a peptide synthesized by non-ribosomal peptide synthases [120], were the compounds related to a decrease in the number of promastigotes from *Le. tropica*.

Antimicrobial activity against clinical and environmentally relevant bacteria has been tested for isolates obtained from the gut of *Pi. evansi* [30, 53, 106] and other insects' guts [121]. There is also interest in examining bacteria from other sources known for their antibacterial and anti-leishmanicidal bioactivities. Recently, extracts from *Streptomyces* sp. demonstrated the ability to inhibit bacterial growth in solid medium and submerged fermentation, with a reported IC_{50} of 9 μg/mL against *Le. amazonensis* [99]. Metabolites isolated from other bacteria, such as valinomycin, staurosporine, violacein, bacteriocin, toxin peptides, and acetogenins, have also been evaluated for their leishmanicidal activity, yielding promising results [122–126]. A wide range of fungal, protozoal, and cyanobacterial compounds further emphasize the relevance of antimicrobial compounds against *Leishmania* (reviewed by [127]). This evidence supports the need for further investigation into the role of microbial metabolism and its interaction with *Leishmania* within sand fly gut microbiota, given the unexplored knowledge available at this level.

6 Depletion of Microbiota

Antibiotics are used to suppress the microbiota and observe their effect in modulating pathogens. The use of antibiotics during in vivo co-infection assays with *Leishmania* and/or in vitro assays with bacteria previously isolated and co-cultured with the parasite are important components of experimental design. However, they do not represent a viable alternative to block transmission, and/or the derived results are not entirely conclusive due to the increase in bacterial strains with antibiotic resistance, which can also impact human and animal health [89, 92]. Moreover, the effect of antibiotics on the microbiota is nonspecific, as it can deplete several bacterial species through antimicrobial treatment.

The impact of antibiotic treatment on the microbiota and *Leishmania* infection in sand flies has been investigated, particularly in *Lu. longipalpis*, *Pi. evansi,* and *Ph. duboscqi* populations, both wild and laboratory-reared [34, 38, 49, 55]. Antibiotic treatment of females during the sucrose feeding often reduces bacterial diversity in the gut microbiota of *Pi. Evansi,* leading to increased loads of metacyclic promastigote infection [34, 55]. Conversely, *Lu. longipalpis* shows higher loads of metacyclic promastigotes on the ninth day in sand flies not treated with antibiotics compared to antibiotic-fed sand flies [49]. Similarly, Louradour et al. [38] observed that antibiotic treatment prevented *Leishmania* development in the midgut of *Ph. duboscqi,* and feeding with bacterial culture supernatants failed to reverse the antibiotic effect.

In other models of vector-borne diseases, such as malaria, Dong et al. [128] demonstrated that clearing the microbiota with antibiotic treatment can significantly enhance mosquito susceptibility to the pathogen. However, it is important to note that similar findings in mosquitoes suggest that manipulating the microbiota may harm the vector and interfere with vector-borne pathogen infectious cycles [92]. The type of antibiotics is a variable that must be evaluated, along with prior knowledge of the bacteriome in insect vectors, as it can markedly reduce or increase the intensities and prevalence of pathogens in key host organs for multiplication and/or replication. For example, the effects of azithromycin and doxycycline on mosquito microbiota produced a differential alteration in the vectorial capacity of *Anopheles*, as azithromycin decreased *Plasmodium falciparum* load, while high concentrations of doxycycline increased *P. falciparum* infection load [25, 129]. Exposing sand fly microbiota to high levels of antibiotics may also contribute to the presence of antibiotic-resistant bacteria and transferable resistance genes (Resistome). Therefore, longitudinal studies are feasible to investigate this altered microbiome's persistence over generations and its interaction with pathogens, highlighting the necessity of further exploration in this area.

7 Microbiota and Environmental Variables

Considering the profound transformations occurring in various ecosystems due to anthropogenic activities and resultant climate change, insects are increasingly challenged by fluctuating temperatures, with occurrences of extreme high temperatures (EHT) becoming more frequent. EHT is defined as the environmental conditions associated with increasing temperatures that trigger a threshold-like biological response [130]. In response to such challenges, insects employ various strategies for survival, including thermoregulation within microclimates, plastic responses such as rapid hardening and transgenerational acclimation occurring over different timescales, an ontogenetic variation that enables insects to adapt to heat stress according to their life stage, and through their ecological networks that involve the association with endosymbionts and microbiota [131].

While evidence of interactions among microbiota, temperature, and sand flies remains limited, studies indicate that bacteria reported in sand flies also exhibit changes in the phenotype expression of reproductive manipulators like *Wolbachia* and *Cardinium* when undergoing temperature fluctuations [132]. Other associations between bacterial abundance, transmission, and response to environmental conditions have been observed in other insects. For example, in *Gryllus veletis* exposed to varying breeding temperatures associated with seasonal changes, the relative abundance of Proteobacteria increased from 7% to 26% during the transition from summer to winter, while Firmicutes decreased from 50% to 38% over the same period [133]. In aphids, the transmission and infection rates of *Arsenophonus* decreased in populations exposed to 30 °C temperatures but increased in environments with 20 °C temperatures [134]. Additionally, in *Bemisia tabaci* insects, *Rickettsia* has been shown to induce the expression of cytoskeleton genes involved in responding to rising temperatures, thereby enhancing the insect's thermotolerance compared to non-infected individuals [135].

Sand flies are ectothermic organisms, meaning their body temperature is largely dependent on the external environment. Both temperature and microbiota play crucial roles in the ecology and physiology of sand flies. Regarding the interaction of microbiota and temperature, our research showed changes in the relative abundances of bacteria in a wild population of *Lu. longipalpis* according to the temperature preferred by the sand fly [21]. Specifically, we observed a relative abundance of *Pseudomonas* of 57.36% at 25–27 °C that decreased in sand flies that preferred warmer temperatures of 29–33 °C. On the contrary, bacteria of *Bacillus* registered lower relative abundances when *Lu. Longipalpis* preferred temperatures of 25–27 °C (1.21%), but increased at 29–33 °C, with positive interactions with other bacteria such as *Arsenophonus.*

Limited studies have investigated temperature preference in sand flies. For instance, Martins et al. [136] observed that females of *Lu. longipalpis* preferred lower temperatures post-blood ingestion, with reduced activity during this stage, whereas they preferred higher temperatures during later stages of blood ingestion when activity increased. Additionally, an increase in the expression of the Heat Shock Protein HSP90(83) during blood feeding suggested that the sand fly response to thermal stress involves both protein expression and behavioral changes.

The heat shock response serves as a defense mechanism against various stressors, including high temperatures, oxidative stress, and cytotoxins [137]. In mosquitoes, this response has been implicated in defense against insecticides [138]. However, to our knowledge, no studies have explored insecticide resistance in sand flies in relation to HSP expression. Most research has focused on resistance surveillance through WHO tubes or CDC bottles bioassays, and biochemical tests, with limited studies investigating resistance mutations [139–144], and even fewer have considered the role of microbiota in insecticide resistance mechanisms in sand flies, as reported by Azarm et al. [20] who found that the rate of *Wolbachia* infection (*w*Pap) in *Ph. papatasi* was significantly higher in susceptible sand flies than in resistant individuals when exposed to Deltamethrin.

This is particularly relevant as different bacteria reported in sand flies, besides *Wolbachia,* have been implicated in resistance or susceptibility to insecticides using different insect models. For instance, some strains of the endosymbiont *Arsenophonus* are thought to be involved in the susceptibility or resistance to chlorate neonicotinoid insecticides of their host [145]. On the other hand, bacteria associated with the degradation of pyrethroid-type insecticides, such as Alpha-cypermethrin, Beta-cypermethrin, Deltamethrin, and Lambda-cyhalothrin, commonly used for sand fly control, include *Serratia*, known for Beta-cypermethrin degradation, and certain species of *Pseudomonas* and *Bacillus,* which exhibit high metabolic activity in the presence of pyrethroids such as Deltamethrin [146, 147].

Finally, by understanding the mechanisms underlying thermotolerance and insecticide resistance, researchers can inform targeted interventions and enhance our ability to control vector populations, ultimately reducing the burden of vector-borne diseases on human health. Therefore, continued interest and investment in studying these complex dynamics are essential for advancing our understanding and improving our capacity to address emerging challenges posed by environmental changes.

8 Paratransgenesis: A Solution to Alter Insect Vector Competence?

Various interactions have been observed among the microbiota of insect vectors and the pathogens that affect them, prompting the proposal of several biological control strategies to mitigate vector-borne diseases on a global scale [92]. One such strategy is paratransgenesis, a technique that involves the use of recombinant DNA to manipulate bacteria residing in the midgut of insects or symbiotic bacteria, in order to insert genes that express molecules capable of altering vector competence, thereby impeding the transmission of pathogens [148]. To achieve the desired outcomes, the selected bacteria must meet specific criteria to ensure the success of paratransgenesis implementation. According to Beard et al. [149], symbiotic bacteria are prime candidates for paratransgenesis due to their established presence in the insect host and their ability to withstand the conditions within the target organism. Furthermore, these symbiotic bacteria should remain stable, ensuring continued production of the desired molecules to impede pathogen development or establishment within the host organism. This contributes to the efficacy of transgene products associated with the symbiont against pathogens [149].

Additional criteria proposed by these authors include the ease of genetic manipulation and culturing of the selected microorganism, considering that this is an ongoing process over time. Moreover, the modified symbionts should be capable of establishing themselves within natural insect populations and persisting across the different insect instars, which means achieving transstadial transmission in order to ensure a successful control strategy (see Table 3). Similarly, the fitness of the symbiont should remain unaffected despite genetic manipulation.

Despite exhibiting attractive phenotypes when present in insect vectors, certain bacteria can be disregarded as candidates for paratransgenesis due to the difficulty of meeting at least one of the aforementioned criteria. For instance, *Wolbachia*, although capable of inducing immune responses in the host through its surface protein (WSP) and enhancing reactive oxygen species (ROS) and the production of antimicrobial peptides, which can be beneficial against pathogens such as *Leishmania* [154, 155], in some cases gets excluded from consideration, attributing this to its status as an obligate intracellular endosymbiont, making it difficult to manipulate and culture in a cell-free media [156]. However, it is worth noting that there are some successful attempts to establish cell lines infected with *Wolbachia*, as reported by Da Silva Gonçalves et al. [97] in the cell lines of *Lu. Longipalpis,* but further studies

Table 3 Identification of different bacteria prospects as paratransgenesis candidates in phlebotomine sand flies

Insect vector	Potential candidate	Findings/characteristics	References
Ph. duboscqi	*Ochrobactrum anthropi*	Capable of transstadial transmission	[40]
Ph. argentipes	*Bacillus megaterium* *Brevibacterium linens*	– Non-pathological to humans – Potential of acquiring the transformed bacteria through the environment.	[150]
	B. subtilis	– Transformed with Green Fluorescent Protein (GFP) gene reporter – Capable of transstadial transmission – Potential of acquiring the transformed bacteria through breeding sites	[151]
	B. cereus *Lysinibacillus fusiformis*	Capable of transstadial transmission	
Ph. papatasi	*Microbacterium* spp.	Capable of transstadial transmission	[39]
	B. pumilus *B. flexus*	– Nonpathogenic bacteria – Common in sand flies collected from different habitats	[43]
	Enterobacter cloacae subsp. dissolvens *B. subtilis*	– Capable of transstadial transmission – Potential of acquiring the transformed bacteria through breeding sites and the environment	[42].
	E. cloacae	– Transformed with red fluorescent protein and defensin – Very low transstadial transmission, therefore only promising for adult stages	[152]
	Serratia AS1 *E. cloacae subsp. dissolvens*	– Transformed with mCherry fluorescent protein gene – Transformed bacteria introduced via rodent food demonstrated transstadial transmission – Potential of acquiring the transformed bacteria through breeding sites and the environment	[153]
	O. intermedium	– Capable of transstadial transmission	[89].
Lu. longipalpis	*Pseudozyma* sp. *Asaia* sp. *O. intermedium*	– Reduction in the establishment of *Le. mexicana* in the sand fly	[91]
	S. marcescens	– Increasing of survival rate of sand flies infected with both the bacteria and *Le. mexicana*	
Pi. evansi	*E. cloacae* *Pantoea ananatis*	– Extracts of these bacteria showed growth inhibitory activity against *Le. mexicana*	[106]
	E. hormaechei	– Metabolic extracts of the bacterium evidenced activity against *Le. mexicana* – Can be acquired through horizontal transmission	[30]

need to be conducted to validate this endosymbiont for sand fly control strategies.

In the context of sand flies, efforts have been made to identify suitable candidates for paratransgenesis against *Leishmania* parasites. These efforts encompass various studies investigating the plethora of microorganisms of several species of phlebotomine sand flies both from the Old and New worlds, and the identification of bacteria genera or species that may fulfill the requirements for the implementation of this technique, as can be seen in Table 3.

Identifying suitable candidates for paratransgenesis in sand flies poses some challenges, particularly due to the presence of pathogenic bacteria such as *S. marcescens, P. agglomerans, O. anthropi,* and *Escherichia coli.* While some of these microorganisms have been investigated for paratransgenesis in other insect vectors like mosquitoes and triatomines, and are reported in sand flies, they fail to meet the aforementioned criteria. Furthermore, several of these bacteria exhibit varying degrees of pathogenicity in humans or insects [37, 40, 157–159].

The case of *Serratia* is particularly noteworthy, as *S. marcescens* is entomopathogenic, and while it is present during the initial larval stages, it does not persist through the metamorphosis of the sand fly [89]. Consequently, it does not meet the criteria for control strategies based on paratransgenesis. However, research has shown that transformed *Serratia* AS1, introduced through rodent food, can be transmitted from rodent feces to larvae and then to adult phlebotomine sand flies [153], indicating differential behavior regarding transstadial transmission among *Serratia* strains. Moreover, studies have demonstrated that sand flies infected with the entomopathogenic bacteria, in the presence of *Leishmania,* exhibit increased survival rates [91] suggesting a tripartite interaction among the insect, the parasite, and the bacteria that requires further investigation. Similar to *Serratia*, while some authors have found that *Microbacterium* sp. are capable of transstadial transmission [39] others have observed their presence only in the larval stages of *P. papatasi* but not in adults [89].

The different findings among researchers regarding bacteria studied for paratransgenesis highlight the need for additional studies to fully comprehend the implications of various symbionts, and the microbiota in general, as potential candidates for the development of biological control strategies. These studies should explore the intricate interactions within the system comprising bacteria, sand fly species, and *Leishmania* parasites.

9 Virome in Sand Flies: A Field in Need of In-Depth Exploration

The study of phlebotomine sand flies holds significance not only due to their pivotal role in transmitting *Leishmania* species, but also because they are capable of transmitting other diseases of viral origin [160, 161]. Specifically, the virome in phlebotomine sand flies, i.e., the set of viruses present within an organism, may influence human and animal health and affect host phenotypes through interactions with the microbiome, thereby impacting vector competence [162, 163] as interactions between the host and virus can be important determinants of the outcome of infection [164].

A search for unique viral sequences reported in phlebotomine sand flies samples (NCBI: txid7198) in the GenBank nucleotide database (Review: 27.06.2022) revealed the presence of 85 RNA virus sequences classified into seven viral families. Among these, 64% of the reported entries belonged to the *Phenuiviridae* family. Other families, such as *Reoviridae*, *Rhabdoviridae*, *Peribunyaviridae*, *Flaviviridae*, *Nodaviridae*, and *Mitoviridae*, were represented at lower frequencies. However, due to the massive high-throughput sequencing efforts, the number of reported viral sequences is constantly increasing [162]. To date, at least four viruses have been reported to be hosted by *Lu. longipalpis*. These include *Lutzomyia reovirus 1* and *Lutzomyia reovirus 2* of the *Reoviridae* family, *Lutzomyia nodavirus* of the *Nodaviridae* family, and *Lutzomyia longipalpis mitovirus 1*, the sole representative of the *Mitoviridae* family reported in phlebotomine sand flies thus far [165, 166].

10 Sand Fly Bites, Saliva, and Gut Microbiota: Key Players in *Leishmania* Infection Development

Microbiota plays a crucial role in the development of *Leishmania* infection, not only in insects but also in mammals. Recent research has revealed that factors derived from both sand flies (such as saliva and gut microbiota) and *Leishmania* (such as the promastigote secretory gel and exosomes) actively contribute to vector transmission, facilitating parasite survival and establishment in the host [167]. Exposure to sand fly bites or saliva also induces antibody production in both human and animal models [168].

Dey et al. [22] reported that sand fly gut microbiota is transferred to the bite site, promoting neutrophil recruitment and parasite dissemination to distal organs, thus expanding the repertoire of bioactive compounds associated with saliva including the commensal gut microbiota of the vector [22, 169]. More recently, Amni et al. [170] demonstrated that bacteria such as *E. cloacae* and *B. subtilis* significantly influence the infection process caused by *Le. major* parasites. Their findings shed light on the natural transmission of infection through sand fly bite, showing that the effects varied based on the type and dose of bacteria inoculum, and resulted in alterations in the local expression of pro- and anti-inflammatory cytokines in mice receiving different inoculations of bacteria. Amni et al. [170] developed an infection model to mimic transmission by sand flies and examine differences in the onset and progression of *Le. major* infection initiated by inoculation with "low" or "high" doses of these bacteria.

In this context, the potential introduction of commensal microbes via saliva to the bite site during blood feeding and their potential impact on *Leishmania* transmission processes represent a novel area of study that may elucidate aspects of the host immune response and generate more precise treatment strategies, given the variability of clinical manifestation.

11 Conclusions

The intestinal microbiome of sand flies represents a critically important area of study for elucidating differences in vectorial competence between species and understanding the factors influencing the establishment and development of *Leishmania* parasites. Additionally, it serves as a valuable source of bacterial and/or molecular resources with high biotechnological potential to impact pathogen development in crucial organs for infection or to directly affect the survival of the insect vector. Systematic and in-depth studies of paratransgenesis, along with both in vitro and in vivo platforms employing co-infection of bacteria and parasites, are imperative. These efforts should be complemented using animal models to evaluate the ultimate impact on *Leishmania* infection and the evolution of the host immune response. Furthermore, there is a pressing need for more robust metagenomics approaches and further exploration of the virome and mycobiomes in the context of parasite establishment and/or interaction with pathogenic bacteria. Finally, the diversity of secondary endosymbionts, including not only of *Wolbachia* but also others such as *Microsporidia,* among others, requires thorough exploration for their potential importance in alternative biological control strategies.

Acknowledgments The authors would like to thank the Universidad Nacional de Colombia Sede Medellín for the permission granted for the use of electronic resources as well as the research group MICROBIOP for all the suggestions made throughout the elaboration of this review.

Funding This work was supported by projects Hermes 57545, "Interdisciplinary Strategic Alliance Between Leticia, Medellín, and La Paz for the Study of the Microbiome of Insect Vectors of Tropical Diseases and Its Relationship with Climate Change and Society," and Hermes 57720, "Assessment of the Insecticide Degradation Capacity of Bacterial Strains from the Gut of Insects of Agronomic and Public Health Interest in Colombia" of the Universidad Nacional de Colombia Sede Medellín.

References

1. Alvar J, Vélez ID, Bern C, Herrero M, Desjeux P, Cano J, Jannin J, Boer MD, the WHO Leishmaniasis Control Team (2012) Leishmaniasis worldwide and global estimates of its incidence. PLoS One 7(5):e35671. https://doi.org/10.1371/journal.pone.0035671
2. World Health Organization (2024) The global health observatory: explore a world of health data. *Leishmaniasis*. https://www.who.int/data/gho/data/themes/topics/topic-details/GHO/leishmaniasis. Accessed March 2024
3. Alvar J, Yactayo S, Bern C (2006) Leishmaniasis and poverty. Trends Parasitol 22(12):552–557. https://doi.org/10.1016/j.pt.2006.09.004
4. World Health Organization (2024) Leishmaniasis. https://www.who.int/news-room/fact-sheets/detail/leishmaniasis
5. El-Sayed A, Kamel M (2020) Climatic changes and their role in emergence and re-emergence of diseases. Environ Sci Pollut Res 27(18):22336–22352. https://doi.org/10.1007/s11356-020-08896-w
6. Muskus CE, Villa MM (2002) Metaciclogénesis: un proceso fundamental en la biología de Leishmania. [Metacyclogenesis: a basic process in the biology of Leishmania]. Biomedica 22(2):167–177
7. Ramalho-Ortigao M, Saraiva EM, Traub-Csekö YM (2010) Sand fly-Leishmania interactions: long relationships are not necessarily easy. Open Parasitol J 4(1):195–204. https://doi.org/10.2174/1874421401004010195
8. Tom A, Kumar NP, Kumar A, Saini P (2024) Interactions between Leishmania parasite and sandfly: a review. Parasitol Res 123(1):6. https://doi.org/10.1007/s00436-023-08043-7
9. Volf P, Myskova J (2007) Sand flies and Leishmania: specific versus permissive vectors. Trends Parasitol 23(3):91–92. https://doi.org/10.1016/j.pt.2006.12.010
10. Cecílio P, Cordeiro-da-Silva A, Oliveira F (2022) Sand flies: basic information on the vectors of leishmaniasis and their interactions with Leishmania parasites. Commun Biol 5(1):305. https://doi.org/10.1038/s42003-022-03240-z
11. Pimenta PF, Saraiva EM, Rowton E, Modi GB, Garraway LA, Beverley SM, Turco SJ, Sacks DL (1994) Evidence that the vectorial competence of phlebotomine sand flies for different species of Leishmania is controlled by structural polymorphisms in the surface lipophosphoglycan. Proc Natl Acad Sci 91(19):9155–9159. https://doi.org/10.1073/pnas.91.19.9155
12. Telleria EL, Martins-da-Silva A, Tempone AJ, Traub-Csekö YM (2018) Leishmania, microbiota and sand fly immunity. Parasitology 145(10):1336–1353. https://doi.org/10.1017/S0031182018001014
13. Sacks DL (2001) Leishmania-sand fly interactions controlling species-specific vector competence. Cell Microbiol 3(4):189–196. https://doi.org/10.1046/j.1462-5822.2001.00115.x
14. Sacks D, Kamhawi S (2001) Molecular aspects of parasite-vector and vector-host interactions in leishmaniasis. Annu Rev Microbiol 55(1):453–483
15. Engel P, Moran NA (2013) The gut microbiota of insects—diversity in structure and function. FEMS Microbiol Rev 37(5):699–735. https://doi.org/10.1111/1574-6976.12025
16. Tabbabi A, Mizushima D, Yamamoto DS, Kato H (2023) Effects of host species on microbiota composition in Phlebotomus and Lutzomyia sand flies. Parasit Vectors 16(1):310. https://doi.org/10.1186/s13071-023-05939-2
17. Monteiro CC, Villegas LEM, Campolina TB, Pires ACMA, Miranda JC, Pimenta PFP, Secundino NFC (2016) Bacterial diversity of the American sand fly Lutzomyia intermedia using high-throughput metagenomic sequencing. Parasit Vectors 9(1):480. https://doi.org/10.1186/s13071-016-1767-z
18. Vivero RJ, Jaramillo NG, Cadavid-Restrepo G, Soto SIU, Herrera CXM (2016) Structural differences in gut bacteria communities in developmental stages of natural populations of Lutzomyia evansi from Colombia's Caribbean coast. Parasit Vectors 9(1):496. https://doi.org/10.1186/s13071-016-1766-0
19. Dillon RJ (2019) A transdisciplinary consideration of sand flies & leishmaniasis. C R Biol 342(7–8):272–273. https://doi.org/10.1016/j.crvi.2019.09.025
20. Azarm A, Koosha M, Dalimi A, Zahraie-Ramazani A, Akhavan AA, Saeidi Z, Mohebali M, Azam K, Vatandoost H, Oshaghi MA (2024) Association between *Wolbachia* infection and susceptibility to Deltamethrin insecticide in *Phlebotomus papatasi* (Diptera: Psychodidae), the main vector of zoonotic cutaneous leishmaniasis. Vector-Borne and Zoonotic Diseases 24(3):159–165. https://doi.org/10.1089/vbz.2023.0085
21. Duque-Granda D, Vivero-Gómez RJ, Junca H, Cadavid-Restrepo G, Moreno-Herrera CX (2024) Interactions and effects of temperature preference under a controlled environment on the diversity and abundance of the microbiome in *Lutzomyia longipalpis* (Diptera: Psychodidae). Biotechnol Rep 44:e00857. https://doi.org/10.1016/j.btre.2024.e00857
22. Dey R, Joshi AB, Oliveira F, Pereira L, Guimarães-Costa AB, Serafim TD et al (2018) Gut microbes egested during bites of infected sand flies augment severity of leishmaniasis via inflammasome-derived IL-1β. Cell Host Microbe 23(1):134–143. https://doi.org/10.1016/j.chom.2017.12.002
23. Amorim CF, Lovins VM, Singh TP, Novais FO, Harris JC, Lago AS, Carvahlo LP, Carvalho M, Beiting DP, Scott P, Grice EA (2023) Multiomic profiling of cutaneous leishmaniasis infections reveals microbiota-driven mechanisms underlying disease severity. Sci Transl Med 15. https://doi.org/10.1126/scitranslmed.adh1469

24. Cansado-Utrilla C, Zhao SY, McCall PJ, Coon KL, Hughes GL (2021) The microbiome and mosquito vectorial capacity: rich potential for discovery and translation. Microbiome 9(1):111. https://doi.org/10.1186/s40168-021-01073-2
25. Maitre A, Wu-Chuang A, Aželytė J, Palinauskas V, Mateos-Hernández L, Obregon D, Hodžić A, Valiente Moro C, Estrada-Peña A, Paoli J-C, Falchi A, Cabezas-Cruz A (2022) Vector microbiota manipulation by host antibodies: the forgotten strategy to develop transmission-blocking vaccines. Parasit Vectors 15(1):4. https://doi.org/10.1186/s13071-021-05122-5
26. Herren JK, Mbaisi L, Mararo E, Makhulu EE, Mobegi VA, Butungi H, Mancini MV, Oundo JW, Teal ET, Pinaud S, Lawniczak MKN, Jabara J, Nattoh G, Sinkins SP (2020) A microsporidian impairs *Plasmodium falciparum* transmission in *Anopheles arabiensis* mosquitoes. Nat Commun 11(1):2187. https://doi.org/10.1038/s41467-020-16121-y
27. Hoffmann AA, Montgomery BL, Popovici J, Iturbe-Ormaetxe I, Johnson PH, Muzzi F, Greenfield M, Durkan M, Leong YS, Dong Y, Cook H, Axford J, Callahan AG, Kenny N, Omodei C, McGraw EA, Ryan PA, Ritchie SA, Turelli M, O'Neill SL (2011) Successful establishment of Wolbachia in Aedes populations to suppress dengue transmission. Nature 476(7361):454–457. https://doi.org/10.1038/nature10356
28. Katak RDM, Cintra AM, Burini BC, Marinotti O, Souza-Neto JA, Rocha EM (2023) Biotechnological potential of microorganisms for mosquito population control and reduction in vector competence. Insects 14(9):718. https://doi.org/10.3390/insects14090718
29. Schlein Y, Polacheck I, Yuval B (1985) Mycoses, bacterial infections and antibacterial activity in sandifies (Psychodidae) and their possible role in the transmission of leishmaniasis. Parasitology 90(1):57–66. https://doi.org/10.1017/S0031182000049015
30. Vivero RJ, Mesa GB, Robledo SM, Herrera CXM, Cadavid-Restrepo G (2019) Enzymatic, antimicrobial, and leishmanicidal bioactivity of gram-negative bacteria strains from the midgut of *Lutzomyia evansi*, an insect vector of leishmaniasis in Colombia. Biotechnol Rep 24:e00379. https://doi.org/10.1016/j.btre.2019.e00379
31. Omondi ZN, Demir S (2021) Bacteria composition and diversity in the gut of sand fly: impact on Leishmania and sand fly development. Int J Trop Insect Sci 41(1):25–32. https://doi.org/10.1007/s42690-020-00184-x
32. Peterkova-Koci K, Robles-Murguia M, Ramalho-Ortigao M, Zurek L (2012) Significance of bacteria in oviposition and larval development of the sand fly Lutzomyia longipalpis. Parasit Vectors 5(1):145. https://doi.org/10.1186/1756-3305-5-145
33. Telleria EL, Sant'Anna MRV, Alkurbi MO, Pitaluga AN, Dillon RJ, Traub-Csekö YM (2013) Bacterial feeding, Leishmania infection and distinct infection routes induce differential defensin expression in Lutzomyia longipalpis. Parasit Vectors 6(1):12. https://doi.org/10.1186/1756-3305-6-12
34. Vivero RJ, Castañeda-Monsalve VA, Atencia MC, Hoyos-Lopez R, Hurst GD, Cadavid-Restrepo G, Moreno-Herrera CX (2021) Molecular phylogeny of heritable symbionts and microbiota diversity analysis in phlebotominae sand flies and *Culex nigripalpus* from Colombia. PLoS Negl Trop Dis 15(12):e0009942. https://doi.org/10.1371/journal.pntd.0009942
35. Heerman M, Weng J-L, Hurwitz I, Durvasula R, Ramalho-Ortigao M (2015) Bacterial infection and immune responses in Lutzomyia longipalpis sand fly larvae midgut. PLoS Negl Trop Dis 9(7):e0003923. https://doi.org/10.1371/journal.pntd.0003923
36. Tinoco-Nunes B, Telleria EL, Da Silva-Neves M, Marques C, Azevedo-Brito DA, Pitaluga AN, Traub-Csekö YM (2016) The sandfly Lutzomyia longipalpis LL5 embryonic cell line has active Toll and Imd pathways and shows immune responses to bacteria, yeast and Leishmania. Parasit Vectors 9(1):222. https://doi.org/10.1186/s13071-016-1507-4
37. Vaselek S (2024) Overview of microbial studies in sandflies and their progress toward development of paratransgenic approach for the control of *Leishmania* sp. Front Trop Dis 5. https://doi.org/10.3389/fitd.2024.1369077
38. Louradour I, Monteiro CC, Inbar E, Ghosh K, Merkhofer R, Lawyer P, Paun A, Smelkinson M, Secundino N, Lewis M, Erram D, Zurek L, Sacks D (2017) The midgut microbiota plays an essential role in sand fly vector competence for *Leishmania major*. Cell Microbiol 19(10):e12755. https://doi.org/10.1111/cmi.12755
39. Guernaoui S, Garcia D, Gazanion E, Ouhdouch Y, Boumezzough A, Pesson B, Fontenille D, Sereno D (2011) Bacterial flora as indicated by PCR-temperature gradient gel electrophoresis (TGGE) of 16S rDNA gene fragments from isolated guts of phlebotomine sand flies (Diptera: Psychodidae). J Vector Ecol 36(Suppl 1):S144–S147. https://doi.org/10.1111/j.1948-7134.2011.00124.x
40. Volf P, Kiewegová A, Nemec A (2002) Bacterial colonisation in the gut of Phlebotomus duboseqi (Diptera: Psychodidae): transtadial passage and the role of female diet. Folia Parasitol 49(1):73–77. https://doi.org/10.14411/fp.2002.014
41. Papadopoulos C, Karas PA, Vasileiadis S, Ligda P, Saratsis A, Sotiraki S, Karpouzas DG (2020) Host species determines the composition of the prokaryotic microbiota in phlebotomus sandflies. Pathogens 9(6):6. https://doi.org/10.3390/pathogens9060428
42. Maleki-Ravasan N, Oshaghi MA, Afshar D, Arandian MH, Hajikhani S, Akhavan AA, Yakhchali B, Shirazi MH, Rassi Y, Jafari R, Aminian K, Fazeli-Varzaneh RA, Durvasula R (2015) Aerobic bacterial flora of biotic and abiotic compartments of a hyperendemic Zoonotic Cutaneous Leishmaniasis (ZCL) focus. Parasit Vectors 8(63). https://doi.org/10.1186/s13071-014-0517-3

43. Mukhopadhyay J, Braig H, Rowton E, Ghosh K (2012) Naturally occurring culturable aerobic gut flora of adult Phlebotomus papatasi, vector of Leishmania major in the Old World. PLoS One 7(5):e35748. https://doi.org/10.1371/journal.pone.0035748
44. Akhoundi M, Bakhtiari R, Guillard T, Baghaei A, Tolouei R, Sereno D, Toubas D, Depaquit J, Abyaneh MR (2012) Diversity of the bacterial and fungal microflora from the midgut and cuticle of phlebotomine sand flies collected in North-Western Iran. PLoS One 7(11):e50259. https://doi.org/10.1371/journal.pone.0050259
45. Dillon RJ, Kordy EE, Shehata M, Lane RP (1996) The prevalence of a microbiota in the digestive tract of Phlebotomus papatasi. Ann Trop Med Parasitol 90(6):669–673. https://doi.org/10.1080/00034983.1996.11813102
46. Fraihi W, Fares W, Perrin P, Dorkeld F, Sereno D, Barhoumi W, Sbissi I, Cherni S, Chelbi I, Durvasula R, Ramalho-Ortigao M, Gtari M, Zhioua E (2017) An integrated overview of the midgut bacterial flora composition of *Phlebotomus perniciosus*, a vector of zoonotic visceral leishmaniasis in the Western Mediterranean Basin. PLoS Negl Trop Dis 11(3):e0005484. https://doi.org/10.1371/journal.pntd.0005484
47. Campolina TB, Villegas LEM, Monteiro CC, Pimenta PFP, Secundino NFC (2020) Tripartite interactions: Leishmania, microbiota and Lutzomyia longipalpis. PLoS Negl Trop Dis 14(10). https://doi.org/10.1371/journal.pntd.0008666
48. Pires ACAM, Villegas LEM, Campolina TB, Orfanó AS, Pimenta PFP, Secundino NFC (2017) Bacterial diversity of wild-caught Lutzomyia longipalpis (a vector of zoonotic visceral leishmaniasis in Brazil) under distinct physiological conditions by metagenomics analysis. Parasit Vectors 10(1):627. https://doi.org/10.1186/s13071-017-2593-7
49. Kelly PH, Bahr SM, Serafim TD, Ajami NJ, Petrosino JF, Meneses C, Kirby JR, Valenzuela JG, Kamhawi S, Wilson ME (2017) The gut microbiome of the vector Lutzomyia longipalpis is essential for survival of Leishmania infantum. mBio 8(1). https://doi.org/10.1128/mbio.01121-16
50. Sant'Anna MRV, Darby AC, Brazil RP, Montoya-Lerma J, Dillon VM, Bates PA, Dillon RJ (2012) Investigation of the bacterial communities associated with females of Lutzomyia sand fly species from South America. PLoS One 7(8):e42531. https://doi.org/10.1371/journal.pone.0042531
51. Pereira de Oliveira SM, de Morais BA, Gonçalves CA, Giordano-Dias CM, Vilela ML, Brazil RP, D'Almeida JM, Asensi MD, Mello RP (2001) Digestive tract microbiota in female Lutzomyia longipalpis (Lutz & Neiva, 1912) (Diptera: Psychodidae) from colonies feeding on blood meal and sucrose plus blood meal. Cad Saude Publica 17(1):229–232
52. de Oliveira SMP, de Moraes BA, Gonçalves CA, Giordano-Dias CM, d'Almeida JM, Asensi MD, Mello RP, Brazil RP (2000) Prevalência da microbiota no trato digestivo de fêmeas de Lutzomyia longipalpis (Lutz & Neiva, 1912) (Diptera: Psychodidae) provenientes do campo. Rev Soc Bras Med Tro 33:319–322. https://doi.org/10.1590/S0037-86822000000300012
53. Vivero RJ, Villegas-Plazas M, Cadavid-Restrepo GE, Herrera CXM, Uribe SI, Junca H (2019) Wild specimens of sand fly phlebotomine Lutzomyia evansi, vector of leishmaniasis, show high abundance of *Methylobacterium* and natural carriage of *Wolbachia* and *Cardinium* types in the midgut microbiome. Sci Rep 9(1):17746. https://doi.org/10.1038/s41598-019-53769-z
54. Tabbabi A, Watanabe S, Mizushima D, Caceres AG, Gomez EA, Yamamoto DS, Cui L, Hashiguchi Y, Kato H (2020) Comparative analysis of bacterial communities in *Lutzomyia ayacuchensis* populations with different vector competence to *Leishmania* Parasites in Ecuador and Peru. Microorganisms 9(1):E68. https://doi.org/10.3390/microorganisms9010068
55. Vivero RJ, Castañeda-Monsalve VA, Romero LR, Hurst GD, Cadavid-Restrepo G, Moreno-Herrera CX (2021) Gut microbiota dynamics in natural populations of Pintomyia evansi under experimental infection with Leishmania infantum. Microorganisms 9(6):6. https://doi.org/10.3390/microorganisms9061214
56. de Oliveira SMP, de Morais BA, Gonçalves CA, Giordano-Dias CM, Vilela ML, Brazil RP, d'Almeida JM, Asensi MD, Mello RP (2001) Microbiota do trato digestivo de fêmeas de Lutzomyia longipalpis (Lutz & Neiva, 1912) (Diptera: Psychodidae) provenientes de colônia alimentadas com sangue e com sangue e sacarose. Cad Saúde Pública 17:229–232. https://doi.org/10.1590/S0102-311X2001000100024
57. Sekiguchi H, Tomioka N, Nakahara T, Uchiyama H (2001) A single band does not always represent single bacterial strains in denaturing gradient gel electrophoresis analysis. Biotechnol Lett 23(15):1205–1208. https://doi.org/10.1023/A:1010517117046
58. Cui H-L, Zhou P-J, Oren A, Liu S-J (2008) Intraspecific polymorphism of 16S rRNA genes in two halophilic archaeal genera, *Haloarcula* and *Halomicrobium*. Extremophiles 13(1):31–37. https://doi.org/10.1007/s00792-008-0194-2
59. Poretsky R, Rodriguez-R LM, Luo C, Tsementzi D, Konstantinidis KT (2014) Strengths and limitations of 16S rRNA gene amplicon sequencing in revealing temporal microbial community dynamics. PLoS One 9(4):e93827. https://doi.org/10.1371/journal.pone.0093827
60. McCarthy CB, Diambra LA, Pomar RVR (2011) Metagenomic analysis of taxa associated with Lutzomyia longipalpis, vector of visceral leishmaniasis, using an unbiased high-throughput approach. PLoS Negl Trop Dis 5(9):e1304. https://doi.org/10.1371/journal.pntd.0001304

61. Romo Bechara N, Wasserberg G, Raymann K (2022) Microbial ecology of sand fly breeding sites: aging and larval conditioning alter the bacterial community composition of rearing substrates. Parasit Vectors 15(1):265. https://doi.org/10.1186/s13071-022-05381-w
62. Wang J, Gou Q, Luo G, Hou X, Liang G, Shi M (2022) Total RNA sequencing of Phlebotomus chinensis sandflies in China revealed viral, bacterial, and eukaryotic microbes potentially pathogenic to humans. Emerg Microbes Infect 11(1):2080–2092. https://doi.org/10.1080/22221751.2022.2109516
63. Keeling PJ, McCutcheon JP (2017) Endosymbiosis: the feeling is not mutual. J Theor Biol 434:75–79. https://doi.org/10.1016/j.jtbi.2017.06.008
64. Wernegreen JJ (2004) Endosymbiosis: lessons in conflict resolution. PLoS Biol 2(3):E68. https://doi.org/10.1371/journal.pbio.0020068
65. Estrela S, Kerr B, Morris JJ (2016) Transitions in individuality through symbiosis. Curr Opin Microbiol 31:191–198. https://doi.org/10.1016/j.mib.2016.04.007
66. Husnik F, Keeling PJ (2019) The fate of obligate endosymbionts: reduction, integration, or extinction. Curr Opin Genet Dev 58–59:1–8. https://doi.org/10.1016/j.gde.2019.07.014
67. Bennet GM, Moran NA (2015) Heritable symbiosis: the advantages and perils of an evolutionary rabbit hole. J Proc Natl Acad Sci 112(33):10169–10176. https://doi.org/10.1073/pnas.1421388112
68. Heyworth ER, Ferrari J (2015) A facultative endosymbiont in aphids can provide diverse ecological benefits. J Evol Biol 28(10):1753–1760. https://doi.org/10.1111/jeb.12705
69. Sanches P, De Moraes CM, Mescher MC (2023) Endosymbionts modulate virus effects on aphid-plant interactions. ISME J 17(1):2441–2451. https://doi.org/10.1038/s41396-023-01549-z
70. Xi Z, Ramirez JL, Dimopoulos G (2008) The *Aedes aegypti* toll pathway controls dengue virus infection. PLoS Pathog 4(7):e1000098. https://doi.org/10.1371/journal.ppat.1000098
71. Ono M, Braig HR, Munstermann LE, Ferro C, O'Neill SL (2001) *Wolbachia* infections of phlebotomine sand flies (Diptera: Psychodidae). J Med Entomol 38(2):237–241. https://doi.org/10.1603/0022-2585-38.2.237
72. Alipour H, Izadpanah L, Azizi K, Shahriari-Namadi M, Kalantari M (2021) Potential co-infection of *Wolbachia* with *Leishmania* among sand fly vectors caught from endemic leishmaniasis foci in Fars province, southern Iran. J Parasit Dis 45(3):817–822. https://doi.org/10.1007/s12639-021-01366-1
73. Duque-Granda D, Moreno-Herrera CX, Cadavid-Restrepo GE, Vivero-Gómez RJ (2023) Molecular detection and phylogenetic analyses of Arsenophonus endosymbiont in wild specimens of phlebotomine sand flies from Colombia. J Asia Pac Entomol 26(1):102023. https://doi.org/10.1016/j.aspen.2022.102023
74. Karatepe B, Aksoy S, Karatepe M (2018) Investigation of Wolbachia spp. and Spiroplasma spp. in Phlebotomus species by molecular methods. Sci Rep 8(1):10616. https://doi.org/10.1038/s41598-018-29031-3
75. Karimian F, Vatandoost H, Rassi Y, Maleki-Ravasan N, Choubdar N, Koosha M, Arzamani K, Moradi-Asl E, Veysi A, Alipour H, Shirani M, Oshaghi MA (2018) wsp-based analysis of Wolbachia strains associated with Phlebotomus papatasi and P. sergenti (Diptera: Psychodidae) main cutaneous leishmaniasis vectors, introduction of a new subgroup wSerg. Pathog Global Health 112(3):152–160. https://doi.org/10.1080/20477724.2018.1471438
76. Li K, Chen H, Jiang J, Li X, Xu J, Ma Y (2016) Diversity of bacteriome associated with Phlebotomus chinensis (Diptera: Psychodidae) sand flies in two wild populations from China. Sci Rep 6:36406. https://doi.org/10.1038/srep36406
77. Lozano-Sardaneta YN, Valderrama A, Sánchez-Montes S, Grostieta E, Colunga-Salas P, Sánchez-Cordero V, Becker I (2021) Rickettsial agents detected in the genus Psathyromyia (Diptera:Phlebotominae) from a Biosphere Reserve of Veracruz, Mexico. Parasitol Int 82:102286. https://doi.org/10.1016/j.parint.2021.102286
78. Lozano-Sardaneta YN, Jacobo-Olvera E, Ruiz-Tovar K, Sánchez-Montes S, Rodríguez-Rojas JJ, Fernández-Figueroa EA, Roldán-Fernández SG, Rodriguez-Martinez LM, Dzul-Manzanilla F, Correa-Morales F, Treviño-Garza N, Díaz-Albíter HM, Zwetsch A, Valadas SYOB, Nilce-Silveira A, Becker I, Huerta H (2022) Detection of *Wolbachia* and *Leishmania* DNA in sand flies (Diptera: Psychodidae, Phlebotominae) from a focus of cutaneous leishmaniasis in Tabasco, Mexico. Parasitol Res 121(2):513–520. https://doi.org/10.1007/s00436-021-07412-4
79. Lozano-Sardaneta YN, Marina CF, Torres-Monzón JA, Sánchez-Cordero V, Becker I (2023) Molecular detection of *Wolbachia* and *Bartonella* as part of the microbiome of phlebotomine sand flies from Chiapas, Mexico. Parasitol Res 122(6):1293–1301. https://doi.org/10.1007/s00436-023-07829-z
80. Matsumoto K, Izri A, Dumon H, Raoult D, Parola P (2008) First detection of Wolbachia spp., including a new genotype, in sand flies collected in Marseille, France. J Med Entomol 45(3):466–469. https://doi.org/10.1603/0022-2585(2008)45[466:fdowsi]2.0.co;2
81. Parvizi P, Fardid F, Soleimani S (2013) Detection of a new strain of Wolbachia pipientis in Phlebotomus perfiliewi transcaucasicus, a potential vector of Visceral Leishmaniasis in North West of Iran, by targeting the major surface protein gene. J Arthropod Borne Dis 7(1):46–55
82. Latrofa MS, Varotto-Boccazzi I, Louzada-Flores VN, Iata R, Mendoza-Roldan JA, Roura X, Zatelli A, Epis S, Bandi C, Otranto D (2023) Interaction between Wolbachia pipientis and Leishmania infantum in heartworm infected dogs. Parasit Vectors 16(77). https://doi.org/10.1186/s13071-023-05662-y

83. Johnson KN (2015) Bacteria and antiviral immunity in insects. Curr Opin Insect Sci 8:97–103. https://doi.org/10.1016/j.cois.2015.01.008
84. Bell-Sakyi L, Beliavskaia A, Hartley CS, Jones L, Luu L, Haines LR, Hamilton JGC, Darby AC, Makepeace BL (2021) Isolation in natural host cell lines of *Wolbachia* strains *w*Pip from the mosquito *Culex pipiens* and *w*Pap from the sand fly *Phlebotomus papatasi*. Insects 12(10):871. https://doi.org/10.3390/insects12100871
85. Ballinger MJ, Perlman SJ (2019) The defensive *Spiroplasma*. Curr Opin Insect Sci 32:36–41. https://doi.org/10.1016/j.cois.2018.10.004
86. Mancini MV, Damiani C, Short SM, Cappelli A, Ulissi U, Capone A, Serrao A, Rossi P, Amici A, Kalogris C, Dimopoulos G, Ricci I, Favia G (2020) Inhibition of *Asaia* in adult mosquitoes causes male-specific mortality and diverse transcriptome changes. Pathogens 9(5):380. https://doi.org/10.3390/pathogens9050380
87. Massey JH, Newton ILG (2022) Diversity and function of arthropod endosymbiont toxins. Trends Microbiol 30(2):185–198. https://doi.org/10.1016/j.tim.2021.06.008
88. Boelaert M, Sundar S (2014) Leishmaniasis. In: Farrar J, Hotez PJ, Junghanss T, Kang G, Lalloo D, White N (eds) Manson's tropical diseases, 23rd edn. Elsevier, p 636
89. Vaselek S, Sarac BE, Uzunkaya AD, Yilmaz A, Karaaslan C, Alten B (2024) Identification of *Ochrobactrum* as a bacteria with transstadial transmission and potential for application in paratransgenic control of leishmaniasis. Parasitol Res 123(82):1–10. https://doi.org/10.1007/s00436-023-08087-9
90. Karimian F, Koosha M, Choubdar N, Oshaghi MA (2022) Comparative analysis of the gut microbiota of sand fly vectors of zoonotic visceral leishmaniasis (ZVL) in Iran; host-environment interplay shapes diversity. PLoS Negl Trop Dis 16(7):e0010609. https://doi.org/10.1371/journal.pntd.0010609
91. Sant'Anna MR, Diaz-Albiter H, Aguiar-Martins K, Al Salem WS, Cavalcante RR, Dillon VM, Bates PA, Genta FA, Dillon RJ (2014) Colonisation resistance in the sand fly gut: Leishmania protects Lutzomyia longipalpis from bacterial infection. Parasit Vectors 7(1):329. https://doi.org/10.1186/1756-3305-7-329
92. Shaw WR, Catteruccia F (2019) Vector biology meets disease control: using basic research to fight vector-borne diseases. Nat Microbiol 4(1):20–34. https://doi.org/10.1038/s41564-018-0214-7
93. Coon KL, Brown MR, Strand MR (2016) Gut bacteria differentially affect egg production in the anautogenous mosquito Aedes aegypti and facultatively autogenous mosquito Aedes atropalpus (Diptera: Culicidae). Parasit Vectors 9(1):375. https://doi.org/10.1186/s13071-016-1660-9
94. Douglas AE (2015) Multiorganismal insects: diversity and function of resident microorganisms. Annu Rev Entomol 60(1):17–34. https://doi.org/10.1146/annurev-ento-010814-020822
95. Kwong WK, Mancenido AL, Moran NA (2017) Immune system stimulation by the native gut microbiota of honey bees. R Soc Open Sci 4(2):170003. https://doi.org/10.1098/rsos.170003
96. Wong AC-N, Dobson AJ, Douglas AE (2014) Gut microbiota dictates the metabolic response of *Drosophila* to diet. J Exp Biol. https://doi.org/10.1242/jeb.101725
97. Da Silva Gonçalves D, Iturbe-Ormaetxe I, Martins-da-Silva A, Telleria EL, Rocha MN, Traub-Csekö YM, O'Neill SL, Sant'Anna MRV, Moreira LA (2019) Wolbachia introduction into Lutzomyia longipalpis (Diptera: Psychodidae) cell lines and its effects on immune-related gene expression and interaction with Leishmania infantum. Parasit Vectors 12(1):33. https://doi.org/10.1186/s13071-018-3227-4
98. Boulanger N, Lowenberger C, Volf P, Ursic R, Sigutova L, Sabatier L, Svobodova M, Beverley SM, Späth G, Brun R, Pesson B, Bulet P (2004) Characterization of a Defensin from the Sand Fly *Phlebotomus duboscqi* induced by challenge with bacteria or the protozoan parasite *Leishmania major*. Infect Immun 72(12):7140–7146. https://doi.org/10.1128/IAI.72.12.7140-7146.2004
99. Santos VLS, Rosa FC, Dias LRL, Batista LA, Melo MC, Sales LAT, Branco ADJM, Araújo TRR, De Miranda RDCM, Aliança ASDS (2024) An evaluation of the antibacterial, antileishmanial, and cytotoxic potential of the secondary metabolites of Streptomyces sp. ARH (A3). Microorganisms 12(3):476. https://doi.org/10.3390/microorganisms12030476
100. Moraes CS, Seabra SH, Castro DP, Brazil RP, De Souza W, Garcia ES, Azambuja P (2008) Leishmania (Leishmania) chagasi interactions with Serratia marcescens: Ultrastructural studies, lysis and carbohydrate effects. Exp Parasitol 118(4):561–568. https://doi.org/10.1016/j.exppara.2007.11.015
101. Suryawanshi RK, Patil CD, Koli SH, Hallsworth JE, Patil SV (2017) Antimicrobial activity of prodigiosin is attributable to plasma-membrane damage. Nat Prod Re 31(5):572–577. https://doi.org/10.1080/14786419.2016.1195380
102. Moraes CS, Seabra SH, Albuquerque-Cunha JM, Castro DP, Genta FA, Souza WD, Brazil RP, Garcia ES, Azambuja P (2009) Prodigiosin is not a determinant factor in lysis of Leishmania (Viannia) braziliensis after interaction with Serratia marcescens d-mannose sensitive fimbriae. Exp Parasitol 122(2):84–90. https://doi.org/10.1016/j.exppara.2009.03.004
103. Williams RP (1973) Biosynthesis of Prodigiosin, a secondary metabolite of Serratia marcescens. Appl Microbiol 25(3):396–402. https://doi.org/10.1128/am.25.3.396-402.1973
104. Krawczyk K, Szabelska-Beręsewicz A, Przemieniecki SW, Szymańczyk M, Obrępalska-Stęplowska A (2022) Insect Gut Bacteria Promoting the Growth of Tomato Plants (Solanum lycopersi-

cum L.). Int J Mol Sci 23(21):13548. https://doi.org/10.3390/ijms232113548
105. Xu Y, Xian Z-N, Yue W, Yin C-F, Zhou N-Y (2023) Degradation of polyvinyl chloride by a bacterial consortium enriched from the gut of Tenebrio molitor larvae. Chemosphere 318:137944. https://doi.org/10.1016/j.chemosphere.2023.137944
106. Vivero RJ, Cadavid Restrepo GE, Moreno Herrera CX, Ospina V, Uribe SI, Robledo SM (2016) Antagonistic effect of bacteria isolated from the digestive tract of Lutzomyia evansi against promastigotes of Leishmania infantum, antimicrobial activities and susceptibility to antibiotics. Adv Microbiol 06(10):760–775. https://doi.org/10.4236/aim.2016.610075
107. Kamhawi S (2006) Phlebotomine sand flies and Leishmania parasites: friends or foes? Trends Parasitol 22(9):439–445. https://doi.org/10.1016/j.pt.2006.06.012
108. Dostálová A, Volf P (2012) Leishmania development in sand flies: parasite-vector interactions overview. Parasit Vectors 5(1):276. https://doi.org/10.1186/1756-3305-5-276
109. Telleria EL, Tinoco-Nunes B, Leštinová T, De Avellar LM, Tempone AJ, Pitaluga AN, Volf P, Traub-Csekö YM (2021) Lutzomyia longipalpis antimicrobial peptides: differential expression during development and potential involvement in vector interaction with microbiota and Leishmania. Microorganisms 9(6):1271. https://doi.org/10.3390/microorganisms9061271
110. Cunningham ML, Titus RG, Turco SJ, Beverley SM (2001) Regulation of differentiation to the infective stage of the protozoan parasite Leishmania major by tetrahydrobiopterin. Science 292(5515):285–287. https://doi.org/10.1126/science.1057740
111. Bates PA (2008) Leishmania sand fly interaction: progress and challenges. Curr Opin Microbiol 11(4):340–344. https://doi.org/10.1016/j.mib.2008.06.003
112. Voronin D, Tran-Van V, Potier P, Mavingui P (2010) Transinfection and growth discrepancy of Drosophila Wolbachia strain wMel in cell lines of the mosquito Aedes albopictus. J Appl Microbiol 108(6):2133–2141. https://doi.org/10.1111/j.1365-2672.2009.04621.x
113. Aliota MT, Peinado SA, Velez ID, Osorio JE (2016) The wMel strain of Wolbachia Reduces Transmission of Zika virus by *Aedes aegypti*. Sci Rep 6(1):28792. https://doi.org/10.1038/srep28792
114. Hughes GL, Koga R, Xue P, Fukatsu T, Rasgon JL (2011) Wolbachia infections are virulent and inhibit the human malaria parasite Plasmodium Falciparum in Anopheles gambiae. PLoS Pathog 7(5):e1002043. https://doi.org/10.1371/journal.ppat.1002043
115. Rancès E, Ye YH, Woolfit M, McGraw EA, O'Neill SL (2012) The relative importance of innate immune priming in Wolbachia-mediated dengue interference. PLoS Pathog 8(2):e1002548. https://doi.org/10.1371/journal.ppat.1002548
116. Iturbe-Ormaetxe I, Walker T, O'Neill SL (2011) Wolbachia and the biological control of mosquito-borne disease. EMBO Rep 12(6):508–518. https://doi.org/10.1038/embor.2011.84
117. Gulsen SH, Tileklioglu E, Bode E, Cimen H, Ertabaklar H, Ulug D, Ertug S, Wenski SL, Touray M, Hazir C, Bilecenoglu DK, Yildiz I, Bode HB, Hazir S (2022) Antiprotozoal activity of different Xenorhabdus and Photorhabdus bacterial secondary metabolites and identification of bioactive compounds using the easyPACId approach. Sci Rep 12(1):10779. https://doi.org/10.1038/s41598-022-13722-z
118. Bode E, Heinrich AK, Hirschmann M, Abebew D, Shi Y-N, Vo TD, Wesche F, Shi Y-M, Grün P, Simonyi S, Keller N, Engel Y, Wenski S, Bennet R, Beyer S, Bischoff I, Buaya A, Brandt S, Cakmak I et al (2019) Promoter activation in Δhfq Mutants as an efficient tool for specialized metabolite production enabling direct bioactivity testing. Angew Chem Int Ed Engl 58(52):18957–18963. https://doi.org/10.1002/anie.201910563
119. Fuchs SW, Grundmann F, Kurz M, Kaiser M, Bode HB (2014) Fabclavines: bioactive peptide-polyketide-polyamino hybrids from *Xenorhabdus*. ChemBioChem 15(4):512–516. https://doi.org/10.1002/cbic.201300802
120. McInerney BV, Taylor WC, Lacey MJ, Akhurst RJ, Gregson RP (1991) Biologically active metabolites from Xenorhabdus spp., Part 2. Benzopyran-1-one derivatives with gastroprotective activity. J Nat Prod 54(3):785–795. https://doi.org/10.1021/np50075a006
121. Kefi S, Essid R, Papetti A, Abid G, Bouslama L, Aouani E, Tabbene O, Limam F (2023) Antioxidant, antibacterial, and antileishmanial potential of Micromeria nervosa extracts and molecular mechanism of action of the bioactive compound. J Appl Microbiol 134(2):lxad007. https://doi.org/10.1093/jambio/lxad007
122. Abengózar MÁ, Cebrián R, Saugar JM et al (2017) Enterocin AS-48 as evidence for the use of bacteriocins as New Leishmanicidal agents. Antimicrob Agents Chemother 61(4):e02288–e02216. https://doi.org/10.1128/AAC.02288-16
123. Antonello AM, Sartori T, Folmer Correa AP, Brandelli A, Heermann R, Rodrigues Júnior LC, Peres A, Romão PRT, Da Silva OS (2018) Entomopathogenic bacteria *Photorhabdus luminescens* as drug source against *Leishmania amazonensis*. Parasitology 145(8):1065–1074. https://doi.org/10.1017/S0031182017002001
124. Awada B, Hamie M, El Hajj R, Derbaj G, Najm R, Makhoul P, Ali DH, Abou Fayad AG, El Hajj H (2022) HAS 1: a natural product from soil-isolated Streptomyces species with potent activity against cutaneous leishmaniasis caused by Leishmania tropica. Front Pharmacol 13:1023114. https://doi.org/10.3389/fphar.2022.1023114
125. Leon LL (2001) Antileishmanial activity of the violacein extracted from Chromobacterium violaceum. J Antimicrob Chemother 48(3):449–450. https://doi.org/10.1093/jac/48.3.449

126. Pimentel-Elardo SM, Kozytska S, Bugni TS, Ireland CM, Moll H, Hentschel U (2010) Antiparasitic compounds from Streptomyces sp. strains isolated from mediterranean sponges. Mar Drugs 8(2):373–380. https://doi.org/10.3390/md8020373
127. da Cunha AFS, Oliveira YLDC, Dolabella SS, Scher R, Souto EB, Lopez JA, Jain S (2021) Antileishmanial compounds from microbial metabolites: a promising source. Appl Microbiol Biotechnol 105(21):8227–8240. https://doi.org/10.1007/s00253-021-11610-6
128. Dong Y, Manfredini F, Dimopoulos G (2009) Implication of the mosquito midgut microbiota in the defense against malaria parasites. PLoS Pathog 5(5):e1000423. https://doi.org/10.1371/journal.ppat.1000423
129. Gendrin M, Yerbanga RS, Ouedraogo JB, Lefèvre T, Cohuet A, Christophides GK (2016) Differential effects of Azithromycin, Doxycycline, and Cotrimoxazole in ingested blood on the vectorial capacity of malaria mosquitoes. Open Forum Infect Dis 3(2):ofw074. https://doi.org/10.1093/ofid/ofw074
130. Ma CS, Ma G, Pincebourde S (2021) Survive a warming climate: insect responses to extreme high temperatures. Annu Rev Entomol 66:163–184. https://doi.org/10.1146/annurev-ento-041520-074454
131. Tougeron K, Iltis C, Rampnoux E, Goerlinger A, Dhondt L, Hance T (2023) Still standing: the heat protection delivered by a facultative symbiont to its aphid host is resilient to repeated thermal stress. Curr Res Insect Sci 3:100061. https://doi.org/10.1016/j.cris.2023.100061
132. Corbin C, Heyworth ER, Ferrari J, Hurst GD (2017) Heritable symbionts in a world of varying temperature. Heredity 118(1):10–20. https://doi.org/10.1038/hdy.2016.71
133. Ferguson LV, Dhakal P, Lebenzon JE, Heinrichs DE, Bucking C, Sinclair BJ (2018) Seasonal shifts in the insect gut microbiome are concurrent with changes in cold tolerance and immunity. Funct Ecol. https://doi.org/10.1111/1365-2435.13153
134. Chang CY, Sun XW, Tian PP, Miao NH, Zhang YL, Liu XD (2022) Plant secondary metabolite and temperature determine the prevalence of Arsenophonus endosymbionts in aphid populations. Environ Microbiol 24(8):3764–3776. https://doi.org/10.1111/1462-2920.15929
135. Brumin M, Kontsedalov S, Ghanim M (2011) *Rickettsia* influences thermotolerance in the whitefly *Bemisia tabaci* B biotype. Insect Sci 18(1):57–66. https://doi.org/10.1111/j.1744-7917.2010.01396.x
136. Martins KA, Morais CS, Broughton SJ, Lazzari CR, Bates PA, Pereira MH, Dillon RJ (2023) Response to thermal and infection stresses in an American vector of visceral leishmaniasis. Med Vet Entomol 37(2):238–251. https://doi.org/10.1111/mve.12626
137. Gong WJ, Golic KG (2006) Loss of Hsp70 in Drosophila is pleiotropic, with effects on thermotolerance, recovery from heat shock and neurodegeneration. Genetics 172(1):275–286. https://doi.org/10.1534/genetics.105.048793
138. Mack LK, Attardo GM (2024) Heat shock proteins, thermotolerance, and insecticide resistance in mosquitoes. Front Insect Sci 4:1309941. https://doi.org/10.3389/finsc.2024.1309941
139. Alexander B, Barros VC, de Souza SF, Barros SS, Teodoro LP, Soares ZR, Gontijo NF, Reithinger R (2009) Susceptibility to chemical insecticides of two Brazilian populations of the visceral leishmaniasis vector Lutzomyia longipalpis (Diptera: Psychodidae). Trop Med Int Health 14(10):1272–1277. https://doi.org/10.1111/j.1365-3156.2009.02371.x
140. Balaska S, Calzolari M, Grisendi A, Scremin M, Dottori M, Mavridis K, Bellini R, Vontas J (2023) Monitoring of insecticide resistance mutations and pathogen circulation in sand flies from Emilia-Romagna, a Leishmaniasis endemic region of Northern Italy. Viruses 15(1):148. https://doi.org/10.3390/v15010148
141. Gomes B, Purkait B, Deb RM, Rama A, Singh RP, Foster GM, Coleman M, Kumar V, Paine M, Das P, Weetman D (2017) Knockdown resistance mutations predict DDT resistance and pyrethroid tolerance in the visceral leishmaniasis vector Phlebotomus argentipes. PLoS Negl Trop Dis 11(4):e0005504. https://doi.org/10.1371/journal.pntd.0005504
142. Hassan MM, Widaa SO, Osman OM, Numiary MSM, Ibrahim MA, Abushama HM (2012) Insecticide resistance in the sand fly, Phlebotomus papatasi from Khartoum State, Sudan. Parasit Vectors 5(46). https://doi.org/10.1186/1756-3305-5-46
143. Pathirage DRK, Karunaratne SHPP, Senanayake SC, Karunaweera ND (2020) Insecticide susceptibility of the sand fly leishmaniasis vector Phlebotomus argentipes in Sri Lanka. Parasit Vectors 13(1):246. https://doi.org/10.1186/s13071-020-04117-y
144. Balaska S, Fotakis EA, Chaskopoulou A, Vontas J (2021) Chemical control and insecticide resistance status of sand fly vectors worldwide. PLoS Negl Trop Dis 15(8):e0009586. https://doi.org/10.1371/journal.pntd.0009586
145. Pang R, Chen M, Yue L, Xing K, Li T, Kang K, Liang Z, Yuan L, Zhang W (2018) A distinct strain of *Arsenophonus* symbiont decreases insecticide resistance in its insect host. PLoS Genet 14(10):e1007725. https://doi.org/10.1371/journal.pgen.1007725
146. Bhatt P, Huang Y, Zhan H, Chen S (2019) Insight into microbial applications for the biodegradation of pyrethroid insecticides. Front Microbiol 10. https://doi.org/10.3389/fmicb.2019.01778
147. Cycoń M, Piotrowska-Seget Z (2016) Pyrethroid-degrading microorganisms and their potential for the bioremediation of contaminated soils: a review. Front Microbiol 7:1463. https://doi.org/10.3389/fmicb.2016.01463
148. Beard CB, O'Neill SL, Tesh RB, Richards FF, Aksoy S (1993) Modification of arthropod vector competence via symbiotic bacteria. Parasitol Today 9(5):179–183. https://doi.org/10.1016/0169-4758(93)90142-3
149. Beard CB, Cordon-Rosales C, Durvasula RV (2002) Bacterial symbionts of the triatominae and their

potential use in control of Chagas Disease transmission. Annu Rev Entomol 47(1):123–141. https://doi.org/10.1146/annurev.ento.47.091201.145144
150. Hillesland H, Read A, Subhadra B, Hurwitz I, McKelvey R, Ghosh K, Das P, Durvasula R (2008) Identification of aerobic gut bacteria from the kala azar vector, *Phlebotomus argentipes*: a platform for potential paratransgenic manipulation of sand flies. Am J Trop Med Hyg 79(6):881–886. https://doi.org/10.4269/ajtmh.2008.79.881
151. Hurwitz I, Hillesland H, Fieck A, Das P, Durvasula R (2011) The paratransgenic sand fly: a platform for control of *Leishmania* transmission. Parasit Vectors 4(82). https://doi.org/10.1186/1756-3305-4-82
152. Abassi R, Akhlaghi M, Oshaghi MA, Akhavan AA, Yaghoobi-Ershadi MR, Bakhtiary R (2019) Dynamics and fitness cost of genetically engineered Entrobacter cloacae expressing defensin for paratransgenesis in Phlebotomus papatasi. J Bacteriol Parasitol 10:349. https://doi.org/10.35248/2155-9597.1000349
153. Ghassemi M, Akhavan AA, Zahraei-Ramazani A, Yakhchali B, Arandian MH, Jafari R, Akhlaghi M, Shirani-Bidabadi L, Azam K, Koosha M, Oshaghi MA (2023) Rodents as vehicle for delivery of transgenic bacteria to make paratransgenic sand fly vectors of cutaneous leishmaniasis in field condition. Sci Rep 13(1):14912. https://doi.org/10.1038/s41598-023-41526-2
154. Brattig NW, Bazzocchi C, Kirschning CJ, Reiling N, Buttner DW, Ceciliani F, Geisinger F, Hochrein H, Ernst M, Wagner H, Bandi C, Hoerauf A (2004) The major surface protein of *Wolbachia* endosymbionts in filarial nematodes elicits immune responses through TLR2 and TLR4. J Immunol 173(1):437–445. https://doi.org/10.4049/jimmunol.173.1.437
155. Pan X, Zhou G, Wu J, Bian G, Lu P, Raikhel AS, Xi Z (2011) *Wolbachia* induces reactive oxygen species (ROS)-dependent activation of the Toll pathway to control dengue virus in the mosquito *Aedes aegypti*. Proc Natl Acad Sci 109(1):E23–E31. https://doi.org/10.1073/pnas.1116932108
156. Varotto-Boccazzi I, Epis S, Arnoldi I, Corbett Y, Gabrieli P, Paroni M, Nodari R, Basilico N, Sacchi L, Gramiccia M, Gradoni L, Tranquillo V, Bandi C (2020) Boosting immunity to treat parasitic infections: *Asaia* bacteria expressing a protein from *Wolbachia* determine M1 macrophage activation and killing of *Leishmania* protozoans. Pharmacol Res 161. https://doi.org/10.1016/j.phrs.2020.105288
157. Karimian F, Vatandoost H, Rassi Y, Maleki-Ravasan N, Mohebali M, Shirazi MH, Koosha M, Choubdar N, Oshaghi MA (2019) Aerobic midgut microbiota of sand fly vectors of zoonotic visceral leishmaniasis from northern Iran, a step toward finding potential paratransgenic candidates. Parasit Vectors 12(10). https://doi.org/10.1186/s13071-018-3273-y
158. Ratcliffe NA, Furtado Pacheco JP, Dyson P, Castro HC, Gonzalez MS, Azambuja P, Mello CB (2022) Overview of paratransgenesis as a strategy to control pathogen transmission by insect vectors. Parasit Vectors 15(112). https://doi.org/10.1186/s13071-021-05132-3
159. Wijerathna T, Gunathunga S, Gunathilaka N (2020) Recent developments and future directions in the paratransgenesis based control of Leishmania transmission. Biol Control 145:104260. https://doi.org/10.1016/j.biocontrol.2020.104260
160. Ma E, Zhu Y, Liu Z, Wei T, Whang P, Cheng G (2021) Interaction of viruses with the insect intestine. Annu Rev 8:115–131. https://doi.org/10.1146/annurev-virology-091919-100543
161. Stürchler D (2019) *Lutzomyia* - just a name, a fly, or a disease vector? Travel Med Infect Dis 32:101530–101530. https://doi.org/10.1016/j.tmaid.2019.101530
162. de Almeida JP, Aguiar ER, Armache JN, Olmo RP, Marques JT (2021) The virome of vector mosquitoes. Curr Opin Virol 49:7–12. https://doi.org/10.1016/j.coviro.2021.04.002
163. Virgin HW (2014) The virome in mammalian physiology and disease. Cell 157(1):142–150. https://doi.org/10.1016/j.cell.2014.02.032
164. Hudson AI, Fleming-Davies AE, Páez DJ, Dwyer G (2016) Genotype-by-genotype interactions between an insect and its pathogen. J Evol Biol 29(12):2480–2490. https://doi.org/10.1111/jeb.12977
165. Aguiar ERGR, Olmo RP, Paro S, Ferreira FV, da Silva de Faria IJ, Todjro YMH, Lobo FP, Kroon EG, Meignin C, Gatherer D, Imler J-L, Marques JT (2016) Sequence-independent characterization of viruses based on the pattern of viral small RNAs produced by the host. Nucleic Acids Res 44(7):3477–3478. https://doi.org/10.1093/nar/gkw044
166. Fonseca P, Ferreira F, Da Silva F, Oliveira LS, Marques JT, Goes-Neto A, Aguiar E, Gruber A (2020) Characterization of a novel mitovirus of the sand fly Lutzomyia longipalpis using genomic and virus–host interaction signatures. Viruses 13(1):9. https://doi.org/10.3390/v13010009
167. Serafim TD, Coutinho-Abreu IV, Dey R, Kissinger R, Valenzuela JG, Oliveira F, Kamhawi S (2021) Leishmaniasis: the act of transmission. Trends Parasitol 37(11):976–987. https://doi.org/10.1016/j.pt.2021.07.003
168. Andrade BB, Teixeira CR (2012) Biomarkers for exposure to sand flies bites as tools to aid control of Leishmaniasis. Front Immunol 3. https://doi.org/10.3389/fimmu.2012.00121
169. Aksoy S (2018) Insect gut microbiota: accessories to the bite. Cell Host Microbe 23(1):8–9. https://doi.org/10.1016/j.chom.2017.12.016
170. Amni F, Maleki-Ravasan N, Nateghi-Rostami M, Hadighi R, Karimian F, Meamar AR, Badirzadeh A, Parvizi P (2023) Co-infection of Phlebotomus papatasi (Diptera: Psychodidae) gut bacteria with Leishmania major exacerbates the pathological responses of BALB/c mice. Front Cell Infect Microbiol 13:1115542. https://doi.org/10.3389/fcimb.2023.1115542

MIX
Papier aus verantwortungsvollen Quellen
Paper from responsible sources
FSC® C105338

If you have any concerns about our products,
you can contact us on
ProductSafety@springernature.com

In case Publisher is established outside the EU,
the EU authorized representative is:
Springer Nature Customer Service Center GmbH
Europaplatz 3, 69115 Heidelberg, Germany

Printed by Libri Plureos GmbH
in Hamburg, Germany